AF360789

Sustainable Production in Food and Agriculture Engineering

Sustainable Production in Food and Agriculture Engineering

Editors

Jolanta B. Królczyk
Pawel Sobczak
Wioletta Żukiewicz-Sobczak

MDPI • Basel • Beijing • Wuhan • Barcelona • Belgrade • Manchester • Tokyo • Cluj • Tianjin

Editors
Jolanta B. Królczyk
Opole University of Technology
Poland

Pawel Sobczak
University of Life Sciences in
Lublin
Poland

Wioletta Żukiewicz-Sobczak
State University of Applied
Sciences in Kalisz
Poland

Editorial Office
MDPI
St. Alban-Anlage 66
4052 Basel, Switzerland

This is a reprint of articles from the Special Issue published online in the open access journal *Sustainability* (ISSN 2071-1050) (available at: https://www.mdpi.com/journal/sustainability/special_issues/Sustainable_Production_Food_Agriculture_Engineering).

For citation purposes, cite each article independently as indicated on the article page online and as indicated below:

LastName, A.A.; LastName, B.B.; LastName, C.C. Article Title. *Journal Name* **Year**, *Volume Number*, Page Range.

ISBN 978-3-03943-763-4 (Hbk)
ISBN 978-3-03943-764-1 (PDF)

Contents

About the Editors

Jolanta B. Królczyk is Associate Professor of the Department of Manufacturing Engineering and Production Automation in Opole University of Technology. She also has associations with industry, where she has held various positions including Environmental Consultant, Project Manager, Product Engineer, and Managing Director of a manufacturing company. Prof. Królczyk is Manager, co-author, and the executor of numerous projects at university as well as in the industry. The area of her scientific research concerns agricultural engineering (mixing of granular materials, agricultural use of biochar, sustainable development) and the construction and mechanical engineering (production optimization, quality management, surface analysis). Within the framework of scientific cooperation, Prof. Królczyk has participated in internships at such institutions as Fachhochschule Trier, University of Applied Science (Germany); International Institute for Sustainability w Rio de Janeiro (Brazil); and Utah State University (USA). In 2013, she participated in the Top 500 Innovators Program in Haas School of Business in University of California, Berkeley (USA). Prof. Królczyk is author or co-author of over 100 scientific publications and implementations to the industry and has scientific membership in TEAM Society—Technique, Education, Agriculture & Management, Polish Society of Agricultural Engineering, and TOP500 Association.

Paweł T. Sobczak graduated from the University of Life Sciences in Lublin, Poland and is currently employed at the Department of Food Engineering and Machines at the University of Life Sciences in Lublin. He was appointed Professor of Engineering and Technical Sciences in 2020. Prof. Sobczak specializes in food and feed engineering, and is an expert appraiser at the Polish Society of Food Processing Engineering and Technology "SPOMASZ" in the food industry. His scientific interests include granulation, mechanical engineering, agriculture, food processing, environmental pollution, renewable energy, and food. He has managed 4 research projects, completed a scientific internship at the University of Bari (Italy), and cooperates with numerous companies from the food industry in Poland. Prof. Sobczak is author or co-author of over 150 original papers, 5 patents, and 6 utility models.

Wioletta Żukiewicz-Sobczak is Associate Professor at State University of Applied Sciences in Kalisz, and graduated from the University of Life Sciences in Lublin. Here, she obtained her Ph.D. in Agricultural Sciences in the field of food and nutrition technology in 2009. In 2014, she carried out a postdoctoral appointment in health sciences at the Medical University of Poznan. She is currently a head of the Department of Food and Nutrition in State University of Applied Sciences in Kalisz. Dr. Zukiewicz-Sobczak conducts research in the field of health sciences, food, nutrition, and sustainable development which has led to over 120 peer-reviewed publications as author or co-author. She currently serves as a reviewer of both international and Polish journals, Deputy Editor in *Health Problems of Civilization*, has completed scientific internships at the University of Bari (Italy) and University in Lund (Sweden), and is manager of 2 research projects

Preface to "Sustainable Production in Food and Agriculture Engineering"

This book highlights recent research on sustainable production in food and agriculture engineering, encompassing applications in food and agriculture engineering; biosystem engineering; plant, animal, and horticultural production engineering; food and agricultural processing engineering; the dehydration and storing industry; economics and management of production and agricultural farms; agricultural machines and devices; eco-power industry; IT for agricultural engineering and ergonomics in agriculture; and bioengineering. This book is a selection of rigorously peer-reviewed papers relating to recent trends in food and agriculture engineering.

Jolanta B. Królczyk, Pawel Sobczak, Wioletta Żukiewicz-Sobczak
Editors

Article

The Effect of Wheat Moisture and Hardness on the Parameters of the Peleg and Normand Model during Relaxation of Single Kernels at Variable Initial Loading

Jawad Kadhim Al Aridhee [1], Grzegorz Łysiak [2,*], Ryszard Kulig [2], Monika Wójcik [2] and Marian Panasiewicz [2]

[1]   College of Agriculture, Al Muthanna University, Al Muthanna 66001, Iraq; jawadaridhee@gmail.com
[2]   Department of Food Engineering and Machines, University of Life Sciences in Lublin, 20-033 Lublin, Poland; ryszard.kulig@up.lublin.pl (R.K.); monika.wojcik@up.lublin.pl (M.W.); marian.panasiewicz@up.lublin.pl (M.P.)
*   Correspondence: grzegorz.lysiak@up.lublin.pl; Tel.: +48-512-247-128

Received: 6 November 2019; Accepted: 10 December 2019; Published: 11 December 2019

Abstract: The aim of the study was to examine the Peleg and Normand model to characterize the overall stress relaxation behavior of wheat kernel at varying load conditions. The relaxation experiments were made with the help of the universal testing machine, Zwick Z020, by subjecting the samples to compression at four distinct initial load levels, i.e., 20 N, 30 N, 40 N, and 50 N. The measurements were made for four wheat varieties (two soft and two hard-type endosperms) and seven levels of moisture content. Relaxation characteristics were approximated with the help of the Peleg and Normand equation. An interactive influence of the load level, moisture, and wheat hardness on the Peleg and Normand constants has been confirmed. For moist kernels, a higher amount of absorbed compression energy was released, since less energy was required to keep the deformation at a constant level. The constants differed depending on wheat hardness. Higher values of k_1 revealed that the initial force decay was slower for hard varieties. This is more characteristic of elastic behavior. Similarly, higher values of k_2 pointed to a larger amount of elastic (recoverable) energy at the end of the relaxation. The initial loading level had no or only a slight effect on the model coefficients ($Y(t)$, k_1, and k_2). The parameters of the Peleg and Normand model decreased with an increase in the water content in the kernels.

Keywords: wheat; stress relaxation; Initial load; Peleg and Normand; compression

---

1. Introduction

Agricultural and food products most often have a complex structure and exhibit complex behavior during processing. These determine the suitability for processing and constitute the basic quality criterion for acceptance by contractors and customers. This also applies to wheat grains. Wheat grain has a multi-layered and complex structure that depends on the genetic characteristics, the environment, and the specific cultivation conditions. The individual components of its structure exhibit different mechanical properties, including resistance to external loads and resistance to the formation and propagation of cracks. It results, on the one hand, in different susceptibilities of the varieties to the production of flour and, consequently of bread, biscuits, or pasta, and on the other hand, in the complexity and diversity of numerous grain processing technologies. Hence, the optimization of wheat breeding programs as well as the appropriate wheat selection for technology for many years required proper assessment of the technological quality of grain. However, the quality of wheat cannot be simply defined, since it changes depending on the farmer, grain dealer, seed company, milling industry, pasta industry, and consumer. For the farmers it can be yield or resistance to disease, for miller's protein content, hardness, or many others. By knowing the quality, processors can avoid purchasing grain that does not meet their needs. There are a lot of methods used to determine the technological quality of kernels. A number of these are time consuming and expensive (farinograph, alveograph), and hence are often impractical to use as the way to pay producers premiums based on kernel quality expectations. Finding measures of wheat which can be conducted quickly and less expensively gives the opportunity to develop new approaches to predict kernel or dough behavior. Between them, the research on mechanical properties of kernels continues to expand.

Mechanical properties of wheat kernels are naturally associated with the conditions of mechanical separation of the endosperm and the outer bran layer, the breaking resistance of the bran or the breaking susceptibility of the endosperm itself, or starch and proteins. Khazaei and Mann [1] stated that the relaxation data could be useful in estimating the susceptibility of materials to damage. Ponce-García et al. [2] successfully applied rheological measurements to distinguish among wheat classes, varieties, and different moisture contents. The test on intact kernels was used by Figueroa et al. [3,4] to establish relationships among protein composition, viscoelasticity of dough, and baking outcomes. However, there are many different conclusions and opinions regarding the scope of application of many studies. The reason for this is undoubtedly the differences arising from the use of different materials, equipment, procedures, detailed measurement conditions, and different methods of interpretation of results.

The interpretation of stress relaxation test results is most often based on rheological models (mechanical analogs) known from the rheology theory. According to these, food materials can be assumed to be composed of springs (representing ideal solids) and dashpots (considered ideal fluids) arranged in different ways. The most frequently used mechanical analogs are the Maxwell model, the Kelvin–Voigt model, and the standard linear model, SLM [5]. The spring constants in the Maxwell model give information on the stiffness, that is, the solid nature of the sample. The dashpots constants represent the relaxation times of individual Newtonian elements. The Maxwell model is suitable for representing stress relaxation data, but it does not consider the equilibrium stress. For this reason, a generalized Maxwell–Wiechert model consisting of a few elements in parallel with a spring is better for describing the viscoelastic behavior of food [5]. It was reported by Sozer and Dalgic [6] that most viscoelastic foods do not follow the simple Maxwell model; therefore, it is necessary to use more complex models to describe their stress relaxation behavior.

The Peleg and Normand model has fewer constants than the Maxwell model, however it is a simple, quick, and effective method to handle stress relaxation data [7]. According to Sozer et al. [8], the Peleg and Normand equation can be a good alternative to the Maxwell model. The authors observed that at large deformations, the Maxwell model did not fit their data very well, and the Peleg and Normand model helped overcome this problem. Apart from its debated mathematical convenience, the Peleg and Normand simulation are easy to perform and analyze the results. Stress relaxation parameters were sensitive enough to account for the textural characteristics of raw, dried, and cooked noodles [9–11]. Lewicki and Łukaszuk [12] concluded that the rheological properties, based on the Peleg model, strongly correlated with the moisture content and reflected the changes in the distribution of components and structure of apple tissue during convective drying. On the other hand, Buňka et al. [13] stated that only the parameter expressing the extent of material relaxation provided an adequate description of the actual changes occurring during the ripening of Edam cheese. The initial rate at which the stress relaxed provided no relevant description. Similarly, Singh et al. [14] and Filipčev [15] support the opinion of the inadequacy of parameter k_1 to follow textural changes or in differentiating various food products. Other authors found, however, that this parameter differed between various samples, such as cooked spaghetti [8], cooked Asian noodles [9], or wheat breads [16].

More studies have reported rheological properties evaluated under large strains (e.g., 5% strain or greater) [17]. This results from the fact that during real processing conditions, large deformations occur, resulting in the expected irreversible deformation, flow, or fracture of the material [18]. Creep and relaxation tests for large deformations were carried out for various food products, but their results indicate different opinions on the adequacy and variability of parameters of various rheological models. Higher applied stress can result in faster relaxation, according to Guo et al. [19], and larger proportion of the unrelaxed stress, according to [20]. The constants of the Peleg and Normand model for steamed bread [7] and bulk relaxation test of Jatropha curcas seeds [21] significantly decreased with larger strains. Karaman et al. [22] indicated a slightly more elastic behavior of cheese at higher compression levels. According to Khazaei and Mann [1], the deformation level had a definite effect on the decay forces for Maxwell elements for buckthorn berries, whereas the effect on relaxation times was not always ambiguous and different for each Maxwell element. Also, Bargale and Irudayaraj [23] stated that the effect of deformation level on relaxation time was stable and did not show a clear tendency. Moreover, Faridi and Faubion [24], following Shelef and Bousso [25], pointed to the independence of the relaxation parameters on the initial stress.

In summary, there are still different opinions on specific rheological measurements, and many particular cases must be treated individually. The goal of this study is the determination of the influence of the initial loading level on relaxation characteristics described by the model of Peleg and Normand. The effect was examined for different wheat types and varying moisture content.

2. Materials and Methods

2.1. Material Selection

Four wheat varieties were used in the experiments, two soft-type endosperms and two hard-type endosperms. Of these, two soft Polish cultivars (Zawisza and Wydma) and one hard SMH87 originated from the plant breeding station HR Smolice (harvested in 2015). The second hard variety was obtained from the pasta production company, Lubella. The names of the wheat varieties are denoted here as: Zawisza—SOFT1, Wydma—SOFT2, SMH87—HARD1, and wheat from Lubella—HARD2. The selection of both the soft-type and hard-type endosperm varieties was motivated by their confirmed significant differences in stress–strain behavior and fragmentation susceptibility.

2.2. Chemical Analyses

Chemical analyses were performed in the Central Agroecological Laboratory of the University of Life Sciences in Lublin, Poland. The Infratec™ 1241 (Foss, Denmark), a whole grain analyzer that uses near-infrared transmittance technology, was used to estimate the amounts of protein, wet gluten, starch, and Zeleny index. The apparatus uses a wavelength in the range of 570–1100 nm. From the batch sample about 200 g, 10 subsamples were scanned. The result of the analysis was the average of 10 measurements.

2.3. Moisture Determination

The initial moisture content of a batch sample of wheat kernels was determined by applying the air oven method. Three 5 g samples of grains were dried for over 2 h at a temperature of 130 °C, in accordance with the Polish Standard, PN-EN ISO 712:2012. The initial moisture content (m_i) was calculated according to the formula (wet basis):

$$m_i = \frac{M_w - M_d}{M_w} \cdot 100\%,\qquad(1)$$

where M_w is mass of the sample before drying and, M_d is mass of the sample after drying.

Different moisture levels, i.e., 8%, 10%, 12%, 14%, 16%, 18%, and 20% (wet basis), were assigned in the experimental plan. To obtain the respective levels, each batch sample with known weight and moisture was dried at 40 °C until its weight corresponding to the driest sample (8%) was obtained. The batch sample was then divided into seven smaller samples, and to each of them, the required amount of water was added. The amount of water was calculated using simple mass balance equations.

2.4. Stress Relaxation Procedures

The measurements were performed in the laboratory of the Department of Food Engineering and Machines. A Zwick Z020 universal testing machine was used for relaxation measurements. It was equipped with 0.1 kN capture. Application of the 0.1 N capture with an accuracy grade of 0.5% resulted in a 0.5 N accuracy of force measurement. The machine was operated using the *testXpert* firmware supplied by Zwick (version 7.1). Measurements were carried out for all four wheat varieties with the seven specified levels of moisture content. A single kernel of wheat was placed with its ventral side on the bottom of the machine's plate and then loaded axially until a required load level was obtained. The constant compression rate during the loading at 10 mm/min was adjusted. The strain (deformation) of the kernel was kept constant during next 300 s. The decreasing value of the loading force as a function of time was measured with the help of the *testXpert* software at a frequency of 100 Hz. Twenty replications were performed.

The experiments were conducted for four distinct initial load levels, i.e., 20 N, 30 N, 40 N, and 50 N. The maximum applied value (50 N) corresponds to about half of the value of the force causing kernel rupture (at ambient conditions) (Figure 1).

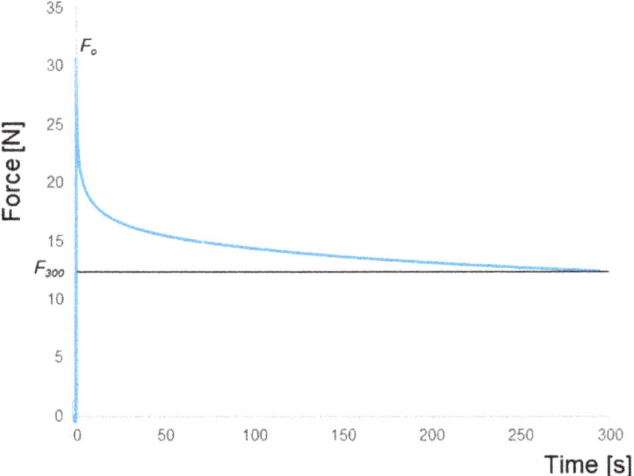

Figure 1. An example of relaxation characteristics of wheat kernel.

2.5. Calculations and Statistical Analysis

Peleg and Normand [26,27] suggested that stress relaxation data could be calculated as normalized stress and fitted to formula (2). The formula has been used by many authors, and depending on the assumptions made, it is based on decay in stress, modulus, or force. Force, which was used in the study, is also a valid criterion [28]. Hence, the relaxation of wheat kernel can be expressed in terms of force decay, as:

$$\frac{F_o\, t}{F_o - F_t} = k_1 + k_2 t, \tag{2}$$

where F_o is the initial loading force in Newtons at time $t = 0$; F_t is the force in Newtons at relaxation time t; t is the time of relaxation in seconds; k_1 and k_2 are constants; the reciprocal of k_1 determines the initial decay rate; and k_2 is the hypothetical value of the asymptotic normalized force.

The decay parameter $Y(t)$ was calculated according to:

$$Y(t) = \frac{F_o}{F_o - F_{(300)}}, \tag{3}$$

where $F_{(300)}$ – is the force established at 300 s.

k_1 indicates how fast the material in question is relaxing in energy (at least initially). If the reciprocal of the k_1 represents the initial decay rate, then its value can be associated with a low rate of decay, indicating a pronounced elastic behavior. A higher value of k_1 suggests a harder, more solid-like material that dissipates less energy, thus needing more force to be compressed [29]. Parameter k_2 represents the degree of solidity, and it varies between 1, for a material that is an ideal liquid, to infinity, for an ideal elastic solid where the stress does not relax at all [27].

The constants of the Peleg and Normand model were estimated by fitting the experimental data to the above formulas using Excel software. Experimental data were analyzed using Statistica, Dell Inc. [30] version 13. In all the analyses, a significance level at 0.05 was acquired.

3. Results

The chemical composition of wheat varieties is presented in Table 1. Hard varieties had higher gluten and protein content. The amount of wet gluten was close to 34–35% for hard wheats and 25–26% for soft types. The protein amounts were close to 15% and 12%, respectively. Higher starch content was characteristic for soft wheats (near 71%) in comparison to hard ones (about 64%). Finally, the ability of the flour proteins to swell in an acid medium expressed by Zeleny index was about two times higher for hard wheats.

The amount of protein in soft, hard, and durum can vary according to the differences in grading systems in different countries. However, within a species, wheat cultivars are further classified in terms of protein content as soft wheats (protein content about 10%), hard (protein content higher than in soft), and durum (are generally high in gluten-producing proteins, 15%). Generally, soft wheat has been bred to yield flour containing less protein than hard wheats, about 8% to 11% versus 10% to 14% protein, respectively [31]. Although the hardness is postulated to be strongly correlated with protein content [32], soft wheats can have a higher protein content than the hard cultivars, or wheat lines with different protein content can have very close hardness scores [33]. This all means that the cause of starch–protein adhesion is not straightforward, and is still being discussed [32–35].

Table 1. Chemical composition of wheat.

Wheat	Parameter	Unit	Result
HARD1	Protein content (d.b.)	%	15.2 ± 0.6
	Starch content (d.b.)	%	63.7 ± 1.9
	Wet gluten content	%	35.2 ± 3.6
	Zeleny index	mL	61.8 ± 10.7
HARD2	Protein content (d.b.)	%	15.0 ± 0.7
	Starch content (d.b.)	%	64.6 ± 2.0
	Wet gluten content	%	33.7 ± 3.4
	Zeleny index	mL	64.6 ± 11.7
SOFT1	Protein content (d.b.)	%	11.8 ± 0.5
	Starch content (d.b.)	%	70.3 ± 2.2
	Wet gluten content	%	26.2 ± 2.6
	Zeleny index	mL	34.0 ± 5.9
SOFT2	Protein content (d.b.)	%	11.9 ± 0.5
	Starch content (d.b.)	%	71.1 ± 2.2
	Wet gluten content	%	25.4 ± 2.6
	Zeleny index	mL	34.5 ± 6.0

d.b.—dry basis.

3.1. The Force Decay

Figure 2 shows the effect of wheat variety and moisture on the force decay rate. The decay $Y(t)$ represents a rate of decline in the force as a result of relaxation. The Tukey test was used for the testing of hypotheses of means equality. Distinct letters denote groups for which the means differed significantly at $p < 0.05$. Four groups were distinguished for studied cultivars. The highest values of the force decay rate were characteristic of hard wheats. The differences between the average values were particularly noticeable for lower grain moistures (up to 14%). For higher water amounts, the average values for hard and soft varieties were close. However, significant differences between soft and hard kernels were confirmed.

A considerable decline in the force decay was caused by increasing the moisture of kernels. Only for 18% and 20% moisture levels were the means were not statistically different, though the slower decline rate was observed already above 16%. The interactive effect between the two factors was also significant and confirmed by ANOVA analysis.

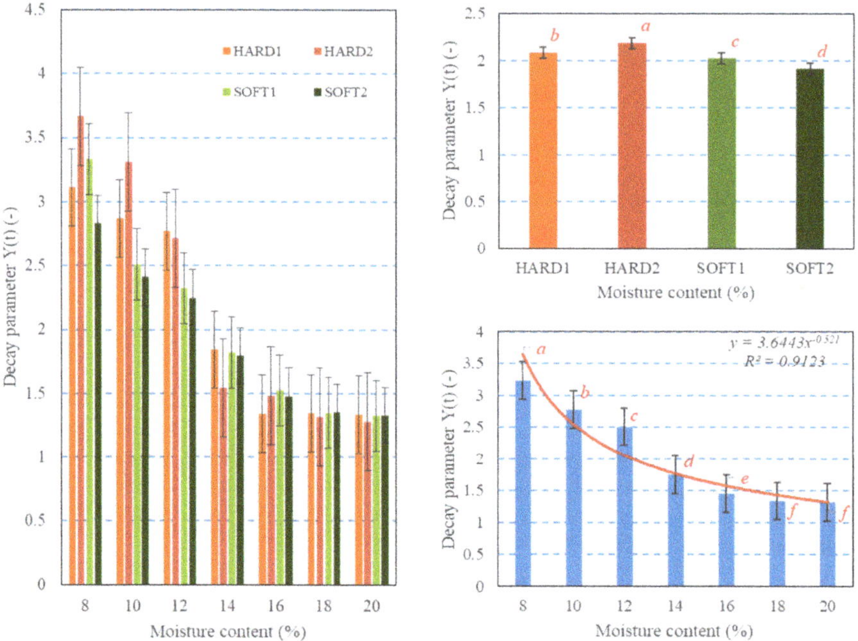

Figure 2. Influence of moisture content and wheat type on the force decay based on the Peleg and Normand equation: *a–d* and *a–f*—homogeneous groups; whiskers denote the standard error of the mean.

Figure 3 shows the effect of initial load level on the decay parameter for varied moisture and wheat type. For individual samples differing in moisture content, the effect of loading was not confirmed, although the significant effects of moisture, initial loads, as well as interaction of the two factors (moisture content × initial load) on the parameter were statistically evidenced by variance analysis. However, the effect of load was unclear both for hard and soft-type endosperm. In this case, two-way ANOVA proved no influence of initial load nor wheat type versus initial load interaction on the force decay rate *Y(t)*.

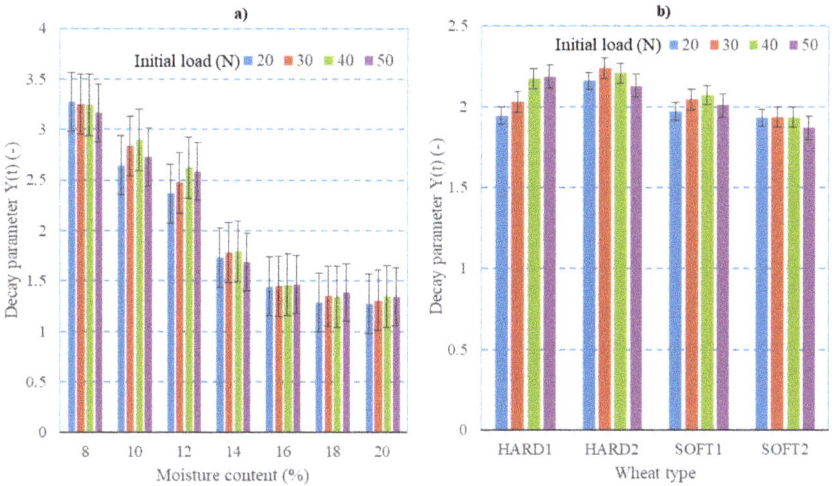

Figure 3. Plot of interactions: (**a**) moisture–initial load and (**b**) wheat type–initial load on the force decay based on the Peleg and Normand equation; whiskers denote the standard error of the mean.

Next, Figure 4 shows the general effect of the initial loading level on the average values and the distribution of the decay parameter. Only slight variations of the parameter were observed for the initial loads applied. The averages ranged from 2.0 to 2.1, although some means were statistically different. The observed small decrease in decay for 50 N loading might be caused by higher plastic deformation for moist kernels. Apart from the averages, the variability of individual results represented by their histograms (Figure 4b) was also similar. The scale parameter of the distributions was found to be also similar and not statistically dependent on the initial loading. The two-fold higher number of observations in the range of 1.0–1.5 was a reason for wheat softness at higher moistures (at 18–20%) and fact that the decay tended to unity for softer (liquid) materials.

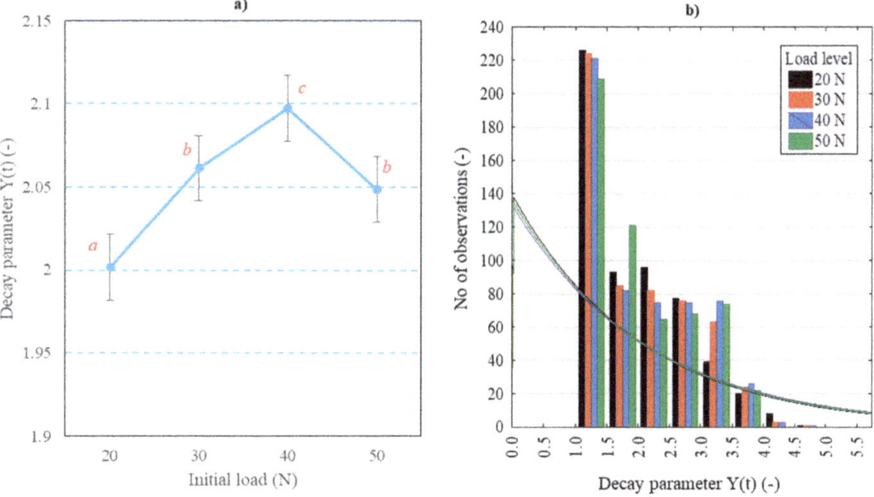

Figure 4. Influence of the initial loading level of wheat kernels on averages (**a**) and histograms (**b**) of *Y(t)*;/ *a–c*/—homogeneous groups; whiskers denote the standard error of the mean.

The results of three-way ANOVA analysis are presented in Table 2. There were seven null hypotheses with the analysis: (1–3) the population means of each of the three independent factors are equal; (4–6) there is no interaction between all possible combinations of two factors; and (7) there is no interaction between all three factors. The F-statistic is the mean square for each main effect, and the interaction effect divided by the within variance is simply a ratio of two variances. A large F ratio means that the null hypothesis is wrong (the data are not sampled from populations with the same mean)—variation among group means is more than you would expect to see by chance. In our case, all seven hypotheses were wrong. From the three analyzed main effects, the initial load was less significant (lowest F = 22.4). The highest effect was evidenced for the kernel moisture contents (largest F = 4783.6). It is necessary to remark that although there was very little or limited influence of loading level on the decay parameter, the latter was interactively dependent on all the two-way and all the three-way interactions, i.e., moisture, wheat variety, and the initial load (Table 2). However, the interaction of the studied initial load with the two other factors was the lowest. The "p" column presents the statistical significance level (i.e., p-value) of the three-way ANOVA. It can be seen that the statistical significance level of all seven hypotheses was close to zero ($p = 0.0000$). This value is <0.05 (i.e., it satisfies $p < 0.05$), which means that there was a statistically significant effect.

Table 2. Analysis of variance of the force decay parameter $Y(t)$.

Effect	Univariate Tests of Significance for $Y(t)$				
	SS	DF	MS	F	p
Intercept	9431.540	1	9431.540	233,734.0	0.0000
Wheat type	21.051	3	7.017	173.9	0.0000
Moisture	1158.090	6	193.015	4783.3	0.0000
Load	2.713	3	0.904	22.4	0.0000
Wheat type × Moisture	72.133	18	4.007	99.3	0.0000
Wheat type × Load	5.558	9	0.618	15.3	0.0000
Moisture × Load	5.316	18	0.295	7.3	0.0000
Wheat type × Moisture × Load	21.858	54	0.405	10.0	0.0000
Error	85.788	2126	0.040		

SS—sum of squares, DF—degrees of freedom, MS—mean square, F—F ratio, p—p-value

3.2. Coefficients k_1 and k_2

Figures 5 and 6 show the effect of moisture content on the coefficients k_1 and k_2 of the model presented by Peleg and Normand. Values of k_1 decreased with an increase in the kernel moisture content from about 17.0 to 5.0. The means of k_1 were significantly different for each moisture level, although the relationship was not proportional. A relatively fast decrease in the parameter with moisture rising up to 14% was noticeably slower for higher water levels, with practically no difference at 18% and 20%. This was observed both for soft and hard type wheats. On the contrary, for low moisture levels, in the 8–12% range, the differences between the kernel types were very clear. The values of the k_1 parameter were evidently higher for the two hard varieties. In turn, for high moisture contents, these differences vanished completely. This is an interesting and noteworthy observation regarding the possibility of using the relaxation test to distinguish varietal differences. According to this, a too high moisture content may be the reason for disappearing differences in the values of Peleg and Normand parameters.

The values of parameter k_2 decreased from about 3.3 to 1.4 together with the rise in the moisture contents used. The impact of both moisture and wheat type was very similar to the parameter k_1 described above.

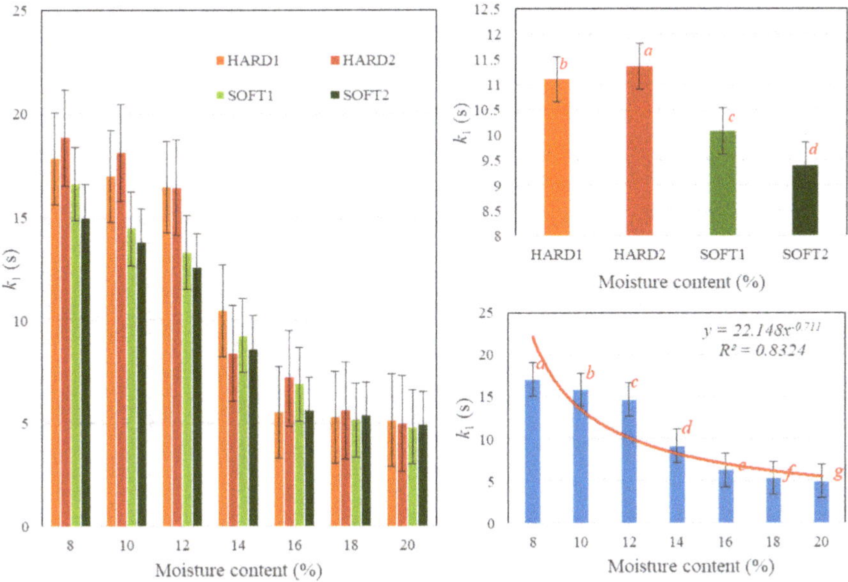

Figure 5. Influence of moisture and wheat type on the coefficient k_1 of the Peleg and Normand model; *a–d* and *a–g*—homogeneous groups, whiskers denote the standard error of the mean.

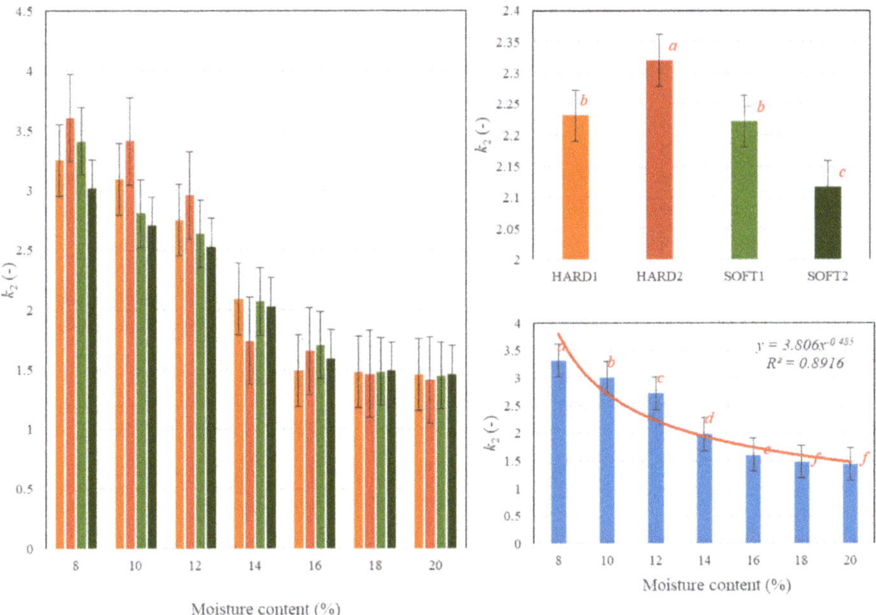

Figure 6. Influence of moisture and wheat type on the coefficient k_2 of the Peleg and Normand model; *a–c* and *a–g*—homogeneous groups, whiskers denote the standard error of the mean.

The effect of the loading level on the coefficients k_1 and k_2 is presented in Figures 7 and 8.

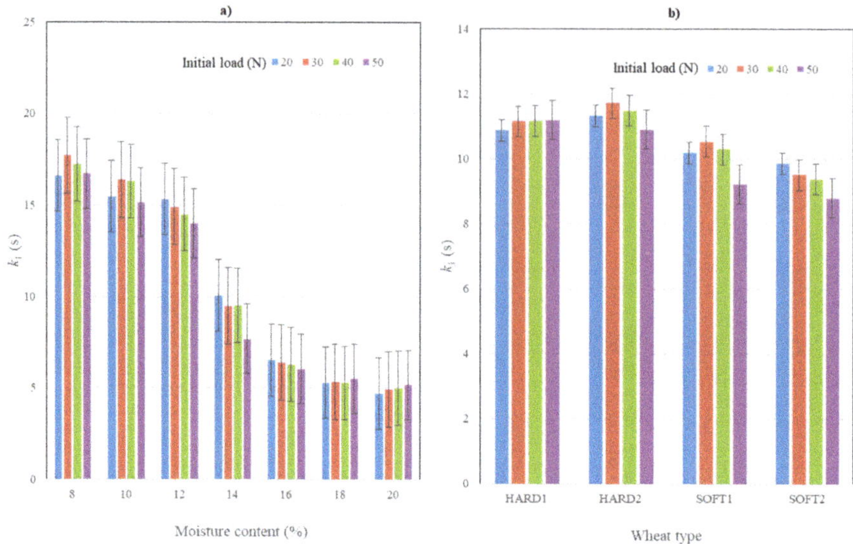

Figure 7. Influence of initial load on the coefficient k_1 of the Peleg and Normand model for different moisture (**a**) and wheat type (**b**), whiskers denote the standard error of the mean.

The analysis for the individual moisture levels showed that the mean values of k_1 did not differ significantly. Some differences were noted for 14% and 20% moisture levels. One-way ANOVA, in which the data were completely randomized, did not confirm any significant effect of the initial load on k_1. However, two-way ANOVA, including both effects of the loading level and moisture level, confirmed their interaction on the values of the coefficient. The interactive effect was also confirmed for load and wheat type interaction. Similarly, three-way ANOVA analysis, which was the most appropriate for our three-factorial experiment, proved the interactive influence of the three analyzed factors, i.e., moisture, wheat variety, and load, on k_1 (Table 3). Similarly, to the above analyzed decay parameter, from three factors, the initial load was less significant (F = 16.66). The highest effect was confirmed for wheat moisture (largest F = 2340.4). As for the previous analyses, the influence of wheat type was better evidenced than that of initial loading. K_1 was interactively influenced by all the two-way and all the three-way interactions, i.e., moisture, wheat variety, and the initial load (Table 2). However, the interaction of the studied initial load with the two other factors was also the lowest.

Table 3. Analysis of variance of the force decay parameter k_1.

Effect	Univariate Tests of Significance for k_1				
	SS	**DF**	**MS**	**F**	**p**
Intercept	1,215,041	1	1,215,041	69,015	0.0000
Wheat type	5419	3	1806	102.61	0.0000
Moisture	247,224	6	41,204	2340.4	0.0000
Load	880	3	293	16.66	0.0000
Wheat type × Moisture	6622	18	368	20.90	0.0000
Wheat type × Load	680	9	76	4.29	0.0000
Moisture × Load	1883	18	105	5.94	0.0000
Wheat type × Moisture × Load	3690	54	68	3.88	0.0000
Error	37,464	2128	18		

SS—sum of squares, DF—degrees of freedom, MS—mean square, F—F ratio, *p*—*p*-value.

In the case of k_2, the changing initial load had a more significant impact. We were able to distinguish two, three, and even four homogeneous groups, depending on the moisture level; however, no straightforward relationship was found. For k_2, main effects, like wheat type and load, were significant, but their interaction was not. Neither the load nor the interactive effect was significant ($p > 0.05$). Finally, no effect of load was observed on the relaxation coefficients for either soft or hard endosperms. Like for k_1, the three-way ANOVA analysis proved the interactive influence of the three analyzed factors on k_2 (Table 4). All the individual factors and interactive effects were confirmed to be statistically significant at very low p-values ($p = 0.0000$). All the observations for k_1 were similar, with two exceptions. The F-statistics for k_2 were higher (stronger effect), and the effect of load in this case was slightly higher (F = 70.6) than that of wheat type (67.8).

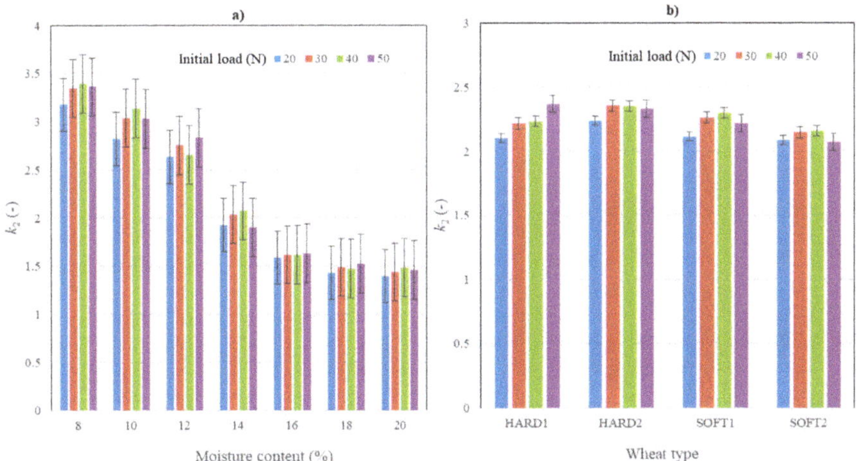

Figure 8. Influence of initial load on the coefficient k_2 of the Peleg and Normand model for different moisture (**a**) and wheat type (**b**), whiskers denote the standard error of the mean.

Table 4. Analysis of variance of the force decay parameter k_2.

Effect	Univariate Tests of Significance for k_2				
	SS	**DF**	**MS**	**F**	**p**
Intercept	7925.104	1	7925.104	249,207.8	0.0000
Wheat type	6.472	3	2.157	67.8	0.0000
Moisture	732.497	6	122.083	3838.9	0.0000
Load	6.734	3	2.245	70.6	0.0000
Wheat type × Moisture	31.723	18	1.762	55.4	0.0000
Wheat type × Load	3.305	9	0.367	11.5	0.0000
Moisture × Load	4.171	18	0.232	7.3	0.0000
Wheat type × Moisture × Load	13.557	54	0.251	7.9	0.0000
Error	67.673	2128	0.032		

SS—sum of squares, DF—degrees of freedom, MS—mean square, F—F ratio, p—p-value.

The overall effect of the initial load no the coefficients k_1 and k_2 is presented in Figure 9. The values of the two constants changed only very little. k_1 ranged from 10.04 to 10.73 and k_2 from 2.14 to 2.27. The lowest and most statistically different value of k_1 was noted at 50 N, whereas in the range from 20 to 30 N, the means were very close. In turn, the lowest values of k_2 were obtained for the 20 N load level. The homogeneous groups marked in the figure were established by the tree-way ANOVA analysis.

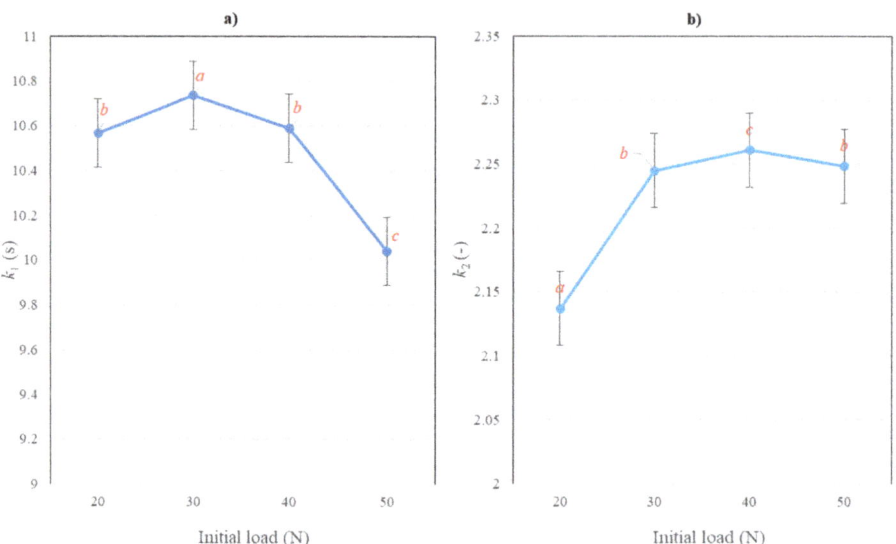

Figure 9. Influence of initial load on the coefficients k_1 (**a**) and k_2 (**b**) of the Peleg and Normand model; *a–c*—homogeneous groups; whiskers denote the standard error of the mean.

Apart from the averages, we analyzed the variability of the individual model coefficients. Their distributions are presented in Figure 10. In both cases the distributions were similar and not statistically dependent upon the initial loading. For example, the scale parameter of the exponential distributions was 0.093–0.099 for k_1 and 0.44–0.47 for k_2. Independent of the distribution fitting model applied (normal, lognormal, Weibull, or others) their parameters were very close and independent of the loading level.

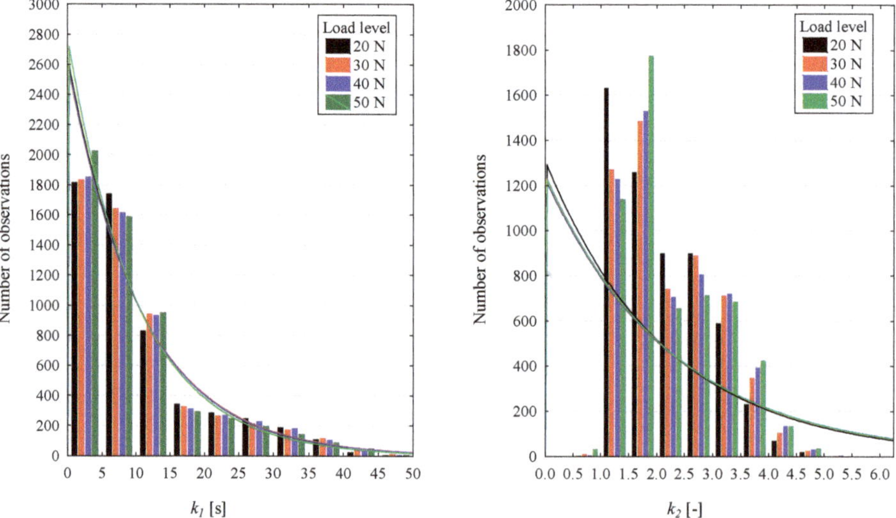

Figure 10. Histograms of the coefficients k_1 and k_2 of the Peleg and Normand model for different initial loading levels.

Finally, the analyzed impact of the initial load on the model parameters can be compared with the wheat type, which, as observed, was very small or insignificant for higher moisture contents. The histograms of the parameters k_1 and k_2 are shown in Figure 11. In the case of k_1, the distributions (medians) were clearly shifted towards higher values for hard wheats, with a slightly larger spreading (scale factor). However, in the case of k_2, the median values for the examined wheats were similar, with a slightly different scale factor.

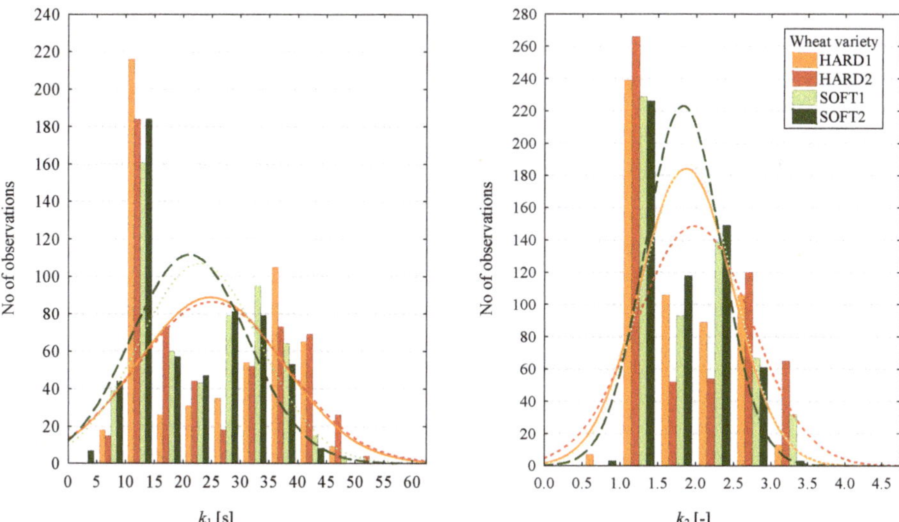

Figure 11. Histograms of the coefficients k_1 and k_2 of the Peleg and Normand model for different wheat hardness.

4. Discussion

The observed decrease in Y(t) with moisture means that for dry kernels, only a small amount of the initial force was relaxed, and the residual force established at 300 s was relatively high. In fact, as moisture content increases, the kernel becomes softer, and consequently, a lower force is required to maintain a certain deformation level [1]. Water influences the rheological properties of both liquid and solid foods. It affects the response of solid foods to force [36]. Ozturk and Takhar [37] explained a decreasing trend in the values of relaxation constants as moisture level increased, with higher resistance to the relaxation of stresses at lower moisture levels. Samples with higher moisture content relaxed faster than the ones with lower moisture, as a result of the softening process. In the study, it was observed that an increase in the moisture content of kernels caused the force to decrease more quickly. This is due to the increased plasticity of the kernel and energy losses in non-recoverable deformations [28,38]. In the model of Peleg and Normand, the coefficient k_1 represents the reciprocal of the initial decay rate. This means that a high value of k_1 corresponds to a slower rate of force relaxation in the first few seconds. The coefficient k_2 represents the hypothetical value of the asymptotic normalized force (not relaxed). With an increase in moisture content, the residual force at the end of relaxation decreased, as did the values of the coefficient. This demonstrates that for moist samples, much more energy was relaxed and less energy was necessary to keep the deformation at a constant level. Similar results have also been reported in other research works [28,37].

Generally, well noticeable differences between hard and soft cultivars were observed and statistically confirmed. All three parameters, Y(t), k_1, and k_2, were statistically higher for the hard type endosperm, but the differences could be confirmed only at lower moisture levels and they diminished with increasing water content in kernels (especially above 16%). The changes in all obtained coefficients of $Y(t)$, k_1, and k_2, for the studied wheat variety were very slight, and no clear tendency was evidenced. According to Edwards [39], the moisture sensitivity of starch strength, storage protein strength, and/or the strength of starch/protein adhesion may be responsible for differences in the strength of wheats with varying moisture contents. Such responses to water may be due to the structure of the endosperm and strength of the cell wall architecture. Soft and hard wheats exhibit the same trend with moisture content; however, they may do so at different response rates. A similar trend for soft and hard wheats was observed in our study, with little difference in rates, mainly for lower moisture contents. Glenn and Johnston [35] pointed to no significant differences in any of the mechanical properties of the starch and protein components, either within a variety or among soft, hard, and durum wheat types, though mean values for each parameter were highest for durum and the lowest soft caryopses. Many studies have underlined the meaning of the overall porosity of the endosperm structure and indicated that endosperm strength is unrelated to the protein content [40,41]. Accordingly, endosperm can be very resistant, either at low protein or high protein content, depending on the adherence between the protein and starch granules. Also, Haddad et al. [42] stated that relaxation tests make it possible to relate endosperm behavior to different types listed in materials science. However, their proposal for wheat classification was based only on the relationship between the rupture energy and the Young modulus. Following Ponce-Garcia [43] the elastic, viscoelastic, and flow properties allow the indirect measurement of wheat characteristics related to the chemical composition, including nongluten components. The elastic behavior of intact kernels was related to the sedimentation volume and composition of glutenin. He demonstrated that hard wheats showed higher plastic deformation work than durum wheat kernels, and therefore had higher elastic properties [44]. Some relationships between wheat hardness and Zeleny index and protein quality were observed in [45]. In our study, we determined the basic chemical composition of wheats, however any conclusions on their relationship to measured rheological parameters were not direct or simply not justified. Some generalization based on wheat hardness may be briefly discussed here; nevertheless, the link between the biochemical studies and mechanical studies is not established.

Irrespective of the wheat softness, no obvious or very little effect (some interactions were observed) of the initial load on the model of Peleg and Normand was demonstrated. Strain level is postulated to be meaningful in relaxation experiments; however, no unique opinion in research literature exists. Some reports showed that the initial stress can influence the amount of unrelaxed energy [20] and the relaxation speed [19], which directly relates to the constants of the relaxation models [21–23]. Lewicki and Spiess [20] showed that the proportion of the unrelaxed stress is larger if the developed stress is higher. Guo et al. [19] stated that higher applied stress can result in higher relaxation speed. The obtained results confirm those of Bargale and Irudayaraj [23] on the uni-axial compression of barley kernels, in which the authors reported that the effect of the deformation level on relaxation time was consistent and did not show a clear trend. Also, according to Faridi and Faubion [24] and Shelef and Bousso [25], stress relaxation can be fairly independent of the initial stress. An analysis of the available research results allows one to state that many conclusions are true for specific circumstances only. These results demonstrate that wheat properties, which may differ very slightly, influence the relaxation characteristics more significantly than the level of loading applied. This allows a statement on its little meaning for wheat compression in the range of loads applied in the study. The results showed that this method could be a useful tool to distinguish among wheat classes, cultivars, and different moisture levels in the kernels.

To the above-mentioned different conclusions, the general opinion about the limited applicability of the Peleg and Normand model must be emphasized. To date, relaxation tests have had few practical applications, however the reasons for this seem to be not in the method itself but in the dependence of the outcomes on a number of experimental conditions (material properties, load, speed, rate of deformation, time) and research works, which cannot be easily compared. Moreover, the presented results underline the meaning of the interactive effect between various measurement conditions. Hence, the obtained parameters of the Peleg and Normand model must be examined or interpreted together with other, often very important, factors. One of these is the time of relaxation used for the model development. Only a few studies have reported the effect of relaxation time on the model parameters [13,46]. The dependence of the model parameters on the relaxation time and varied applicability to differentiate cheese samples was also reported in studies of Bunka et al. [13]. Morales et al. [46] observed that both k_1 and k_2 changed with relaxation time. The time significantly influenced the accuracy of both the rate the force decreased at the initial relaxation phase and the value of residual force. Depending on the relaxation time, the parameters were over or underestimated. Similar conclusions can be derived from the work of Al Aridhee [47]. He demonstrated very strong dependences of the both model parameters on the relaxation holding time—k_1 significantly and linearly increased with time, whereas for longer relaxation periods k_2 decreased exponentially. For times approaching zero, k_2 increased to infinity. It must be added that the effect of time was quantitatively and qualitatively comparable to that of moisture contents.

5. Conclusions

The study led to a clear identification of the effect of wheat moisture and confirmed the influence of the wheat type on the Peleg and Normand model coefficients. The parameters k_1 and k_2, and the force decay $Y(t)$, decreased with the increase in water content in the kernels. The relatively fast decrease in the parameter with moisture rising up to 14% was noticeably slower for higher water levels, practically no difference at 18% and 20%. This was observed both for soft and hard type wheats.

The constants differed depending on the wheat hardness. The highest values of the force decay rate were characteristic of hard cultivars. For higher water amounts, the average values for hard and soft varieties were similar. On the contrary, for low moisture levels, in the 8–12% range, the differences between the grain types were very clear. This is an interesting and important observation regarding the possibility of using the relaxation test to distinguish varietal differences. According to this, too high moisture levels may be the reason for disappearing differences in the values of Peleg and Normand parameters.

The initial loading level had no effect or only a slight effect on the parameters of the Peleg and Normand model. The values of all three constants changed only very little. The effect was similar for all parameters, and none of the three analyzed parameters were influenced more than others. Apart from the averages, the distributions were very close and independent of the loading level. It must be added, however, that an interactive influence of moisture, wheat variety, and load level on the Peleg and Normand constants was statistically confirmed.

The research showed that in comparison to wheat moisture and wheat type, the influence of initial loading was very weak, ambiguous, or simply not present. Very small differences between the parameters of the Peleg and Norman model show the very weak influence of the initial load on wheat relaxation experiments, and the practical application of different loads will probably lead to similar results. It is reasonable to state that experiments on larger loads than those applied here may lead to the development and propagation of cracks, making the relaxation tests more difficult to control and interpret.

The grain protein content and protein quality aspects together with grain hardness explain most of the variation in wheat behavior. However, endosperm can be very resistant, either at low or high protein content. Thus, any relationships between wheat chemical composition and measured rheological parameters are not straightforward, and the link between the biochemical and mechanical studies has not been established yet.

Attempts to apply the Peleg and Normand model into practice need further study, including the influence of all relevant factors.

Author Contributions: Conceptualization: J.K.A.A., G.Ł.; methodology: J.K.A.A., G.Ł.; software: J.K.A.A.; validation: J.K.A.A., G.Ł.; formal analysis: G.Ł.; investigation: J.K.A.A.; resources: J.K.A.A., G.Ł., M.W., M.P.; data curation: J.K.A.A., G.Ł.; writing—original draft preparation: J.K.A.A.; writing:— J.K.A.A., Gr.Ł.; writing - review

& editing: G.Ł., M.W., M.P.; visualization: J.K.A.A.; supervision: G.Ł., R.K.; project administration: J.K.A.A., G.Ł., R.K.; funding acquisition: J.K.A.A., G.Ł., M.W., M.P.

Acknowledgments: I deeply appreciate the scholarship from the Iraqi Ministry of the Higher Education and Scientific Research.

References

1. Khazaei, J.; Mann, D. Effects of moisture content and number of loadings on force relaxation behavior of chickpea kernels. *Int. Agrophys.* **2005**, *19*, 305–313.
2. Ponce-García, N.; Ramírez-Wong, B.; Torres-Chávez, P.I.; Figueroa-Cárdenas, J.D.; Serna Saldívar, S.O.; Cortez-Rocha, M.O. Effect of moisture content on the viscoelastic properties of individual wheat kernels evaluated by the uniaxial compression test under small strain. *Cereal Chem.* **2013**, *90*, 558–563. [CrossRef]
3. Figueroa, J.D.C.; Hernández, Z.J.E.; Véles, J.J.; Rayas-Duarte, P.; Martínez-Flores, H.E.; Ponce García, N. Evaluation of degree of elasticity and other mechanical properties of wheat kernels. *Cereal Chem.* **2011**, *88*, 12–18. [CrossRef]
4. Figueroa, J.D.C.; Ramírez-Wong, B.; Peña, J.; Khan, K.; Rayas-Duarte, P. Potential use of the elastic properties of intact wheat kernels to estimate millings, rheological and bread—making quality of wheat. In *Wheat Science Dynamics: Challenges and Opportunities*; Chibbar, R.N., Dexter, J., Eds.; Agrobios: Jodhpur, India, 2011; pp. 317–325.
5. Steffe, J.F. *Rheological Methods in Food Process Engineering*, 2nd ed.; Freeman press: East Lansing, MI, USA, 1996.
6. Sozer, N.; Dalgic, A.C. Modeling of rheological characteristics of various spaghetti types. *Eur. Food Res. Technol.* **2007**, *225*, 183–190. [CrossRef]
7. Wu, M.Y.; Chang, Y.H.; Shiau, S.Y.; Chen, C.C. Rheology of fiber—Enriched steamed bread: Stress relaxation and texture profile analysis. *J. Food Drug Anal.* **2012**, *20*, 133–142.
8. Sozer, N.; Kaya, A.; Dalgic, A.C. The effect of resistant starch addition on viscoelastic properties of cooked spaghetti. *J. Texture Stud.* **2008**, *39*, 1–16. [CrossRef]
9. Hatcher, D.W.; Bellido, G.G.; Dexter, J.E.; Anderson, M.J.; Fu, B.X. Investigation of uniaxial stress relaxation parameters to characterize the texture of yellow alkaline noodles made from durum and common wheats. *J. Texture Stud.* **2008**, *39*, 695–708. [CrossRef]
10. Shiau, S.Y.; Wu, T.T.; Liu, Y.L. Effect of the amount and particle size of wheat fiber on textural and rheological properties of raw, dried and cooked noodles. *J. Food Qual.* **2012**, *35*, 207–216. [CrossRef]
11. Shiau, S.Y.; Chang, Y.H. Instrumental textural and rheological properties of raw, dried, and cooked noodles with transglutaminase. *Int. J. Food Prop.* **2013**, *16*, 1429–1441. [CrossRef]
12. Lewicki, P.; Łukaszuk, A. Changes of rheological properties of apple tissue undergoing convective drying. *Dry. Technol.* **2000**, *18*, 707–722. [CrossRef]
13. Buňka, F.; Pachlová, V.; Pernická, L.; Burešová, I.; Kráčmar, S.; Lošák, T. The dependence of Peleg's coefficients on selected conditions of a relaxation test in model samples of edam cheese. *J. Texture Stud.* **2013**, *44*, 187–195. [CrossRef]
14. Singh, H.; Rockall, A.; Martin, C.R.; Chung, O.K.; Lookhart, G.L. The analysis of stress relaxation data of some viscoelastic foods using a texture analyzer. *J. Texture Stud.* **2006**, *37*, 383–392. [CrossRef]
15. Filipčev, B.V. Texture and stress relaxation of spelt—amaranth composite breads. *Food Feed Res.* **2014**, *41*, 1–9. [CrossRef]
16. Mandala, I.; Karabela, D.; Kostaropoulos, A. Physical properties of breads containing hydrocolloids stored at low temperature. I. Effect of chilling. *Food Hydrocoll.* **2007**, *21*, 1397–1406. [CrossRef]
17. ASAE S368.4. *Compression Test of Food Materials of Convex Shape*; American Society of Agricultural and Biological Engineers: Saint Joseph, MI, USA, 2006; pp. 554–556.
18. Karim, A.A.; Norziah, M.H.; Seow, C.C. Methods for the study of starch retrogradation. *Food Chem.* **2000**, *71*, 9–36. [CrossRef]

19. Guo, L.; Wang, D.; Tabil, L.G.; Wang, G. Compression and relaxation properties of selected biomass for briquetting. *Biosyst. Eng.* **2016**, *148*, 101–110. [CrossRef]

20. Lewicki, P.; Spiess, W.E. Rheological properties of raisins: Part, I. Compression test. *J. Food Eng.* **1995**, *24*, 321–338. [CrossRef]

21. Herak, D.; Kabutey, A.; Choteborsky, R.; Petru, M.; Sigalingging, R. Mathematical models describing the relaxation behaviour of Jatropha curcas L. bulk seeds under axial compression. *Biosyst. Eng.* **2015**, *131*, 77–83. [CrossRef]

22. Karaman, S.; Yilmaz, M.T.; Toker, O.S.; Dogan, M. Stress relaxation/creep compliance behaviour of kashar cheese: Scanning electron microscopy observations. *Int. J. Dairy Technol.* **2016**, *69*, 254–261. [CrossRef]

23. Bargale, P.C.; Irudayaraj, J. Mechanical strength and rheological behavior of barley kernels. *Int. J. Food Sci. Technol.* **1995**, *30*, 609–623. [CrossRef]

24. Faridi, H.; Faubion, J.M. *Dough Rheology and Baked Product Texture*; Springer: Berlin/Heidelberg, Germany, 2012.

25. Shelef, L.; Bousso, D. A new instrument for measuring relaxation in flour dough. *Rheol. Acta* **1964**, *3*, 168–172. [CrossRef]

26. Peleg, M. Linearization of relaxation and creep curves of solid biological materials. *J. Rheol.* **1980**, *24*, 451–463. [CrossRef]

27. Peleg, M.; Normand, M.D. Comparison of two methods for stress relaxation data presentation of solid foods. *Rheol. Acta* **1983**, *22*, 108–113. [CrossRef]

28. Łysiak, G. Influence in of moisture on rheological characteristics of the kernel of wheat. *Acta Agrophys* **2007**, *9*, 91–97.

29. Guo, Z.; Castell-Perez, M.E.; Moreira, R.G. Characterization of masa and low-moisture corn tortilla using stress relaxation methods. *J. Texture Stud.* **1999**, *30*, 197–215. [CrossRef]

30. Dell Inc. Dell Statistica (data analysis software system), version 13. 2016. Available online: software.dell.com (accessed on 6 December 2019).

31. Delcour, J.A.; Joye, I.J.; Pareyt, B.; Wilderjans, E.; Brijs, K.; Lagrain, B. Wheat gluten functionality as a quality determinant in cereal-based food products. *Annu. Rev. Food. Sci. Technol.* **2012**, *3*, 1–24. [CrossRef]

32. Mikulíková, D. The effect of friabilin on wheat grain hardness: A review. *Czech J. Gen. Plant Breed.* **2018**, *43*, 35–43. [CrossRef]

33. Turnbull, K.M.; Rahman, S. Endosperm texture in wheat. *J. Cereal Sci.* **2002**, *36*, 327–337. [CrossRef]

34. Pasha, I.; Anjum, F.M.; Morris, C.F. Grain Hardness: A Major Determinant of Wheat Quality. *Food Sci. Tech. Int.* **2010**, *16*, 511–522. [CrossRef]

35. Glenn, G.M.; Johnston, R.K. Mechanical properties of starch, protein and endosperm and their relationship to hardness in wheat. *Food Struct.* **1992**, *11*, 187–199.

36. Lewicki, P. Water as the determinant of food engineering properties. A review. *J. Food Eng.* **2004**, *61*, 483–495. [CrossRef]

37. Ozturk, O.K.; Takhar, P.S. Stress relaxation behavior of oat flakes. *J. Cereal Sci.* **2017**, *77*, 84–89. [CrossRef]

38. Al Aridhee, J.; Łysiak, G. Stress relaxation characteristics of wheat kernels at different moisture. *Acta Sci. Pol. Tech. Agrar.* **2015**, *14*, 3–10.

39. Edwards, M.A. Morphological Features of Wheat Grain and Genotype Affecting Flour Yield. Ph.D. Thesis, Southern Cross University, Lismore, NSW, Australia, 2010.

40. Dexter, J.E.; Marchylo, B.A.; MacGregor, A.W.; Tkachuk, R. 1989. The structure and protein composition of vitreous, piebald and starchy durum wheat kernels. *J. Cereal Sci.* **1989**, *10*, 19–32. [CrossRef]

41. Dobraszczyk, B.J.; Schofield, J.D. Rapid Assessment and Prediction of Wheat and Gluten Baking Quality With the 2-g Direct Drive Mixograph Using Multivariate Statistical Analysis. *Cereal Chem.* **2002**, *79*, 607–612. [CrossRef]

42. Haddad, Y.; Benet, J.C.; Delenne, J.Y.; Mermet, A.; Abecassis, J. Rheological Behaviour of Wheat Endosperm—Proposal for Classification Based on the Rheological Characteristics of Endosperm Test Samples. *J. Cereal Sci.* **2001**, *34*, 105–113. [CrossRef]

43. Ponce-García, N.; Ramírez-Wong, B.; Escalante-Aburto, A.; Torres-Chávez, P.I.; Figueroa, J.D.C. Mechanical Properties in Wheat (*Triticum aestivum*) Kernels Evaluated by Compression Tests: A Review. In *Viscoelastic and Viscoplastic Materials*; El-Amin, M., Ed.; IntechOpen: London, UK, 2016.

44. Ponce-García, N.; Figueroa, J.D.C.; López-Huape, G.A.; Martínez, H.E.; Martínez-Peniche, R. Study of viscoelastic properties of wheat kernels using compression load method. *Cereal Chem.* **2008**, *85*, 667–672. [CrossRef]
45. Hrušková, M.; Švec, I. Wheat hardness in relation to other quality factors. *Czech J. Food Sci.* **2009**, *27*, 240–248. [CrossRef]
46. Morales, R.; Arnau, J.; Serra, X.; Guerrero, L.; Gou, P. Texture changes in dry-cured ham pieces by mild thermal treatments at the end of the drying process. *Meat Sci.* **2008**, *80*, 231–238. [CrossRef]
47. Al Aridhee, J. Investigation and Modelling of Stress Relaxation of Wheat Kernels in View of Grinding Prediction. Ph.D. Thesis, University of Life Sciences, Lublin, Poland, 2018.

 *sustainability*

Article

Numerical Modeling of the Shape of Agricultural Products on the Example of Cucumber Fruits

Andrzej Anders *, Dariusz Choszcz, Piotr Markowski, Adam Józef Lipiński, Zdzisław Kaliniewicz and Elwira Ślesicka

Department of Heavy Duty Machines and Research Methodology, University of Warmia and Mazury in Olsztyn, Olsztyn 10-957, Poland; choszczd@uwm.edu.pl (D.C.); pitermar@uwm.edu.pl (P.M.); adam.lipinski@uwm.edu.pl (A.J.L.); arne@uwm.edu.pl (Z.K.); elwira.slesicka@uwm.edu.pl (E.Ś.)
* Correspondence: andrzej.anders@uwm.edu.pl; Tel.: +48-089-523-4534

Received: 24 April 2019; Accepted: 14 May 2019; Published: 16 May 2019

Abstract: The aim of the study was to build numerical models of cucumbers cv. *Śremski* with the use of a 3D scanner and to analyze selected geometric parameters of cucumber fruits based on the developed models. The basic dimensions of cucumber fruits–length, width and thickness—were measured with an electronic caliper with an accuracy of $d = 0.01$ mm, and the surface area and volume of fruits were determined by 3D scanning. Cucumber fruits were scanned with an accuracy of $d = 0.13$ mm. Six models approximating the shape of cucumber fruits were developed with the use of six geometric figures and their combinations to calculate the surface area and volume of the analyzed agricultural products were identified. The surface area and volume of cucumber fruits calculated by 3D scanning and mathematical formulas were compared. The surface area calculated with the model combining two truncated cones and two hemispheres with different diameters, joined base-to-base, was characterized by the smallest relative error of 3%. Fruit volume should be determined with the use of mathematical formulas derived for a model composed of an ellipsoid and a spheroid. The proposed geometric models can be used in research and design.

Keywords: 3D scanner; geometric model; reverse engineering; fruit; cucumber

1. Introduction

Advanced measurement techniques and software supporting complex simulations of selected technological processes are required to introduce new products and technologies on the market and to improve product quality. Models of agricultural products should account for the designed technological processes and should accurately reflect the products' shape [1]. A 3D model that accurately describes a product's geometric and physical parameters can be used in the design process. A traditional approach to modeling relies on the assumption that agri-food products are homogeneous and isotropic, and the modeled objects are assigned regular shapes (e.g., cylinder, sphere, cone, etc.) Computer-Aided Design (CAD) and Computational Fluid Dynamics (CFD) software can be applied to simulate complex processes that occur during the processing of agri-food products [2]. The development of a model that closely approximates the shape of the original agricultural product and can be used in computer simulations poses the key challenge in the research and design of food processing equipment. Numerical modeling based on traditional methods is a laborious and difficult task, in particular when the studied objects have irregular shape [3]. In the process of measuring fruits and seeds, many researchers rely solely on image analysis tools and measuring devices such as calipers and micrometers [4,5]. In the literature, traditional methods have been used to determine the geometric parameters of soybeans (*Glycine max* L. Merr.) [6], sunflower seeds (*Helianthus annuus* L.) [7], oilseed rape seeds (*Brassica napus* L.) [8,9], mustard seeds (*Sinapis alba*) [10] and flax seeds (*Linnum usitatissimum* L.) [11]. In small objects such as seeds, only basic dimensions can be measured with a caliper or a micrometer. In larger products such as fruits and vegetables, the analyzed parameters can be measured with a caliper or a micrometer at any point on the object's surface.

In the literature, traditional and advanced measuring techniques have been deployed to accurately render the shape of the analyzed products. Erdogdu et al. [12] relied on a machine vision system designed by Luzuriaga et al. [13] to determine the geometric parameters of shrimp cross-sections and to develop mathematical models of the thermal processing of shrimp. Crocombe et al. [14] analyzed the surface of meat pieces by laser scanning to develop a numerical model and simulate meat refrigeration time. Jancsok et al. [15] used a machine vision system to build numerical models of pears cv. Konferencja. Borsa et al. [16] performed computed tomography scans and calculated the radiation dose absorbed by the examined food products. Sabliov et al. [17] proposed an image analysis method for measuring the volume and surface area of axially symmetric agricultural products. Zapotoczny [18] developed a test stand for measuring the geometric parameters of cucumber fruits with the use of digital image analysis. The cited author registered changes in the shape and size of greenhouse-grown cucumbers during storage. Scheerlinck et al. [19] relied on a machine vision system to develop a 3D model of strawberries and a thermal system for disinfecting fruit surfaces. Du and Sun [20] and Zheng et al. [21] developed an image analysis technique for measuring the surface area and volume of beef loin and beef joints. Kim et al. [22] generated 3D geometric models of food products with a complex shape with the use of computed tomography. Goni et al. [23] modeled the geometric properties of the studied objects with the involvement of magnetic resonance imaging. Siripon et al. [24] analyzed chicken half-carcasses with a 3D scanner (Atos, GOM, Germany) and used the results to simulate cooking processes. Mieszkalski [25,26] developed computer models of carrots, apples cv. Jonagored and chicken eggs. The shape of biological objects was described with Bézier curves. The resulting mathematical models were used to generate 3D figures that accurately rendered the shape and basic dimensions of the studied products. Balcerzak et al. [27] modeled the geometric parameters of corn and oat kernels in the 3ds Max environment. Images of kernel cross-sections were used to acquire geometric data, generate meshes and determine nodal coordinates. Ho Q. T. and others used multiscale modeling in food engineering. Multiscale models support evaluations of the phenomena occurring inside agricultural raw materials on a micro and macro scale. The authors relied on X-ray tomography to generate multiscale models [28]. The volume of agricultural raw materials can also be determined by water displacement. However, this method cannot be applied to materials that easily absorb water [29].

The majority of methods require complex and expensive measuring devices and software. A thorough knowledge of various imaging techniques is required to model irregularly shaped objects. Models that accurately render the shape of the analyzed products can be developed with the use of a 3D scanner. This technique is considerably simpler, but it is not yet widely used. 3D models can be used to analyze the shape of whole products or their fragments [30,31].

The dimensions and basic geographic parameters of agricultural materials have been long determined with the use of simple measuring devices, including analog and digital calipers, micrometers and dial indicators. The main limitation of conventional measuring techniques is that they investigate only characteristic points in the examined objects, and the measured values can be used to calculate selected parameters, such as surface area and volume, with mathematical formulas [29]. In contrast, indirect methods rely on the acquisition of images of the investigated object and digital image analysis. The advances made in digital technology and computing power have contributed to the widespread popularity of indirect measuring methods. Indirect measurements produce linear dimensions as well as images of the analyzed surfaces. The main advantage of indirect methods is that measurements are rapid, whereas the main limitation stems from the fact that measurements are performed along the contours of the acquired image, which are projected onto a plane [9]. A relatively new method has been proposed for registering the shape of a sample as a cloud of points. The location of every point in the modeled sample is determined with the use of 3D scanners, which register the position of the laser beam, a structured light source. The points registered by a 3D scanner support the development of a numerical model, which can be used in metrological analyses. The development of a numerical model with the described method is time-consuming, but the results can be stored in computer memory [32,33].

The presented methods for measuring the geometric properties of objects produce highly similar results, provided that the required precision thresholds are met. However, the time and conditions of measurement can vary. Approximation formulas are widely applied to calculate volume and area. The main problem is the selection of the optimal model for determining the above parameters with the required accuracy. The aim of this study was to compare selected geometric parameters of cucumber fruits acquired from 3D models and models based on basic geometric figures and direct caliper measurements.

2. Materials and Methods

The experiment was performed on cucumber fruits cv. *Śremski* stored indoors at a constant temperature of 18.1 °C and 60% humidity. Cucumbers were purchased from the Pozorty Production and Experimental Station in Olsztyn. Fifty whole cucumber fruits without visible signs of damage were randomly selected for the experiment. Cucumbers were purchased on five occasions in the second half of August 2018, and 10 cucumbers were purchased each time. The length, width and thickness of cucumber fruits were measured with an electronic caliper with an accuracy of $d = 0.01$ mm. Each fruit was additionally measured with an electronic caliper at the points presented in Figure 1. Cucumbers were scanned with the Nextengine 3D scanner with a resolution of 15 points per mm^2. Scanning precision was 0.13 mm. Cucumbers were mounted on a turntable. Individual images were combined in the ScanStudio HD PRO program [34]. The developed numerical models were used to determine the surface area and volume of cucumber fruits. The above parameters were measured in the MeshLab program [35].

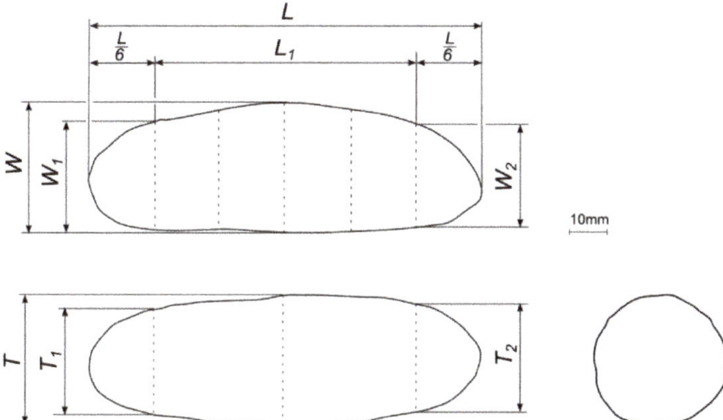

Figure 1. Shape of a selected cucumber fruit: *L*—length (mm), *W*—width (mm), *T*—thickness (mm), L_1—length of the middle section (mm), W_1, W_2—width of the terminal section (mm), T_1, T_2—thickness of the terminal section (mm).

The measured dimensions were used to build six geometric models whose shape resembled the shape of cucumber fruits. The surface area and volume of fruits were calculated from the developed models. Geometric models were built based on basic geometric figures, including an ellipsoid, cylinder, hemisphere, truncated cone and a combination of selected figures. The analyzed geometric models are presented in Figure 2.

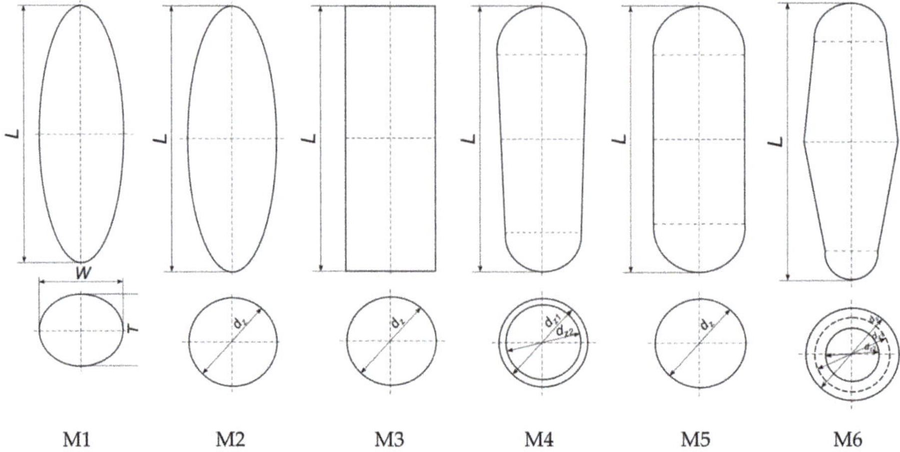

M1   M2   M3   M4   M5   M6

Figure 2. Models of cucumber fruits: M1—ellipsoid, M2—spheroid, M3—cylinder, M4—truncated cone and two hemispheres, M5—cylinder and two hemispheres, M6—two truncated cones and two hemispheres.

Mathematical formulas were derived for every geometric model and were used to calculate the surface area and volume of cucumbers [36,37]:

ellipsoid model (M1):

$$A_{M1} = 2 \cdot \pi \cdot \left(\left(\frac{L}{2} \right)^2 + \frac{\frac{T}{2} \cdot \left(\frac{L}{2} \right)^2}{\sqrt{\left(\frac{W}{2} \right)^2 - \left(\frac{L}{2} \right)^2}} \cdot F(\Theta, m) + \frac{T}{2} \cdot \sqrt{\left(\frac{W}{2} \right)^2 - \left(\frac{L}{2} \right)^2} \cdot E(\Theta, m) \right) \tag{1}$$

where:

$$m = \frac{\left(\frac{L}{2} \right)^2 \cdot \left(\left(\frac{T}{2} \right)^2 - \left(\frac{L}{2} \right)^2 \right)}{\left(\frac{T}{2} \right)^2 \cdot \left(\left(\frac{W}{2} \right)^2 - \left(\frac{L}{2} \right)^2 \right)} = \frac{L^2 \cdot T^2 - L^4}{T^2 \cdot W^2 - L^2 \cdot T^2} \tag{2}$$

$$\Theta = \arcsin \left(\sqrt{\frac{\sqrt{W^2 - L^2}}{|W|}} \right) \tag{3}$$

and where $F(\Theta, m)$ and $E(\Theta, m)$ are incomplete elliptic integrals of the first and second kind [37].

$$V_{M1} = \frac{\pi \cdot T \cdot W \cdot L}{6} \tag{4}$$

spheroid model (M2), when: $\frac{L}{2} > \frac{d_z}{2}$, then:

$$A_{M2} = 2 \cdot \pi \cdot \left(\frac{d_z}{2} \right)^2 \cdot \left(1 + \frac{\frac{L}{2}}{\frac{d_z}{2} \cdot e} \cdot \arcsin(e) \right) = \frac{4 \cdot \pi \cdot d_z^2 + \pi \cdot L \cdot d_z \cdot e \cdot \arcsin(e)}{8} \tag{5}$$

where:

$$e = \sqrt{1 - \frac{d_z^2 \cdot L^2}{16}} \tag{6}$$

$$V_{M2} = \frac{\pi \cdot d_z^2 \cdot L}{6} \tag{7}$$

cylinder model (M3):

$$A_{M3} = \pi \cdot d_z \cdot L + 2 \cdot \pi \cdot \left(\frac{d_z}{2}\right)^2 \tag{8}$$

$$V_{M3} = \frac{\pi \cdot d_z^2 \cdot L}{4} \tag{9}$$

model combining a truncated cone and two hemispheres (M4)

$$A_{M4} = \frac{\pi}{2} \cdot \left(d_{z1}^2 + d_{z2}^2\right) + \pi \cdot \sqrt{\left(\frac{d_{z1}}{2}\right)^2 + L_1^2} \cdot \left(\frac{d_{z1}}{2} + \frac{d_{z2}}{2}\right) \tag{10}$$

$$V_{M4} = \frac{\pi}{12} \cdot \left(d_{z1}^3 + d_{z2}^3 + L_1 \cdot \left(d_{z1}^2 + d_{z1} \cdot d_{z2} + d_{z2}^2\right)\right) \tag{11}$$

model combining a cylinder and two hemispheres (M5)

$$A_{M5} = \pi \cdot d_w \cdot \left(\frac{d_w}{2} + \frac{d_w}{2} + L_1\right) \tag{12}$$

$$V_{M5} = \pi \cdot d_w^2 \cdot \left(\frac{d_w}{6} + \frac{L_1}{4}\right) \tag{13}$$

model combining two truncated cones and two hemispheres (M6)

$$A_{M6} = \frac{(\pi \cdot d_{z1} + \pi \cdot d_z) \cdot \sqrt{d_{z1}^2 + L_1^2} + 2 \cdot \pi \cdot d_{z1}^2 + (\pi \cdot d_{z2} + \pi \cdot d_z) \cdot \sqrt{d_{z2}^2 + L_1^2} + 2 \cdot \pi \cdot d_{z2}^2}{4} \tag{14}$$

$$V_{M6} = \frac{2 \cdot \pi \cdot d_{z2}^3 + \pi \cdot L_1 \cdot d_{z2}^2 + \pi \cdot L_1 \cdot d_z \cdot d_{z2} + 2 \cdot \pi \cdot d_{z1}^3 + \pi \cdot L_1 \cdot d_{z1}^2 + \pi \cdot L_1 \cdot d_z \cdot d_{z1} + 2 \cdot \pi \cdot L_1 \cdot d_z^2}{24} \tag{15}$$

In models M2, M3, M4, M5 and M6, geometric mean diameter was calculated with the following formulas:

$$d_w = \frac{W_1 + W_2 + T_1 + T_2}{4} \tag{16}$$

$$d_z = \frac{W + T}{2} \tag{17}$$

$$d_{z1} = \frac{W_1 + T_1}{2} \tag{18}$$

$$d_{z2} = \frac{W_2 + T_2}{2} \tag{19}$$

every cucumber fruit was weighed on the Radwag WAA 100/C/2 electronic scale to the nearest 0.001 g. The significance of differences between the mean values of the measured parameters was determined in the Kruskal-Wallis test with multiple comparisons of mean ranks. The aim of the analysis was to identify homogeneous groups. The results were processed statistically in the Statistica 13.3 PL program at a significance level of $\alpha = 0.05$.

3. Results and Discussion

Cucumber fruits (*Cucumis sativus* L.) cv. *Śremski* are botanical berries with a more or less elongated shape, varied size, smooth or spiny skin. Cucumbers are filled with seeds, and their color ranges from dark green to yellow. At harvest maturity, cucumbers are cylindrical in shape, without a neck, with a gently tapering end at the flower base and a small seed chamber. The smallest of the examined cucumbers weighed 43.05 g, and the largest 123.70 g. The surface area of cucumbers determined in the 3D scanner ranged from 74.84 cm^2 do 145.38 cm^2, with an average of 111.25 cm^2. Based on the generated 3D images, the volume of cucumbers was determined in the range of 46.65 cm^3 to 127.38 cm^3, with an average of 77.26 cm^3 (Table 1). Exemplary 3D models of cucumber fruits are presented in Figures 3 and 4.

Table 1. Geometric parameters of cucumber fruits.

Variable1	Mean	Range	Standard Deviation
L (mm)	113.14	39.10	9.94
W (mm)	37.23	13.04	3.28
T (mm)	35.47	14.82	3.31
A^{3D} (mm^2)	111.25	70.54	16.12
V^{3D} (mm^3)	77.26	80.73	18.89

^{3D}-3D scan

Figure 3. 3D model of a cucumber fruit with a texture overlay.

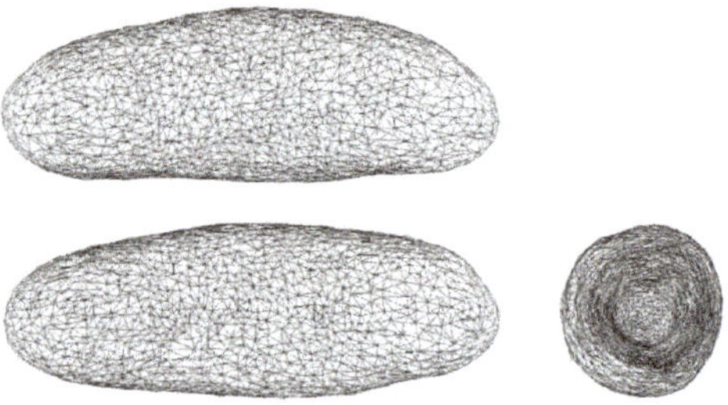

Figure 4. 3D model of a cucumber fruit represented by a triangle mesh.

The mean dimensions, surface area and volume of the analyzed cucumber fruits are presented in Table 1.

The significance of differences between the mean surface area and mean volume of cucumbers was determined in the Kruskal-Wallis nonparametric test. The significance of differences between the parameters acquired by 3D scanning and the parameters calculated with mathematical formulas is presented in Tables 2 and 3. The mean surface area of cucumber fruits calculated from the 3D model did not differ significantly from the mean surface area calculated from the spheroid model (M2—formula 5) and the model combining two truncated cones and two hemispheres with different diameters (M6—formula 14).

The mean volume of cucumber fruits calculated from the 3D model did not differ significantly from the mean volume calculated from the ellipsoid model (M1—formula 4), spheroid model (M2—formula 7) and the geometric model combining two truncated cones and two hemispheres with different diameters (M6—formula 15).

Table 2. The significance of differences between the mean surface area of cucumber fruits.

Surface Area A (Kruskal-Wallis Test) H(6, N = 350) = 132.2065; p = 0.000				
Probability of Multiple Comparisons				
Measurement Method	**Number of Observations N**	**Rank Sum**	**Mean Rank**	**Mean**
3D	50	9288.50	185.77	111.25 [bc]
M1	50	6939.50	138.79	101.71 [a]
M2	50	7863.00	157.26	105.93 [ab]
M3	50	15,564.00	311.28	150.45 [d]
M4	50	6181.00	123.62	100.17[a]
M5	50	5737.00	114.74	98.57 [a]
M6	50	9852.00	197.04	114.06 [c]

Values marked with the same letters in columns do not differ significantly; [a,b,c,d] ($p \leq 0.05$).

Table 3. The significance of differences between the mean volume of cucumber fruits.

Volume V (Kruskal-Wallis Test) H(6, N = 350) = 124.2550; p = 0.000				
Probability of Multiple Comparisons				
Measurement Method	**Number of Observations N**	**Rank Sum**	**Mean Rank**	**Mean**
3D	50	8301.00	166.02	77.26 [a]
M1	50	8910.00	178.20	79.21 [a]
M2	50	8982.00	179.64	79.29 [a]
M3	50	15,085.00	301.70	118.93 [c]
M4	50	5492.00	109.84	65.85 [b]
M5	50	5262.00	105.24	65.16 [b]
M6	50	9393.00	187.86	81.27 [a]

Values marked with the same letters in columns do not differ significantly; [a,b,c] ($p \leq 0.05$).

The distribution of surface area values computed from the 3D model and the proposed geometric models is presented in Figure 5. The distribution of volume values computed from the same models is presented in Figure 6.

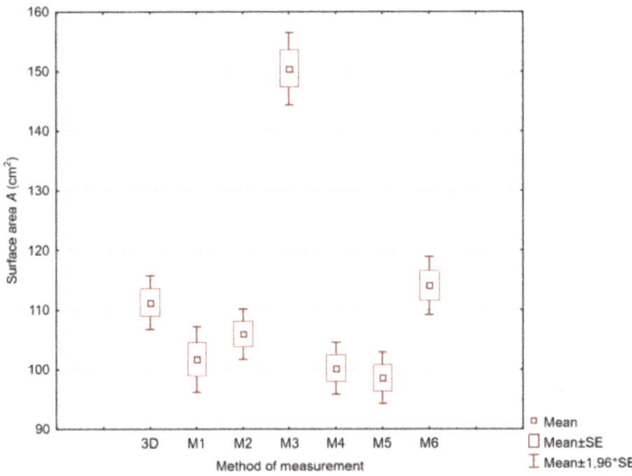

Figure 5. Parameters of normal distribution of cucumber surface area.

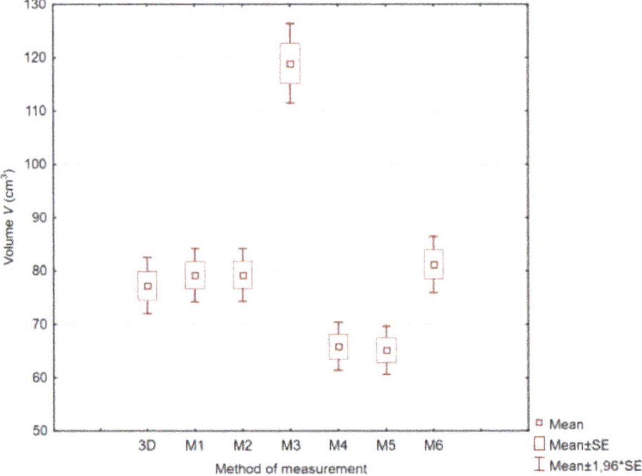

Figure 6. Parameters of normal distribution of cucumber volume.

If we assume that fruit dimensions acquired from 3D scans are burdened by a small error, these parameters can be used as a reference to compare the results of caliper measurements and to describe the shape of cucumber fruits with selected geometric figures. The relative error between the values acquired from 3D scans and direct measurements was regarded as the error of the method. The data presented in Figure 7 indicate that the error in direct measurements of cucumber surface area was smallest for the model combining two truncated cones and two hemispheres with different diameters (M6) where it did not exceed 3%. The error was estimated at 5% when model M2 and formula 4 were used. The data presented in Figure 8 indicate that the error in direct measurements of cucumber volume was smallest for the ellipsoid model (M1), the spheroid model (M2), and the model combining two truncated cones and two hemispheres with different diameters (M6). The error did not exceed 6% when ellipsoids were used, and it was estimated at 6% when model M6 was used.

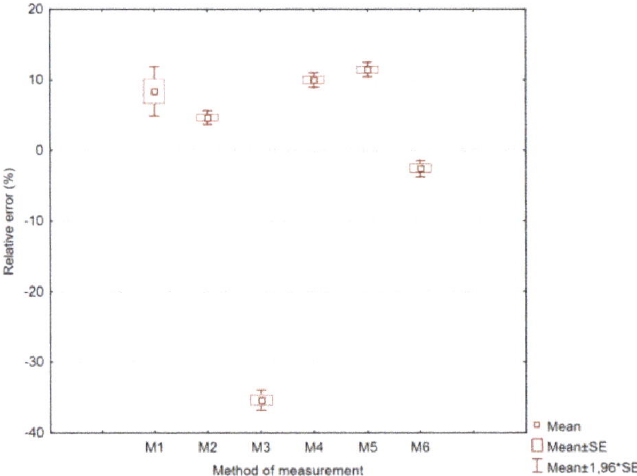

Figure 7. Relative error of cucumber surface area determined with geometric models and the 3D model.

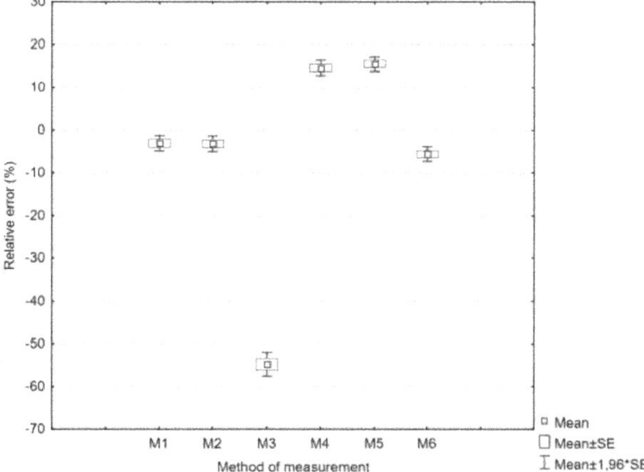

Figure 8. Relative error of cucumber volume determined with geometric models and the 3D model.

The results of this study were compared with the findings of other authors. Zapotoczny (2002) investigated the geometrical parameters of greenhouse-grown cucumbers under laboratory conditions with the use of image analysis methods. The cited author analyzed 2D images of 27 greenhouse-grown cucumbers and determined their average length at 163.17 mm, average width at 32.00 mm, and average projected area at 51.20 cm^2. The multiscale modeling approach deployed by Ho et al. (2013) supports the description of the phenomena occurring inside agricultural raw materials. Multiscale models consist of interconnected sub-models that describe the behavior of raw material in different spatial scales. This approach supports the prediction of processes and phenomena occurring inside raw materials. However, multiscale modeling is relatively complex, and not widely used. Rahmi and Ferruh (2009) described the applicability of 3D models for processing agricultural raw materials and for food production. The presented models were generated based on 3D scans of selected materials, including chicken egg, pear fruit, strawberry fruit, banana and apple. The authors modeled the cooling process in pear fruit and compared the results with experimental findings. Cucumber fruits have never been analyzed in studies on modeling and determination of geometrical parameters of agricultural raw materials.

4. Conclusions

1. Geometric models and direct measurements of the geometric parameters of agricultural products facilitate the planning of spraying, sorting and packaging operations. These methods enable small-scale farmers to easily determine the geometric parameters (volume, surface area) of raw materials without the use of expensive and sophisticated devices such as 3D scanners. Direct measurements of the geometric parameters of agricultural raw materials are consistent with sustainable development principles and can be applied on a large scale.
2. Models where the relative error of measurement does not exceed 5% are recommended when the surface area of cucumbers is calculated with an electronic caliper and mathematical formulas of the presented geometric models. The above condition was fulfilled by the spheroid model (M2) and the model combining two truncated cones and two hemispheres with different diameters (M6). Relative error was higher in the range of 8% to 12% when the surface area of cucumbers was determined with the ellipsoid model (M1), the model combining a truncated cone and two hemispheres (M4) and the model combining a cylinder and two hemispheres (M5). The surface area of cucumbers should not be calculated with the cylinder model (M3) where relative error reached 35%.

3. The volume of fruits can be calculated with the use of the ellipsoid model (M1), the spheroid model (M2) and, similarly to surface area measurements, the model combining two truncated cones and two hemispheres with different diameters (M6). The relative error of the above geometric models did not exceed 5.5%. Relative error was higher in the range of 14% to 16% when cucumber volume was determined with the model combining a truncated cone and two hemispheres (M4) and the model combining a cylinder and two hemispheres (M5). The relative error of the cylinder model (M3) was determined at 54%.

4. The significance of differences between the mean values of surface area was determined in the Kruskal-Wallis test, and no significant differences were observed in models M1, M2, M4 and M5. However, models M1, M4 and M5 cannot be used to determine the surface area of cucumber fruit due to high mean relative error at 8.37%, 9.98% and 11.44%, respectively.

5. In the literature, the mathematical formula for calculating the volume of an ellipsoid (M1) is often used to determine the volume of agricultural products with an ellipsoidal shape. Relative error is estimated at 3% when the volume of ellipsoidal fruits is calculated with the above mathematical formula.

6. In the group of the evaluated methods for determining the geometric parameters of agricultural materials, 3D scanning is the most informative approach. Numerical models support the determination of a full range of geometric parameters (dimensions, area, volume) of entire objects and their fragments. The shape of the analyzed object is stored in computer memory as a cloud of points, and can be used to measure volume without the involvement of displacement methods where the sample is immersed in liquid. Numerical models can also be archived and used for future research.

7. The measurable result of the study was the development of models supporting the determination of the geometric parameters (surface area, volume) of agricultural materials based on their basic dimensions (length, width, thickness). In most cases, the proposed models support the determination of the above geometric parameters with a relative error below 5% within a short period time. Therefore, they can be used in the research and design of new cucumber processing equipment.

8. Further research should focus on the development of models of agricultural raw materials that facilitate the determination of geometric parameters for planning and performing of production processes in agriculture.

Author Contributions: A.A. developed the concept and design of the study; A.A., P.M. and Z.K. conducted the experiments; A.A., Z.K., P.M. and E.Ś. contributed to the literature study; D.C., P.M., Z.K., A.A. and A.J.L. analyzed the data and made final calculations; A.A., Z.K. and P.M. wrote the paper; A.A., P.M. and Z.K. critically revised it.

Funding: This research received no external funding.

Conflicts of Interest: The authors declare no conflict of interest.

References

1. Datta, A.K.; Halder, A. Status of food process modeling and where do we go from here (synthesis of the outcome from brainstorming). *Compr. Rev. Food Sci. Food Saf.* **2008**, *7*, 117–120. [CrossRef]

2. Verboven, P.; De Baerdemaeker, J.; Nicolai, B.M. Using computational fluid dynamics to optimize thermal processes. In *Improving the Thermal Processing of Foods*; Richardson, P., Ed.; CRC Press: Boca Raton, FL, USA, 2004; pp. 82–102.

3. Goni, S.M.; Purlis, E.; Salvadori, V.O. Three-dimensional reconstruction of irregular foodstuffs. *J. Food Eng.* **2007**, *82*, 536–547. [CrossRef]

4. Frączek, J.; Wróbel, M. Methodic aspects of seed shape assessment. *Inżynieria Rolnicza* **2006**, *12*, 155–163. (In Polish)

5. Szwedziak, K.; Rut, J. Assessment of pollutants of the grain corn with the help of computer analysis of the image. *Postępy Techniki Przetwórstwa Spożywczego* **2008**, *1*, 14–15. (In Polish)

6. Deshpande, S.D.; Bal, S.; Ojha, T.P. Physical properties of soybean. *J. Agric. Eng. Res.* **1993**, *56*, 89–98. [CrossRef]

7. Gupta, R.K.; Das, S.K. Physical properties of Sunflower seeds. *J. Agric. Eng. Res.* **1997**, *66*, 1–8. [CrossRef]

8. Cahsir, S.; Marakoglu, T.; Ogut, H.; Ozturk, O. Physical properties of rapeseed (*Brassica napus oleifera* L.). *J. Food Eng.* **2005**, *69*, 61–66.

9. Tańska, M.; Rotkiewicz, D.; Kozirok, W.; Konopka, I. Measurement of the geometrical features and surface color of rapeseeds using digital image analysis. *Food Res. Int.* **2005**, *38*, 741–750. [CrossRef]

10. Jadwisieńczak, K.; Kaliniewicz, Z. Analysis of the mustard seeds cleaning process. Part 1. Physical properties of seeds. *Inżynieria Rolnicza* **2011**, *9*, 57–64. (In Polish)

11. Coskuner, Y.; Karababa, E. Some physical properties of flaxseed (*Linum usitatissimum* L.). *J. Food Eng.* **2007**, *78*, 1067–1073. [CrossRef]

12. Erdogdu, F.; Balaban, M.O.; Chau, K.V. Modeling of heat conduction in elliptical cross-section: II, Adaptation to thermal processing of shrimp. *J. Food Eng.* **1998**, *38*, 241–258. [CrossRef]

13. Luzuriaga, D.A.; Balaban, M.O.; Yeralan, S. Analysis of visual quality attributes of white shrimp by machine vision. *J. Food Sci.* **1997**, *62*, 113–118. [CrossRef]

14. Crocombe, J.P.; Lovatt, S.J.; Clarke, R.D. Evaluation of chilling time shape factors through the use of three-dimensional surface modeling. In Proceedings of the 20th International Congress of Refrigeration 1999: IIR/IIF, Sydney, Australia, 19–24 September 1999; p. 353.

15. Jancsok, P.T.; Clijmans, L.; Nicolai, B.M.; De Baerdemaeker, J. Investigation of the effect of shape on the acoustic response of 'conference' pears by finite element modeling. *Postharvest Biol. Technol.* **2001**, *23*, 1–12. [CrossRef]

16. Borsa, J.; Chu, R.; Sun, J.; Linton, N.; Hunter, C. Use of CT scans and treatment planning software for validation of the dose component of food irradiation protocols. *Radiat. Phys. Chem.* **2002**, *63*, 271–275. [CrossRef]

17. Sabliov, C.M.; Bolder, D.; Keener, K.M.; Farkas, B.E. Image processing method to determine surface area and volume of axi-symmetric agricultural products. *Int. J. Food Prop.* **2002**, *5*, 641–653. [CrossRef]

18. Zapotoczny, P. Measuring geometrical parameters of cucumbers fruits using computer image analysis. *Problemy Inżynierii Rolniczej* **2002**, *4*, 57–64. (In Polish)

19. Scheerlinck, N.; Marquenie, D.; Jancsok, P.T.; Verboven, P.; Moles, C.G.; Banga, J.R.; Nicolai, B.M. A model-based approach to develop periodic thermal treatments for surface decontamination of strawberries. *Postharvest Biol. Technol.* **2004**, *34*, 39–52. [CrossRef]

20. Du, C.; Sun, D.W. Estimating the surface area and volume of ellipsoidal ham using computer vision. *J. Food Eng.* **2006**, *73*, 260–268. [CrossRef]

21. Zheng, C.; Sun, D.W.; Du, C.J. Estimating shrinkage of large cooked beef joints during air-blast cooling by computer vision. *J. Food Eng.* **2006**, *72*, 56–62. [CrossRef]

22. Kim, J.; Moreira, R.G.; Huang, Y.; Castell-Perez, M.E. 3-D dose distributions for optimum radiation treatment planning of complex foods. *J. Food Eng.* **2007**, *79*, 312–321. [CrossRef]

23. Goni, S.M.; Purlis, E.; Salvadori, V.O. Geometry modeling of food materials from magnetic resonance imaging. *J. Food Eng.* **2008**, *88*, 561–567. [CrossRef]

24. Siripon, K.; Tansakul, A.; Mittal, G.S. Heat transfer modeling of chicken cooking in hot water. *Food Res. Int.* **2007**, *40*, 923–930. [CrossRef]

25. Mieszkalski, L. Computer-aiding of mathematical modeling of the carrot (*Daucus carota* L.) root shape. *Ann. Wars. Univ. Life Sci. SGGW Agric.* **2013**, *61*, 17–23.

26. Mieszkalski, L. Bezier curves in modeling the shapes of biological objects. *Ann. Wars. Univ. Life Sci. SGGW Agric.* **2014**, *64*, 117–128.

27. Balcerzak, K.; Weres, J.; Górna, K.; Idziaszek, P. Modeling of agri-food products on the basis of solid geometry with examples in AutoDesk 3ds Max and finite element mesh generation. *J. Res. Appl. Agric. Eng.* **2015**, *60*, 5–8.

28. Ho, Q.T.; Carmeliet, J.; Datta, A.K.; Defraeye, T.; Delele, M.A.; Herremans, E.; Opara, L.; Ramon, H.; Tijskens, E.; Sman, R.; et al. Multiscale modeling in food engineering. *J. Food Eng.* **2013**, *114*, 279–291. [CrossRef]

29. Anders, A. *Determination of the Geometric Parameters of Seeds with Different Methods*; Wydawnictwo Uniwersytetu Warmińsko-Mazurskiego w Olsztynie: Olsztyn, Poland, 2019; ISBN 978-83-8100-163-2. (In Polish)

30. Rahmi, U.; Ferruh, E. Potential use of 3-dimensional scanners for food process modeling. *J. Food Eng.* **2009**, *93*, 337–343.
31. Anders, A.; Markowski, P.; Kaliniewicz, Z. Numerical modelling of agricultural products on the example of bean and yellow lupine seeds. *Int. Agrophys.* **2015**, *29*, 397–403. [CrossRef]
32. Anders, A.; Markowski, P.; Kaliniewicz, Z. Evaluation of geometric and physical properties of chosen pear cultivars based on numerical models obtained by a 3D scanner. *Zeszyty Problemowe Postępów Nauk Rolniczych* **2014**, *577*, 3–12. (In Polish)
33. Anders, A.; Markowski, P.; Kaliniewicz, Z. The application of a 3D scanner for the evaluation of geometric properties of *Cannabis sativa* L. seeds. *Acta Agrophys.* **2014**, *21*, 391–402. (In Polish)
34. NextEngine User Manual. 2010. Available online: http://www.nextengine.com (accessed on 1 July 2018).
35. MeshLab Visual Computing Lab—ISTI—CNR. 2013. Available online: http://meshlab.sourceforge.net (accessed on 1 July 2018).
36. Gastón, A.L.; Abalone, R.M.; Giner, S.A. Wheat drying kinetics. Diffusivities for sphere and ellipsoid by finite elements. *J. Food Eng.* **2002**, *52*, 313–322. [CrossRef]
37. Bronsztejn, I.N.; Siemiendiajew, K.A.; Musiol, G.; Muhling, H. *Nowoczesne Kompendium Matematyki*; PWN: Warszawa, Poland, 2009. (In Polish)

 *sustainability*

Article

The Effect of Selected Parameters on Spelt Dehulling in a Wire Mesh Cylinder

Andrzej Anders, Ewelina Kolankowska, Dariusz Jan Choszcz, Stanisław Konopka and Zdzisław Kaliniewicz *

Department of Heavy Duty Machines and Research Methodology, University of Warmia and Mazury in Olsztyn, 10-719 Olsztyn, Poland; andrzej.anders@uwm.edu.pl (A.A.); ewelina.kolankowska@uwm.edu.pl (E.K.); dariusz.choszcz@uwm.edu.pl (D.J.C.); stanislaw.konopka@uwm.edu.pl (S.K.)
* Correspondence: zdzislaw.kaliniewicz@uwm.edu.pl; Tel.: +48-089-523-39-34

Received: 29 October 2019; Accepted: 14 December 2019; Published: 19 December 2019

Abstract: A spelt dehuller with an innovative structural design is described in the study. In the developed solution, spelt kernels are separated by the mechanical impact of friction in a wire mesh cylinder with 4 × 4 mm openings. The dehuller is powered by a motor with a rotating impeller with an adjustable blade angle. In the experimental part of the study, spelt kernels are dehulled at five rotational speeds of the shaft: 160 to 400 rpm at intervals of 60 rpm, and five rotor blade angles: 50° to 90° at intervals of 10°. The efficiency of spelt dehulling (removal of glumes and glumelles) is evaluated based on kernel separation efficiency, husk separation efficiency, and the proportion of damaged kernels.

Keywords: spelt; threshing; dehuller

---

1. Introduction

Sustainable agriculture is the production of healthy, high-quality foods in a way that protects the environment and provides economic benefits to farmers. Spelt as well as emmer and einkorn are ancient wheat species. Spelt is a relict species that has recently been "rediscovered" and is being widely used in the production of flour, groats, flakes, pasta, bread, vodka, and beer [1,2]. The growing popularity of spelt on the consumer market can be attributed to its unique flavor, health-promoting properties, environmental benefits, and high content of biologically active compounds including essential nutrients. Spelt grain contains high-quality protein; unsaturated fatty acids; B complex vitamins; PP vitamin (niacin, vitamin B3, and pellagra-preventing factor); and minerals such as zinc, potassium, calcium, and iron [3–6]. Spelt kernels are enveloped by tough husks that protect this grain against atmospheric pollution and radiation. Tough husks are difficult to separate, which poses a considerable problem during threshing and limits the processing suitability of spelt grain. Spelt is difficult to harvest because grain is not effectively separated from spikelets by combine harvesters. Spelt spikes are hard, awned or awnless, and nonfree-threshing. Loose spikes have a brittle rachis, which is broken during threshing into several fragments, and each fragment contains one spikelet. Spikelets typically have two flowers with two kernels per spikelet, and kernels are tightly enclosed by four glumelles and two glumes, which makes the species nonfree-threshing. Spelt spikelets rarely contain three kernels. Spelt kernels are vitreous, white to red in color, with distinctive brush hairs in the apical part [7,8].

The harvested grain cannot be directly used in the food processing industry due to considerable contamination with chaff. After harvest, spikelets require additional treatment in a process that is commonly referred to as threshing. The harvested material is a mixture of grain and spikelets. This mixture has to be separated before threshing, because repeated threshing only increases energy consumption, contributes to grain damage, and increases grain loss [9].

The popularity of spelt is on the rise in the agricultural sector, in particular in organic farms. According to Mieczysław Babalski, one of the leading spelt producers in Poland, and Józef Tyburski,

PhD, spelt can be effectively dehulled in a modified thresher, an abrasive discdehuller, or specialist equipment manufactured in Western Europe [7,8,10]. Shelling machines and tangential dehullers can also be used for this purpose [11]. Most commercially available dehulling machines feature a cylindrical sieve and a rotating impeller with beaters. However, their efficiency is not highly satisfactory because spelt grain is encased by tightly adhering husks. Dehulling machines are also very expensive, and many farmers make attempts to build their own dehulling equipment. For instance, Tudor (2012) designed and built a dehulling machine as part of the Farmer FNE11–731 project [12]. It should also be noted that the grain and spikelet mixture has to be cleaned several times during supplementary threshing.

Dehulling devices are not highly efficient in separating spelt grain from husks, and they are very expensive. In Polish farms, modified clover hullers are often used for the supplementary threshing of spelt. However, spelt grain is not effectively separated by modified clover hullers, and the processed material requires further cleaning and sorting.

To address these concerns, a spelt dehuller with an innovative structural design is proposed in this study [13]. In the developed solution, spelt kernels are separated by the mechanical impact of friction in a wire mesh cylinder with 4 × 4 mm openings. The dehuller is powered by a motor with a rotating impeller, and the blade angle can be adjusted from 50° to 90°, at intervals of 10° [14]. The aim of the study is to determine the effect of selected parameters (rotor blade angle and rotational speed of the shaft) on the efficiency of glume and glumelle removal from spelt kernels.

2. Materials and Methods

The experimental material comprised spikelets of spelt cv. *Schwabenkorn* purchased in an organic farm in Praslity (54°01′55″ N 20°21′29″ E), municipality of Dobre Miasto, Region of Warmia and Mazury in Poland. The experimental material (spelt) was purchased from an organic farm (not from a commercial seed center) immediately after harvest. In order to remove impurities, weeds, and kernels of other cereal species, the seeds were cleaned using screens/sieves and a pneumatic separator. The percentage share of seeds of the main species in an average sample of input material (purity of the examined material) was determined at 85%, and its relative moisture content was determined at 11.56 ± 2.00%.

The proposed dehuller (Figure 1) features a wire mesh cylinder with 4 × 4 mm openings (Figure 2) and an open area factor (as a ratio (percentage) of the area available for the material to pass (openings) to the total area of the screen) of 0.38. Cylindrical mesh sieves are generally applied in specialist machines for separating spelt grain from chaff and in modified clover hullers.

The experiment was conducted in two stages. The process of removing husks from spelt kernels was examined in the first stage, and the separation of grain from threshed spikelets was analyzed in the second stage.

The following parameters were adopted in the first stage of the experiment:

1. Fixed parameters:

    –   Width of the grain inlet s_z = 10 mm;
    –   Sample mass m_p = 300 g;
    –   Relative moisture content of the sample—11.56 ± 2.00%;
    –   Rubber impeller blades (properties T-REX40 red natural rubber (T-Rex Rubber International, Netherlands): hardness: 40 ± 5° ShA, min. 600% elongation at break, tensile strength—16 MPa, max. abrasion resistance—210 mm^3);
    (a) Distance between the cylinder and beaters s_r = 2 mm.

2. Independent variables:

    –   Rotor blade angle αw—50° ÷ 90°, at intervals of 10°;
    –   Rotational speed of the shaft nw—160 ÷ 400 rpm, at intervals of 60 rpm;

3. Dependent variables:

- Kernel separation efficiency η_z, %;
- Husk separation efficiency η_p, %;
- Proportion of damaged kernels U_z, %.

Figure 1. Prototype of a spelt dehuller: 1—dehulling chamber, 2—wire mesh cylinder, 3—rotating impeller, 4—receiving outlet, 5—gearmotor, 6—radial fan, and 7—impeller blades.

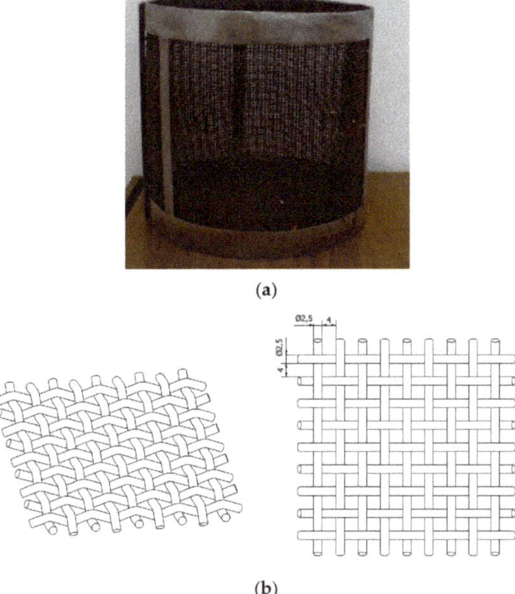

Figure 2. Wire mesh cylinder with 4×4 mm openings [15]: (a) cylinder view, (b) sieve structure.

Kernel and husk separation efficiency η and the proportion of damaged kernels U_z were determined based on the mass of the mixture components separated in each sample in the first stage of the experiment [16].

In each sample, kernel separation efficiency η_z was calculated with the use of the following formula:

$$\eta_z = \frac{M_z}{M_c} \cdot 100\% \tag{1}$$

where

M_z—mass of separated kernels (g),
M_c—mass share of kernels in a sample (g).
In each sample, husk separation efficiency η_p was calculated with the use of the following formula:

$$\eta_p = \frac{M_p}{M_m} \cdot 100\% \tag{2}$$

where

M_p—husk mass (g),
M_c—mass share of husks in a sample (g).
The proportion of damaged kernels U_z was calculated with the use of the following formula:

$$U_z = \frac{M_u}{M_c} \cdot 100\% \tag{3}$$

where

M_u—mass of damaged kernels (g).
In the first stage of the experiment, the dehuller's operating parameters were set before every measurement. Spelt spikelets were fed into the hopper, and the dehuller was turned on. Machine start-up time was around 30 s. The sliding gate in the hopper was opened, and husks were removed from spikelets by the mechanical impact of friction. The processed material was weighed on a laboratory scale to the nearest 0.01 g. A preweighed sample of spelt spikelets was always present in the feeder hopper to ensure the repeatability of measurements.

In the second stage of the experiment, the threshed material was separated into whole kernels, damaged kernels, spikelets, and husks in a pneumatic separator ([17], Figure 3) and a Fritsch Analysette 3 (Fritsch GmbH, Germany) vibratory sieve shaker. Based on initial measurements, the air flow rate in the pneumatic separator was set at 7.5 m·s^{-1}. The vibratory sieve shaker consisted of 5 mesh screens with longitudinal openings (4.0 × 20 mm, 3.5 × 20 mm, 3.0 × 20 mm; 2.5 × 20 mm, and 2.0 × 20 mm). Processing time was 3 min, and the amplitude of vibration was 0.1 mm. Spelt kernels were considered damaged when they were broken, cracked, had a interior visible, or had damaged seed coats.

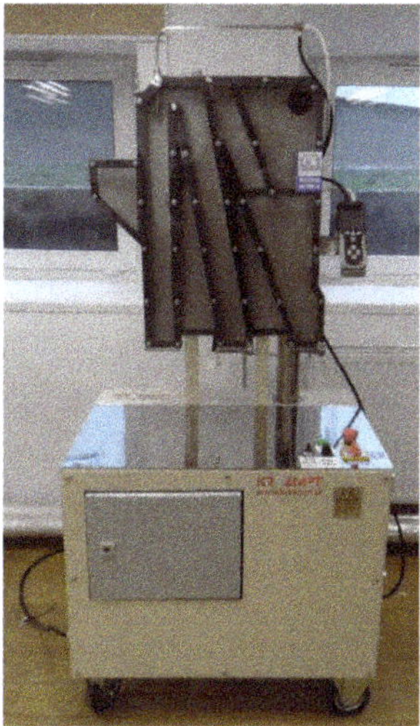

Figure 3. W.124984 pneumatic separator [15].

In all sub-stages of the experiment, the measurements for each combination of independent variables were performed in triplicate.

The results of all measurements were processed statistically by correlation analysis and stepwise multivariate polynomial regression analysis for a second-order polynomial model. Statistical analyses were conducted in Statistica PL v. 13.1 [18].

3. Results and Discussion

The results of the statistical analysis of the proportion of spelt kernels damaged at different rotor blade angles and rotational speeds are presented in Table 1.

The results of the multiple regression analysis indicate that the correlation coefficient for the proportion of spelt kernels damaged during threshing in a wire mesh cylinder with 4 × 4 mm openings at the tested rotor blade angles and rotational speeds ranged from 0.32 to 0.58. The multiple regression equations describing the proportion of damaged kernels generally fit the empirical data well, and the coefficient of determination for the wire mesh cylinder reached 0.70.

The equations describing the percentage of kernels damaged at different rotor blade angles and rotational speeds are presented graphically in Figure 4. In the wire mesh cylinder with 4 × 4 mm openings, the proportion of damaged kernels increased with a rise in rotational speed in the following range of rotor blade angles: $\alpha_w = 70° \div 80°$.

In the tested wire mesh cylinder, the efficiency of kernel separation was highly correlated with rotational speed, and the value of the correlation coefficient was determined at 0.84 (Table 2). The efficiency of kernel separation was less correlated with the rotor blade angle, and the value of the correlation coefficient was determined at 0.23. The multiple regression equation describing the efficiency of kernel separation was characterized by good and very good fit to empirical data. In the wire mesh cylinder with 4 × 4 mm openings, the coefficient of determination after stepwise elimination of non-significant variables was determined at 0.88.

The quadratic equation describing kernel separation efficiency at different rotor blade angles and rotational speeds is presented graphically in Figure 5. In the wire mesh cylinder with 4 × 4 mm openings, kernel separation efficiency increased with a rise in rotational speed. The rotor blade angle exerted a smaller and more ambiguous effect on kernel separation efficiency. Kernels were most efficiently separated at a rotational speed of 400 rpm and a rotor blade angle of 90°.

Table 1. A correlation analysis of the proportion of spelt kernels damaged during threshing in a wire mesh cylinder with 4×4 mm openings.

	General Data:			
	Correlation Coefficients are Significant at $\alpha < 0.05$			
	N = 25			
No.	**Variable**	**Mean**	**Standard Deviation**	**Coefficient of Variation (%)**
1.	Rotor blade angle α_w (°)	70.00	14.43	20.62
2.	Rotational speed n_w (rpm)	280.00	86.60	30.93
3.	Proportion of damaged kernels U_z (%)	4.74	3.96	83.60

	Correlation Matrix			
		α_w	n_w	U_z
	α_w	1	0	0.584
	n_w		1	0.320
	U_z			1

Variable	**F-Statistic**	**Coefficient of Determination R^2**	**Standard Error of the Estimate**	**t-Statistic**
Free term				−3.18
α_w				3.76
α_w^2	8.81	0.70	2.44	−3.84
n_w				−0.30
n_w^2				−0.04
$\alpha_w \cdot n_w$				1.14

Quadratic equation for two independent variables:

$$U_z = 1.601419 \cdot \alpha_w - 0.011224 \cdot \alpha_w^2 - 0.016039 \cdot n_w - 0.000003 \cdot n_w^2 + 0.000465 \cdot \alpha_w \cdot n_w - 54.454576$$

Stepwise regression did not decrease the degree of the polynomial function for two independent variables

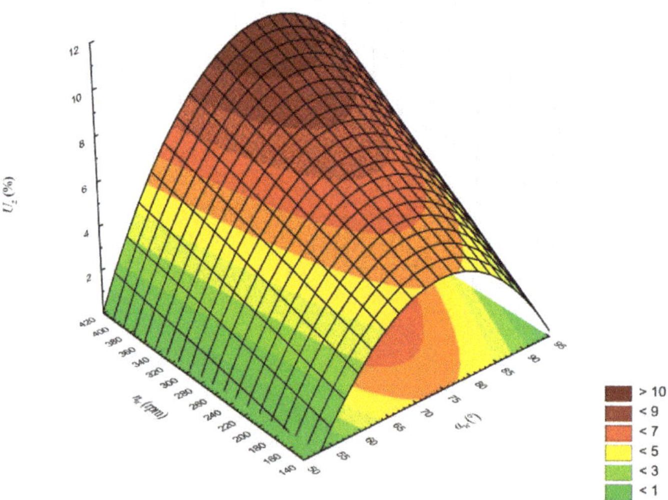

Figure 4. Proportion of kernels damaged at different rotor blade angles α_w and rotational speeds n_w in a wire mesh cylinder with 4×4 mm openings.

Table 2. A correlation analysis of kernel separation efficiency during threshing in a wire mesh cylinder with 4×4 mm openings.

General Data: Correlation Coefficients are Significant at $\alpha < 0.05$ N = 25				
No.	**Variable**	**Mean**	**Standard Deviation**	**Coefficient of Variation (%)**
1.	Rotor blade angle α_w (°)	70.00	14.43	20.62
2.	Rotational speed n_w (rpm)	280.00	86.60	30.93
3.	Kernel separation efficiency η_z (%)	52.14	8.58	16.46
Correlation Matrix				
		α_w	n_w	η_z
	α_w	1	0	0.226
	n_w		1	0.845
	η_z			1
Variable	**F-Statistic**	**Coefficient of Determination R^2**	**Standard Error of the Estimate**	**t-Statistic**
Free term				5.65
α_w	53.77	0.88	3.11	−4.41
α_w^2				4.68
n_w				11.41
Quadratic equation for two independent variables: $\eta_z = -2.3039{\cdot}\alpha_w + 0.0174{\cdot}\alpha_w^2 + 0.0837{\cdot}n_w + 101.1434$				

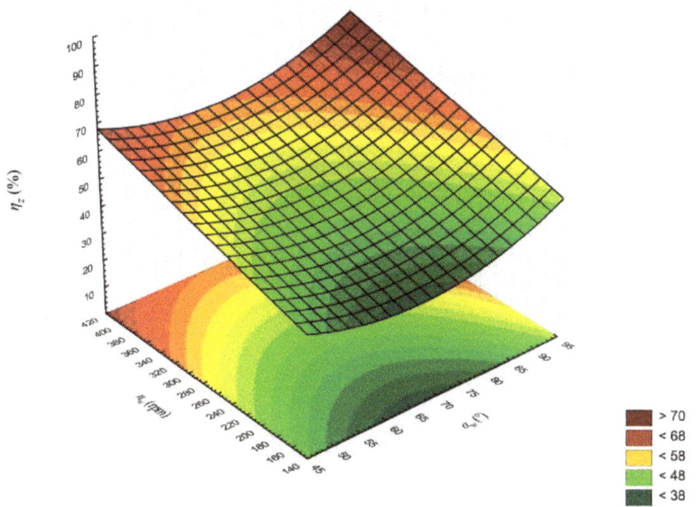

Figure 5. Kernel separation efficiency at different rotor blade angles α_w and rotational speeds n_w in a wire mesh cylinder with 4×4 mm openings.

Husk separation efficiency was measured at different rotor blade angles and rotational speeds; the results were processed statistically and are presented in Table 3. In the wire mesh cylinder with 4×4 mm openings, the correlation coefficient between husk separation efficiency and rotor blade angle was determined at 0.67, and the correlation coefficient between husk separation efficiency and rotational speed was determined at 0.54. The multiple regression equation describing husk separation efficiency was characterized by a very good fit to empirical data. The value of the coefficient of determination for the tested mesh cylinder was 0.89.

Table 3. A correlation analysis of husk separation efficiency during threshing in a wire mesh cylinder with 4 × 4 mm openings.

	General Data: Correlation Coefficients are Significant at $\alpha < 0.05$ N = 25			
No.	**Variable**	**Mean**	**Standard Deviation**	**Coefficient of Variation (%)**
1.	Rotor blade angle α_w (°)	70.00	14.43	20.62
2.	Rotational speed n_w (rpm)	280.00	86.60	30.93
3.	Husk separation efficiency η_p (%)	42.42	9.52	22.45
Correlation Matrix				
		α_w	n_w	η_p
	α_w	1	0	0.673
	n_w		1	0.541
	η_p			1
Variable	**F-Statistic**	**Coefficient of Determination R^2**	**Standard Error of the Estimate**	**t-Statistic**
Free term				5.85
α_w	54.86	0.89	3.42	−4.95
α_w^2				5.36
$\alpha_w \cdot n_w$.				7.20
Quadratic equation for two independent variables: $\eta_p = -2.8529 \cdot \alpha_w + 0.0219 \cdot \alpha_w^2 + 0.0008 \cdot \alpha_w \cdot n_w + 114.3675$				

The quadratic equation describing husk separation efficiency at different rotor blade angles and rotational speeds is presented graphically in Figure 6. In the wire mesh cylinder with 4 × 4 mm openings, husks were most effectively separated at the rotational speed of 400 rpm and the rotor blade angle of 90°.

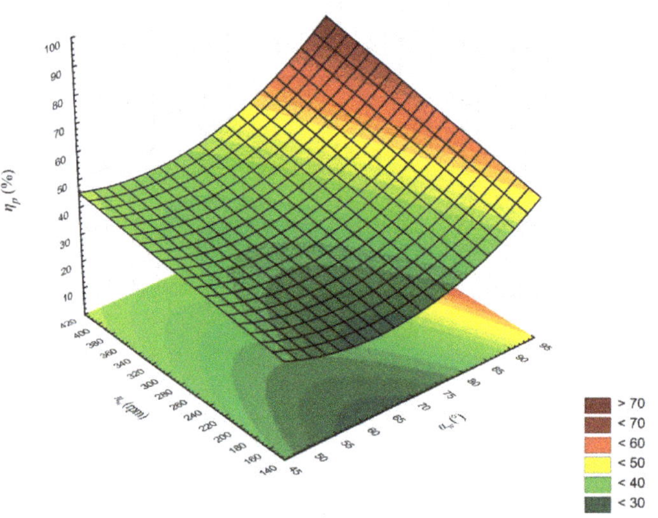

Figure 6. Husk separation efficiency at different rotor blade angles α_w and rotational speeds n_w in a wire mesh cylinder with 4 × 4 mm openings.

4. Conclusions

In the proposed spelt dehuller with an adjustable blade angle, where spelt kernels were separated by the mechanical impact of friction in a wire mesh cylinder with 4×4 mm openings, the highest proportion of damaged kernels U_z (13.77%) was noted at a rotational speed of 400 rpm and a rotor blade angle of 80°. Kernel separation efficiency η_z was highest (68.72%) at a rotational speed of 400 rpm and a rotor blade angle of 50°. Husk separation efficiency η_p was highest (60.53%) at a rotational speed of 400 rpm and a rotor blade angle of 90°.

The proportion of damaged kernels, kernel separation efficiency, and husk separation efficiency in the proposed spelt dehuller can be described by linear equations or stepwise regression for a second-order polynomial model with the elimination of non-significant variables, where the rotor blade angle and rotational speed are the independent variables. The multiple regression equations for the separation process in a wire mesh cylinder with 4×4 mm openings were characterized by good and very good fit to empirical data. The values of the coefficient of determination R^2 ranged from approximately 0.70 to 0.90.

It can be concluded that the dehulling efficiency of spelt is significantly influenced by both tested variables, i.e., the rotational speed of the impeller n_w and the rotor blade angle α_w.

Grain processing is particularly important in the case of spelt, which has high nutritional value and can be used in the food industry, thus contributing to the development of sustainable agriculture and the extensive use of agricultural resources.

Author Contributions: E.K. and D.J.C. conceived and designed the experiments; E.K. performed the experiments; A.A., D.J.C., S.K., and Z.K. contributed to the literature study; E.K., D.J.C., and S.K. analyzed the data; A.A., E.K., and Z.K. wrote the paper; A.A., D.J.C., and S.K. critically revised the manuscript.

Funding: This research received no external funding.

Conflicts of Interest: The authors declare no conflict of interest.

References

1. Cegielska, A.; Gromulska, W. Różnorodność produktów z orkiszu (Diverse spelt products). *Przegląd Zbożowo-Młynarski* **2008**, *5*, 30–31. (In Polish)
2. Majewska, K.; Dąbkowska, E.; Żuk-Gołaszewska, K.; Tyburski, J. Baking quality of flour obtained from grain of chosen spelt varieties (*Triticum spelta* L.). *Żywność Nauka Technol. Jakość* **2007**, *2*, 60–71.
3. Capouchová, I. Technological quality of spelt (*Triticum spelta* L.) from ecological growing system. *Sci. Agric. Biochem.* **2001**, *32*, 307–322.
4. Kohajdova, Z.; Karovicova, J. Nutritional Value and Baking Applications of Spelt Wheat. *Acta Sci. Pol. Technol. Aliment.* **2008**, *7*, 5–14.
5. Wieser, H. Comparative investigations of gluten proteins from different wheat species. III. N–terminal amino acid sequences of a–gliadins potentially toxic for celiac patients. *Eur. Food Res. Technol.* **2001**, *213*, 183–186.
6. Abdel-Aal, E.S.M.; Hucl, P. Amino acid composition and in vitro protein digestibility of selected ancient wheats and their end products. *J. Food Comp. Anal.* **2002**, *15*, 737–747. [CrossRef]
7. Babalski, M.; Przybylak, Z.; Przybylak, K. *Uzdrawiające Ziarna Zbóż (Cereal Grains with Healing Properties)*; Eko Media: Bydgoszcz, Poland, 2013; pp. 1–191. ISBN 9788363537128. (In Polish)
8. Tyburski, J.; Babalski, M. *Uprawa Pszenicy Orkisz (Spelt Cultivation)*; Centrum Doradztwa Rolniczego w Brwinowie, Oddział w Radomiu: Radom, Poland, 2006; pp. 1–25. ISBN 8360185263. (In Polish)
9. Frączek, J.; Reguła, T. Method of evaluation of susceptibility of spelt grains to mechanical damages during the threshing process. *Inżynieria Rol.* **2010**, *4*, 51–58.
10. Choszcz, D.J.; Konopka, S.; Zalewska, K. Characteristics of physical properties of selected varieties of spelt. *Inżynieria Rol.* **2010**, *4*, 23–28.
11. Budzyński, W. (Ed.) *Pszenice—Zwyczajna, Orkisz, Twarda: Uprawa i Zastosowanie (Common Wheat, Spelt and Durum Wheat: Cultivation and Applications)*; Powszechne Wydawnictwo Rolnicze i Leśne: Poznań, Poland, 2012; pp. 1–328. ISBN 9788309011354. (In Polish)

12. Tudor, N. Farmer Built Spelt Dehuller. USDA Sustainable Agriculture Research and Education Report. 2012. Available online: http://mysare.sare.org/MySare/ProjectReport.aspx?do=viewRept&pn=FNE11--731&t=1&y=2014 (accessed on 18 November 2015).
13. Kolankowska, E.; Choszcz, D. Urządzenie do Usuwania Plew z Ziarna Orkiszu (Device for Removing Chaff from Spelt Grain). Patent PL408757, 29 November 2019. (In Polish).
14. Kolankowska, E.; Choszcz, D. Urządzenie do Usuwania Plew z Ziarna Orkiszu (Device for Removing Chaff from Spelt Grain). Utility Model Application No W123522, 31 May 2017. (In Polish).
15. Kolankowska, E. Doskonalenie Procesu Usuwania Plew z Ziarna Orkiszu (Optimization of the Dehulling Process for Spelt Kernels). Ph.D. Thesis, Uniwersytet Warmińsko-Mazurki w Olsztynie, Wydział Nauk Technicznych, Olsztyn, Poland, 2019. (In Polish).
16. Grochowicz, J. *Maszyny do Czyszczenia i Sortowania Nasion (Seed Cleaning and Sorting Machines)*; Wydawnictwo Akademii Rolniczej: Lublin, Poland, 1994; pp. 1–326. ISBN 839016129X. (In Polish)
17. Kolankowska, E.; Choszcz, D. Separator powietrzny (Air Separator). Utility Model Application No W124984, 31 December 2018. (In Polish).
18. Stanisz, A. *Przystępny kurs Statystyki w Oparciu o Program STATISTICA PL na Przykładach z Medycyny. Tom 1. Statystyki Podstawowe (Accessible Course in Statistics Based on the STATISTICA PL Software on Examples from Medicine. Tome 1. Basic Statistics)*; StatSoft Polska: Kraków, Poland, 2007; pp. 1–532. ISBN 8388724185. (In Polish)

 *sustainability*

Article

Assessment of the Potential Use of Young Barley Shoots and Leaves for the Production of Green Juices

Agata Blicharz-Kania [1], Dariusz Andrejko [1], Franciszek Kluza [1], Leszek Rydzak [1,*] and Zbigniew Kobus [2]

1   Department of Biological Bases of Food and Feed Technologies, University of Life Sciences in Lublin, 20-612 Lublin, Poland
2   Department of Technology Fundamentals, University of Life Sciences in Lublin, 20-612 Lublin, Poland
*   Correspondence: leszek.rydzak@up.lublin.pl; Tel.: 81-531-9650

Received: 9 June 2019; Accepted: 19 July 2019; Published: 21 July 2019

Abstract: It is possible to use the aboveground parts of barley, which are cultivated as a forecrop. They are often simply composted or dried for bedding. It is worth trying other more effective methods of processing aboveground biomass. The aim of this study was to preliminary investigate the possibility of using young barley leaves and shoots for the production of green juice with potential health properties. The material was collected at days 7, 14, 21, and 28 after plant emergence. The length and strength of the shoots were measured and the pressing yield was calculated. The pH value and the content of protein, chlorides, and reducing sugars were also determined. The juice was additionally subjected to pasteurisation and freezing, and changes in pH and chlorophyll content occurring during storage were determined. The pressing yield of young barley leaves and shoots was estimated to be between 69% and 73%. The product was characterised by a high content of total protein (34.45%–51.81%$_{d.w.}$) and chlorophylls (6.62 mg·g^{-1}). The chlorophyll content declined during barley juice storage. Pasteurisation of the juice from young barley leaves does not induce statistically significant changes in the pH of the juice, but reduces the chlorophyll content. Our results revealed that the most effective way to preserve the green juice is by freezing. This process does not induce changes in juice acidity and only slightly reduces the chlorophyll content during storage of the product.

Keywords: juice; barley; pressing; protein; chlorophylls; green food

1. Introduction

In order to improve the soil quality and protect it from weeds and, consequently, obtain a better crop of the cultivated plant, producers often use forecrop cultivation. Depending on what kind of plant will be grown, an appropriate forecrop will be selected. For cereals, the best options are root crops (mainly potatoes), legumes, or rapeseed. However, when growing vegetables, such as tomatoes, carrots, or white cabbage, it is better to use a cereal forecrop, and rye, wheat, or barley can be used [1]. These cereals can be used via two techniques. The first method involves ploughing whole plants and using them as a fertilizer. Plants that are a forecrop can also be cut. Then, only roots and postharvest residues are ploughed. In the case of the latter method, it is possible to continue using the green parts of the plants, which are often simply composted or dried for bedding.

Recently, products from young barley leaves and shoots have gained popularity. They are available in a variety of forms, i.e., juice, tablets, or powder. Young parts of plants are a source of phenolic acids and many vitamins, e.g., C, E, and B group vitamins. Additionally, barley grass contains substantial amounts of carotenoids, folic acid, calcium, iron, magnesium, potassium, zinc, and copper. Importantly, products obtained from cereal leaves and shoots can be used as supplements in a high-protein diet. Barley grass contains approximately 30% of protein in dry matter [2–4]. The chemical index of nutritional value (Chemical Score, CS) is 41.44% (Methionine) [5]. Barley shoots and leaves are also a source of chlorides. Naturally occurring chlorides exert a beneficial effect on organism function. Chloride ions participate in the regulation of water, as well as electrolyte metabolism and maintaining acid–base balance [6].

One of the most important active compounds in "green food" is chlorophyll. Seed plants contain type A and B chlorophylls, which differ according to the type of substituent on the second pyrrole ring. However, these pigments are unstable. Many factors, e.g., UV radiation or pH and temperature changes, cause chlorophyll degradation in food products. Chlorophylls are a valuable source of magnesium; they can also improve metabolism and eliminate unpleasant mouth odours. Additionally, they have antibacterial and anti-inflammatory properties, remove toxins from the liver and blood [7–11], and even act as a haemoglobin substitute [12].

The consumption of green juices has therapeutic properties: it exerts an antidiabetic effect, regulates blood pressure, strengthens immunity, protects the liver, and has anti-acne and antidepressant activities. It also improves the function of the digestive tract and prevents hypoxia, cardiovascular diseases, fatigue, and constipation. Additionally, it alleviates atopic dermatitis and has anti-inflammatory, antioxidant, and anticancer effects [4]. Kubatka et al. [13] have demonstrated positive changes in tumour cells in rats treated with juice from young barley leaves. A significantly more pronounced effect of the therapeutic treatment was observed in a study group receiving a diet supplemented with the juice. It has also been confirmed that supplementation of the diet with barley leaf powder can relieve the clinical symptoms of diabetes [14]. Additionally, barley grass contains substantial amounts of dietary fibre (mainly an insoluble fraction), which has a positive effect on metabolism through regulating the appetite and, thus preventing the development of overweight. Son et al. [3] recommend enrichment of the diets of young children using valuable nutrients from young barley leaves.

Products from young barley leaves have been used in East Asia for a long time. Currently, they are available in supermarkets, as well as online, in the United States and many European countries. The increase in consumer awareness has contributed to an appreciation of the health benefits of "green food" [8]. A number of products based on young cereal leaves are recommended as dietary supplements and, hence, are currently being manufactured to be purchased in pharmacies.

There is a paucity of scientific publications confirming the health-promoting properties of the juice from young barley leaves. There are also no preliminary investigations describing the impact of production and processing on the quality of green juices. Furthermore, the relationship between the date of raw material harvesting and the pressing process, as well as the content of nutrients—including proteins—in the juice, has not yet been analysed. Another unexplored area is the changes occurring in the chlorophyll content of barley juice, depending on the thermal treatment. Knowledge of these relationships has great practical significance as it provides information on methods for the acquisition of a product with a high nutritional value and, at the same time, ensures the longest possible growth period for the plant (and, thus, the greatest mass of roots to be used as forecrop).

The aim of this work was to preliminary examine the efficiency of the process of pressing green juice from young barley leaves and shoots and to determine the chemical composition of the product obtained. An additional goal of the research was to compare the effect of preservation methods like pasteurization and freezing on the chlorophyll content and pH of barley juice.

2. Materials and Methods

2.1. Research Material

Barley grain cv. Kangoo was used for the investigations. The seeds, weighing 5 kg, were sown under laboratory conditions (Figure 1). No fertilisation was applied during the cultivation. The material was collected at days 7, 14, 21, and 28 after plant emergence. The crop area was divided into four parts, and these four parts were then divided again into 12 smaller ones. Three parts from the whole field were chosen at random for each series of investigation (different harvest time: 7, 14, 21, and 28 days). Each sample of shoots and leaves of young barley for juice production weighed 200 g.

Figure 1. Cultivation of barley.

2.2. Measurement of Physical Properties

Samples without mechanical damage were selected for determination of the length and strength of the barley. The length of five barley shoots was measured with the use of a calliper to the nearest 0.01 mm. The shoot strength was determined by applying a uniaxial tensile test to the leaves. The test was carried out using a Zwick/Roell Z0.5 materials testing machine (Zwick.Roell AG, BT1-FR0.5TN.D14, Ulm, Germany) equipped with a measuring head at a maximum force of 50 N (travel speed = 50 mm·min^{-1}). Tensile strength was applied to the material until rupture. TestXpert II software (Zwick/Roell AG, Ulm, Germany) was used to assess the force required for destruction of the barley grass.

2.3. Pressing

The juice was squeezed from barley grass (with a weight of 200 g for each repetition) using a press designed for pressing green plant leaves, i.e., Manual Juicer BL-30 (BioChef, Byron Bay, NS, Australia).

The pressing efficiency was calculated using the following equation:

$$W_j(\%) = \frac{M_j}{M_i} \cdot 100 \tag{1}$$

where:

W_j—pressing yield, %;
M_j—mass of juice after pressing, kg;
M_i—mass of input material, kg.

2.4. Preserving Juice

The juice obtained in the first harvest term (at seven days after plant emergence) was additionally subjected to pasteurisation for 10 min at 75 °C and freezing. Immediately after the thermal treatment, the product was cooled (in a blast chiller–freezer) to 3 °C. The other batch of juice was blast-frozen to a temperature of −18 °C. The material was stored under appropriate conditions, i.e., 4 °C in the case of the unprocessed (UJ) and pasteurised (PJ) juice and −18 °C in the case of the frozen juice (FJ). The study material was analysed after three (UJ and PJ) and seven (UJ, PJ, and FJ) storage days.

2.5. Determination of pH

Juice samples were placed in 50 mL beakers and the pH was recorded with a pH meter (model 780, Metrohm AG, Herisau Switzerland).

2.6. Measurement of Total Protein Content

The determinations were carried out using the Kjeldahl method and a Foss Kjeltec 8400 automatic distiller (Foss Anatytical AB, Höganäs, Sweden). The total protein content was calculated using a 6.25 conversion factor.

2.7. Determination of the Chloride Content

The chloride content was determined with the Mohr method using a TitraLab AT1000 Series automatic titrator (HACH Company, Willstätterstraße, Germany). The solution was titrated with a 0.1 N silver nitrate solution. The chloride content was given as g in 100 $g_{f.j.}$ (of fresh juice).

2.8. Determination of the Content of Reducing Sugars

The Lane–Eynon method was used to determine the content of reducing sugars. The material was extracted and deproteinised. The content of reducing sugars was determined in the obtained liquid by direct hot titration of a specific copper salt with the analysed sugar solution (against methylene blue as an indicator of the end of the reaction) [15,16].

2.9. Determination of Dry Matter Content

The moisture content of the research material was measured by drying 3 g of juice at 105 °C for 3 h. The measurements were carried out in triplicate.

2.10. Measurement of Chlorophyll Content

The juice was analysed for the content of chlorophylls A and B. The pigments were extracted with methyl alcohol. The chlorophyll content was measured using a UV–vis Helios Omega 3 spectrophotometer (Thermo Scientific, England). The measurement consisted of the determination of the absorbance (A) at different wavelengths (λ): 650 and 665 nm [17]. Next, the content of chlorophylls A and B and the total chlorophyll content were calculated with Equations (2)–(4):

Chlorophyll A content ($C_{chl(a)}$):

$$C_{chl(a)} = 16.5 \cdot A_{(665)} - 8.3 \cdot A_{(650)} \tag{2}$$

Chlorophyll B content ($C_{chl(b)}$):

$$C_{chl(b)} = 33.8 \cdot A_{(650)} - 12.5 \cdot A_{(665)} \tag{3}$$

Total chlorophyll content (C):

$$C = 4.0 \cdot A_{(665)} + 25.5 \cdot A_{(650)} \tag{4}$$

where:
$A_{(650)}$ = absorbance at a 650 nm wavelength;
$A_{(665)}$ = absorbance at a 665 nm wavelength.
The chlorophyll content was calculated in $mg \cdot g^{-1}$, taking into account the sample weight.

2.11. Statistical Analysis

The data were analysed statistically. A significance level of $\alpha = 0.05$ was assumed for inference. The analysis was carried out using ANOVA (StatSoft Polska, Poland) with post hoc tests for homogeneous groups based on Tukey's test. These groups comprised means between which no statistically significant difference was found at the assumed significance level, α.

The determinations were carried out in triplicate, except for leaf length and strength tests, which were repeated five times.

3. Results and Discussion

3.1. Characterisation of the Physical Traits of the Raw Material

Changes in the length of the barley shoots relative to the harvest date are shown in Table 1.

Table 1. Properties of the raw material relative to the harvest time of barley leaf harvesting.

Harvest Time (Day)	7	14	21	28	*p*-Value
Length of leaves (mm)	9.08 ± 1.14 [a]	15.48 ± 1.26 [b]	17.58 ± 1.44 [bc]	19.53 ± 1.35 [c]	<0.0001
Strength of leaves (N)	3.75 ± 0.25 [a]	3.27 ± 0.45 [ab]	3.20 ± 0.39 [ab]	3.01 ± 0.69 [b]	0.0144

[a,b,c,d] Means in the same line denoted by different letters were significantly different. The results are expressed as mean ± SD ($n = 5$).

The largest gain in the length of barley shoots was noted within seven days after emergence. In the following days, the rate of shoot growth declined. There were no statistically significant differences in the length of shoots between the material harvested at day 21 and day 28 after emergence. The length of shoots collected on day 21 and day 28 was 17.58 and 19.53 cm, respectively. The height of the plants was characteristic of unfertilised crops [18–20].

The strength of the barley shoots decreased over time. However, there were only significant changes in the tensile strength of the shoots in the material collected on day 28 of growth (in comparison to the material collected on day 7). Changes in the strength of cereal shoots are associated with the chemical composition, which is modified during plant growth [18].

3.2. The Pressing Efficiency

The effect of the harvest date on the pressing yield is shown in Figure 2.

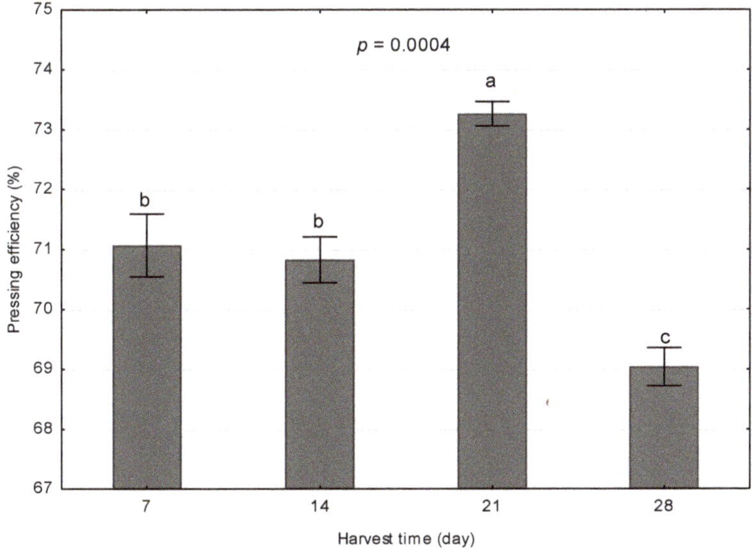

Figure 2. Pressing yield of the barley shoot and leaf juice relative to the harvest date.

The pressing yield ranged from 69.04% to 73.26%. It was obtained from 137 to 146 mL of juice (depending on the time of harvest). The highest value was noted during the processing of material harvested on cultivation day 21. The longer period of cultivation was associated with a statistically significant drop in the pressing yield. The pressing yield of barley leaves collected on day 28 was estimated at 69.4%. These results are consistent with data obtained by other authors. Paulíčková et al. [2] reported a pressing yield of 68% in a study that involved the extraction of juice from barley shoots. The decline in the pressing yield accompanying the longer harvesting period is probably caused by changes in the chemical composition, which lead to an increase in the fibre content.

3.3. Juice Acidity

Irrespective of the harvest date, the juice from the leaves and shoots of young barley had an acidic reaction (changes in pH are shown in Figure 3), shown by the significant decrease in pH values over time. The pH of the products pressed from leaves and shoots collected on day 7–28 ranged from 5.71 to 5.95. These values are similar to the pH of vegetable juices such as carrot or beetroot juice [21–23].

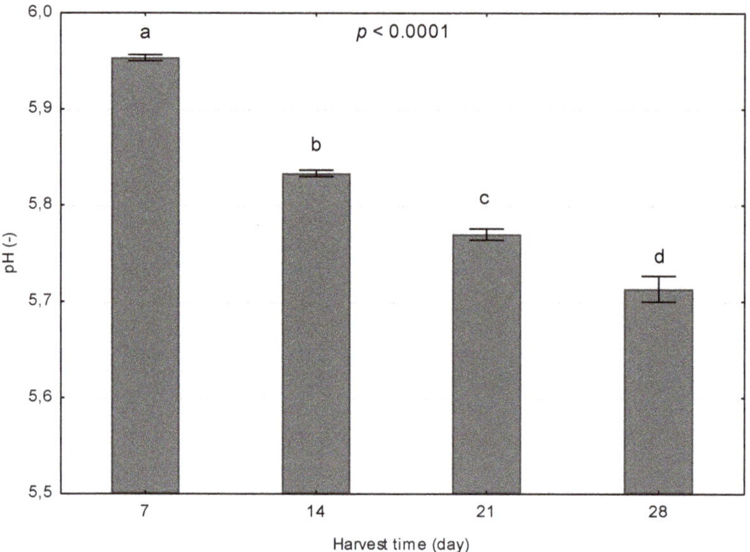

Figure 3. Impact of barley shoot harvesting time on juice pH.

The pH values of the juice from young barley leaves did not change significantly during storage (Figure 4). The statistical analysis only revealed significant differences in the pH value in the case of juice refrigerated for seven days. The pasteurisation and freezing processes did not change this parameter significantly. Juice acidity has an important effect on pigments and other ingredients (e.g., chlorophyll, carotenoids, anthocyanins, myoglobin, etc.) responsible for the colour of fruits, vegetables, and meat [21,24–26].

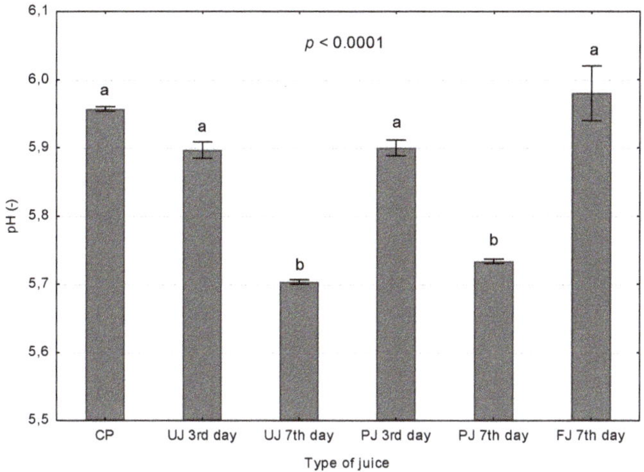

Figure 4. Impact of the conditions and length of storage on barley juice pH. (CP—control probe for fresh juice, UJ—unprocessed juice, PJ—pasteurised juice, FJ—frozen juice; 3rd day—after three storage days, 7th day—after seven storage days).

3.4. Total Protein Content

Protein content was expressed as a percentage of dry matter, which was 4.71% on average. Changes in the protein content of the analysed juices are shown in Figure 5. The total protein content in the juice increased along with the barley harvest date. The differences in the protein content between the harvest dates were statistically significant. The highest protein content, i.e., 51%, was determined in samples collected on day 21. The statistical analysis revealed that the content of this component in the juice produced after the next harvest (day 28) was significantly lower (42.68%). As demonstrated by Paulíčková et al. [2], the total amino acid content in juice from barley leaves and shoots decreased over time. The highest protein content recorded for the Malz cultivar (collected in phase I—DC 29) was $30.44\%_{d.m.}$. However, it should be noted that the authors collected the raw material at a later stage of barley growth. Therefore, these results may explain the lower protein content in the juice from barley leaves and shoots harvested on day 28 of growth (in comparison with the material obtained on day 21).

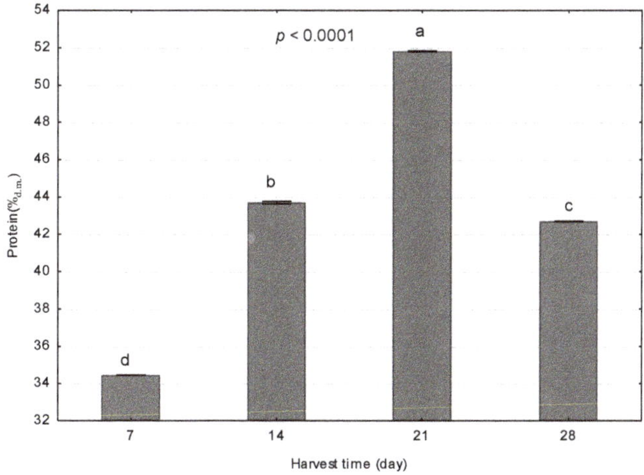

Figure 5. Changes in the protein content in barley shoot juice relative to the harvest date.

3.5. Chloride Content

The changes in chloride content in the analysed juices are shown in Figure 6. The chloride content in the juices was positively correlated with the length of barley growth. The statistical analysis demonstrated statistically significant differences in this parameter between the harvest dates. The chloride content ranged from 0.021 to 0.117 g·100 g $_{f.j.}$ $^{-1}$ for juice pressed from barley leaves and shoots collected on days 7–28. The increase in the chloride content was clearly correlated with a decrease in the pH of the juice. Park et al. [27] also showed that the amount of chlorides in the aboveground parts of plants may depend on the type of fertilization used.

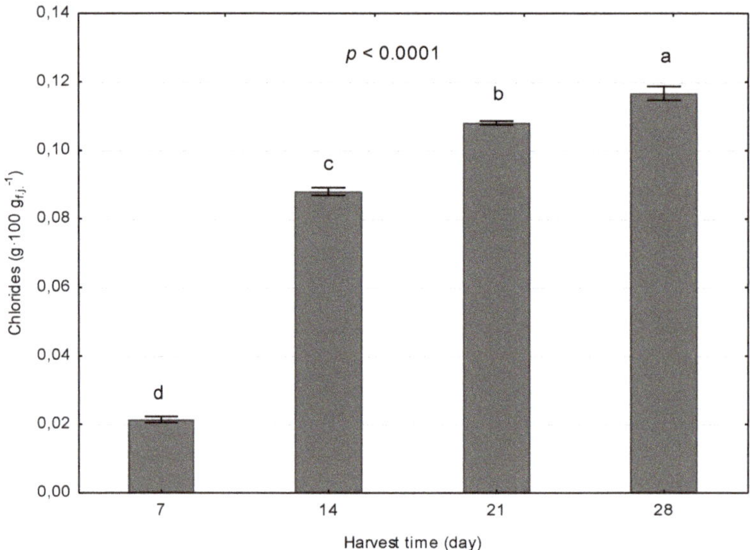

Figure 6. Changes in the chloride content in barley shoot juice relative to the harvest date.

3.6. Content of Reducing Sugars

Changes in the content of reducing sugars are shown in Figure 7. Their highest content was determined in the juice from barley leaves and shoots collected on day 21 of growth (8.20 g·100 g$_{d.m.}$ $^{-1}$). The differences in the value of this parameter between products obtained from shoots collected on days 14 and 28 were not statistically significant. The statistical analysis confirmed the lower content of reducing sugars only for juice made from raw material collected on day 7. The study conducted by Paulíčková et al. [2] showed that the content of simple sugars varied, depending on the plant growth phase. Other factors that significantly determined the changes in the analysed parameter include the conditions of the soil and the variety of barley. Paulíčková et al. [2] demonstrated that the sugar content in most barley varieties steadily decreased throughout the growing season.

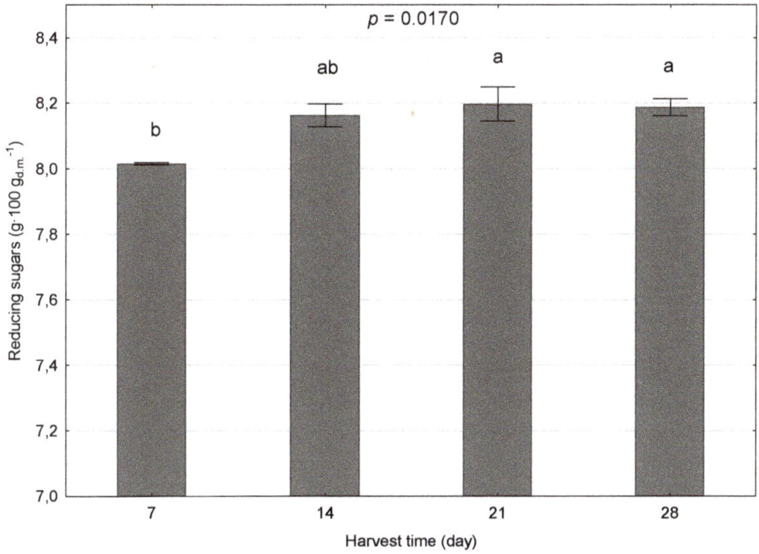

Figure 7. Changes in the content of reducing sugars in barley shoot juice, depending on the harvest date.

3.7. Chlorophyll Content

The chlorophyll content in the barley juice and the impact of the thermal treatment methods on changes in this parameter are shown in Table 2. The content of chlorophyll A and B and the total chlorophyll content changed significantly during storage. The highest determined content of total chlorophylls, i.e., 6.62 mg/g, was found in fresh juice. The chlorophyll contents in fresh raw material obtained by other authors ranged from 10.15 to 19.62 mg/g [28,29]. After seven days of storage, the total chlorophyll content was reduced by 37.9% (UJ), 42.12% (PJ), and 2.43% (FJ), in comparison with the untreated juice. It can thus be concluded from the present investigations that chlorophyll A is more sensitive to heat than chlorophyll B. Similar findings were reported by Weemaes et al. [30], who analysed the kinetics of chlorophyll degradation in thermally treated broccoli juice.

Table 2. Changes in the chlorophyll content in juice from young barley leaves during storage in various conditions ($p < 0.0001$).

Type of Heat Treatment/Ti-Me of Storage	Fresh (Control Probe)	Unprocessed (UJ) 3rd Day	7th Day	Pasteurised (PJ) 3rd Day	7th Day	Frozen (FJ)
Chlorophyll A (mg·g^{-1})	4.80 ± 0.01 [a]	4.15 ± 0.02 [c]	2.99 ± 0.01 [e]	3.80 ± 0.00 [d]	2.83 ± 0.015 [f]	4.66 ± 0.01 [b]
Chlorophyll B (mg·g^{-1})	1.53 ± 0.00 [a]	1.51 ± 0.00 [bc]	0.85 ± 0.01 [d]	1.50 ± 0.00 [c]	0.83 ± 0.00 [e]	1.51 ± 0.00 [b]
Total chlorophylls (mg·g^{-1})	6.62 ± 0.01 [a]	6.02 ± 0.04 [c]	4.11 ± 0.04 [e]	5.72 ± 0.02 [d]	3.83 ± 0.007 [f]	6.46 ± 0.04 [b]

[a,b,c,d,e,f] Means in the same line denoted by different letters were significantly different. The results are expressed as mean ± SD ($n = 3$).

The process of freezing had the lowest effect on changes in the content of chlorophyll A and total chlorophylls. By contrast, in the case of chlorophyll B, the least significant changes were noted for the unpasteurised product refrigerated for one day and in the case of the frozen juice stored for seven days. Paulíčková et al. [2] also confirmed the significant effect of thermal treatment of barley leaf juice (freezing, drying, and freeze-drying) on changes in the nutrient content. The analysis of their results also allows for the conclusion that freezing exerts the weakest effect on the quality of the product. Koca et al. [25] demonstrated a faster rate of degradation of chlorophylls at a lower pH value (from 5.5 to 7.5). Hence, the significantly lower chlorophyll content of juice stored for seven days may be caused by, e.g., changes in the pH of the product.

4. Conclusions

The present investigations confirm the potential benefits of consuming green juice from young barley shoots and leaves as part of a daily diet. The product obtained from the material harvested after 7, 14, 21, and 28 days of growth contains many valuable nutrients, e.g., a high level of total protein (with CS = 41.44 Meth). Additionally, high pressing yields of approximately 70% can be achieved.

The study has demonstrated that barley leaves and shoots harvested on day 21 of plant growth are the best raw material for the production of juice. The process of pressing material collected at this time exhibits the highest efficiency, and the juice contains the highest levels of protein and reducing sugars, as well as a high chloride content.

The most effective way to preserve the juice from young barley leaves is freezing. This process does not induce changes in juice acidity and only slightly reduces the content of chlorophylls A and B during storage of the product. Pasteurisation of the juice significantly reduces the chlorophyll content, but does not induce statistically significant changes in the pH of the juice.

The reported results are a preliminary study on the topic. However, it is necessary to continue research to determine the impact of other factors (e.g., barley varieties, growing conditions) on the quality of juices from young barley shoots and leaves. It is also possible scale-up the experiment using a higher number of samples, filed conditions, etc.

In addition, this way of using shoots and leaves of young barley will have economic importance. The costs of obtaining raw materials for juice production will be reduced. The use of shoots and leaves of young barley, which is grown as a forecrop, for the production of green juices, may have a beneficial effect on the development of sustainable crop production. This will allow more efficient use of the plants.

Author Contributions: Conceptualization, A.B.-K., D.A., F.K. and L.R.; methodology, A.B.-K. and D.A.; formal analysis, A.B.-K. and F.K.; investigation, A.B.-K. and L.R.; data curation, A.B.-K. and L.R.; writing—original draft preparation, A.B.-K. and Z.K.; writing—review and editing, D.A.; visualization, A.B.-K.; supervision, D.A.

Funding: This research received no external funding.

Conflicts of Interest: The authors declare no conflicts of interest.

References

1. Jabłońska-Ceglarek, R.; Rosa, R. Forecrop green manures and the size and quality of white cabbage yield. *Electron. J. Pol. Agric. Univ.* **2003**, *6*. Available online: http://www.ejpau.media.pl/volume6/issue1/horticulture/art-08.html (accessed on 21 July 2019).

2. PaulíčkoVá, I.; EhrENbErgEroVá, J.; FIEdlEroVá, V.; Gabrovska, D.; Havlova, P.; Holasova, M.; Vaculová, K. Evaluation of barley grass as a potential source of some nutritional substances. *Czech J. Food Sci.* **2007**, *25*, 65–72. [CrossRef]

3. Son, H.-K.; Lee, Y.-M.; Park, Y.-H.; Lee, J.-J. Effect of Young Barley Leaf Powder on Glucose Control in the Diabetic Rats. *Korean J. Community Living Sci.* **2016**, *27*, 19–29. [CrossRef]

4. Zeng, Y.; Pu, X.; Yang, J.; Du, J.; Yang, X.; Li, X.; Li, L.; Zhou, Y.; Yang, T. Preventive and Therapeutic Role of Functional Ingredients of Barley Grass for Chronic Diseases in Human Beings. *Oxidative Med. Cell. Longev.* **2018**, *2018*, 3232080. [CrossRef] [PubMed]

5. Barczak, B.; Nowak, K. Skład aminokwasowy białka biomasy jęczmienia ozimego (Hordeum vulgare L.) w zależności od stadium rozwoju rośliny i nawożenia azotem. *Acta. Sci. Pol. Agric.* **2008**, *7*, 3–15.

6. Jarosz, M.; Szponar, L.; Rychlik, E.; Wierzejska, R. Woda i elektrolity. In *Normy Żywienia Dla Popul. Pol. Nowelizacja*, 1st ed.; Jarosz, M., Ed.; Instytut Żywności i Żywienia: Warsaw, Poland, 2012; pp. 143–151.

7. García-Caparrós, P.; Almansa, E.M.; Chica, R.M.; Lao, M.T. Effects of Artificial Light Treatments on Growth, Mineral Composition, Physiology, and Pigment Concentration in Dieffenbachia maculata "Compacta" Plants. *Sustainability* **2019**, *11*, 2867. [CrossRef]

8. Kawka, K.; Lemieszek, M.K. Prozdrowotne właściwości młodego jęczmienia. *Med. Ogólna I Nauk. O Zdrowiu* **2017**, *23*, 7–12. [CrossRef]

9. Polak, R. Chlorofile jako naturalne źródło energii biomasy. *Przemysł Chem.* **2019**, *1*, 138–142. [CrossRef]

10. Pradhan, J.; Das, S.; Das, B.K. Antibacterial activity of freshwater microalgae: A review. *Afr. J. Pharm. Pharmacol.* **2014**, *8*, 809–818.

11. Kandhasamy, S.; Chin, N.L.; Yusof, Y.A.; Lai, L.L.; Mustapha, W.A.W. Effect of Blender and Blending Time on Color and Aroma Characteristics of Juice and Its Freeze-Dried Powder of Pandanus amaryllifolius Roxb. Leaves (Pandan). *Int. J. Food Eng.* **2016**, *12*, 75–81.

12. Qamar, A.; Saeed, F.; Nadeem, M.T.; Hussain, A.I.; Khan, M.A.; Niaz, B. Probing the storage stability and sensorial characteristics of wheat and barley grasses juice. *Food Sci. Nutr.* **2019**, *7*, 554–562. [CrossRef] [PubMed]

13. Kubatka, P.; Kello, M.; Kajo, K.; Kruzliak, P.; Výbohová, D.; Šmejkal, K.; Maršík, P.; Zulli, A.; Gönciová, G.; Mojžiš, J.; et al. Young barley indicates antitumor effects in experimental breast cancer in vivo and in vitro. *Nutr. Cancer* **2016**, *68*, 611–621. [CrossRef] [PubMed]

14. Son, H.-K.; Lee, Y.-M.; Lee, J.-J. Nutrient Composition and Antioxidative Effects of Young Barley Leaf. *Korean J. Community Living Sci.* **2016**, *27*, 851–862. [CrossRef]

15. Majumdar, T.; Wadikar, D.; Vasudish, C.; Premavalli, K.; Bawa, A. Effect of Storage on Physico-Chemical, Microbiological and Sensory Quality of Bottlegourd-Basil Leaves Juice. *Am. J. Food Technol.* **2011**, *6*, 226–234. [CrossRef]

16. Teixeira, E.M.B.; Carvalho, M.R.B.; Neves, V.A.; Silva, M.A.; Arantes-Pereira, L. Chemical characteristics and fractionation of proteins from Moringa oleifera Lam. leaves. *Food Chem.* **2014**, *147*, 51–54. [CrossRef]

17. Pielesz, A. Skład chemiczny algi brązowej Fucus vesiculosus L. *Postępy Fitoter.* **2011**, *1*, 9–17.

18. Briggs, D.E. *Barley*; Springer Science and Business Media: Berlin/Heidelberg, Germany, 2012; pp. 14–24.

19. Frank, A.B.; Bauer, A.; Black, A.L. Effects of Air Temperature and Fertilizer Nitrogen on Spike Development in Spring Barley. *Crop. Sci.* **1992**, *32*, 793–797. [CrossRef]

20. McMaster, G.S.; Wilhelm, W.W.; Frank, A.B. Developmental sequences for simulating crop phenology for water-limiting conditions. *Aust. J. Agric. Res.* **2005**, *56*, 1277–1288. [CrossRef]

21. Kırca, A.; Özkan, M.; Cemeroğlu, B.; Kirca, A.; Cemeroğlu, B. Effects of temperature, solid content and pH on the stability of black carrot anthocyanins. *Food Chem.* **2007**, *101*, 212–218. [CrossRef]

22. Nabrdalik, M.; Świsłowski, P. Microbiological Evaluation of Unpasterized Fruit and Vegetable Juices. In Proceedings of the ECOpole, Zakopane, Poland, 5–8 October 2016.

23. Yoon, K.Y.; Woodams, E.E.; Hang, Y.D. Fermentation of beet juice by beneficial lactic acid bacteria. *LWT Food Sci. Technol.* **2005**, *38*, 73–75. [CrossRef]

24. Czapski, J. Heat stability of betacyanins in red beet juice and in betanin solutions. *Eur. Food Res. Technol.* **1990**, *191*, 275–278. [CrossRef]

25. Koca, N.; Karadeniz, F.; Burdurlu, H.S. Effect of pH on chlorophyll degradation and colour loss in blanched green peas. *Food Chem.* **2007**, *100*, 609–615. [CrossRef]

26. Saguy, I. Thermostability of red beet pigments (betanine and vulgaxanthin–I): Influence of pH and temperature. *J. Food Sci.* **1979**, *44*, 1554–1555. [CrossRef]

27. Park, J.; Cho, K.H.; Ligaray, M.; Choi, M.-J. Organic Matter Composition of Manure and Its Potential Impact on Plant Growth. *Sustainability* **2019**, *11*, 2346. [CrossRef]

28. Cao, X.; Zhong, Q.; Wang, Z.; Zhang, M.; Mujumdar, A.S. Effect of microwave freeze drying on quality and energy supply in drying of barley grass. *J. Sci. Food Agric.* **2017**, *98*, 1599–1605. [CrossRef]

29. Cao, X.; Zhang, M.; Mujumdar, A.S.; Zhong, Q.; Wang, Z. Effect of nano-scale powder processing on physicochemical and nutritional properties of barley grass. *Powder Technol.* **2018**, *336*, 161–167. [CrossRef]

30. Weemaes, C.A.; Ooms, V.; Van Loey, A.M.; Hendrickx, M.E. Kinetics of Chlorophyll Degradation and Color Loss in Heated Broccoli Juice. *J. Agric. Food Chem.* **1999**, *47*, 2404–2409. [CrossRef]

Article

Application of Non-Parametric Bootstrap Confidence Intervals for Evaluation of the Expected Value of the Droplet Stain Diameter Following the Spraying Process

Andrzej Bochniak [1], Paweł Artur Kluza [1,*], Izabela Kuna-Broniowska [1] and Milan Koszel [2,*]

[1] Department of Applied Mathematics and Computer Science, University of Life Sciences in Lublin, 20-950 Lublin, Poland; andrzej.bochniak@up.lublin.pl (A.B.); izabela.kuna@up.lublin.pl (I.K.-B.)

[2] Department of Machinery Exploitation and Management of Production Processes, University of Life Sciences in Lublin, 20-950 Lublin, Poland

* Correspondence: pawel.kluza@up.lublin.pl (P.A.K.); milan.koszel@up.lublin.pl (M.K.); Tel.: +48-81-531-9619 (P.A.K.); +48-81-531-9732 (M.K.)

Received: 6 November 2019; Accepted: 5 December 2019; Published: 9 December 2019

Abstract: In the era of sustainable agriculture, the issue of proper and precise implementation of agrotechnical operations, without harmful effects on the natural environment, begins to play an important role. Statistical tools also become important, for example, when assessing the malfunction of plant cultivation equipment. The study presents a comparison of six nonparametric bootstrap methods used for construction of confidence intervals for the expected value of an average diameter of droplet stains following the spraying process. The simulation tests were carried out based on experiment with nozzle sprayer Lechler 110-03 using two spray nozzles: a new one and an old one. It was assumed that the distribution of the droplet stain size was consistent with the lognormal distribution. The paper considers the influence of the sample size, mean value and standard deviation of the droplet stain diameter on the interval range as well as on the estimated coverage probabilities of the confidence intervals. It was shown that in general these methods can be applied for this purpose. For the double bootstrap method and the studentized method, the empirical confidence levels of the constructed intervals turned out to be less distinct than the assumed level but the lengths of these intervals were greater than the lengths of intervals obtained using the other four methods.

Keywords: bootstrap methods; confidence intervals; lognormal distribution; sprayer; droplet diameters; sustainable agriculture

---

1. Introduction

Sustainable agriculture aims to promote a sustainable farming system that relies on the rational use of natural resources, which allows reducing the negative impact of agriculture on the environment and prevents the loss of organic matter in the soil. It is part of a larger movement towards sustainable development, which suggests that natural resources are limited, recognizes the constraints of economic growth and encourages equality in resource allocation. Sustainable agricultural development has been proposed as one of the alternatives to reverse the spiral of resource degradation and poverty. The goal of sustainable agriculture is to protect natural resources [1,2]. Sustainable development is one of the principles on which the European Union (EU) policy should be based, including the EU's common agricultural policy. The common agricultural policy is the main institutional structure and the main factor of reforms affecting the development of sustainable agriculture [3].

Spraying is the basic and the most widely used treatment in plant protection but improperly preformed or defective technical equipment may degrade the quality of agricultural raw materials and pose a threat to humans and the environment. Recently, plant protection has been severely criticized for creating environmental and human hazards. The use of pesticides, in addition to unquestionable effectiveness, has serious disadvantages. Creation of resistant biotypes, which are difficult to combat later, can be another possible threat. In addition, chemicals change crop biochemistry, pollute the environment and often destroy natural enemies who would limit the host population [4].

However, the use of herbicides, despite many controversies, is one of the basic methods of protecting plants against weeds in modern agriculture. The use of plant protection products (generally known as pesticides or agrochemicals) makes it possible to increase food production by destroying weeds and pests that attack crops. As a result, productivity per hectare increases and also reduces losses during transport and storage of food [5]. In addition, the use of pesticides gives the farmer the opportunity to intervene quickly when the crop is directly threatened by diseases and pests. This is especially important in the case of mass and sudden outbreaks of pests. Some chemicals, for example, seed mortars, have a preventive effect. In turn, granulated pesticides introduced into the soil and taken up by plant roots operate for the whole period or a significant part of the growing season [6].

The effect of herbicides depends on both the technical and technological processes that make up the spraying procedure, as well as the behaviour of the active substance on the sprayed surface of the plants and inside them. These processes include the production of spray liquid droplets (atomization), transfer from the atomizer to the plant surface, droplet retention on plants, formation of sediments on the leaf surface after evaporation of water from the spray liquid, up-taking of the active substance (absorption) and its translocation and metabolism [7].

A number of technological, technical and climatic factors influence the quality of the sprayer work. The most important of them include the type of machine, choice of nozzles, spray parameters, temperature and humidity as well as compliance with the instructions of producers of plant agents [8]. It should be noted that the nozzle wear degree has a decisive influence on spray quality. Application of excessive numbers of droplets on the protected surface causes their merging, which worsens the quality of spraying. Gravity makes droplets on the surface of plants penetrate the soil and reach groundwater [5,9]. Besides, if nozzles generate very small droplets, they are drifted away by wind or the liquid evaporates before falling on protected plants.

In spraying operations, it is important that as much of the spray liquid as possible settles on the surface of the plant. This is the basic condition for achieving high effectiveness of the procedure, as well as reducing losses that burden the environment. Liquid settlement on plants depends mainly on the development phase of the plant, crop density, dose, type of usable liquid and operation of the spraying apparatus [10]. It was stated that in the total spray liquid expenditure balance (during wheat spraying) 52.4% were losses in the air (as a result of drift and evaporation) and as a result of falling on the soil, while 47.6% sprayed liquid stopped on the plants [11]. These are significant losses that can cause environmental pollution.

A uniform distribution of the spray liquid has a significant influence on the environmental impact of agrochemicals. Uneven application means that a more larger dose of the agent is applied over a large field area than it results from actual needs. It is not only a waste but it also carries the risk of the accumulation of residues of plant protection agents in agricultural raw materials [12].

For these reasons, statistical analysis of the droplet stains after the spraying process is important for adjustment of the operating parameters of the sprayer. One of the statistical methods that can be used is the analysis of the droplet stain distribution, especially point or interval estimation of the parameters of their distribution. Naturally occurring droplets are usually irregular in shape and different in size. This resulted in the introduction of different mean droplet diameter definitions along with statistical description of the droplet size distribution (linear, surface and volume (Sauter)).

The conducted research regarding the estimation of droplet size and quantity involves the division into raindrops and droplets following the spraying process, where various types of probability distribution can be applied to describe each of the two cases presented above. As regards empirical data obtained for raindrops, the following distributions are used most frequently—exponential [13], gamma [14–16], Gauss [17], lognormal [18,19], Weibull [20,21] and Poisson [22]. The gamma distribution is used to describe the droplet size distribution following the spraying process [23]. Equivalently, the following distributions are also used for the same type of data—upper truncated lognormal [24,25], Rosin–Rammler and Nukiyama–Tanasawa [26] and Weibull [18,27]. The lognormal distributions with the upper truncation [15,23] are applicable for the distribution of droplet diameters because the measured droplet diameters are in a certain interval ranging from zero to a definite upper limit, which in an extreme case can be a maximum measured droplet size.

A number of parametric and non-parametric methods have been designed for interval estimation of the expected value of a random variable with a lognormal distribution. A few such procedures were described and compared in Reference [28], including the Cox method for a large sample size presented in Reference [29], a conservative method described in Reference [30] and a parametric bootstrap method [31]. Based on numerical results, it was demonstrated [28] that all these procedures, with the exception of the Cox method, are too conservative or too liberal disregarding the sample size. Therefore, the methods of construction confidence intervals presented above are not satisfactory for small sample sizes. This problem was solved in Reference [32], where the method was developed for construction of exact confidence intervals for small sample sizes using the idea of a generalized p-value and generalized confidence intervals.

The bootstrap method was designed by Bradley Efron [23,24,26,33]. The group of bootstrap methods includes parametric and non-parametric methods, including inter alia—the basic, percentile, bias-corrected, bias-corrected and accelerated, studentized and double bootstrap methods. These methods differ as regards the selection of appropriate quantiles of the distribution of the estimators from bootstrap samples. Bootstrap is a relatively new method but in the computer era it is frequently used for obtaining confidence intervals and test hypotheses in situations when the sample sizes are small, the distribution of estimators is unknown or when the fulfilment of relevant assumptions is doubtful. The statistical research carried out in Reference [27], showed that the information on the asymmetry of the probability distribution of a random variable or initial estimation of the asymmetry coefficient can be helpful while determining the sample size both in bootstrap tests and classical significance tests.

The aim of this study was to investigating the possibility of using various bootstrap methods to generate confidence intervals to assess the average droplet trace diameter, which can be characterized by an asymmetric probability distribution set close to the lognormal one. The simulations were limited to the ranges of droplet trace sizes obtained in experimental experiments using the Lechler 110-03 sprayer. The usefulness of particular methods is indicated depending on the size of the random sample and the different degree of variability of the obtained droplet trace sizes. The tests carried out are aimed at determining the methods of assessing the degree of wear of the sprayer nozzle.

2. Materials and Methods

2.1. Material

The analysis carried out in the research concerns the estimation of the mean value of droplet stain diameters following the spraying process carried out in an experimental study on a Lechler 110-03 spray nozzle type. The nominal flow rate for this type of the spray nozzle is $1.17 \, \text{L} \cdot \text{min}^{-1}$. The spray boom shift during spraying was $9 \, \text{km} \cdot \text{h}^{-1}$ ($2.5 \, \text{m} \cdot \text{s}^{-1}$) and the pressure was 5 bar (0.5 MPa). After the droplets applied on each piece of foil dried, scans of 5×5 cm images were obtained. The scanner resolution during scanning was 300 dpi. The size of droplet stains and the number of droplets was calculated using the computer software Image Pro+ by Media Cybernetics. The new spray nozzle and older one were examined for comparison in a research experiment. The older nozzle was obtained as a result of laboratory wearing using Kaolin KOM water slime from SURMIN-KAOLIN. The slime was produced by adding 9.1 kg of kaolin to 150 L of water [34]. After reaching 10% wearing, the liquid outflow rate was measured at 1.29 L · min^{-1}. In Figure 1 examples of scans obtained from the spraying process using new spray nozzle (a) and older one (b) are shown.

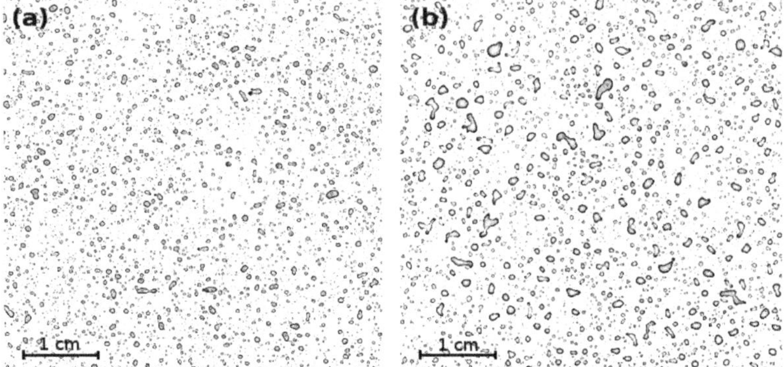

Figure 1. Example scans of droplet stains from: (**a**) new spray nozzle (**b**) spray nozzle in longer use.

2.2. Mathematical Background

The density function $f(x)$ of the lognormal distribution for the diameter size of the splashed droplets following the spraying process is given by:

$$f(x) = \frac{1}{x\sigma\sqrt{2\pi}} exp\left(-\frac{(log(x) - \mu)^2}{2\sigma^2}\right),$$

(1)

where:

$x \in (0, +\infty)$ – droplet diameter [μm],
$\mu \in< 0, +\infty)$ – expected value of variable ln(X), location parameter,
$\sigma > 0$ – standard deviation, scale parameter,
$log(\cdot)$ – natural logarithm.
$Y = log(X) \sim N(\mu, \sigma)$.

The parameters of the lognormal distribution turn out to be the functions of both parameters μ and σ^2 and this dependence poses difficulties in obtaining exact and optimal tests and confidence intervals. In particular, the expected value of lognormal random variable X depends on μ and σ^2:

$$E(X) = E(exp(Y)) = exp\left(\mu + \frac{1}{2}\sigma^2\right).$$

(2)

The overlapping of droplet stains following the spraying process is an additional difficulty in using these stains to estimate the distribution of the spraying parameters. The poor accuracy in point estimation of the mean value and standard deviation has an impact on the interval estimation of these parameters. Taking into account the above difficulties, six non-parametric bootstrap methods were used to construct the confidence intervals—basic, percentile, bias-corrected, bias-corrected and accelerated, studentized and double bootstrap methods.

During the research work, the sensitivity of the lengths and coverage of intervals constructed using the discussed bootstrap methods were compared, given the changes in the sample sizes and changes in the values of standard deviation. The consistency of empirical coverages was also compared with the assumed confidence level.

Let $x = (x_1, x_2, \ldots, x_n)$ denote an n-element random sample, whereas $R(x, F)$ is some statistics determined on the sample space with distribution F, θ is the unknown parameter and σ is the standard deviation of this parameter. The bootstrap sample is an n-element realization of the variable X,

$\mathbf{x}^* = (x_1^*, x_2^*, \ldots, x_n^*)$ and its elements are selected at random with replacement from sample x. In order to obtain bootstrap samples, independent B-fold ($B \geq 1000$) generation of n-element data sequences $\mathbf{x}^* = (x_1^*, x_2^*, \ldots, x_n^*)$ is carried out. These data sequences are used to determine the estimator $\hat{\theta}$ of the θ parameter. As a result, B-element sequence of estimators $\hat{\theta}_j^* = R\left(x_j^*\right), j = 1, 2, \ldots, n$ is obtained. Out of the sequence of estimators $\hat{\theta}_j^*$, the quantiles of a fixed rank are calculated, which are then used to obtain the confidence interval for parameter $\theta = R(x)$.

In the simplest basic method (B), the $(1 - \alpha)\%$ confidence interval for parameter θ is obtained according to the formula:

$$\left(2\hat{\theta} - \hat{\theta}^*_{\left(1 - \frac{\alpha}{2}\right)}, 2\hat{\theta} - \hat{\theta}^*_{\left(\frac{\alpha}{2}\right)}\right), \tag{3}$$

where:

$\hat{\theta}^*_{(\alpha)}$ denotes α^{th} quantile from the sequence of values $\hat{\theta}_j^*$,

$\hat{\theta}$ denotes the estimator from the original sample $\mathbf{x} = (x_1, x_2, \ldots, x_n)$.

The percentile (P), bias-corrected (BC) and bias-corrected and accelerated (BCa) methods confidence intervals do not use estimator $\hat{\theta}$ but are calculated exclusively based on quantiles of a relevant rank from the sequence of estimators:

$$\left(\hat{\theta}^*_{(\alpha_L)}, \hat{\theta}^*_{(\alpha_U)}\right), \tag{4}$$

where the ranks α_L and α_U of appropriate percentiles can be calculated according obtained sequence of estimators $\hat{\theta}_j^* = R\left(x_j^*\right), j = 1, 2, \ldots, n$ depending on selected method.

The bootstrap method with bias correction [35] contributes to improvement of the precision of the percentile method in the case when the median of estimator values for all bootstrap samples is not equal to the value of the estimator for the original sample. This method adds two steps after estimation of the distribution of the considered parameter for bootstrap samples:

- calculation of the fraction k of bootstrap samples for which the estimator value of the parameter is lower than the estimator calculated on the basis of the original sample,
- calculation of interval bounds regarding the calculated fraction of samples in order to correct the bias.

Similar to the previous method, the bias-corrected and accelerated bootstrap method (BCa) [24,35] takes into account the fact that estimators of the parameter for bootstrap samples can be biased and the impact of individual values in a sample on the value of the estimator is considered as well. When the distribution is skewed, some adjustment needs to be made and BCa method gives such a possibility.

The ranks α_L and α_U of percentiles in BCa method depend on two constants a and z_0 called the acceleration and bias correction coefficient. The standard error of $\hat{\theta}$ is stabilized by coefficient a and z_0 measures the median bias of $\hat{\theta}^*$. This results in faster convergence of the estimator confidence limit into its exact value than in the case of the quantile method. Thus, the probability of coverage for the BCa method is closer to the nominal probability than in the percentile method.

There are different ways to determine the value of coefficient a. One of them uses the jackknife method [24], which explores the impact of n individual elements of a sample by using n special sample size $n - 1$. Each of the n new samples is obtained through deleting each time only one element from the original sample. Based on such samples, n estimators of the θ parameter are calculated.

In the studentized method, sometimes referred to as the bootstrap-t method, the appropriate statistic quantiles are used in the following form:

$$t^* = \frac{\hat{\theta}^* - \hat{\theta}}{\hat{\sigma}\left(\hat{\theta}^*\right)}, \tag{5}$$

where $\hat{\sigma}\left(\hat{\theta}^*\right)$ is the estimator of the standard error of the estimator $\hat{\theta}^*$ determined based on the bootstrap sample $x^* = (x_1^*, x_2^*, \ldots, x_n^*)$.

The double bootstrap method is a modification of the previous method [36], where the standard error $\sigma\left(\hat{\theta}^*\right)$ of $\hat{\theta}^*$ is estimated on bases of additional B_1 bootstrap samples $\mathbf{x}^{**} = (x_1^{**}, x_2^{**}, \ldots, x_n^{**})$ generated from the $\mathbf{x}^*$ sample, where $\bar{\theta}^{**} = \frac{1}{B_1}\sum_{i=1}^{B_1}\hat{\theta}_i^{**}$.

This section outlines the six bootstrap methods, which were used in this study. More details and appropriate mathematical formulas of individual methods can be found in numerous studies available in the cited literature.

2.3. Computer Simulations

On the basis of the experimental tests, the ranges of the average diameters of droplet traces obtained in the spraying process using the tested nozzles were estimated. The determined ranges of means as well as standard deviations obtained for the droplet stains were used as the basis for conducting computer simulations using the six bootstrap methods described in the previous section. The simulations were carried out taking into account different sample sizes (10–600 elements). For each combination of two values of the analyzed mean and standard deviation, ten thousand samples were generated consisting of equal elements each, which were used for generating bootstrap samples. For each random sample x, the total of $B = 2000$ bootstrap samples were generated, based on which the relevant percentiles were estimated. In the case of a double bootstrap, the standard deviation was estimated based on $B_1 = 250$ samples. For each generated sample, all six types of confidence intervals were calculated and the assumed confidence level was $\alpha = 0.05$. Computer simulations were carried out using software MATLAB R2014a by employing our own code and embedded procedures on the computer with the operating system Windows 10 Professional equipped with an Intel Xeon E5-2609 2.40 GHz processor.

3. Results

The analysis took into account the mean values and standard deviation values estimated on the basis of the measurements performed with the Lechler 110-03 sprayer. The first step involved testing the consistency of the distribution of size of droplets stain diameters from the measurements with the lognormal distribution and determining its parameters.

The considered sample scans, which were previously presented on Figure 1, resulted in a highly asymmetrical distribution of droplet stain diameters. The distribution of droplet diameters obtained while spraying using a new and old nozzle are presented in Table 1. The drops traces were divided into 5 size classes—0–150 μm, 150–250 μm, 250–350 μm, 350–450 μm and above 450 μm. The estimates were obtained using computer software Image Pro+ by Media Cybernetics. The table includes the number of objects assigned to classes, the percentage of objects, the percentage of surface covered and the average drop values and standard deviations in each class. The data compiled in Table 1 show a strongly asymmetrical distribution.

In addition, to determine the approximate parameters of the droplet size distribution, a division into a larger number of size classes was made (Figure 2). For both nozzles (new and old), using the MatLab function *fitnlm*, parameters of log-normal distribution were determined, which are best fitted to the obtained histograms for selected droplet size classes. For the new nozzle, the average is $\mu_x = 160.3$ μm and the standard deviation $\sigma_x = 65.8$ μm, while for the older nozzle the average is $\mu_x = 186.7$ μm and the standard deviation $\sigma_x = 81.2$ μm.

Due to the overlapping of droplet stains following the spraying process and the difficulties in construction of the confidence intervals for the expected value of the lognormal distribution, we chose the bootstrap methods [24]. As already mentioned, the confidence intervals, constructed using the discussed bootstrap methods were compared in terms of the sensitivity of their lengths and probabilities of the expected value coverage to changes in sample sizes and in values of standard deviation. The consistency of empirical coverage probabilities with the assumed confidence level was also compared.

Table 1. Droplet traces size distribution in an example datasets.

Stain Class	No. Objects	% Objects	% Area	Diameter Mean	Diameter Std
New nozzle					
<150 μm	511	62.01	43.96	123.89	19.19
150–250 μm	242	29.37	36.72	184.29	22.61
250–350 μm	61	7.40	15.95	293.17	52.33
350–450 μm	9	1.09	3.17	382.51	69.80
>450 μm	1	0.12	0.19	471.00	0.00
Older nozzle					
<150 μm	323	51.43	37.49	132.34	14.37
150–250 μm	204	32.48	35.71	196.42	15.69
250–350 μm	80	12.74	19.09	295.28	24.28
350–450 μm	12	1.91	3.69	412.07	39.19
>450 μm	9	1.43	4.01	521.74	47.16

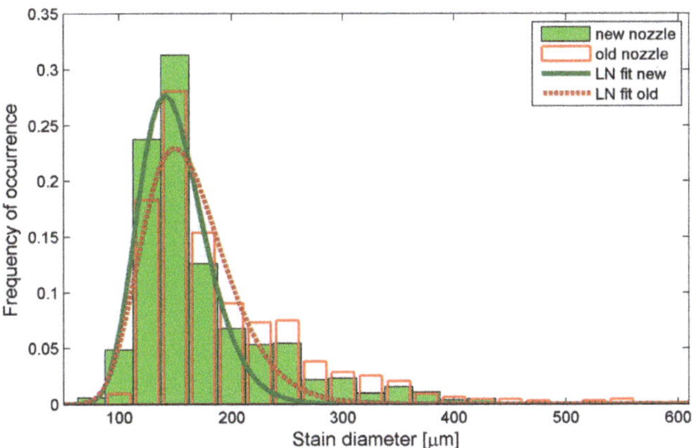

Figure 2. Histograms of the droplet stain diameters from the experiment and fitted lognormal distribution for new and older nozzles.

The analysis took into account the mean values and standard deviation values estimated based on the measurements performed with the Lechler 110-03 sprayer. In order to generate samples from the lognormal distribution for the population mean, the range from 90 μm to 210 μm was taken with a step of ten units, whereas for the standard deviation the range was from 30 μm to 180 μm with a step of ten units. For each combination of two values of the analyzed mean and standard deviation, ten thousand samples were generated consisting of thirty elements each, which were random samples used for generating bootstrap samples. For each random sample x, the total of $B = 2000$ bootstrap samples were generated, based on which the relevant percentiles were estimated. In the case of a double bootstrap, the standard deviation was estimated based on $B_1 = 250$ samples. The values of B and B_1 were determined according to Reference [36]. For each generated sample, all six types of confidence intervals were calculated and the assumed confidence level was $\alpha = 0.05$.

The largest differences between the results obtained by different methods occurred for small samples. Figure 3 presents numerical results of the coverage of the calculated confidence intervals for the assumed 95% confidence levels using the six bootstrap methods with division to the selected lengths of the mean value and standard deviation value (μ_x, σ_x) of variable X. In a majority of cases, all the considered methods give the coverage probabilities lower than the assumed level of 95% for sample sizes of 30. The confidence intervals constructed with the studentized and double bootstrap methods give the coverage probabilities not greater than 94% for each considered value of the mean; noteworthy, the variation of the feature may be high and, in a particular case, the variation coefficient may even be equal to 80%. The other four methods allow obtaining coverage probabilities not greater than 93%; however, only for the variation coefficient not higher than 34% (Figure 3).

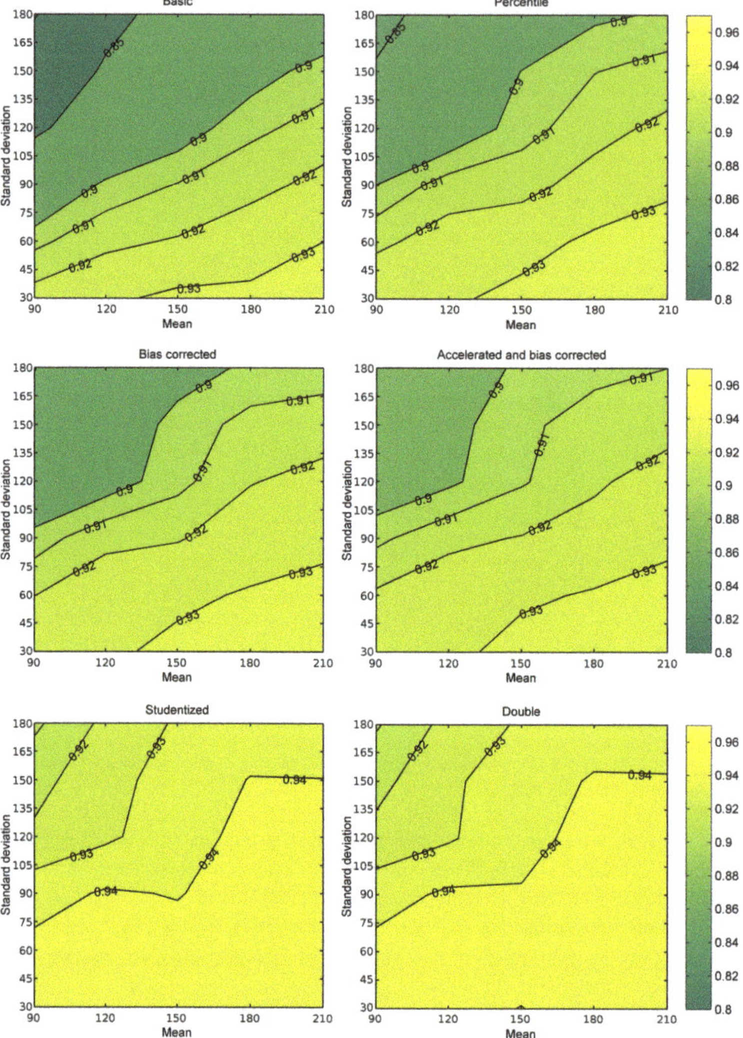

Figure 3. Coverage probabilities of the mean value by confidence intervals constructed by the six types of bootstrap methods ($n = 30$).

In order to study the impact of the sample size on the lengths of confidence intervals for the examined bootstrap methods, the sample sizes considered in the simulations were 10, 20, 30, 50, 80, 100, 150, 200, 400 and 600. Since the interval lengths for larger sample sizes overlap, the figure presents only the lengths for the smaller sample sizes (up to 100 elements). The samples were generated from a population for which the mean value was 160 μm and the standard deviation ranged from 40 to

120 µm. For each of the sizes, five thousand samples were generated and the length of the confidence interval calculated with the six studied methods was estimated based on each sample. The lengths of the intervals obtained were averaged for each method separately. From the graph presented in Figure 4 (for the standard deviation of 80 µm, close to the sample obtained from the experiment), it is evident that the intervals for the studentized and double bootstrap methods are wider, compared with the basic, percentile, bias correction and accelerated methods. Concurrently, it can be noted that the lengths of the confidence intervals for the four methods mentioned above are fairly equal (Figure 4).

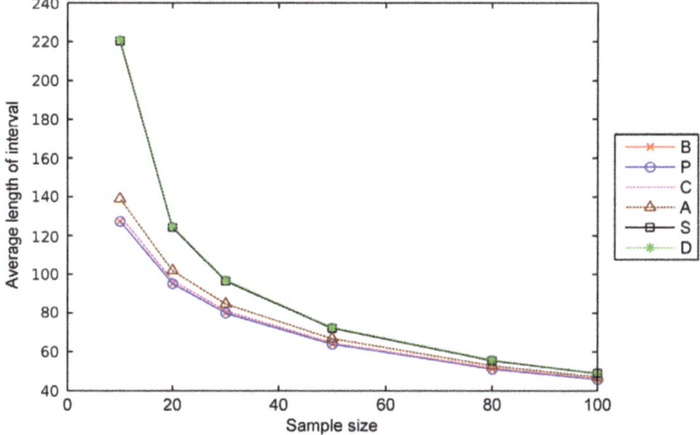

Figure 4. Changes in the length of confidence intervals constructed using nonparametric bootstrap methods depending on the sample size; $\mu_x = 160$ µm, $\sigma_x = 65$ µm (B—basic, P—percentile, C—bias corrected, A—accelerated, S—studentized, D—double bootstrap).

The four methods—namely, basic, percentile, bias correction and accelerated methods—give intervals characterized by similar length but different from the results obtained using the studentized and double bootstrap methods, which are mutually comparable (Hall, 1992). Hence, the graph shows the comparison of the lengths of the confidence intervals only for two methods—the percentile and double bootstrap methods (Figure 4). For small sample sizes and large standard deviation, the lengths of the intervals obtained using the double bootstrap method are even approximately 3 times longer than those obtained using the percentile method. For larger sample sizes, these differences are less visible.

In the study of the influence of the sample size on the coverage probabilities of confidence intervals, obtained in particular for the smaller sample sizes, the follwing sample sizes were included in the simulation—10, 20, 30, 50, 80, 100, 150, 200, 400 and 600 (Figure 5). For each sample size, five thousand samples were generated from the same population as previously (mean value 160 µm and standard deviation from 40 to 120 µm) and the percentage of covering of the assumed value of 160 was checked.

The intervals obtained using the analyzed methods give a similar covering of the true value of the population mean only for samples with a size of at least 100–200 elements. For smaller sample sizes, we obtained a clear division of the discussed methods into three groups. Given the interval lenghts discussed above, the covering is closest to the assumed confidence level for the studentized and double bootstrap intervals, followed by the bias correction, percentile, accelerated and basic methods. According to the considered criterion, the worst covering is provided by the basic method (Figure 6).

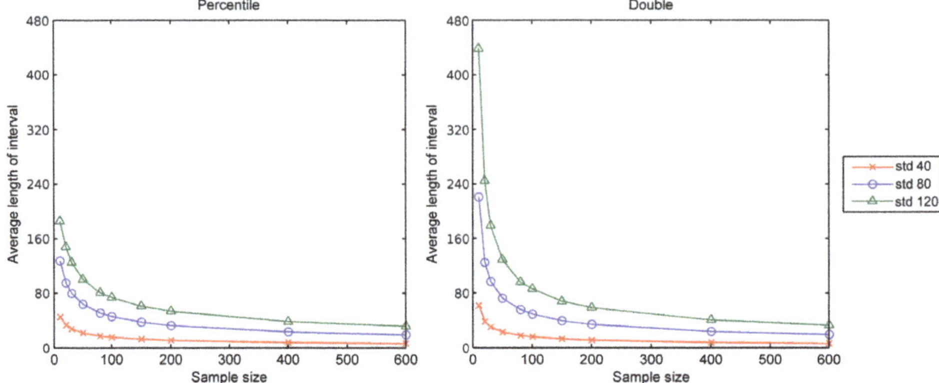

Figure 5. Changes in the lengths of confidence intervals constructed using nonparametric bootstrap methods (percentile and double bootstrap) depending on the sample size and different values of standard deviation; $\mu_x = 160\,\mu\mathrm{m}$.

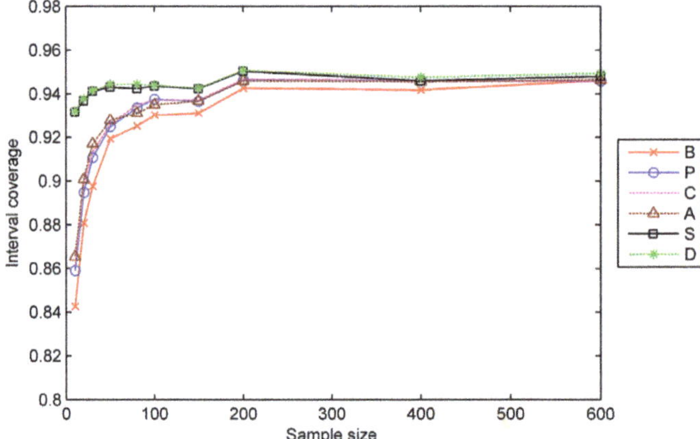

Figure 6. Changes in the empirical confidence level of intervals obtained using the nonparametric bootstrap methods depending on the sample size; $\mu_x = 160\ \mu\mathrm{m}$, $\sigma_x = 65\ \mu\mathrm{m}$ (B—basic, P—percentile, C—bias corrected, A—accelerated, S—studentized, D—double bootstrap).

The graph presented in Figure 7 and Table 2 show the influence of the standard deviation on the coverage of confidence intervals. As in the case of the comparison of the interval lengths, the graph shows a comparison of only the percentile method and the double bootstrap method (Figure 7, Table 2). The difference between the coverages of the intervals obtained using these methods are visible mainly for small sample sizes.

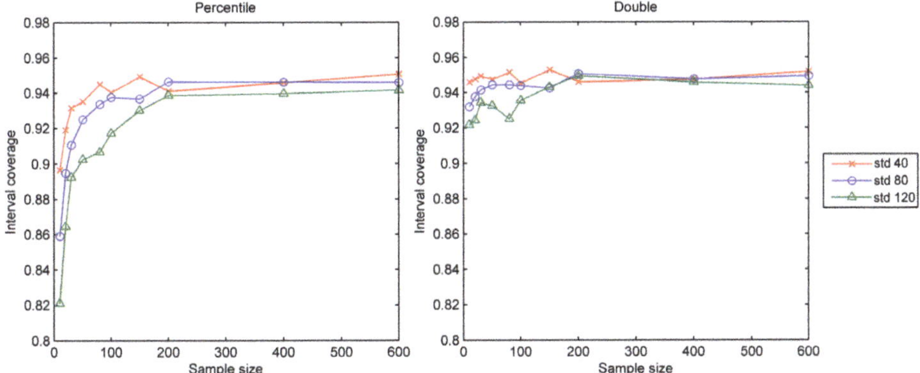

Figure 7. Changes in the empirical confidence level of intervals constructed using nonparametric bootstrap methods (percentile and double bootstrap) depending on the sample size and different values of standard deviation; $\mu_x = 160$ μm.

Table 2. Coverage probabilities of the expected value by confidence intervals constructed using nonparametric bootstrap methods depending the sample size and different values of standard deviation, $\mu_x = 160$ μm.

Sample Size	Standard Deviation	B	P	BC	BCa	S	D
10	40	0.897	0.897	0.895	0.895	0.947	0.946
	80	0.843	0.859	0.861	0.865	0.932	0.932
	120	0.789	0.821	0.828	0.847	0.921	0.922
20	40	0.918	0.919	0.919	0.917	0.946	0.948
	80	0.881	0.895	0.899	0.901	0.937	0.938
	120	0.838	0.864	0.870	0.882	0.924	0.924
30	40	0.928	0.932	0.933	0.934	0.950	0.949
	80	0.898	0.911	0.913	0.917	0.941	0.942
	120	0.869	0.892	0.897	0.905	0.934	0.934
50	40	0.935	0.935	0.936	0.939	0.947	0.947
	80	0.919	0.925	0.926	0.928	0.943	0.944
	120	0.883	0.902	0.907	0.914	0.932	0.933
80	40	0.944	0.945	0.946	0.946	0.950	0.952
	80	0.925	0.934	0.935	0.931	0.942	0.944
	120	0.895	0.907	0.908	0.910	0.926	0.925
100	40	0.943	0.940	0.941	0.941	0.944	0.945
	80	0.930	0.938	0.937	0.935	0.944	0.944
	120	0.905	0.917	0.919	0.920	0.934	0.936
150	40	0.948	0.949	0.948	0.948	0.952	0.953
	80	0.931	0.937	0.937	0.937	0.942	0.943
	120	0.921	0.930	0.932	0.932	0.942	0.943
200	40	0.941	0.941	0.940	0.941	0.945	0.946
	80	0.943	0.946	0.947	0.946	0.950	0.951
	120	0.928	0.939	0.941	0.939	0.948	0.949
400	40	0.947	0.946	0.946	0.945	0.947	0.947
	80	0.942	0.946	0.946	0.946	0.946	0.948
	120	0.936	0.940	0.939	0.941	0.944	0.946
600	40	0.949	0.951	0.949	0.949	0.952	0.952
	80	0.946	0.946	0.946	0.946	0.948	0.949
	120	0.940	0.942	0.941	0.939	0.942	0.944

4. Discussion

Agricultural ecosystems have been designed to provide food, fibre and fuel. Ecosystem services such as nutrient recirculation, pollination and water flow are often referred to as endangered resources in modern societies. Agricultural systems can therefore be seen as heterogeneous but nested socio-ecological systems requiring an institutional environment [3].

Man-made climate change has created major challenges to achieve sustainable agriculture by depleting ecological and natural resources. Existing climate change has an impact on water resources. Some countries are already experiencing severe water scarcity or are reaching their limits [37]. The most promising strategy for achieving sustainable development is to replace hazardous agrochemicals with environmentally friendly preparations from symbiotic microorganisms that could provide protection against biotic and abiotic substances [38].

The main conflict in the pursuit of sustainable development is between economic maximization of growth and environmental protection [39]. On a global scale, agriculture is one of the main sources of degradation of ecosystem services. Also in Europe, most of today's agriculture has a severe negative impact on ecosystems, for example contributing to global carbon cycle disorders, loss of biodiversity and increasing eutrophication. The transition to more sustainable agriculture is necessary to avoid further degradation of ecosystems [3]. The presented assessments of various authors regarding the impact of pesticide use on the natural environment, agricultural production and human health indicate that an important element is proper plant protection technique and compliance with spraying parameters. These parameters can have a significant impact on the quality of the plant protection treatment and the state of the environment.

To reduce the harmful effects of plant protection agents on the environment, we should aim to reduce herbicide doses. In Reference [40], authors based on their own research, say that the dose can be reduced from 12.5% to 50%. In their view, this does not significantly reduce crop yield, while maintaining the required herbicidal effectiveness. The use of herbicides in reduced doses brings the best results in spring cereals, because weeds in these crops are in the early stages of development and thus are easier to eliminate. Very good results can be obtained in plantations with low weed intensity and in the case of controlling species sensitive to a given herbicide. The weaker effect of reduced doses is obtained in winter cereals and especially barley. This fact is caused by the fact that weeds are more advanced in development. Producers of plant protection products are also trying to reduce the content of the active substance, contributing to the reduction of production costs and demonstrating their concern for the state of the environment [8].

Another way to protect the environment is to reduce the number of treatments through the combined use of agrochemicals. Appropriate mixtures of plant protection products give the possibility of wider protection of cereals and affect the increase of yields. The effectiveness of mixtures often does not differ from the effectiveness of preparations used separately and in some cases is even greater. It should be remembered that only well-tested and tested combinations of preparations can be used together [41].

It was therefore considered that it would be more reasonable to systematically implement an integrated model for controlling harmful organisms than to use only chemical methods. The idea of sustainable development combines economic growth with environmental protection and the global balance of ecosystems. According to this concept, the increase in agricultural production can only occur through the increase of resource productivity and thus the introduction of technologies that simultaneously protect resources and maintain their high quality for future generations. The need to develop environmentally friendly agriculture is increasingly accepted in Poland. An important element of it is the change of some agrotechnical methods due to the need to protect ground and surface waters. Restricting the use of chemicals is justified because they cause a number of negative effects [42]—they contribute to the emergence of pesticide-resistant breeds of pests, diseases and weeds, destroy beneficial organisms by reducing their numbers and can even lead to their complete elimination from ecosystems, reduce the value of the environment by contaminating soil, surface and groundwater, lead to a decrease in the nutritional value of plant and animal products as a result of persistence of their residues or metabolites, negatively affect human health.

Ensuring the quality of spraying is a key process in plant protection, which is to ensure high yields while minimizing the adverse impact on the environment [43]. Proper spraying of field crops is influenced by many factors that can be grouped into three categories—1. Equipment and techniques used; 2. Spray quality. 3. Operator skills [44]. All leading atomizer manufacturers offer "low-drift" nozzles that produce droplets less prone to drift [44]. Currently produced atomizers and in particular Venturi nozzles, are designed to produce larger diameter droplets. These designs were created because droplets with a smaller diameter (<200 μm) are more likely to drift [45]. The quality of the plant protection treatment is influenced by the height of the boom position above the sprayed surface [45].

One of the methods used for improvement of the properties of liquids used for plant protection is usage of adjuvants. They improve spraying efficiency by equalizing droplet size and reducing wind drift. They reduce the level of the surface tension of liquid, so they enable coverage more surface of the leaf blade and at the same time they increase absorbing properties of plants. An additional benefit of adjuvants is availability of combining various types of plant protection agents and their use during one treatment [46].

According to Directive 2009/128/EC of the European Parliament establishing a framework for the sustainable use of pesticides, ways to reduce the impact of pesticides on the environment are determined and thus also maintaining the safety of field crops. Particularly great importance is attached to the precise use of pesticides on protected plants. An important issue is the application of the pesticide in such a way that it is effective in use and at the same time safe for the environment. The safety of pesticide use in agriculture depends on the techniques used. At present, great importance is attached to eliminating drift of liquid used during plant protection treatment. The quality of the plant protection treatment depends, among others on atomizer type, working pressure and working speed. The type of atomizer used and the operating pressure determine the droplet size produced by the atomizer. Sprayers wear out during operation and the parameters of their work change. These changes are expressed by the increase in the flow rate of the liquid as well as the spectrum of droplets produced (VMD) [46].

The research topics discussed in the papers make it important to determine the most precise determination of the distribution of droplets generated by sprayers. Because the statistical probability distribution for droplet trace diameters is not fully specified, as presented in the introduction, computer methods become helpful when analysing these problems. Their example are the bootstrap methods, which allows, for example, to generate confidence intervals for a specific parameter or statistics. As the simulation studies show, they do not always give the right results. Results are influenced by among others size of the random sample or too large variability of the examined feature. However, simulation studies have shown that under certain conditions, these methods give reliable results and allow the study properties of variables with asymmetrical statistical distributions, which is confirmed by Reference [27].

Despite some advantages, bootstrap confidence intervals may be heavily distorted and therefore may be misleading. This problem is especially noticeable for simple bootstrap methods and small samples. This is confirmed by the results of other authors who also received inaccurate estimates of confidence intervals in their studies [47]. In addition, bootstrap methods are time consuming, and not always more complex methods give better results, and significantly increase the cost of calculations [48]. Despite these drawbacks, bootstrap methods are a powerful tool for creating confidence intervals. They do not need sophisticated mathematical models and are often used in the absence of information about statistical distributions of the studied features. In addition, for samples of over 200 elements, all examined bootstrap methods give similar results, so in such cases the percentile method seems sufficient. It is the least time-consuming than other more advanced bootstrap methods. For smaller samples studentized and double bootstrap gives more proper estimates.

The applied method of assessing the size of sprayed droplets obtained from sprayers by means of measuring the size of traces left on the surface is burdened with a certain error. Some drops fall in the same place or close to each other, as a result of which some traces are enlarged compared to real ones. For larger drops, the most commonly left traces do not have a circular shape, which causes additional inaccuracies in the estimation of the size of traces left by the computer image analysis program. However, this method, apart from determining the droplet size distribution, also allows the surface coverage of falling drops to be assessed.

In future research, the authors intend to focus on further analysis of spray droplet distribution using more accurate droplet size measurement methods (using in-flight laser measurement). Additionally, the use of the bootstrap method to estimate other parameters of the droplet size distribution obtained from the sprayer will allow for a more precise assessment of sprayer nozzle wear and unevenness of the generated droplets.

5. Conclusions

The comparison of the empirical probabilities of coverage of the expected value by the examined six bootstrap confidence intervals used for estimation of the mean droplet stain diameter following the spraying process performed with the Lechler 110-03 sprayer nozzle leads to the following conclusions:

- The size distribution of traces of droplets obtained from sprayers is asymmetrical, similar to the log-normal distribution.
- The six bootstrap methods compared give confidence intervals that in general do not hold the assumed confidence level especially for small samples.
- The bootstrap methods are generally useful for constructing confidence intervals for the expected value of droplet diameter but not all methods always give confidence intervals that in general hold the assumed confidence level.
- The simulation studies conducted here can be used in practice with the interval estimation of the expected value of droplet stain diameters. An adequate method should be selected depending on the sample size, in particular in the case of smaller sizes (below 100) and depending on the variability of data.
- The studentized and double bootstrap methods allow obtaining less distinct coverage than the assumed confidence level, compared with the other four methods.
- For small sample sizes, the lengths of the confidence intervals obtained using the studentized and double bootstrap methods are similar, greater than the intervals from the other four methods.
- Although the confidence intervals obtained by the studentized and double bootstrap methods maintain the assumed coverage level, however, for small samples the estimated confidence intervals are too wide, which makes it impossible to use them in practice. It is recommended to use samples with at least 100-200 elements for usefull confidence intervals.
- The coverage of intervals obtained using the studentized and double bootstrap methods is less sensitive to the variability of the feature, compared with the basic and percentile methods. For small sample sizes, a confidence level of 93% or more can be obtained if the coefficient of variation does not exceed certain values. For example, if the sample size is ca. 30 elements, the standard deviation of the sample cannot be higher than 22%–33%, 25%–36%, 25%–34%, 23%–33%, 109%–123% and 110%–123% of the sample mean for the basic, percentile, bias-corrected, bias-corrected and accelerated, studentized and double bootstrap methods, respectively.
- Determining the confidence interval for the expected value of droplet size using bootstrap methods requires knowledge of the exact diameters of individual traces, that is, using image analysis or another measurement method, such as laser droplet size measurement. However, the method presented in this work allows the analysis of surface coverage.

Author Contributions: Conceptualization, A.B. and I.K-.B.; methodology, A.B. and I.K-.B.; software, A.B..; validation, A.B., P.A.K., I.K-.B. and M.K.; formal analysis, P.A.K. and I.K-.B.; investigation, M.K.; resources, M.K..; data curation, P.A.K. and M.K.; writing—original draft preparation, A.B., P.A.K., I.K-.B. and M.K; writing—review and editing, A.B., P.A.K., I.K-.B. and M.K; visualization, A.B. and P.A.K.; supervision, A.B. and I.K-.B.; project administration, A.B. and I.K-.B.

Funding: This research received no external funding.

Conflicts of Interest: The authors declare no conflict of interest.

References

1. Cheng, X.; Shuai, C.; Liu, J.; Wang, J.; Liu, Y.; Li, W.; Shuai, J. Modelling environment and poverty factors for sustainable agriculture in the Three Gorges Reservoir of China. *Land Degrad. Dev.* **2018**, *29*, 3940–3953. [CrossRef]

2. Kiełbasa, B.; Pietrzak, S.; Ulén, B.; Drangert, J.O.; Tonderski, K. Sustainable agriculture: The study on farmers' perception and practices regarding nutrient management and limiting losses. *J. Water Land Dev.* **2018**, *36*, 67–75, [CrossRef]

3. Öhlund, E.; Zurek, K.; Hammer, M. Towards sustainable agriculture? The EU framework and local adaptation in Sweden and Poland. *Environ. Policy Gov.* **2015**, *25*, 270–287, [CrossRef]

4. Piesik, D. Biologiczna walka z chwastami na przykładzie Rumex confertus Willd. *Postępy Nauk Rolniczych* **2001**, *3*, 85–98.

5. Biziuk, M.; Hupka, J.; Warendecki, W.; Zygmunt, B.; Siłwoiecki, A.; Zelechowska, A.; Dąbrowski, Ł.; Wiergowski, M.; Zaleska, A.; Tyszkiewicz, H. *Pestycydy: Występowanie, Oznaczanie i Unieszkodliwianie*; Wydawnictwo Naukowo-Techniczne: Warszawa, Poland, 2001.

6. Lipa, J.J. Zwalczanie szkodników, chwastów i patogenów. *Zesz. Probl. Postęp. Nauk Rol., Rol. Ekol.* **1987**, *324*, 131–155.

7. Miłkowski, P.; Woźnica, Z. Zachowanie się kropel opryskowych na powierzchni roślin a skuteczność chwastobójcza Glifosatu. *Pr. Kom. Nauk Roln. Kom. Nauk Lesn., Poznan. Tow. Przyj. Nauk* **2001**, *91*, 77–85.

8. Sawa, J.; Huyghebaert, B.; Koszel, M. Parametry pracy opryskiwaczy a jakość oprysku. In Proceedings of 4th Conference: "Racjonalna Technika Ochrony Roślin", Skierniewice, Poland, 15–16 October 2003; pp. 84–89.

9. Miczulski B. *Podstawy Praktycznej Ochrony Roślin*; Wydawnictwo Akademii Rolniczej w Lublinie: Lublin, Poland, 1991.

10. Rogalski, L.; Konopka, W. Wybrane charakterystyki opryskiwania pszenicy w zależności od wartości dawki cieczy użytkowej. *Zesz. Probl. Postęp. Nauk Rol.* **2002**, *486*, 367–373. Available online: http://agro.icm.edu.pl/agro/element/bwmeta1.element.agro-c9aa0840-51cb-4418-a22e-1e912970f2f1/c/367-373.pdf (accessed on 25 October 2019).

11. Rogalski, L.; Konopka, W. Bilansowanie rozchodu masy oprysku w łanie pszenicy w zależności od rodzaju cieczy użytkowej. *Zesz. Probl. Postęp. Nauk Rol.* **2002**, *486*, 375–380.

12. Nilars, T.; Taylor, B.; Kappel, D. Wpływ rozpylaczy na jakość i bezpieczeństwo opryskiwania. In Proceeding of 3rd Conference: "Racjonalna Technika Ochrony Roślin". Skierniewice, Poland, 16–17 October 2002; pp. 121–134.

13. Marshall, J.S.; Palmer, W.M.K. The distribution of raindrops with size. *J. Meteorol.* **1948**, *5*, 165–166. [CrossRef]

14. Kozu, T.; Nakamura, K. Rainfall parameter estimation from dual-radar measurements combining reflectivity profiles and path-integrated attenuation. *J. Atmos. Ocean. Technol.* **1991**, *8*, 259–270, [CrossRef]

15. Su, C.L.; Chu, Y.H. Analysis of terminal velocity and VHF backscatter of precipitation particles using Chung-Li VHF radar combined with ground-based disdrometer. *Terr. Atmos. Ocean. Sci.* **2007**, *18*, 97–116. [CrossRef]

16. Ulbrich, C.W. Natural variations in the analytical form of the raindrop size distribution. *J. Appl. Meteorol.* **1983**, *22*, 1764–1775. [CrossRef]

17. Maguire, W.B.; Avery, S.K. Retrieval of raindrop size distribution using two Doppler wind profilers: Model sensitivity testing. *J. Appl. Meteorol. Climatol.* **1994**, *33*, 1623–1635. [CrossRef]

18. Feingold, G.; Levin, Z. Application of the lognormal raindrop distribution to differential reflectivity radar measurement (ZDR). *J. Atmos. Ocean. Technol.* **1987**, *4*, 377–382. [CrossRef]

19. Meneghini, R.; Rincon, R.; Liao, L. On the use of the lognormal particle size distribution to characterize global rain. In Proceedings of the International Geoscience and Remote Sensing Symposium, Toulouse, France, 21–25 July 2003; pp. 1707–1709, [CrossRef]

20. Kuna-Broniowski, M.; Kuna-Broniowska, I. The use of comparators for automatic classification of the splashed rain drops. *Electron. J. Pol. Agric. Univ. Agric. Eng.* **2001**, *4*, #08. Available online: http://www.ejpau.media.pl/volume4/issue2/engineering/art-08.html (accessed on 25 October 2019).

21. Wilks, D.S. Rainfall intensity, the Weibull distribution, and estimation of daily surface runoff. *J. Appl. Meteorol.* **1989**, *28*, 52–58. [CrossRef]

22. Joss, J.; Waldvogel, A. Raindrop size distribution and sampling size errors. *J. Atmos. Sci.* **1969**, *26*, 566–569. [CrossRef]
23. Efron, B.; Tibshirani R.J. Statistical data analysis in the computer age. *Science* **1991**, *253*, 390–395. [CrossRef]
24. Efron, B.; Tibshirani R.J. *An introduction to the Bootstrap*; Chapman & Hall: New York, NY, USA, 1993; [CrossRef]
25. Kamińska, J.; Machowczyk, A.; Szewrański, S. The variation of drop size gamma distribution parameters for different natural rainfall intensity, Wydawnictwo ITP. *Woda-Środ.-Obsz. Wiej.* **2010**, *10*, 95–102. Available online: http://yadda.icm.edu.pl/baztech/element/bwmeta1.element.baztech-article-BATC-0005-0103 (accessed on 25 October 2019).
26. Hall, P. *The Bootstrap and Edgeworth Expansion*; Springer: New York, NY, USA, 1992; [CrossRef]
27. Pekasiewicz, D. Bootstrapowa weryfikacja hipotez o wartości oczekiwanej populacji o rozkładzie asymetrycznym. *Acta Univ. Lodz. Folia Oeconomica* **2012**, *271*, 151–159. Available online: http://cejsh.icm.edu.pl/cejsh/element/bwmeta1.element.hdl_11089_1928/c/Pekasiewicz_151-159.pdf (accessed on 25 October 2019).
28. Zhou, X.; Gao, S. Confidence intervals for the log-normal mean. *Stat. Med.* **1997**, *16*, 783–790. [CrossRef]
29. Land, C.E. An evaluation of approximate confidence interval estimation methods for lognormal means. *Technometrics* **1972**, *14*, 145–158. [CrossRef]
30. Angus, J.E. Inferences on the lognormal mean for complete samples. *Commun. Stat. Simul. Comput.* **1988**, *17*, 1307–1331. [CrossRef]
31. Angus, J.E. Bootstrap one-sided confidence intervals for the lognormal mean. *Statistician* **1994**, *43*, 395–401. [CrossRef]
32. Krishnamoorthy, K.; Mathew, T.P. Inferences on the means of lognormal distributions using generalized p-values and generalized confidence intervals. *J. Stat. Plan. Inference* **2003**, *115*, 103–121. [CrossRef]
33. Efron, B. *The Jackknife, the Bootstrap, and Other Resampling Plans*; Society for Industrial and Applied Mathematics: Philadelphia, PA, USA, 1982. [CrossRef]
34. Ozkan, H.E.; Reichard, D.L.; Ackerman, K.D. Effect of orifice wear on spray patterns from fan nozzles. *Trans. ASAE* **1992**, *35*, 1091–1096. [CrossRef]
35. Efron, B. Better Bootstrap Confidence Intervals. *J. Am. Stat. Assoc.* **1987**, *82*, 171–185. [CrossRef]
36. McCullough, B.D.; Vinod, H.D. Implementing the Double Bootstrap. *Comput. Econ.* **1998**, *12*, 79–95. [CrossRef]
37. Maleksaeidi, H.; Karami, E. Social-ecological resilience and sustainable agriculture under water scarcity. *Agroecol. Sustain. Food Syst.* **2013**, *37*, 262–290. [CrossRef]
38. Tikhonovich, I.A.; Provorov, N.A. Microbiology Is a basis of sustainable agriculture: An opinion. *Ann. Appl. Biol.* **2011**, *159*, 155–168. [CrossRef]
39. McNeill, D. The contested discourse of sustainable agriculture. *Glob. Policy* **2019**, *10* (Suppl. 1). 1758–5899. [CrossRef]
40. Domardzki, K.; Rola, H. Efektywność stosowania niższych dawek herbicydów w zbożach. *Pamiętnik Puławski Materiały Konferencji* **2000**, *120*, 53–63.
41. Głazek, M.; Mrówczyński, M. Łączne stosowanie agrochemikaliów w nowoczesnej technologii produkcji zbóż. *Pamiętnik Puławski Materiały Konferencji* **1999**, *114*, 119–126.
42. Szymańska, E. Zużycie chemicznych środków ochrony roślin i możliwości jego ograniczenia w zrównoważonym systemie produkcji zbóż. *Pamiętnik Puławski Materiały Konferencji* **2000**, *120*, 439–444.
43. De Cock, N.; Massinon, M.; Salah, S.O.T.; Lebeau, F. Investigation on optimal spray properties for ground based agricultural applications using deposition and retention models. *Biosyst. Eng.* **2017**, *162*, 99–111. [CrossRef]
44. Arvidsson, T.; Bergström, L.; Kreuger, J. Spray drift as influenced by meteorological and technical factors. *Pest. Manag. Sci* **2011**, *67*, 586–598. [CrossRef]
45. Butts, T.R.; Luck, J.D.; Fritz, B.K.; Hofmann, W.C.; Kruger, G.R.. Evaluation of spray pattern uniformity using three unique analyses as impacted by nozzle, pressure, and pulse-width modulation duty cycle. *Pest. Manag. Sci.* **2019**, *75*, 1875–1886. [CrossRef]
46. Parafiniuk, S.; Milanowski, M.; Subr A.K. The influence of the water quality of the droplet spectrum produced by agricultural nozzles. *Agric. Agric. Sci. Procedia* **2015**, *7*, 203–208. [CrossRef]

Sustainability **2019**, *11*, 7037

47. Benkwitz, A.; Lütkepohl, H; Wolters, J. Comparison of bootstrap confidence intervals for impulse responses of German monetary systems. *Macroecon. Dyn.* **2001**, *5*, 81–100. Available online: https://www.researchgate.net/publication/23775280_Comparison_of_Bootstrap_Confidence_Intervals_for_Impulse_Responses_of_German_Monetary_Systems (accessed on 4 December 2019). [CrossRef]

48. Trichakis, I.; Nikolos, I; Karatzas, G.P.; Comparison of bootstrap confidence intervals for an ANN model of a karstic aquifer response. *Hydrol. Process.* **2011**, *25*, 2827–2836. [CrossRef]

sustainability

Article

Modeling the Dependency between Extreme Prices of Selected Agricultural Products on the Derivatives Market Using the Linkage Function

Zofia Gródek-Szostak [1,*], Gabriela Malik [2], Danuta Kajrunajtys [3], Anna Szeląg-Sikora [4], Jakub Sikora [4], Maciej Kuboń [4], Marcin Niemiec [5] and Joanna Kapusta-Duch [6]

[1] Department of Economics and Organization of Enterprises, Cracow University of Economics, Krakow 31-510, Poland
[2] Higher School of Economics and Computer Science in Krakow, Kraków 31-510, Poland
[3] Department of International Management, Cracow University of Economics, Krakow 31-510, Poland
[4] Institute of Agricultural Engineering and Informatics, University of Agriculture in Krakow, Krakow 30-149, Poland
[5] Department of Agricultural and Environmental Chemistry, University of Agriculture in Krakow, Krakow 31-120, Poland
[6] Department of Human Nutrition, Faculty of Food Technology, University of Agriculture in Krakow, Krakow 30-149, Poland
* Correspondence: grodekz@uek.krakow.pl; Tel.: +48-12-293-52-29

Received: 4 June 2019; Accepted: 30 July 2019; Published: 1 August 2019

Abstract: The purpose of the article is to identify and estimate the dependency model for the extreme prices of agricultural products listed on the Chicago Mercantile Exchange. The article presents the results of the first stage of research covering the time interval 1975–2010. The selected products are: Corn, soybean and wheat. The analysis of the dependency between extreme price values on the selected futures was based on the estimation of five models of two-dimensional extreme value copulas, namely, the Galambos copula, the Gumbel copula, the Husler–Reiss copula, the Tawn asymmetric copula and the t-EV copula. The next stage of the analysis was to test whether the structure of the dependency described with the estimated copulas is a sufficient approximation of reality, and whether it is suitable for modeling empirical data. The quality of matching the estimated copulas to empirical data of return rates of agricultural products was assessed. For this purpose, the Kendall coefficient was calculated, and the methodology of the empirical combining function was used. The conducted research allowed for the determination of the conduct for this kind of phenomena as it is crucial in the process of investing in derivatives markets. The analyzed phenomena are highly dependent on e.g., financial crises, war, or market speculation but also on drought, fires, rainfall, or even crop oversupply. The conducted analysis is of key importance in terms of balancing agricultural production on a global scale. It should be emphasized that conducting market analysis of agricultural products at the Chicago Mercantile Exchange in the context of competition with the agricultural market of the European Union is of significant importance.

Keywords: agricultural product; price; modeling; management

---

1. Introduction

Sustainable agricultural production is crucial for balancing the needs of current and future populations. The benefits of modern agriculture are significant, allowing food production go hand in hand with an increase in the population [1,2]. Cooperation between business and the agricultural production circles are of particular interest at both the national and international level [3,4]. Demographic forecasts indicate that by 2050, the world's population will reach 9.7 billion. This will lead to an increase in demand for food, further aggravating environmental problems related to intensive agricultural production. For this reason, one of the main challenges is to achieve global food security and sustainable agriculture. While food security aims at ensuring sustainable and healthy food supply over time, sustainable agriculture plays a key role in maintaining resilient agro-ecosystems [5]. The above-mentioned concept of sustainable farming is fostered by ecological farming. Its development is accompanied by a high growth rate of the organic food market and a significant growth of sales of organic food. Even with strict adherence to production practices and increasing their availability, most consumers still are unaware of organic production methods. The customers' awareness of organic products does not necessarily translate into real consumption [6,7].

According to the United States Department of Agriculture (USDA), the projected global wheat trade in the 2019/2020 season is expected to increase by 6.7 million tonnes, i.e., by 4% and reach 184.6 million tonnes in the case of larger export deliveries and lower expected export prices.

The global perspective of the feed grain market in the 2019/2020 season relates to record production and consumption, as well as lower final inventories. Global production of corn is projected to increase, mostly in the USA, South Africa, Russia, Canada, India, and Brazil, compensating for slightly lower crops in China and Ukraine. The global use of corn is expected to increase by 1%, while global corn imports are expected to increase by 2%. Global soy production in the 2018/2019 season increased in relation to the May forecast by 0.7 million tonnes, reaching 355.2 million tonnes [8–11].

In the decision-making processes, as well as in many scientific disciplines, the study of the dependency of extreme values is of great importance. In many situations, their proper identification enables avoiding wrong investment decisions. Studying the dependency between extreme prices of agricultural products is of key importance in the process of analyzing the structure of dependency. This is especially true when it is important to model the dependency between maintaining the maximum and minimum values of the analyzed observations. Dependencies for extremely small or extremely large values may have different characteristics than those determined based on the entire sample. The reason for fluctuations in the prices of agricultural products are, e.g., financial crises, wars, market speculation, but also droughts, fires, rains, or even crop oversupply. For this reason, the difference in price behavior during periods of extreme change, i.e., unusual and rarely observable market events require careful monitoring because it can lead to above-average losses or profits in the process of investing on derivatives markets.

A very useful tool in the process of modeling dependencies between extreme values of time series is the functions of relations of extreme values, also called extreme value copulas. Their main advantage is the possibility of analyzing above-average losses or profits in the field of finance, insurance, but also in the case of examining futures for agricultural products.

One of the first applications of the theory of extreme value based on a two-dimensional copula appeared in the subject literature in 1964 in the study of Gumbel and Goldstein [12]. These copulas were also used in many other fields of science, e.g., in the process of analyzing the return rates of exchange rates [13], the study of capital markets characterized by high volatility [14], in the portfolio analysis and portfolio risk estimation [15] or in the field of insurance [16], and in hydrology [17].

Extreme value copulas were created as possible limit copulas in relation to the maximum i.i.d. (independent and identically distributed) distribution, i.e., a distribution in which all random variables are independent and have the same distribution of probability. Since two-dimensional copulas were used in the empirical part of this article, the basic definitions and conclusions regarding extreme value copulas presented below will be demonstrated for the two-dimensional case. More detailed information on extreme value copulas, together with an extension for a multidimensional case can be found, e.g., in a monograph by R.B. Nelsen [18], a monograph by H. Joe [19], books by J. Segers [20] and G. Gudendorf and J. Segers [21], or in the following papers: C. Genest and J. Seger [22], M. Ribatet and M. Sedki [23].

As a supplier of raw food materials, the agricultural market is of strategic importance from the point of view of food security. Therefore, it is so important to strive towards risk-reducing solutions, including models, and inference based on the statistical analyzes, as exemplified by those proposed by the authors.

In order to tackle the growing challenges of sustainable agricultural production, a better understanding of complex agricultural ecosystems is needed. This can be done using modern digital technologies that constantly monitor the physical environment, producing large amounts of data at an unprecedented rate. The analysis of the big data will enable farmers and enterprises to leverage its value and improve their efficiency. Although the analysis of large data sets leads to progress in various industries, it is not yet widely used in agriculture [24,25].

2. Materials and Methods

The source of data used for the analysis in the empirical part of the paper is the Chicago Mercantile Exchange. The choice of this exchange was motivated by its importance in the international market of agricultural products. The research sample was created using daily quotations of nominal prices of futures for three agricultural products, i.e., corn, wheat and soybean. The value of the contract is expressed in the price of a bushel, the transaction unit of a commodity in US dollars. The data includes the closing price of the contract with the shortest expiration date, so that a series of quotations could be treated as a forward price with the shortest possible execution time. The choice of products was dictated by the importance of trading on the futures market and the availability of appropriately long time series. Empirical data come from the years 1975–2010. A single time series includes 9070 observations. They have been checked for possible discontinuities and errors. In order to minimize the impact of arbitrary interference on the obtained results, no procedures have been applied to correct or supplement empirical data. The selected combining function is one, for which the distance from the empirical combining function is the smallest.

2.1. Theoretical Models of Extreme Value Copulas, the Inference Functions for Margins (IFM) Estimation Method, and the Empirical Combining Function Method

The notion of the linkage function, also often called the copula function, appeared first in subject literature in 1959 in the work of A. Sklar [26]. However, it was not until the 1990s and the monograph of H. Joe [19] and R.B. Nelsen [18] that these functions have gained immense popularity, mainly due to the presentation of their broad practical application in dependence modeling. In the general, and at the same time the simplest view, the copula is a function that allows distinguishing the component describing only the structure of dependence from the cumulative distribution function of the total distribution of a random vector. In other words, it can be said that the copula functions are functions that combine a multidimensional cumulative distribution function of a random variable with its one-dimensional limit cumulative distribution function [19–26].

Among the copula functions, one can distinguish their particularly important class, namely, the copula of extreme values [27–33]. The use of two-dimensional extreme value copulas in empirical studies has been greatly simplified by using the representation introduced into the subject literature by J. Pickands in 1981 [34]. This issue was also an essential element of research conducted in 1977 by A.A. Balkema and S.I. Resnick [35] and L. de Haan and S. I. Resnick [36]. Observations of the extremes were classified in accordance with the Extreme Value Theory (EVT), which is used to describe the behavior of limit properties of extreme values. It consists of analyzing the tails of distributions, which constitute only a small part of the entire distribution examined. Its purpose is not to describe the usual behavior of stochastic phenomena, but the unusual and rarely observed events. The Extreme Value Theory is widely used wherever modeling relationships between the behaviors of the maximum and minimum values of the analyzed observations is particularly important [37–39]. The classic approach for modeling extreme values is based on the block maxima model. This method is used for a large number of observations that have been selected from a large sample. Modeling the behavior of the extreme values of independent random variables with an identical probability distribution is in fact using the maxima or minima of the observations in fixed time blocks. These blocks are designated by means of separate time intervals of equal length, most often months, quarters or years [30].

The next part of this article will present extreme values copulas used in empirical research.

2.1.1. Theory of Extreme Copulas

Among the copula functions, one can distinguish a particularly important class, i.e., the extreme value copula. A copula is a function that allows to distinguish a component describing only the structure of dependence from a total random vector distribution. In other words, the copula functions are functions that combine a multidimensional cumulative distribution of a random variable with its one-dimensional limit distributors [40].

The definition of a copula for a two-dimensional case is as follows:

A two-dimensional copula (2-copula) is each function $C : [0, 1]^2 \rightarrow [0, 1]$ that meets the following conditions:

For each $u_1, u_2 \in [0, 1]$ there is:

$$C(u_1, 0) = C(0, u_2) = 0 \tag{1}$$

1. For each $u_1, u_2 \in [0, 1]$ there is:

$$C(u_1, 1) = u_1 \text{ and } C(1, u_2) = u_2 \tag{2}$$

2. For each $u_1, u_2, v_1, v_2 \in [0, 1]$, such that $u_1 \leq u_2$ and $v_1 \leq v_2$, there is:

$$C(u_2, v_2) - C(u_2, v_1) - C(u_1, v_2) + C(u_1, v_1) \geq 0 \tag{3}$$

For basic information in terms of modeling the price dependence of futures on agricultural products listed on the Chicago Mercantile Exchange using the copula function, let $(X_1, Y_1), (X_2, Y_2), \ldots, (X_n, Y_n)$ be independent pairs of random variables with the same distributions and common copula C. Moreover, let $C_{(n)}$ be a copula with respect to the maximum components, namely $X_n = \max\{X_i\}$, $Y_n = \max\{Y_i\}$, $i = 1, \ldots, n$. From Sklar's theorem [26], it follows that $C_{(n)}(u, v) = C^n\left(u^{\frac{1}{n}}, v^{\frac{1}{n}}\right)$ for every $u, v \in [0, 1]$ The limit of the sequence $\{C_{(n)}\}$ naturally leads to the concept of an extreme value copula. Therefore, the copula C_E is called an extreme value copula if there is a copula C, such that:

$$\forall u, v \in [0, 1] \ C_E(u, v) = \lim_{n \rightarrow \infty} C^n\left(u^{\frac{1}{n}}, v^{\frac{1}{n}}\right) \tag{4}$$

2.1.2. Galambos' Copula

This copula was introduced to subject literature in 1975 by J. Galambos [41]. It is an example of a symmetrical copula, described with the formula below:

$$C_\theta^{Ga}(u, v; \theta) = uv \exp\left\{\left[(-\ln u)^{-\theta} + (-\ln v)^{-\theta}\right]^{-1/\theta}\right\} \tag{5}$$

where $\theta > 0$.

The dependence function of the discussed copula is presented using the following formula:

$$A^{Ga}(t) = 1 - \left(t^{-\theta} + (1 - t)^{-\theta}\right)^{-1/\theta} \tag{6}$$

2.1.3. The Gumbel Copula

This copula was introduced to literature in 1960 by E.J. Gumbel and is also called the Gumbel–Hougaard copula [42]. It is described by the following formula:

$$C_\theta^{Gu}(u, v; \theta) = \exp\left\{-\left[(-\ln u)^\theta + (-\ln v)^\theta\right]^{1/\theta}\right\} \tag{7}$$

where $\theta \geq 1$.

The independence function is described by the formula:

$$A^{Gu}(t) = \left(t^\theta + (1 - t)^\theta\right)^{1/\theta} \tag{8}$$

It can be observed that for $\theta \to \infty$ there is an excellent dependence, while independence occurs in the case of $\theta = 1$.

2.1.4. The Husler–Reiss Copula

The name of the copula comes from its authors, who presented it for the first time in a paper in 1989 [43]. The Husler–Reiss copula is defined by the following formula:

$$C_\theta^{HR}(u, v; \theta) = \exp\left\{-\widetilde{u}\Phi\left[\frac{1}{\theta} + \frac{1}{2}\theta\ln\left(\frac{\widetilde{u}}{\widetilde{v}}\right)\right] - \widetilde{v}\Phi\left[\frac{1}{\theta} + \frac{1}{2}\theta\ln\left(\frac{\widetilde{v}}{\widetilde{u}}\right)\right]\right\} \tag{9}$$

wherein $\theta \geq 0$, $\widetilde{u} = -\ln u$, $\widetilde{v} = -\ln v$. On the other hand, Φ is a cumulative distribution function of a standardized normal distribution.

The dependence function of the discussed copula is determined by the formula:

$$A^{HR}(t) = t\Phi\left[\theta^{-1} + \frac{1}{2}\theta\ln\left(\frac{t}{1-t}\right)\right] + (1 - t)\Phi\left[\theta^{-1} - \frac{1}{2}\theta\ln\left(\frac{t}{1-t}\right)\right] \tag{10}$$

while the parameter θ for the Husler–Reiss copula measures the degree of dependence. This means that starting from $\theta = \infty$, one is dealing with independence until the situation in which $\theta = 0$, which brings total dependence.

2.1.5. Tawn Copula

Introduced in the subject literature in 1988 by J. Tawn, this copula is an asymmetric extension of the Gumbel cupola [44]. The model of the discussed copula is presented below:

$$C_\theta^{Ta}(u, v; \theta) = uv\exp\left\{\theta\frac{(\ln u)(\ln v)}{\ln u + \ln v}\right\} \tag{11}$$

where $0 \leq \theta \leq 1$.

In turn, the independence function of the Tawn copula is given by the formula:

$$A^{Ta}(t) = \theta t^2 - \theta t + 1 \tag{12}$$

2.1.6. The *t*-EV Copula

This copula is known as the *t*-Student extreme copula. For a two-dimensional case, the λ degrees of freedom and the correlation coefficient $\rho \in (-1, 1)$, the *t*-EV copula is determined by the following formula:

$$C_{\lambda, \rho}^{tEV}(u, v; \rho) = \exp\left\{T_{\lambda+1}\left[-\frac{\rho}{\theta} + \frac{1}{\theta}\left(\frac{\ln u}{\ln v}\right)^{1/\lambda}\right]\ln u + T_{\lambda+1}\left[-\frac{\rho}{\theta} + \frac{1}{\theta}\left(\frac{\ln u}{\ln v}\right)^{1/\lambda}\right]\ln v\right\} \tag{13}$$

where T_λ is a cumulative t-Student distribution function with λ degrees of freedom, while the parameter θ depends on the value of the correlation coefficient ρ and the degrees of freedom λ, and is described by the equation: $\theta^2 = \frac{1-\rho^2}{\lambda+1}$ [40].

The function of t-Student extreme dependence copula is presented using the following formula:

$$A_{\lambda,\rho}^{tEV}(t) = tT_{\lambda+1}\left(\frac{\left(\frac{t}{1-t}\right)^{1/\lambda} - \rho}{\sqrt{1-\rho^2}}\sqrt{1+\lambda}\right) + (1-t)T_{\lambda+1}\left(\frac{\left(\frac{1-t}{t}\right)^{1/\lambda} - \rho}{\sqrt{1-\rho^2}}\sqrt{1+\lambda}\right) \tag{14}$$

Estimation of five models of two-dimensional extreme value copulas was carried out based on the two-stage method of maximum likelihood, i.e., the IFM method. It is the inference function method for limit distributions, also called the two-stage maximum likelihood estimation method due to the two-stage estimation process. It was proposed for the first time in the works of H. Joe and J.J. Xu [27] and H. Joe [19]. The authors suggested to divide the set of estimated parameters into two subsets, so as to first estimate the parameters associated with the limit distributions, and then find the estimator responsible for the combined distribution, i.e., the copula. The empirical combining function method is used to assess the quality of matching the parameters of the extreme value copula [35–37]. The principle of selecting an appropriate combining function boils down to selecting the best function from the finite set of candidates, which is a subset of the set of all possible combining functions. There are various ways to define the empirical combining function. In this paper, a practical formula based on the values of the combining function defined on the grid will be used [45]. Since the price series of financial instruments belong to the group of non-stationary processes, daily constant return rates were determined based on the price series to conduct statistical analysis between the prices of the agricultural products. The use of logarithmic return rates is meaningful for the properties of the data series under investigation. As one of the Box–Cox transformations, logarithmation is known to stabilize the variance of the series [38].

$$R_t = \ln\left(\frac{X_t}{X_{t-1}}\right) \tag{15}$$

where X_t is the value of the futures on the day t.

Based on the daily closing prices, the logarithmic return rates for individual agricultural products were calculated, in accordance with the formula presented above.

2.1.7. The Empirical Combining Function Method

This method is used to assess the quality of matching the parameters of the extreme value copula, based on the empirical link function [39–44]. Let the theoretical structure of the dependence be described by the combining function C_θ, dependent on the value of the parameter θ. It is then assumed that the null hypothesis H_0 means that the structure of the dependence is determined using the combining function C_θ, against the alternative hypothesis H_1, which is a negation of hypothesis H_0. The principle of selecting the appropriate combining function comes down to selecting the best function from the finite set of candidates C, which is a subset of the set of all possible combining functions. The combining function selected from the set C is that for which the distance from the empirical copula is the smallest. There are various ways to define the empirical copula. In this work a practical formula was used, based on the values of the combining function defined on the following grid:

$$L = \left\{\left(\frac{i}{m}, \frac{j}{m}\right) : i, j = 0, 1, \ldots, m\right\} \tag{16}$$

The empirical values of the combining function at the points of the L-grid are determined by the following formula:

$$C_m\left(\frac{i}{m}, \frac{j}{m}\right) = \frac{1}{m}\sum_{k=1}^{m} \mathbf{1}(R_k \leq i)\mathbf{1}(S_k \leq j) \tag{17}$$

where R_i, S_i are ranks of variables X and Y.

On the other hand, the distance measured between the combining functions is based on the standard L^2, and in the discrete version takes the form:

$$d_L(C_m, C) = \sqrt{\sum_{i=1}^{m}\sum_{j=1}^{m}\left(C_m\left(\frac{i}{m}, \frac{j}{m}\right) - C\left(\frac{i}{m}, \frac{j}{m}\right)\right)^2} \tag{18}$$

where:

C_m—empirical copula,

C—function of the copula from the distinguished set of copulas C.

2.1.8. Kendall's Correlation Coefficient

Empirical studies often use rank-based correlation coefficients. One of the most popular is the Kendall rank correlation coefficient. It is usually used when there are restrictions related to the use of Pearson's linear correlation coefficient. Its advantage is that, as a representative of the measure of compatibility, it does not depend on limit distributions. Although the limit distributions and the correlation matrix do not determine the form of the combined distribution, similar to the linear correlation coefficient, for any continuous edge distributions, a combined distribution with a given rank correlation coefficient can still be constructed from the entire range $[-1; 1]$.

In order to define the Kendall rank correlation coefficient, a less formal definition of compatibility should first be introduced, as follows [46].

Two different realizations (x_1, y_1) and (x_2, y_2) of the random vector (X, Y) are compatible if

$$(x_1 - x_2)(y_1 - y_2) > 0 \text{ (i.e., } x_1 > x_2 \text{ and } y_1 > y_2, \text{ or } x_1 < x_2 \text{ and } y_1 < y_2)$$

and non-compatible, if

$$(x_1 - x_2)(y_1 - y_2) < 0 \text{ (i.e., } x_1 > x_2 \text{ and } y_1 < y_2 \text{ or } x_1 < x_2 \text{ and } y_1 > y_2)$$

Let (X_1, Y_1) and (X_2, Y_2) be independent random vectors with identical distribution. Kendall's correlation coefficient τ is defined as follows:

$$\tau(X_1, Y_1) = P[(X_1 - X_2)(Y_1 - Y_2) > 0] - P[(X_1 - X_2)(Y_1 - Y_2) < 0] \tag{19}$$

Using the above terminology, it can be concluded that the Kendall's coefficient for the vector (X_1, Y_1) is the probability of compatible realizations of the random vector (X_1, Y_1) minus the likelihood of realizing the non-compatible of the vector.

However, if $\{(x_1, y_1), \ldots, (x_n, y_n)\}$ is an n-element sample of a vector (X, Y) of continuous random variables, then there are $\binom{n}{2} = \frac{n(n-1)}{2}$ different pairs (x_i, y_i) and (x_j, y_j) that are compatible or inconsistent. The Kendall factor τ for the sample version, which is designated to be distinguished from the theoretical coefficient through $\hat{\tau}$, can be calculated based on the following formula:

$$\hat{\tau} = \frac{c - d}{c + d} \tag{20}$$

where c is the number of compliant pairs, and d the number of incompatible pairs.

3. Results

Sustainability is a human-centered concept that comprises multiple aspects and objectives of different interest groups. It is not readily measurable, except as a compromise between different parts of society, of which some may try to represent future generations of mankind [47]. In order to analyze the dependence of extreme price values on selected futures listed on the Chicago Mercantile Exchange, five selected models of two-dimensional extreme values copulas defined in the previous chapter, were used. As said before, the following copulas were used in empirical research: the Gumbel copula, the Galambos copula, the Husler–Reiss copula, the *t*-EV copula and the Tawn asymmetric copula. The estimation of parameters of selected extreme values copulas was carried out based on the two-stage maximum likelihood estimation method, i.e., the IFM method for two different limit distributions, namely, for the normal distribution and the *t*-Student distribution. The selection of distributions was motivated by the results recommending their use, presented widely in the subject literature [48–52]. Results for each of the three agricultural product pairs in question are presented in Tables 1 and 2. In addition to the values of the estimated parameters and their average estimation errors given in brackets, the likelihood function (LLF) values can also be found there. In addition, the extreme t -Student dome has an additional parameter λ, responsible for the number of degrees of freedom, the estimations of which are also included in the tables.

Table 1. Results of the estimation of the parameters of extreme values copulas with the normal limit distribution.

Copula	Parameter	Maize-Soy	LLF	Corn-Wheat	LLF	Soybean-Wheat	LLF
Galambos	θ	0.897 *** (0.013)	1831	0.674 *** (0.011)	1031	0.571 *** (0.010)	693.3
Gumbel	θ	1.628 *** (0.013)	1871	1.411 *** (0.011)	1050	1.316 *** (0.010)	710.2
Hüsler–Reiss	θ	1.268 *** (0.018)	1711	1.039 *** (0.012)	981.9	0.924 *** (0.011)	660.9
Tawn	θ	0.928 *** (0.009)	1883	0.758 *** (0.013)	1033	0.633 *** (0.015)	686.4
t-EV	ρ	0.786 *** (0.012)	1885	0.672 *** (0.018)	1053	0.589 *** (0.016)	711.8
	λ	4.000 ** (1.876)		4.000 ** (1.882)		4.000 ** (1.843)	

* significance at the level of 10%, ** significance at the level of 5%, *** significance at the level of 1%. Source: [own study].

Table 2. Results of the estimation of the parameters of extreme values copulas with *t*-Student limit distribution.

Copula	Parameter	Corn–Soybean	LLF	Corn–Wheat	LLF	Soybean–Wheat	LLF
Galambos	θ	0.895 *** (0.013)	1824	0.673 *** (0.011)	1028	0.570 *** (0.010)	691.9
Gumbel	θ	1.626 *** (0.013)	1864	1.411 *** (0.011)	1047	1.315 ** (0.010)	708.9
Hüsler-Reiss	θ	1.267 *** (0.018)	1706	1.038 *** (0.012)	977.1	0.923 *** (0.011)	658.8
Tawn	θ	0.927 *** (0.009)	1875	0.757 *** (0.013)	1032	0.633 *** (0.015)	685.5
t-EV	ρ	0.785 *** (0.012)	1878	0.671 *** (0.018)	1051	0.588 *** (0.016)	710.9
	λ	4.000 ** (1.876)		4.000 ** (1.882)		4.000 ** (1.843)	

* significance at the level of 10%, ** significance at the level of 5%, *** significance at the level of 1%. Source: [own study].

The picture emerging from the analysis of the results presented in Tables 1 and 2 shows that very similar results have been obtained irrespective of the adopted limit distribution. The copula, which best describes the relationship between the extreme values of all researched agricultural product pairs, is the extreme *t*-Student copula. When comparing the obtained logarithm values of the likelihood function for individual copulas, it can also be observed that the Tawn copula may be useful for describing the relationship between extreme observations for the corn–soybean pair. A minor difference in the logarithm values of the credibility function between the t-EV copula and the Tawn copula, may suggest that the pair corn–soybean is characterized by a small degree of asymmetry, with more observations deviating in plus.

On the other hand, for pairs of agricultural products—corn–wheat and soybean–wheat—the second copula of extreme values, which may be helpful to describe the studied dependencies, with small differences in the value of the credibility function, turned out to be the Gumbel copula. Please note that in the case of an extreme values Gumbel copula, the parameter θ takes on values greater than or equal to 1; the higher the value of this parameter, the stronger the dependence of maximum losses between the analyzed observations. The results given in Tables 1 and 2 indicate a minor correlation between the extreme observations of the researched pairs of return rates of agricultural products quoted on the Chicago Mercantile Exchange. This means that unlike stock exchange indices for which, as it is well known and documented in extensive empirical studies, the occurrence of extremely negative return rates is much more likely than of extremely high rates. In the case of futures, one is not dealing with such strong dependence. The obtained results may rather point to a possible symmetry or very minor asymmetry in the tails of the return rates for the analyzed agricultural product pairs.

The next stage of the analysis was to test whether the structure of the relationship described with the estimated extreme values copulas is a sufficient approximation of reality, and whether it is suitable for modeling empirical data. In order to assess the quality of matching the estimated copulas to the empirical data on return rates of agricultural products, the methodology presented [37–44] was used. It says that one of the methods of checking the quality of matching the copula parameters is to compare the coefficients implied by the selected copula with empirical Kendall coefficients $\hat{\tau}$. Estimation of the Kendall coefficient τ was obtained for all extreme copulas using a simulation method. The results are presented in Tables 3 and 4. On the other hand, Table 5 contains the values of the Kendall correlation coefficient $\hat{\tau}$ calculated for the sample version. All estimated correlations are positive and statistically significant. Regardless of the pair of agricultural products, the strongest dependence is demonstrated by the pair corn–soy, while the weakest dependence is characterized by the pair soybean–wheat. Estimates of the Kendall coefficient for all extreme value copulas were obtained using the simulation method. The results are presented in Tables 3 and 4. Table 5, in turn, contains the values of the Kendall correlation coefficient calculated for the sample version.

Table 3. Values of Kendall's tau coefficient in the case of normal limit distribution.

Type of Copula	Corn–Soybean	Corn–Wheat	Soybean–Wheat
The Galambos' copula	0.382	0.289	0.236
The Gumbel copula	0.386	0.291	0.240
The Husler–Reiss copula	0.357	0.271	0.223
The Tawn copula	0.387	0.290	0.239
The t-EV copula	0.390	0.293	0.242

Source: [own study].

Table 4. Values of Kendall's tau coefficient in the case of t-Student distribution.

Type of Copula	Corn–Soybean	Corn–Wheat	Soybean–Wheat
The Galambos' copula	0.381	0.288	0.235
The Gumbel copula	0.385	0.291	0.239
The Husler-Reiss copula	0.356	0.270	0.222
The Tawn copula	0.386	0.289	0.239
The t-EV copula	0.388	0.293	0.241

Source: [own study].

Table 5. Sample Kendall correlation coefficients between the daily return rates of the surveyed agricultural products.

	Corn	Soybean	Wheat
Corn	1.000	0.417	0.332
Soybean		1.000	0.270
Wheat			1.000

Source: [own study].

The comparison of the empirical values of Kendall coefficients with the corresponding theoretical values demonstrates that for all three pairs of agricultural products, for both applied limit distributions, the best studied extreme dependences is described by the *t*-EV copula. In the case of the pair corn–soybean, the Tawn asymmetrical copula may also be helpful, while for the pairs corn–wheat and soybean–wheat, the Gumbel copula may also be useful. Please also note the Tawn copula, as the Kendall tau coefficient for this copula is quite satisfactory, and compared to the best-rated extreme *t*-Student copula, the differences are minimal.

An alternative method used to assess the quality of matching the parameters of the extreme value copula, is based on an empirical combining function. In order to choose the best copula function, the estimated parameters of the extreme value combining function were compared to the empirical combining function defined for the grid for $m = 303$. The selection of this particular grid size is a natural reference to the analyzed time series, which 303 sub-periods correspond to the subsequent months. The tables below demonstrate the distances of the estimated extreme value copulas from the empirical combining functions, depending on the selected limit distribution.

Analysis of the results listed in Tables 6 and 7 confirms the results obtained to assess the goodness of the adjustment of the extreme value copulas obtained with a methodology using the Kendall coefficient τ. The minor difference between the distances from the empirical joining function for the Tawn copula and the *t*-Student extreme copula proves that both functions may prove equally useful to describe the relationship between extreme values of all researched agricultural product pairs, regardless of the adopted limit distribution. In order to capture the dependence between extreme observations for the pair corn–soybean, the *t*-EV copula is the best, but the Gumbel copula may turn out equally important. On the other hand, for the pair corn–wheat and soybean–wheat, the Tawn copula seems the most accurate, while the *t*-EV copula ranks second. However, the differences are minor and they result from the selection of the size of the grid.

Table 6. Values of distance from the empirical combining function for selected extreme value copulas for normal limit distribution.

	Corn–Soybean	Corn–Wheat	Soybean–Wheat
The Galambos' copula	0.221	0.211	0.188
The Gumbel copula	0.209	0.197	0.174
The Husler–Reiss copula	0.285	0.252	0.219
The Tawn copula	0.218	0.180	0.163
The *t*-EV copula	0.203	0.195	0.174

Source: [own study].

Table 7. Values of distance from the empirical combining function for selected extreme value copulas for *t*-Student limit distribution.

	Corn–Soybean	Corn–Wheat	Soybean–Wheat
The Galambos' copula	0.222	0.211	0.188
The Gumbel copula	0.210	0.197	0.174
The Husler–Reiss copula	0.286	0.252	0.219
The Tawn copula	0.218	0.179	0.162
The *t*-EV copula	0.203	0.195	0.173

Source: [own study].

4. Conclusions

Sustainable agriculture is a key issue for environmentally friendly agriculture: Effective, economically viable and socially desirable. In addition, conservation of resources, environmental protection and agricultural stewardship, i.e., all requirements of sustainability, will increase and not reduce global food production. Other issues, such as the links between sustainable agriculture and the rest of the food and agri-food industry, as well as the consequences of sustainable development for rural communities and society as a whole, have not yet been finally resolved [53].

The empirical results demonstrated that the extreme values copula, which best describes the dependence between the extreme values of all researched pairs of agricultural products, irrespective of the limit distribution adopted, is the extreme *t*-Student, a.k.a. the *t*-EV copula. Moreover, in the case of the pair corn–soybean, the Tawn copula may be useful, and for corn–wheat and soybean–wheat, the Gumbel copula. The obtained results were confirmed by the assessment of the goodness of matching the parameters of the extreme value copulas. It can be observed that the results obtained in order to check the goodness of matching the estimated copula to empirical data of return rates of agricultural products reflect the results obtained in the estimation process. The final conclusion is that the return rate of the analyzed agricultural product pairs may be characterized by a minor degree of asymmetry, with the right tail being particularly heavy, which means higher probability of extreme observations than in the case of normal distribution.

The economic determinants of agricultural production have become the main reason for both producers and agri-food processors to search for exchange-based instruments to reduce the risk associated with adverse changes in product prices. A tool that is perfectly suited for this purpose is futures, which allow transferring the risk relatively cheaply to third parties. Thanks to their effectiveness, futures markets started to increase their turnover steadily. Participation of entities operating on the agricultural products market became common in countries with advanced stock exchange systems, at the same time becoming an irreplaceable link of production, processing, and trade in the agri-food industry. Analyzing agricultural markets is particularly important in terms of the need for ideation and creation of key development strategies for the agricultural sector, both at global and EU level, including the coordinated goals of the Common Agricultural Policy (CAP) [54–57].

In conclusion, our review highlights the vast possibilities of analyzing large data sets in agriculture towards optimizing management decisions based on statistical modeling. It demonstrates that the availability, techniques, and methods for analyzing large data sets, as well as the growing openness of large data sources, will encourage more academic research, public sector initiatives, and business ventures in the agricultural sector. This practice is still at an early stage of development and many barriers have to be overcome.

In practice, the results allow for rationalization of decisions of companies interested in intervention on the futures market for agricultural products, because investing there not only reduces the risk of financial losses, but also generates a high return rate.

The use of these models allows the simulation of the possible behavior of prices of agricultural products, including identification of extreme deviations. This allows programming decisions regarding possible public intervention in agricultural markets. In extraordinary situations—force majeure, weather anomalies, natural disasters—decision modeling can be used to assess the scale of damage for the agricultural market, and also to develop plans for the protection of agricultural producers within the framework of policies such as the CAP. As a consequence, they allow for securing food resources on deficit markets, e.g., with respect to geographical location.

Author Contributions: Conceptualization, Z.G.-S., D.K. and M.N.; methodology, G.M. and A.S.-S. and J.K.-D.; resources, G.M., M.K., A.S.-S. and J.S.; formal analysis, Z.G.-S., D.K. and M.K.; investigation, D.K., Z.G.-S., M.N. and J.S.; resources, M.K., G.M. and J.K.-D.; data curation, A.S.-S., D.K. and M.N.; writing—Z.G.-S., G.M., A.S.-S. and M.K.; visualization, D.K., J.S. and M.N.; funding acquisition, Z.G.-S.

Funding: The Project has been financed by the Ministry of Science and Higher Education within "Regional Initiative of Excellence" Programme for 2019–2022. Project no.: 021/RID/2018/19. Total financing: 11 897 131,40 PLN.

Conflicts of Interest: The authors declare no conflict of interest. The funders had no role in the design of the study; in the collection, analyses, or interpretation of data; in the writing of the manuscript, or in the decision to publish the results.

References

1. Erbaugha, J.; Bierbaum, R.; Castillejac, G.; da Fonsecad, G.A.B.; Cole, S.; Hansend, B. Toward sustainable agriculture in the tropics. *World Dev.* **2019**, *121*, 158–162. [CrossRef]
2. Szeląg-Sikora, A.; Niemiec, M.; Sikora, J.; Chowaniak, M. Possibilities of designating swards of grasses and small-seed legumes from selected organic farms in Poland for feed. In Proceedings of the IX International Scientific Symposium "Farm Machinery and Processes Management in Sustainable Agriculture", Lublin, Poland, 22–24 November 2017; pp. 365–370. [CrossRef]
3. Gródek-Szostak, Z.; Szeląg-Sikora, A.; Sikora, J.; Korenko, M. Prerequisites for the cooperation between enterprises and business supportinstitutions for technological development. *Bus. Non-profit Organ. Facing Increased Compet. Grow. Cust. Demands* **2017**, *16*, 427–439.
4. Gródek-Szostak, Z.; Luc, M.; Szeląg-Sikora, A.; Niemiec, M.; Kajrunajtys, D. Economic Missions and Brokerage Events as an Instrument for Support of International Technological Cooperation between Companies of the Agricultural and Food Sector. *Infrastruct. Environ.* **2019**, 303–308. [CrossRef]
5. Rigby, D.; Cáceresb, D. Organic farming and the sustainability of agricultural systems. *Agric. Syst.* **2001**, *68*, 21–40. [CrossRef]
6. Skafa, L.; Buonocorea, E.; Dumonteta, S.; Capone, R.; Franzesea, P.P. Food security and sustainable agriculture in Lebanon: An environmental accounting framework. *J. Clean. Prod.* **2019**, *209*, 1025–1032. [CrossRef]
7. Yu, J.; Wu, J. The Sustainability of Agricultural Development in China: The Agriculture–Environment Nexus. *Sustainability* **2018**, *10*, 1776. [CrossRef]
8. USDA. *Grain: World Markets and Trade*; USDA Office of Global Analysis: Washington, DC, USA, 2019.
9. USDA. *Oilseeds: Worlds Market and Trade*; USDA Office of Global Analysis: Washington, DC, USA, 2019.
10. USDA. *World Agricultural Production*; USDA Office of Global Analysis: Washington, DC, USA, 2019.
11. EC. Short-Term Outlook for EU Agricultural Markets in 2018 and 2019. In *Agriculture and Rural Development*; Spring: Brussels, Belgium, 2019.
12. Gumbel, E.J.; Goldstein, N. Analysis of empirical bivariate extremal distribution. *J. Am. Stat. Assoc.* **1964**, *59*, 794–816. [CrossRef]
13. Starica, C. Multivariate extremes for models with constant conditional correlations. *J. Empir. Financ.* **1999**, *6*, 515–553. [CrossRef]
14. Longin, F.; Solnik, B. Extreme correlations in international equity markets. *J. Financ.* **2001**, *56*, 649–676. [CrossRef]
15. Hsu, C.P.; Huang, C.W.; Chiou, W.J.P. Effectiveness of copula—Extreme value theory in estimating value-at-risk: Empirical evidence from Asian emerging markets. *Rev. Quant. Financ. Account.* **2012**, *39*, 447–468. [CrossRef]
16. Cebrian, A.; Denuit, M.; Lambert, P. Analysis of bivariate tail dependence using extreme values copulas: An application to the SOA medical large claims database. *Belg. Actuar. J.* **2003**, *3*, 33–41.
17. Renard, B.; Lang, M. Use of a Gaussian copula for multivariate extreme value analysis: Some case studies in hydrology. *Adv. Water Resour.* **2007**, *30*, 897–912. [CrossRef]
18. Nelsen, R.B. *An Introduction to Copulas*, 2nd ed.; Springer Verlag: New York, NY, USA, 2006; pp. 97–101.
19. Joe, H. *Multivariate Models and Dependence Concepts*; Chapman & Hall: London, UK, 1997; pp. 139–168.
20. Segers, J. Nonparametric inference for bivariate extreme-value copulas. In *Topics in Extreme Values*; Ahsanullah, M., Kirmani, S.N.U.A., Eds.; Nova Science Publishers: New York, NY, USA, 2007; pp. 181–203.
21. Gudendorf, G.; Segers, J. Extreme-value copulas. In *Copula Theory and its Applications: Proceedings of the Workshop Held in Warsaw 25–26 September 2009*; Jaworski, P., Durante, F., Härdle, W., Rychlik, T., Eds.; Springer Verlag: New York, NY, USA, 2010; pp. 127–145.
22. Genest, C.; Segers, J. Rank-based inference for bivariate extreme-value copulas. *Ann. Stat.* **2009**, *37*, 2597–3097. [CrossRef]

23. Ribatet, M.; Sedki, M. Extreme value copulas and max-stable processes. *J. Société Fr. Stat.* **2012**, *153*, 138–150.
24. FAO. *The Future of Food and Agriculture: Trends and Challenges*; Food and Agriculture Organization of the United Nations: Roma, Italy, 2017.
25. Wrzaszcz, W.; Zegar, J. Challenges for Sustainable Development of Agricultural Holdings. *Econ. Environ. Stud.* **2016**, *16*, 377–402.
26. Sklar, A. Fonctions de répartition á n dimensions et leurs marges. *Publ. l'Institut Stat. l'Université Paris* **1959**, *8*, 229–231.
27. Joe, H.; Xu, J.J. The estimation method of inference function for margins for multivariate models, Department of Statistics, University of British Columbia. *Tech. Rep.* **1996**, *166*. [CrossRef]
28. Ji, Q.; Bouri, E.; Roubaud, D.; Jawad, S.; Shahzad, H. Risk spillover between energy and agricultural commodity markets: A dependence-switching CoVaR-copula model. *Energy Econ.* **2018**, *75*, 14–27. [CrossRef]
29. Lia, R.L.; Wang, D.H.; Tu, J.Q.; Li, S.P. Correlation between agricultural markets in dynamic perspective—Evidence from China and the US futures markets. *Phys. A Stat. Mech. Appl.* **2016**, *464*, 83–92. [CrossRef]
30. Hill, J.; Schneeweis, T.; Yau, J. International trading/non-trading time effects on risk estimation in futures markets. *J. Futures Mark.* **1990**, *10*, 407–423. [CrossRef]
31. Yang, K.; Tian, F.; Chen, L.; Li, S. Realized volatility forecast of agricultural futures using the HAR models with bagging and combination approaches. *Int. Rev. Econ. Financ.* **2017**, *49*, 276–291. [CrossRef]
32. Cheng, B.; Nikitopoulos, C.S.; Schlögl, E. Pricing of long-dated commodity derivatives: Do stochastic interest rates matter? *J. Bank. Financ.* **2018**, *95*, 148–166. [CrossRef]
33. Greiner, R.; Puig, J.; Huchery, C.; Collier, N.; Garnett, S.T. Scenario modelling to support industry strategic planning and decision making. *Environ. Model. Softw.* **2014**, *55*, 120–131. [CrossRef]
34. Pickands, J. Multivariate extreme value distributions. *Bull. Int. Stat. Inst.* **1981**, *2*, 859–878.
35. Balkema, A.A.; Resnick, S.I. Max-infinite divisibility. *J. Appl. Probab.* **1977**, *14*, 309–319. [CrossRef]
36. De Haan, L.; Resnick, S.I. Limit theory for multivariate sample extremes. *Z. Wahrscheinlichkeitstheorie Verwandte Geb.* **1977**, *40*, 317–337. [CrossRef]
37. Davison, A.C.; Smith, R.L. Models for Exceedances over High Thresholds. *J. R. Stat. Soc. Ser. B (Methodological)* **1990**, *52*, 393–442. [CrossRef]
38. Katz, R.W.; Parlange, M.B.; Naveau, P. Statistics of extremes in hydrology. *Adv. Water Resour.* **2002**, *25*, 1287–1304. [CrossRef]
39. McNeil, A.J. *Extreme Value Theory for Risk Managers, Internal Modelling and CAD II*; RISK Books: London, UK, 1999.
40. Demarta, S.; McNeil, A.J. The t copula and related copulas. *Int. Stat. Rev.* **2005**, *73*, 111–129. [CrossRef]
41. Galambos, J. Order statistics of samples from multivariate distributions. *J. Am. Stat. Assoc.* **1975**, *9*, 674–680.
42. Gumbel, E.J. Bivariate exponential distributions. *J. Am. Stat. Assoc.* **1960**, *55*, 698–707. [CrossRef]
43. Husler, J.; Reiss, R.D. Maxima of normal random vectors: Between independence and complete dependence. *Stat. Probab. Lett.* **1989**, *7*, 283–286. [CrossRef]
44. Tawn, J.A. Bivariate extreme value theory: Models and estimation. *Biometrika* **1988**, *75*, 397–415. [CrossRef]
45. Heilpern, S. *Funkcje Łączące, Wydawnictwo Akademii Ekonomicznej im*; Oskara Langego we Wrocławiu: Wroclaw, Poland, 2007.
46. Hsieh, J.J. Estimation of Kendall's tau from censored data. *Comput. Stat. Data Anal.* **2010**, *54*, 1613–1621. [CrossRef]
47. Fisher, J.; Rucki, K. Re-conceptualizing the Science of Sustainability: A Dynamical Systems Approach to Understanding the Nexus of Conflict, Development and the Environment. *Sustain. Dev.* **2017**, *25*, 267–275. [CrossRef]
48. Durrleman, V.; Nikeghbali, A.; Roncalli, T. *Which Copula Is the Right One?* Working Paper; Groupe de Recherche Opérationnelle, Crédit Lyonnais: Lyon, France, 2000.
49. Kamnitui, N.; Genest, C.; Jaworski, P.; Trutschnig, W. On the size of the class of bivariate extreme-value copulas with a fixed value of Spearman's rho or Kendall's tau. *J. Math. Anal. Appl.* **2019**, *472*, 920–936. [CrossRef]
50. Li, F.; Zhou, J.; Liu, C.h. Statistical modelling of extreme storms using copulas: A comparison study. *Coast. Eng.* **2018**, *142*, 52–61. [CrossRef]

51. Gudendorf, G.; Segers, J. Nonparametric estimation of multivariate extreme-value copulas. *J. Stat. Plan. Inference* **2012**, *142*, 3073–3085. [CrossRef]
52. Sriboonchitta, S.; Nguyen, H.T.; Wiboonpongse, A.; Liu, J. Modeling volatility and dependency of agricultural price and production indices of Thailand: Static versus time-varying copulas. *Int. J. Approx. Reason.* **2013**, *54*, 793–808. [CrossRef]
53. Fousekis, P.; Tzaferi, D. Price returns and trading volume changes in agricultural futures markets: An empirical analysis with quantile regressions. *J. Econ. Asymmetries* **2019**, *19*, e00116. [CrossRef]
54. Sreekumar, S.; Sharma, K.C.; Bhakar, R. Gumbel copula based multi interval ramp product for power system flexibility enhancement. *Int. J. Electr. Power Energy Syst.* **2019**, *112*, 417–427. [CrossRef]
55. Keya, N.; Anowar, S.; Eluru, N. Joint model of freight mode choice and shipment size: A copula-based random regret minimization framework. *Transp. Res. Part E Logist. Transp. Rev.* **2019**, *125*, 97–115. [CrossRef]
56. Niemiec, M.; Komorowska, M.; Szeląg-Sikora, A.; Sikora, J.; Kuboń, M.; Gródek-Szostak, Z.; Kapusta-Duch, J. Risk Assessment for Social Practices in Small Vegetable farms in Poland as a Tool for the Optimization of Quality Management Systems. *Sustainability* **2019**, *11*, 3913. [CrossRef]
57. Kapusta-Duch, J.; Szeląg-Sikora, A.; Sikora, J.; Niemiec, M.; Gródek-Szostak, Z.; Kuboń, M.; Leszczyńska, T.; Borczak, B. Health-Promoting Properties of Fresh and Processed Purple Cauliflower. *Sustainability* **2019**, *11*, 4008. [CrossRef]

Article

The Effect of Biochar Used as Soil Amendment on Morphological Diversity of Collembola

Iwona Gruss [1,*], Jacek P. Twardowski [1], Agnieszka Latawiec [2,3], Jolanta Królczyk [4] and Agnieszka Medyńska-Juraszek [5]

1   Department of Plant Protection, Wroclaw University of Environmental and Life Sciences, 50-363 Wrocław, Poland
2   Institute of Agricultural Engineering and Informatics, University of Agriculture in Kraków, 30-149 Kraków, Poland
3   Department of Geography and the Environment, Rio Conservation and Sustainability Science Centre, Pontifícia Universidade Católica, Rio de Janeiro 22453-900, Brazil
4   Department of Manufacturing Engineering and Production Automation, Faculty of Mechanical Engineering, Opole University of Technology, 45-271 Opole, Poland
5   Institute of Soil Sciences and Environmental Protection, Wroclaw University of Environmental and Life Sciences, 53, 50-357 Wrocław, Poland
*   Correspondence: iwona.gruss@upwr.edu.pl; Tel.: +48-71-320-1754

Received: 24 July 2019; Accepted: 15 September 2019; Published: 19 September 2019

Abstract: Biochar was reported to improve the chemical and physical properties of soil. The use of biochar as a soil amendment have been found to improve the soil structure, increase the porosity, decrease bulk density, as well increase aggregation and water retention. Knowing that springtails (Collembola) are closely related to soil properties, the effect of biochar on morphological diversity of these organisms was evaluated. The main concept was the classification of springtails to the life-form groups and estimation of QBS-c index (biological quality index based on Collembola species). We conducted the field experiment where biochar was used as soil amendment in oilseed rape and maize crops. Wood-chip biochar from low-temperature (300 °C) flash pyrolysis was free from PAH (polycyclic aromatic hydrocarbon) and other toxic components. Results showed that all springtail life-form groups (epedaphic, hemiedaphic, and euedaphic) were positively affected after biochar application. The QBS-c index, which relates to springtails' adaptation to living in the soil, was higher in treatments where biochar was applied. We can recommend the use of Collembola's morphological diversity as a good tool for the bioindication of soil health.

Keywords: biochar; biological soil quality; Collembola life-form groups; QBS-c index

---

1. Introduction

One of the major threats to global agriculture is soil degradation, including decreased fertility and increased erosion [1]. The common problem is acidification and soil organic matter depletion, which decreases soil aggregate stability [2]. Therefore, the development of methods is needed to sustain soil resources by different remediation strategies. The application of organic materials like manure, compost, and biomass waste seems very promising, but a lot of attention has been paid to stable forms of organic carbon like biochars [3,4]. The main feature of biochar is the porous carbonaceous structure, which can contain amounts of extractable humic-like and fluvic-like substances [5]. Biochar was reported to improve the chemical and physical properties of soil [6]. The use of biochar as a soil amendment has been found to improve the soil structure, increase the porosity, decrease bulk density, as well increase aggregation and water retention [7–9]. On the other hand, the main concerns with respect to biochar use as a soil amendment is its potential contamination with heavy metals (HMs) and polycyclic aromatic hydrocarbons (PAHs) [10].

Springtails (Hexapoda: Collembola) are a key group of soil arthropods with densities often reaching thousands of individuals per square meter [11,12]. They contribute mainly in substrate decomposition and nutrient cycling [12,13]. Moreover, these organisms are sensitive to environmental changes in soil and are therefore often used as indicators of soil quality [14]. For bioindication, Collembola species diversity is used [15,16]. The disadvantage of this method is the difficulty in the determination to the species level. An alternative could be the QBS-c index (biological quality index based on Collembola), which responds to the morphological diversity of springtails [17]. Using this index, each individual is evaluated in terms of different morphological traits, e.g., for antennal length, size of furca, presence of ocelli pigmentation, and the presence of hairs and/or scales along the body. The principle of this index is that the presence of individuals with better adaptation to live in soil (with reduced appendages or less pigmented) indicates better soil quality [17,18]. Also, on the basis of morphological traits, springtails can be divided into three main life-forms [19]. First, epedaphic Collembola are adapted to live on the soil surface. The major features of this group are a pigmented body, well developed eyes and appendages, as well a fast dispersal ability. In contrast to them, soil dwelling species (euedaphic), with a relatively small, less pigmented body and reduced eyes. Their dispersal ability is limited. Species showing adaptations between epedaphic and euedaphic species are classified as belonging to the hemiedaphic group [20]. The vertical stratification of springtails reflects their function in the ecosystem. For instance, only epedaphic springtails contribute in the early stages of organic matter decomposition [19]. Ponge et al. [21] suggested that Collembola living in the soil characterized by limited active dispersal, may suffer more from land use intensification, than species living on the soil surface. In contrast, Ellers et al. [22] showed stronger effects of intensive land use on epedaphic than on euedaphic Collembola. The majority of studies on biochar effect on springtails were conducted in laboratory conditions on one model species [23–25]. Considering the impact of biochar under field conditions some experiments have been made also on nematodes [26] and earthworms [27].

The potential of biochar for pH and nutrient availability changes or improvement of some physical properties like porosity, water retention, or temperature and impact on soil microbial life, are well documented [28–30]. Therefore, springtails can be affected directly by the changes in soil chemical properties [31,32] or indirectly from biochar-induced changes in microorganisms' biomass [33]. It has been reported that many Collembola species feed on bacteria or fungi [34,35]. Biochar particles might be considered analogous to soil aggregates in that their large internal surface areas and pores could be important for biological processes [36].

Within the presented study we aimed to estimate the effect of biochar on the morphological diversity of Collembola species and evaluate its potential for field application.

It was hypothesized that:

1. Biochar will increase the soil biological quality mainly through improvement of the physical and chemical soil properties.
2. From the analyzed life-form groups of springtails, the response of euedaphic assemblages will be most distinct after biochar amendment. On the other hand, springtails living on the soil surface and in the litter layer (epigeic and hemiedaphic) will be more sensitive to cover plants.
3. The QBS-c index will show higher values in crops where biochar was applied.

2. Materials and Methods

2.1. Experimental Design

The field experiment was set up in mid-April 2014 in the south of Poland (50.5740 N, 17.8908 E) and continued until October 2016. The soil type was poor (sandy and weakly acidic) agricultural soil [37]. The climate of the area is moderately warm with an average annual temperature of 8.4 °C and an average annual rainfall of 611 mm. The biochar effect on soil dwelling springtails was explored in two crops (oilseed rape and maize) compared to control (two crops with no biochar application). Within each treatment (plot), three replicates (subplots) were established. The area of each subplot was 3×3 m. The research area was previously (before 2014) used as conventionally agricultural field. The forecrop was maize. Biochar was applied up to a depth of 30 cm at a rate of 50 t/ha and was mixed by ploughing. No chemical protection was applied before or during experiment period. Weeds were removed manually upon occurrence. Weeds were harvested manually a few times during the vegetation season. The maize variety in two years of the study was P8745 (FAO 250, Pioneer Company) and oilseed rape variety Monolit. The only fertilizer used in oilseed rape was ammonium sulfate 34% in a dose of 300 kg/ha, and in maize ammonium phosphate (Polydap) in a dose of 25 kg/ha. The same amount of fertilizers was applied in biochar and control treatments.

2.2. Biochar Characteristic and Soil Properties

The biochar used in the experiment was industrial produced by Fluid S.A. Company (Poland). It was produced in the low-temperature flash pyrolysis (300 °C) of pine and spruce wood chips. Its heating value was 25 MJ/kg. During the experiment selected properties of biochar (pH, organic carbon content, cation exchange capacity, heavy metal content and total PAH's) were analyzed according to International Biochar Initiative (IBI) Standard Product Definition and Product Testing Guidelines [38]. The particle size fraction of biochar applied on the field was more than 2 mm (sieve method).

The tested biochar was alkaline (pH 8.2) and had 52.3% of carbon (Table 1). The surface area of the tested biochar was low (only 16.5 m^2/g) and cation exchange capacity was also lower—39.5 cmol/kg, compared with biochars produced at higher temperatures and from other feedstock, like wheat straw, giant miscanthus, rice husk, or sewage sludge [39,40]. It was free from polycyclic aromatic hydrocarbon and the concentration of all tested toxic compounds was very low or even under the level of detection, passing fixed recommendations for acceptable levels [38].

Table 1. The chemical characteristic of biochar properties used in the experiment (sourced from Gruss et al. [41]).

Parameter	Value
pH H_2O	8.2
CEC (cmol/kg)	16.8
Carbon content (% of DM)	52.3
H/Corg ratio	0.026
Pb (g/t DM)	1.57
Mn (g/t DM)	29.7
Cu (g/t DM)	0.50
Hg (g/t DM)	0.32
Zn (g/t DM)	13.04

For physicochemical analysis, soil samples were collected twice a year from topsoil, before each crop in rotation in five replicates from each plot. The pH, total organic carbon, CEC, exchangeable acidity, and water properties were measured. Soil was classified as Cambisol [38], with a typical sandy loam texture with the addition of medium fine gravel. Application of biochar significantly increased CEC values in both trials, due to the increase of exchangeable Ca2+, Mg2+, and H+ + Al3+ (exchangeable acidity) in the soil sorption complex and total organic carbon in biochar trials (Table 2).

Table 2. Soil properties after biochar application in oilseed rape and maize (sourced from Gruss et al. [41]).

Parameter	Oilseed Rape		Maize	
	Biochar	Control	Biochar	Control
C_{org} (%)	0.94	0.92	0.94	0.78
pH H_2O	6.88	7.26	6.49	7.22
Na^+ (cmol/kg)	0.20 *	0.12	0.12	0.18
Mg^{2+} (cmol/kg)	3.14	1.13	2.76	0.86
K^+ (cmol/kg)	0.30	0.19	0.34	0.31
Ca^{2+} (cmol/kg)	5.12	2.29	5.72	2.28
CEC (cmol/kg)	8.76	3.74	8.98	4.73

* The values in bold font differ significantly between treatments.

2.3. Collembola Studies

Soil samples for Collembola analysis were taken three times during each of the vegetation season (from May to July) in 2015 and 2016. The growth stages according to the BBCH (growth stages of plants) scale [42] in the sampling dates were: maize: 10–15, 32–37, and 61–67; oilseed rape: 60–69, 72–79, and 83–89. On each date, 12 samples were taken from each subplot (36 samples from one plot), and transported to the laboratory. The samples were taken with the use of a soil sampler (diameter 5 cm and depth 10 cm). The volume of one sample was 196 cm^2. Collembola were extracted over 24 h from the soil samples with the use of Tullgren funnels modified by Murphy [43]. After the extraction the springtails were kept in 75% ethyl alcohol.

Springtails from each sample were counted and identified to the species or genus. Each individual was placed on permanent microscope slide and determined to the species level with the use of following keys [44–46]. Springtails were classified to three life-form groups (euedaphic, hemiedaphic and epigeic) according to Karaban [20]. Epedaphic forms have strong pigmentation, fully developed furca and other appendages, and pigmented eyes (8 + 8). Hemiedaphic have reduced body pigmentation, eye numbers, and a reduced furca. Euedaphic forms are characterized by an unpigmented body (or eyes' pigmentation) with eyes and furca not developed. The QBS-c (biological quality index based on Collembola species) is calculated as the sum of EMI values in each sample (Table 3). The springtails species were evaluated for seven morphometric traits according to the scale. The results were the sums of scores (EMI) obtained for each trait. Species which are well adopted to live in soil obtain more EMI scores in comparison to those with adaptation to live on the soil layer [17].

Table 3. Description of the Collembola species identified in the experiment including their morphological description.

Species	Abbr. on the CCA Biplot	Life-form Group	Size *	Pigmentation	Structures on Cuticle	Ocelli	Antennae	Legs	Furcula	EMI Scores (QBS-c)
Bourietiella hortensis (Fitch)	Bou_hor	Epedaphic	4	0	1	0	0	0	0	5
Brachystomella parvula (Schaffer)	Bra_par	Hemiedaphic	4	0	1	0	3	0	3	14
Caprainea marginata (Schoett)	Cap_mar	Epedaphic	4	0	0	0	0	0	0	4
Desoria multisetis (Carpenter & Phillips)	Des_mul	Epedaphic	4	3	0	0	0	0	0	7
Desoria tigrina (Nicolet)	Des_tig	Hemiedaphic	4	6	0	0	0	0	0	10
Folsomia sexuolata (Tullberg)	Fol_sex	Hemiedaphic	4	6	3	6	2	2	2	25
Folsomides angularis (Axelson)	Fol_ang	Hemiedaphic	4	6	3	6	3	2	3	27
Folsomides parvulus (Stach)	Fol_par	Hemiedaphic	4	6	3	3	3	3	3	25
Friesea mirabilis (Tullberg)	Fri_mir	Hemiedaphic	4	3	1	0	2	2	2	14
Hypogastrura spp.	Hopogast	Hemiedaphic	4	0	3	0	2	2	2	13
Isotoma anglicana (Lubbock)	Iso_ang	Epedaphic	4	3	1	0	0	0	0	8
Isotoma antennalis (Bagnall)	Iso_ant	Epedaphic	4	1	0	0	0	0	0	5
Isotoma viridis (Bourlet)	Iso_vir	Epedaphic	4	3	0	0	0	0	0	7
Isotomiella minor (Schaeffer)	Iso_min	Hemiedaphic	4	6	3	3	3	3	3	25
Isotomodes productus (Axelson)	Iso_pro	Hemiedaphic	4	6	3	6	2	3	2	26
Isotomurus palustris (Mueller)	Iso_pal	Epedaphic	4	1	0	0	0	0	0	5
Isotomutus gallicus (Carapelli et al.)	Iso_gal	Epedaphic	4	1	0	0	0	0	0	5
Lepidocyrtus violaceus (Fourcroy)	Lep_vio	Epedaphic	4	0	0	0	0	0	0	4
Mesaphorura spp.	Mesaphor	Euedaphic	4	6	3	6	3	3	6	31
Parisotoma notabilis (Schaeffer)	Par_not	Hemiedaphic	4	6	1	3	0	2	0	16
Proisotoma minima (Absolon)	Pro_mini	Hemiedaphic	4	3	1	3	2	2	2	17
Proisotoma minuta (Tullberg)	Pro_minu	Hemiedaphic	4	3	1	0	2	2	2	14
Proisotoma tenella (Reuter)	Pro_ten	Hemiedaphic	4	3	1	0	2	2	2	14
Protaphorura spp.	Protapho	Euedaphic	4	6	3	6	3	3	6	31
Pseudosinnela sexoculata (Schott)	Pse_sex	Hemiedaphic	4	6	1	3	0	0	0	14
Sminthurides parvulus (Krausbauer)	Smi_par	Epedaphic	4	0	0	0	0	0	0	4
Sminthurinus alpinus (Gisin)	Smi_alp	Epedaphic	4	0	1	0	2	0	0	7
Sphaeridia pumilis (Krausbauer)	Sph_pum	Epedaphic	4	0	0	0	0	0	0	4
Stenacidia violacea (Reuter)	Ste_vio	Epedaphic	4	0	0	0	0	0	0	4
Stenaphorura spp.	Stenapho	Euedaphic	4	6	3	6	3	3	6	31

* Size: >3 mm = 0; 2–3 mm = 2; <2 mm = 4.

Pigmentation: Fully pigmented=0; only strips on the body = 1; r = Reduced to appendages = 3; none = 6.

Structures on cuticle: Well developed chaeta or scales, present trichobothria = 0; relatively low number of structures on cuticle=1; Reduced number of chaetae, presence of PSO (pseudocelli) on cuticle = 3; Low number of chaeteae, other structures present only in selected parts of the body = 6.

Number of ocelli in the eye spot: 8 + 8 = 0; 6 + 6 = 2; form 5 + 5 to 1 + 1 = 3; absence of ocelli = 6.

Antennae: antennae longer than the head = 0; antennae more or less the same length as the head = 2; antennae shorter than the head = 3; antennae much shorter than the head = 6.

Legs: Well developed = 0, Medium developed = 2; Short = 3; Reduced or with reduced claw and mucro =6.

Furcula: Well developed = 0; Medium developed = 2; Short with reduced number of chaetae = 3; the absence of mucro and modification of manubrium = 5; Furcula reduced in residual form = 6.

2.4. Data Analysis

The effect on springtails life-form groups and QBS-c index was analyzed with the mixed model in SAS University Edition (proc Mixed). In the analysis, the effect of crop and treatment, as well their interaction, were included. The year and term of the study were the random factors. The abundance of springtails per sample was relatively low (with the number of individuals of 10.5 per sample). Therefore, the abundance of springtails was calculated for 1 m^2, knowing that the area of one sample was 0.000785 m^2 (5 cm diameter).

The springtails abundance as well morphometric trails in relation to experimental treatments was analyzed using canonical correspondence analysis (CCA) in Canoco, Version 4.5 (Ithaca, New York, USA). Significance of the first canonical axis and all axes together was calculated with Monte Carlo Test.

3. Results

The abundance of Collembola per m^2 and the QBS-c index differed significantly between all tested factors (Table 4). Generally, the most abundant group was epedaphic, then hemiedaphic, and the least, euedaphic Collembola (Figure 1). Within the epedaphic and hemiedaphic groups, both the plant ($p < 0.0001$) and biochar ($p = 0.0009, 0.0058$) significantly differed with respect to springtails abundance. Springtails were significantly more abundant in oilseed rape in comparison to maize crop. At the same time more Collembola were found in biochar, but only in oilseed rape. The abundance of euedaphic Collembola differed between biochar and control plots ($p = 0.003$). In both crops, significantly more individuals were found in biochar treated soil.

Table 4. Results of repeated ANOVA (GML, $p \leq 0.05$) considering effects on treatment, plant and year, and its interactive effects on Collembola life-form groups and QBS-c index.

Dependent Variable	Treatment		Plant		Treatment × Plant	
	F *	p	F *	p	F	p
Epedaphic	11.10	0.0009	20.53	<0.0001	12.43	0.0004
Hemiedaphic	7.65	0.0058	22.04	<0.0001	11.53	0.0007
Euedaphic	13.09	0.0003	0.04	0.8501	2.15	0.1425
QBS-c	6.16	0.01132	4.77	0.0292	0.02	0.8943

* F = ratio of two different measure of variance for the data.

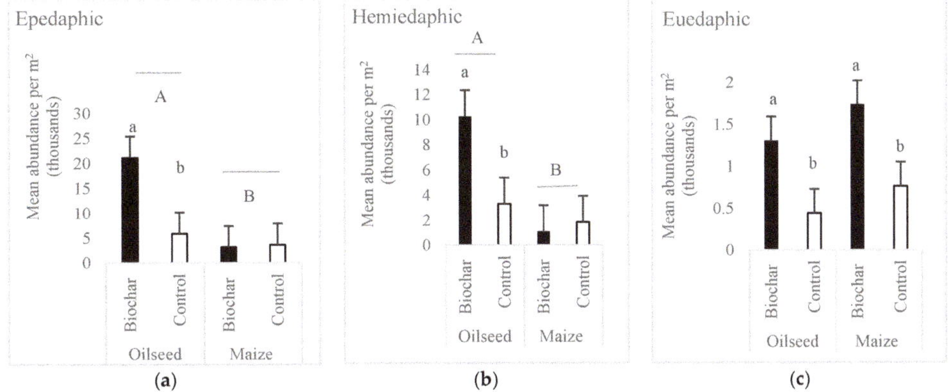

Figure 1. Effect of crop and biochar application on different Collembola life-form groups. Note: (a), (b), (c); indicate significant differences (p ≤ 0.05) between biochar and control; A, B indicate significant differences (p ≤ 0.05) between plants; the results of repeated ANOVA used for data analysis are given in the Table 4.

Considering the QBS-c index, it was significantly higher in oilseed rape crop in comparison to maize (Figure 2) (p = 0.0292). In both plants, the index was significantly higher in treatments, where biochar was applied (p = 0.01132). As shown by the life-form groups, significantly higher QBS-c index was found in oilseed rape only.

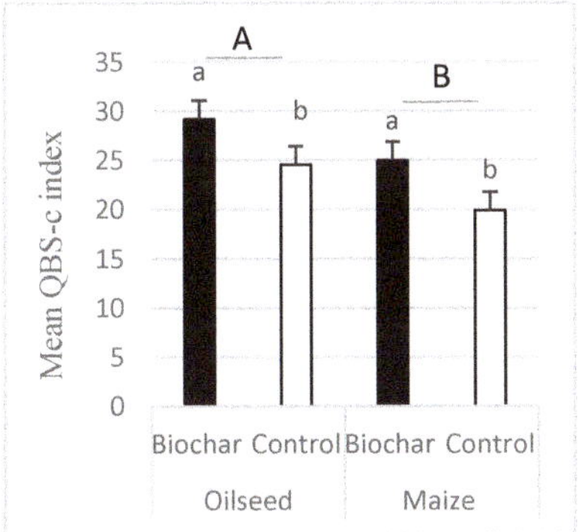

Figure 2. Effect of crop and biochar application on the QBS-c index. Note: (a), (b), (c); indicate significant differences (p ≤ 0.05) between biochar and control; A, B indicate significant differences (p ≤ 0.05) between plants; the results of repeated ANOVA used for data analysis are given in the Table 4.

The morphometric traits used for the calculation of QBS-C index were correlated with experimental treatments (Figure 3). The significance of the first canonical axis (CCA1), as well all axes together was p = 0.002 (Table 5). Biochar (in oilseed rape) was positively correlated with reduced legs, antennae and furcula, as well absence of specific structures on cuticle. In maize, where biochar was applied, springtails were characterized by a reduced number of ocelli. Size and pigmentation were positively correlated with oilseed rape and maize, both without biochar.

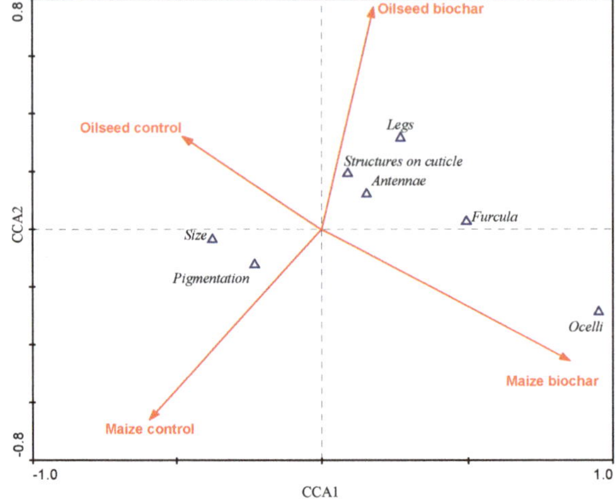

Figure 3. Canonical correspondence analysis (CCA) biplot on Collembola morphomentric traits in relation to experimental treatments. Notes: The morphometric traits are described in more detail in Table 1.

Table 5. Results of Canonical correspondence analysis (CCA) of Collembola morphometric traits correlated with experimental treatments.

CCA Axes	1	2	3	4
Eigenvalues	0.013	0.002	0.000	0.103
Morphometric traits-environment correlations	0.013	0.002	0.000	0.103
Significance of the first canonical axis	F = 26.362, p = 0.002			
Significance of all canonical axes	F = 10.296, p = 0.002			

The eigenvalues of the first two CCA axes were 0.0102 and 0.041, respectively (Table 6). Both the first canonical axis (CCA1), as well all axes were significant (p = 0.002, Monte Carlo test). As shown on the CCA biplot (Figure 4), the Collembola community was affected more by crop than by biochar. There was only minor effect of biochar in oilseed. In maize the group of species related to the control site (e.g., *Desoria tigrina* and *Pseusinella sexoculata*) differed from the species which preferred biochar (e.g., *Stenaphorura* spp., *Isotoma antenalis*). Considering life-form groups, most of the hemiedaphic species were found in the oilseed crops to oil seed (with similar effect for biochar and control). The epedaphic species were frequently found in all of the treatments, while euedaphic mostly in maize with biochar.

Table 6. Results of Canonical correspondence analysis (CCA) of Collembola species in relation to experimental treatment.

CCA Axes	1	2	3	4
Eigenvalues	0.102	0.041	0.013	0.476
Species-environment correlations	0.563	0.388	0.270	0.000
Significance of the first canonical axis	F = 7.754, p = 0.002			
Significance of all canonical axes	F = 4.016, p = 0.002			

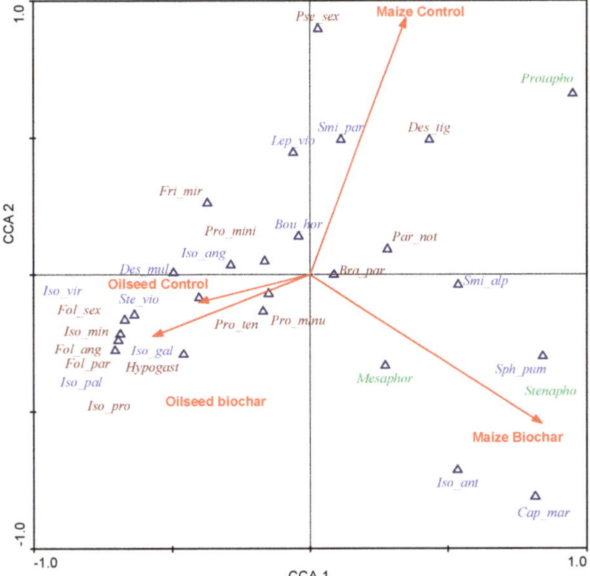

Figure 4. Canonical correspondence analysis (CCA) biplot on Collembola species in relations to the crop and biochar application. * Designation of the life-form groups: epedaphic, hemiedaphic, euedaphic. Note: the detailed description of the species is included in Table 1.

4. Discussion

The low temperature pine/wood chip biochar used in the experiment is characterized by the high carbon content but low surface area, and ability for nutrient storage, therefore the effect of its application to soil had little effect on soil properties. Tested pined wood biochar, was free from PAH and the concentration of all tested toxic compounds was very low or even under the level of detection. The release of toxic compounds like PAH's and heavy metals after biochar application seems to be crucial for survival and reproduction of soil fauna [47,48]. Some of the soil properties were improved, like total organic carbon content (only in maize trials) or CEC, as reported by other authors [49–53]. The liming effect, which is mostly expected [54,55], when biochar is applied to soil was not determined in our experimental trials.

Our estimations showed a higher QBS-c index value and higher individual number of particular life-form groups in biochar amended soils, confirming the hypothesis of soil biological properties improvement. We state that the positive response of Collembola to biochar addition was the result of improved soil properties. Some authors have found that soil mesofauna abundances are closely related to soil conditions, especially soil pH and organic matter content [56,57]. To compare, the significant increase of fungivorous nematodes was found after biochar application in the rates from 12 to 48 t/ha [26]. In contrast, no response of soil faunal feeding activity to biochar addition was found in the rates range from 0 to 30 t/ha [58]. Also, Castracani et al. [59] did not find any interaction between biochar and the abundance of epigeic macroarthropods in the rate of 14 t/ha.

The response of euedaphic springtails was predicted to be most distinct after biochar amendment. This would result from low dispersal ability and living in deeper soil layers [60]. Therefore, euedaphic springtails would be more sensitive to changes in the soil environment [61]. In our experiment the abundance of euedaphic springtails was relatively low compared to hemiedaphic and epedaphic groups. However, in both analyzed crops, its number increased after biochar application. Considering the two other groups living on upper soil layers, the effect was significant only for oilseed.

The main concept of the QBS-c index is that soil quality is positively correlated with the number of Collembola species that are well adapted to soil habitats [17]. Otherwise, numerous occurrences of euedaphic forms of springtails in a given habitat can indicate the better biological quality of the soil [43]. Based on the results, we can agree with the hypothesis that the QBS-c index will have higher values in crops where biochar was applied. For instance, in the study of Twardowski et al. [61], the QBS-c showed higher soil quality in the potato crop rotation in comparison to monoculture. Jacomini et al. [62] found decreased QBS-c index values in degraded soils. To confirm our last hypothesis, two springtails life-form groups (epigeic and hemiedaphic) differed significantly between crops. More springtails were found in oilseed rape in comparison to maize crop. Considering that maize and oilseed rape are plants which differ in their development, this should also affect organisms living on the soil surface. Similarly, Op Akkerhuis [63] found differences in mesofauna abundance between crops, i.e., specifically the mesofauna groups were much more abundant in cereals than in root and tuber crops. We state also that cover plant might modify the effect of biochar on soil fauna.

5. Conclusions

The main findings of the present study are that:

(1) Crop affected more Collembola community than the biochar application. More springtails occurred in oilseed rape.

(2) In each of the life-form groups, biochar caused a significant increase in individual number of Collembola in comparison to the no-biochar treatment. The effect was significant mainly for the oilseed rape crop.

(3) The QBS-C index (biological quality index based on Collembola species) was higher in treatments where biochar was applied.

(4) Collembola related to biochar were characterized by reduced appendages and the absence of specific structures on the cuticle, what indicates better adaptation to live in soil.

To conclude, biochar was found to increase springtails abundance and diversity in field conditions. A greater occurrence of species better adopted to life in the soil with biochar use indicated better soil quality. Thus, we can recommend the use of the morphological diversity of Collembola as a good tool for the bioindication of soil health.

Author Contributions: Conceptualization, I.G. and J.P.T.; Methodology, A.L., J.K. and A.M.-J.; Formal analysis, I.G. and J.P.T.; Writing, Original Draft Preparation, Review and Edition, I.G., and J.P.T.

Funding: This research received no external funding.

Acknowledgments: We kindly thank the Fluid S.A. Company for free-of-charge delivery of biochar for the experiment. We are greatly indebted to Kamila Twardowska and Joanna Magiera-Dulewicz for support in laboratory experiments.

Conflicts of Interest: The authors declare no conflicts of interest.

References

1. Gregory, A.S.; Ritz, K.; McGrath, S.P.; Quinton, J.N.; Goulding, K.W.; Jones, R.J.; Harris, J.A.; Bol, R.; Wallace, P.; Pilgrim, E.S.; et al. A review of the impacts of degradation threats on soil properties in the UK. *Soil Use Manag.* **2015**, *31*, 1–15. [CrossRef] [PubMed]

2. De Meyer, A.; Poesen, J.; Isabirye, M.; Deckers, J.; Rates, D. Soil erosion rate in tropical villages: A case study from Lake Victoria Basin, Uganda. *Catena* **2011**, *84*, 89–98. [CrossRef]

3. Lehmann, J.; Joseph, S. *Biochar for Environmental Management Science and Technology*, 1st ed.; Earthscan: London, UK, 2009; pp. 1–944.

4. Latawiec, A.E.; Królczyk, J.B.; Kubon, M.; Szwedziak, K.; Drosik, A.; Polańczyk, E.; Grotkiewicz, K.; Strassburg, B. Willingness to Adopt Biochar in Agriculture: The Producer's Perspective. *Sustainability* **2017**, *9*, 655. [CrossRef]

5. Lin, Y.; Munroe, P.; Joseph, S.; Henderson, R.; Ziolkowski, A. Water extractable organic carbon in untreated and chemical treated biochars. *Chemosphere* **2012**, *87*, 151–157. [CrossRef]

6. Herath, H.M.; Camps-Arbestain, M.; Hedley, M. Effect of biochar on soil physical properties in two contrasting soils: An Alfisol and an Andisol. *Geoderma* **2013**, *209–210*, 188–197. [CrossRef]

7. Baiamonte, G.; De Pasquale, C.; Marsala, V.; Cimò, G.; Alonzo, G.; Crescimanno, G.; Conte, P. Structure alteration of a sandy-clay soil by biochar amendments. *J. Soils Sediments* **2015**, *15*, 816–824. [CrossRef]

8. Ding, Y.; Liu, Y.; Liu, S.; Li, Z.; Tan, X.; Huang, X.; Zeng, G.; Zhou, L.; Zheng, B. Biochar to improve soil fertility. A review. *Agron. Sustain. Dev.* **2016**, *36*, 36. [CrossRef]

9. Latawiec, A.E.; Peake, L.; Baxter, H.; Cornelissen, G.; Grotkiewicz, K.; Hale, S.; Królczyk, J.B.; Kubon, M.; Łopatka, A.; Medynska-Juraszek, A.; et al. A reconnaissance-scale GIS-based multicriteria decision analysis to support sustainable biochar use: Poland as a case study. *J. Environ. Eng. Landsc.* **2017**, *25*, 208–222. [CrossRef]

10. Freddo, A.; Cai, C.; Reid, B.J. Environmental contextualization of potential toxic elements and polycyclic aromatic hydrocarbons in biochar. *Environ. Pollut.* **2012**, *171*, 18–24. [CrossRef]

11. Hopkin, S.P. *Biology of the Springtails (Insecta: Collembola)*; Oxford University Press: Oxford, UK, 1997; pp. 1–330.

12. Rusek, J. Biodiversity of Collembola and their functional role in the ecosystem. *Biodivers. Conserv.* **1998**, *7*, 1207–1219. [CrossRef]

13. Filser, J. The role of Collembola in carbon and nitrogen cycling in soil. *Pedobiologia* **2012**, *46*, 234–245. [CrossRef]

14. Frampton, G.K. The potential of Collembola as indicators of pesticide usage: Evidence and methods from the UK arable ecosystem. *Pedobiologia* **1997**, *41*, 179–184.

15. Fiera, C. Application of stable isotopes and lipid analysis to understand trophic interactions in springtails. *North West. J. Zool.* **2014**, *10*, 227–235.

16. Sousa, J.P.; Bolger, T.; da Gama, M.M.; Lukkari, T.; Ponge, J.-F.; Simón, C.; Traser, G.; Vanbergen, A.J.; Brennan, A.; Dubs, F.; et al. Changes in Collembola richness and diversity along a gradient of land-use intensity: A pan European study. *Pedobiologia* **2006**, *50*, 147–156. [CrossRef]

17. Parisi, V. The biological soil quality, a method based on microarthropods. *Acta Naturalia de L'Ateneo Parmense* **2001**, *37*, 97–106.

18. Machado, J.S.; Oliveira, F.L.; Sants, J.C.; Paulino, A.T.; Baretta, D. Morphological diversity of springtails (Hexapoda: Collembola) as soil quality bioindicators in land use systems. *Biota Neotrop.* **2019**, *19*, e20180618. [CrossRef]

19. Potapov, A.; Semeninaa, E.; Korotkevich, A.; Kuznetsova, N.; Tiunova, A. Connecting taxonomy and ecology: Trophic niches of collembolans as related to taxonomic identity and life forms. *Soil Biol. Biochem.* **2016**, *110*, 20–31. [CrossRef]

20. Karaban, K.; Karaban, E.; Uvarov, A. Determination of life form spectra in soil Collembola communities: A comparison of two methods. *Pol. J. Ecol.* **2012**, *59*, 381–389.

21. Ponge, J.F.; Dubs, F.; Gillet, S.; Sousa, J.P.; Lavelle, P. Decreased biodiversity in soil springtail communities: The importance of dispersal and land use history in heterogeneous landscapes. *Soil Biol. Biochem.* **2006**, *38*, 1158–1161. [CrossRef]

22. Ellers, J.; Berg, M.P.; Dias, A.T.; Fontana, S.; Ooms, A.; Moretti, M. Diversity in form and function: Vertical distribution of soil fauna mediates multidimensional trait variation. *J. Anim. Ecol.* **2018**, *87*, 933–944. [CrossRef]

23. Greenslade, P.; Vaughan, G.T. A comparison of Collembola species for toxicity testing of Australian soils. *Pedobiologia* **2003**, *47*, 171–179. [CrossRef]

24. Marks, E.A.N.; Mattana, S.; Alcañiz, J.M.; Domene, X. Biochars provoke diverse soil mesofauna reproductive responses in laboratory bioassays. *Eur. J. Soil Biol.* **2014**, *60*, 104–111. [CrossRef]

25. Domene, X.; Hanley, K.; Enders, A.; Lehmann, J. Short-term mesofauna responses to soil additions of corn stover biochar and the role of microbial biomass. *Appl. Soil Ecol.* **2015**, *89*, 10–17. [CrossRef]

26. Zhang, X.; Li, Q.; Liang, W.; Zhang, M.; Bao, X.; Xie, Z. Soil nematode response to biochar addition in a Chinese wheat field. *Pedosphere* **2013**, *23*, 98–103. [CrossRef]

27. Tammeorg, P.; Parviainen, T.; Nuutinen, V.; Simojoki, A.; Vaara, E.; Helenius, J. Effects of biochar on earthworms in arable soil: Avoidance test and field trial in boreal loamy sand. *Agric. Ecosyst. Environ.* **2014**, *191*, 150–157. [CrossRef]

28. Kolton, M.; Harel, Y.M.; Pasternak, Z.; Graber, E.R.; Elad, Y.; Cytryn, E. Impact of biochar application to soil on the root-associated bacterial community structure of fully developed greenhouse pepper plants. *Appl. Environ. Microbiol.* **2011**, *77*, 4924–4930. [CrossRef] [PubMed]

29. Lehmann, J.; Rillig, M.C.; Thies, J.; Masiello, C.A.; Hockaday, W.C.; Crowley, D. Biochar effects on soil biota—A review. *Soil Biol. Biochem.* **2011**, *43*, 1812–1836. [CrossRef]

30. Cole, E.; Zandvakili, O.R.; Blanchard, J.; Xing, B.; Hashemi, M.; Etemadi, F. Investigating responses of soil bacterial community composition to hardwood biochar amendment using high-throughput PCR sequencing. *Appl. Soil Ecol.* **2019**, *136*, 80–85. [CrossRef]

31. Van Straalen, N.M.; Verhoef, H.A. The development of a bioindicator system for soil acidity based on arthropod pH preferences. *J. Appl. Ecol.* **1997**, *34*, 217–232. [CrossRef]

32. Bardgett, R. *The Biology of Soil: A Community and Ecosystem Approach*; Oxford University Press: Oxford, UK, 2005; pp. 1–254.

33. Lehmann, J.; Gaunt, J.; Rondon, M. Bio-char Sequestration in Terrestrial Ecosystems—A Review. *MITIG ADAPT STRAT GL* **2006**, *11*, 403–427. [CrossRef]

34. Berg, M.P.; Stoffer, M.; van den Heuvel, H.H. Feeding guilds in Collembola based on digestive enzymes. *Pedobiologia* **2004**, *48*, 589–601. [CrossRef]

35. Chahartaghi, M.; Scheu, S.; Ruess, L. Sex Ratio and Mode of Reproduction in Collembola in an Oak-Beech Forest. *Pedobiologia* **2006**, *50*, 331–340. [CrossRef]

36. Tisdall, J.M.; Oades, J.M. Organic Matter and Water-Stable Aggregates in Soils. *Eur. J. Soil Sci.* **1982**, *33*, 141–163. [CrossRef]

37. FAO-WRB. *World Reference Base for Soil Resources. International Soil Classification System for Naming Soils and Creating Legends for Soil Maps*; World Soil Resources Reports 106; Food and Agriculture Organization of the United Nations: Rome, Italy, 2014.

38. *Standardized Product Definition and Product Testing Guidelines for Biochar that is Used in Soil*; IBI-STD-2.1; IBI (International Biochar Initiative): Canandaigua, NY, USA, 2014.

39. Liang, B.; Lehmann, J.; Solomon, D.; Kinyangi, J.; Grossman, J.; O'Neill, B.; Skjemstad, J.O.; Thies, J.; Luizao, F.J.; Petersen, J.; et al. Black carbon increases cation exchange capacity in soils. *Soil Sci. Soc. Am. J.* **2006**, *70*, 1719–1730. [CrossRef]

40. Liang, B.; Lehmann, J.; Sohi, S.P.; Thies, J.; O'Neill, B.; Truillo, L.; Gaunt, J.; Solomon, D.; Grossman, J.; Neves, E.G.; et al. Black carbon affects the cycling of non-black carbon in soil. *Org. Geochem.* **2010**, *41*, 206–213. [CrossRef]

41. Gruss, I.; Twardowski, J.P.; Latawiec, A.; Medyńska-Juraszek, A.; Królczyk, J. Risk assessment of low temperature biochar used as soil amendment on soil mesofauna. *Environ. Sci. Pollut. Res.* **2019**, *26*, 18230–18239. [CrossRef] [PubMed]

42. Meier, U. *Growth Stages of Mono-and Dicotyledonous Plants*; Federal Biological Research Centre for Agriculture and Forestry: Berlin, Germany, 2001; pp. 1–204.

43. Murphy, P.W. Extraction methods for soil animals. I. Dynamic methods with particular reference to funnel processes. In *Progress in Soil Zoology*; Butterworths: London, UK, 1962; pp. 75–114.

44. Zimdars, B.; Dunger, W. *Synopses on Palaearctic Collembola Part I. Tullbergiinae Bagnall, 1935*; Abhandlungen und Berichte des Naturkundemuseums: Görlitz, Germany, 1994; pp. 1–71.

45. Fjellberg, A. *The Collembola of Fennoscandia and Denmark. Part II: Entomobryomorpha and Symphypleona*; Fauna Entomologica Scandinavica; Brill Publishers: Leiden, The Netherlands, 2007; pp. 1–264.

46. Hopkin, S.P. *A Key to the Springtails (Collembola) of Britain and Ireland*; Field Studies Council (AIDGAP Project): London, UK, 2007; pp. 1–245.

47. Conti, F.; Visioli, G.; Malcevschi, A.; Menta, C. Safety assessment of gasification biochars using *Folsomia candida* (Collembola) ecotoxicological bioassays. *ESPR* **2017**, *25*, 6668–6679. [CrossRef] [PubMed]

48. Liesch, A.M.; Weyers, S.L.; Gaskin, J.W.; Das, K.C. Impact of two different biochars on earthworm growth and survival. *Ann. Environ. Sci.* **2010**, *4*, 1–9.

49. Sohi, S.P.; Krull, E.; Lopez-Capel, E.; Bol, R. *A Review of Biochar and Its Use and Function in Soil. Advances in Agronomy*; Academic Press: Cambridge, MA, USA, 2010; pp. 47–82.

50. Xie, T.; Reddy, K.R.; Wang, C.; Yargicoglu, E.; Spokas, K. Characteristics and applications of biochar for environmental remediation: A review. *Crit. Rev. Environ. Sci. Technol.* **2015**, *45*, 939–969. [CrossRef]

51. Vaccari, F.P.; Maienza, A.; Miglietta, F.; Baronti, S.; Di Lonardo, S.; Giagnoni, L.; Lagomarsino, L.; Pozzi, A.; Pusceddu, E.; Ranieri, R.; et al. Biochar stimulates plant growth but not fruit yield of processing tomato in a fertile soil. *Agric. Ecosyst. Environ.* **2015**, *207*, 163–170. [CrossRef]

52. Głąb, T.; Palmowska, J.; Zaleski, T.; Gondek, K. Effect of biochar application on soil hydrological properties and physical quality of sandy soil. *Geoderma* **2016**, *28*, 11–20. [CrossRef]

53. Yang, D.; Yunguo, L.; Liu, S.; Huang, X.; Li, Z.; Tan, X.; Zeng, G.; Zhou, L. Potential benefits of biochar in agricultural soils: A review. *Pedosphere* **2017**, *27*, 645–661.

54. Baronti, S.; Vaccari, F.P.; Miglietta, F.; Calzolari, C.; Lugato, E.; Orlandini, S.; Pini, R.; Zulian Genesio, C.L. Impact of biochar application on plant water relations in *Vitis vinifera* (L.). *Eur. J. Agron.* **2014**, *53*, 38–44. [CrossRef]

55. Wang, L.; Butterly, C.R.; Wang, Y.; Herath, H.M.S.K.; Xi, Y.G.; Xiao, X.J. Effect of crop residue biochar on soil acidity amelioration in strongly acidic tea garden soils. *Soil Use Manag.* **2013**, *30*, 119–138. [CrossRef]

56. Obia, A.; Cornelissen, G.; Mulder, J.; Dörsch, P. Effect of Soil pH Increase by Biochar on NO, N_2O and N_2 Production during Denitrification in Acid Soils. *PLoS ONE* **2015**, *10*, e0138781. [CrossRef] [PubMed]

57. Hågvar, S. Reactions to soil acidification in microarthropods: Is competition a key factor? *Biol. Fertil. Soils* **1990**, *9*, 178–181. [CrossRef]

58. Domene, X.; Mattana, S.; Hanley, K.; Enders, A. Medium-term effects of corn biochar addition on soil biota activities and functions in a temperate soil cropped to corn. *Soil Biol. Biochem.* **2014**, *72*, 152–162. [CrossRef]

59. Castracani, C.; Maienza, A.; Grasso, D.A.; Genesio, L.; Malcevschi, A.; Miglietta, F.; Vaccari, F.P.; Mori, A. Biochar–macrofauna interplay: Searching for new bioindicators. *Sci. Total Environ.* **2015**, *536*, 449–456. [CrossRef]

60. Larsen, T.; Schjønning, P.; Axelsen, J. The impact of soil compaction on euedaphic Collembola. *Appl. Soil Ecol.* **2004**, *26*, 273–281. [CrossRef]

61. Twardowski, J.P.; Hurej, M.; Gruss, I. Diversity and abundance of springtails (Hexapoda: Collembola) in soil under 90-year potato monoculture in relation to crop rotation. *Arch. Agron. Soil Sci.* **2016**, *62*, 1158–1168. [CrossRef]

62. Jacomini, C.; Nappi, P.; Sbrilli, G.; Mancini, L. *Indicatori ed Indici Ecotossicologicie Biologici Applicati al Suolo: Stato Dell'arte*; Agenzia Nazionale per la Protezione dell'Ambiente (ANPA): Roma, Italy, 2000.

63. Op Akkerhuis, G.J.; de Ley, F.; Zwetsloot, H.; Ponge, J.-F.; Brussaard, L. Soil microarthropods (Acari and Collembola) in two crop rotations on a heavy marine clay soil. *Rev. Ecol. Biol. Sol.* **1988**, *25*, 175–202.

 *sustainability*

Article

Method for the Reduction of Natural Losses of Potato Tubers During their Long-Term Storage

Tomasz Jakubowski [1] **and Jolanta B. Królczyk** [2,*]

1   Department of Machinery, University of Agriculture, 30-149 Krakow, Balicka 116B, Poland; tomasz.jakubowski@ur.krakow.pl
2   Faculty of Mechanical Engineering, Opole University of Technology, 5 Mikolajczyka Street, 45-271 Opole, Poland
*   Correspondence: j.krolczyk@po.opole.pl

Received: 25 November 2019; Accepted: 29 January 2020; Published: 2 February 2020

Abstract: The purpose of the study was to establish whether UV-C radiation applied to potato tubers prior to their storage affected their natural losses over a long period of time. A custom-built UV-C radiation stand constructed for the purpose of this experiment was equipped with a UV-C NBV15 radiator generating a 253.7 nm long wave with power density of 80 to 100 μW·cm^{-2}. Three varieties of edible medium late potatoes, Jelly, Syrena, and Fianna, were the objects of the research. The measurement of tightly controlled storage conditions was carried out over three seasons between 2016/2017 and 2018/2019, in a professional agricultural cold store with automated adjustment of interior microclimate parameters. The obtained data were processed using the variance analysis ($\alpha = 0.05$). There was a statistically significant reduction in transpiration- and respiration-caused losses in the UV-C radiated potato tubers in comparison to those of the control sample. Additionally, the Jelly variety reacted to UV-C radiation demonstrating a reduction in sprout weight.

Keywords: potato; tuber; storage losses; UV-C

---

1. Introduction

The biological effects of UV-C (ultraviolet light) on the preservation of fresh fruits and vegetables is well researched. Treatment with UV-C radiation is one of the methods used to reduce the number of pathogens on the surface of fresh fruits and vegetables [1]. It can be an alternative to other traditional methods, such as disinfectants (chlorine, chlorine dioxide, bromine, iodine, trisodium phosphate, sodium chlorite, sodium hypochlorite, quaternary ammonium compounds, acids, hydrogen peroxide, ozone, permanganate salts) [2–5], modified atmosphere packaging [6–11], low temperature storage [11–13], or the use of edible films [14–18]. The alternative methods mentioned above are selective in reducing the number of pathogens on the surface of fresh fruits and vegetables, whereas the UV-C is a nonselective method. UV-C has a germicidal effect, but it is strongly dependent on the natural resistance of the microorganisms to UV-C [19], and the surface topography on which the microorganisms are attached [20]. By delaying the ripening process, the UV-C treatment extends the shelf life of fruits and vegetables [21–24]. A number of studies showed that UV light can be used to control the fungal decay of citrus fruits [25], kumquats [26], carrots [27,28], apples [29], strawberries [30–33], sweet cherries [30], mandarins [34], bell peppers [35], mangos [36,37], blueberries [38], grapes [39], and persimmon fruits [40]. Recent publications also describe similar effects on potatoes. Pristijono et al. [41] describe their preliminary research on the effect of UV-C irradiation on the sprouting of stored potatoes. Rocha et al. [42] present the use of UV-C radiation and fluorescent light to control postharvest soft rot in potato seed tubers. According to Stevens et al. [43] irradiation of potato tubers (*Ipomea batatas* L.) with UV-C increases their resistance to rot caused by *Fusarium solani* fungi.

Long-term storage of potato tuber crop is associated with quantitative changes involving, inter alia, reduction of the original weight of tubers due to respiration, transpiration, and sprouting. These processes, termed tuber weight losses, are unavoidable, but their extent can be minimized. Respiration, transpiration, and sprouting processes are mainly determined by the temperature and relative humidity of air during storage. Meteorological conditions during the vegetation season, genetic characteristics of varieties, and mechanical damage to tubers caused during harvesting and post-harvest treatment may also be responsible for natural losses of the potato tubers [44–49]. Several researchers [50–52] recommend using physical methods to improve crop condition during the storage alongside the biological and chemical methods of protection. Various systems using the UV-C light exposure have appeared in the literature for crops other than potatoes [53,54]. Zhu et al. [53] evaluated the effect of three UV-C wavelengths (222, 254, and 282 nm) on degradation of the mycotoxin patulin introduced into apple juice and apple cider. Falguera et al. [54] investigated the influence of ultraviolet irradiation (UV) on some quality attributes (color, pH, soluble solids content, formol index, total phenolics, sugars, and vitamin C) and enzymatic activities (polyphenol oxidase, peroxidase, and pectinolytic enzymes) of fresh apple juice. Another pilot experiment carried out by Jakubowski [55] analyzed the effect of UV-C radiation on the possible reduction of storage losses caused by transpiration and respiration of potato tubers of the following varieties: Lord, Vineta, Owacja, Ditta, Finezja, and Tajfun. It was assumed that ultraviolet radiation in the C band (253.7 nm) applied to potato tubers before storage would cause a reduction of pests present on the potato periderm. This would indirectly cut down the overall storage losses. It was also assumed that following the exposure of potato tubers to UV-C radiation, the population of pathogens typically present during storage processes would also be limited. This, in turn, could lead to a more effective repair of the periderm damage caused by mechanical harvesting, transport, and initial crop storage. In order to verify the research hypothesis, the tubers of selected varieties were initially mechanically damaged under laboratory conditions, so that the continuity of periderm was broken. An MTS Insight 2 strength testing machine (a device allowing for damage of the potato tuber pulp so that the shape and size of the damage are identical in the tested material regardless of the size, weight, or mechanical properties of the tuber) was used to damage the tubers. The results of the pilot research proved a reduction in the natural loss of potato compared to the control sample across all the used varieties. For Vineta and Ditta, these differences were statistically significant ($\alpha = 0.05$). The experiment described above was conducted over a 5-month storage period, and the tuber weight loss (represented by very early, early, and medium-early varieties) was analyzed only before and after the storage time. It is therefore reasonable to measure the reaction of stored potato tubers to UV-C radiation in subsequent stages during their storage, focusing on the losses caused by respiration and transpiration (particularly after a period of physiological dormancy when the sprouting process begins, which is accompanied by an increased transpiration process).

The purpose of this project was to determine the impact of pre-storage UV-C irradiation of potato tubers on potato tuber natural losses measured over a long period of storage.

2. Materials and Methods

2.1. Potato Tubers

The medium-late varieties of edible potatoes used in the research were Jelly, Syrena, and Fianna. Each tuber went through an identical annual agrotechnical procedure. The tubers for analysis, under the stage of full technical maturity and mechanically undamaged, were randomly selected from the crop commercial fraction ($\Phi = 35–55$ mm). The selection of tubers size was necessary to provide uniform UV-C exposure doses in the research procedure. When determining the minimum sample size, the t-test was applied for a single sample (at population mean and standard deviation values from pilot studies), the test target power was assumed to be equal 0.9, and the probability of a type I error $\alpha = 0.05$. Each experiment combination involved three replications (with 30 pcs for single replication).

The scope of the research encompassed the measurements of tuber weight losses arising from the processes of respiration, transpiration, and sprouting. The tuber weight was determined before storage (M0), during the initial storage period corresponding to the cooling stage (M1), during the proper storage period (M2), during the first signs of sprouting (M3), and immediately after the storage process (M4). The tuber weight losses were calculated as the difference between M0 and Mn, n = 1–4. Immediately after the storage process, the potato tuber sprout weight and number (Mk, Lk) were also determined.

2.2. Period of Trials and Conditions

Accurate storage experiments were carried out over three seasons between 2016/2017 and 2018/2019 in a professional agricultural cold store with automated adjustment of interior microclimate parameters. After the crop was harvested in the second to third decades of September, the selected tubers underwent weight examination and ultraviolet irradiation in the C band. Then, they were placed in a cold store for initial storage. To standardize the conditions of heat and mass exchange with the environment while storing and to minimize the impact of possible temperature and humidity differentiation, the tubers were stored in wooden cases in single layers and the free spaces between the cases were of similar size. Initial storage lasted for approximately 10 days at a temperature of 15 °C and a relative humidity of 90%–95%. After this period, the storage temperature was gradually reduced to 7 °C. This process took approximately 14 days. The air relative humidity during this period amounted to 92%–95%. From the second to third decades of October to the first decade of March, tubers were stored at a temperature of approximately 7 °C and a relative humidity of 92%–95%. At 10 days before the expected end of the storage period, mid-March, the temperature in the cold chamber was increased to 10 °C.

2.3. Equipment

Potato tuber weight and sprout weight were determined using the AS310.R2 analytical scale (d = 0.1 mg) with the RS232 interface. Potato tubers underwent ultraviolet irradiation (Figure 1) in the C band for 900 s at a constant height of the UV-C radiator (0.7 m) above the surface of the rollers rotating at a constant speed of 25 rpm (exposure time and working parameters of the station were selected based on the results of pilot studies [55]). In order to expose potato tubers to UV-C, the custom-built stand illustrated in Figure 1 was used. A UV-C NBV15 radiator was used (light wave length 253.7 nm, power 15 W, and power density from 80 to 100 μW·cm^{-2}), equipped with a precise timer (AURATON 100). The lifetime of the NBV15 radiator applied in the research, guaranteeing stability of its operational parameters (UV-C radiation intensity of 0.9 W·m^{-2} at a distance of 1 m from the radiator), is 8000 h. The radiator is equipped with a reflector made of high quality aluminum with a high reflection coefficient (similar to the coefficient of a mirror). The potato tuber radiation stand is equipped with a system of exchangeable, parallel, and sliding rollers acting as the bottom of the chamber. The rollers, with a diameter of 45 to 55 mm, are installed on a rail, and the distances between them range from 15 to 25 mm. They are driven electrically, with rotational speed control ranging from 20 to 35 rpm. The speed range was selected so that the potato tubers placed on the rollers were set in rotation, but at the same time they were not displaced along the chamber, which allowed for equal irradiation of the entire surface of the potato tubers. The presented test stand, together with the technology allowing for limitation of storage losses of potato tubers, was submitted as an invention to the Patent Office of the Republic of Poland (P.419392, P.425887: More details about patents are mentioned below in the Patents section).

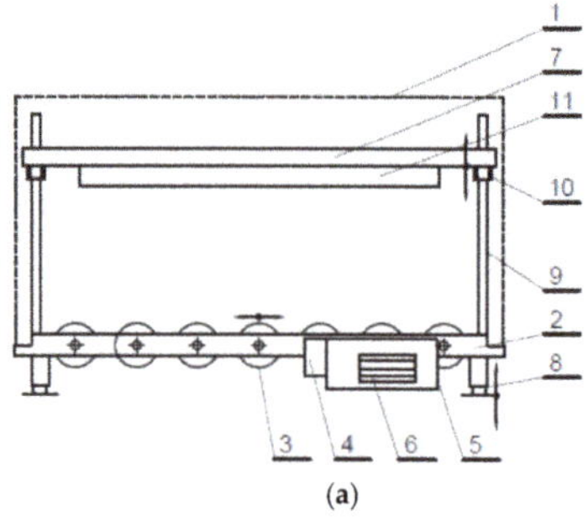 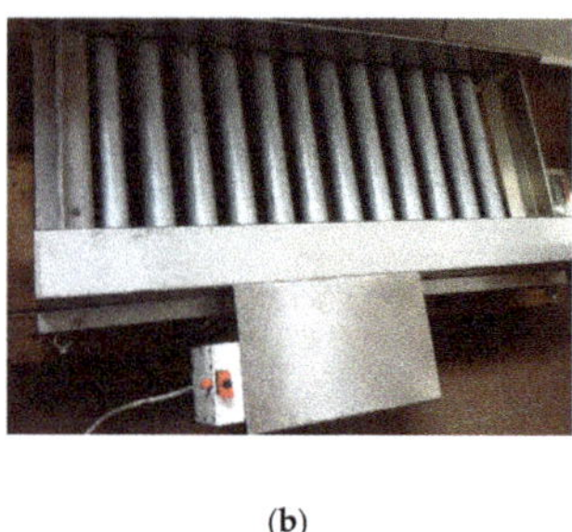

(a) (b)

Figure 1. Stand for potato tubers UV-C radiation: (**a**) stand layout, (**b**) view of rotating rollers. L: 1: chamber housing, 2: device frame, 3: rotating rollers, 4: mechanism for gradeless regulation of roller rotation speed with the device switch, 5: engine, 6: engine cooling system, 7: radiator frame, 8: height-adjustable foot (screw mechanism for leveling the device), 9: radiator frame guide, 10: gradeless UV-C radiator control above the bottom of the chamber (rollers), 11: UV-C radiators.

2.4. Statistical Analysis

The analysis of variance, preceded by the test regarding the distribution normality in the samples (Kolmogorov–Smirnov test) and the variance homogeneity test (Levene's test), was carried out. The zero hypothesis was verified based on the F-Snedecor test. When examining tuber weight data, the variance was analyzed for arrangements with repeated measurements (with Mauchley sphericity test), and for sprouting data, the variance was analyzed for main effects. Due to the biological nature of the examined object, a possibility of a lack of parity in the samples was assumed a priori. Differences between statistically significant averages were examined using the Spjotvoll–Stoline multiple comparison test (generalization of the Tukey procedure for samples with different N). Groups of homogeneous variables were set. In the analysis of data presented in graphical form (Figures 2–4), the Student's t-test was used for related variables. The obtained results were analyzed at the significance level $\alpha = 0.05$ using the STATISTICA 13.3 package.

3. Results

The results of the experiment are presented in Tables 1–3 and in Figures 2–4. Test results from Kolmogorov–Smirnov, Levene, and Mauchley tests allowed for the variance analysis presented in the methodology. The variance analysis proved a significant effect of potato tuber UV-C irradiation before storage on weight loss after storage (predictor: "UV-C exposition {3}", received values: $F = 6.868$; $p = 0.0089$) (Table 1). The variance analysis for arrangements with repeated measurements proved a significant impact of repeated measurement (time: measurement in subsequent storage stages; predictor: "Time {4}", received values: $F = 4498.591$; $p = 0.0000$) and its interaction with other qualitative predictors (predictor: "Time {4} x Variety {2} x UV-C exposition {3}"; received values: $F = 2.329$; $p = 0.0301$) (Table 1). In all the experiment variants, smaller natural losses of potato tuber weight (caused by transpiration and respiration) were observed for the UV-C radiated tubers in comparison to the control sample (Figure 2). Multiple comparisons of average pairs and homogeneous-variables groups determined on this basis (Table 2) showed significant differences in the size of potato tuber natural weight loss (caused by transpiration and respiration) between subsequent storage stages. However, in stages I-III no differences resulting from the variety nor UV-C radiation were shown (letters "a", "b", "c": no statistical differences) (Figure 2). Significant differences in the size of potato tuber natural weight loss (caused by transpiration and respiration) were observed in stage IV for the Jelly variety (3.787 g·g^{-1} for sample Jelly+UV-C and 4.308 g·g^{-1} for control sample; letters "d" and "e") (Table 2). The variance analysis for the impact of the year of the experiment, variety, and UV-C radiation on the weight and number of potato tuber sprouts after the storage period proved a significant impact (Table 3) on the last two quality predictors and their interactions on dependent variables (predictors: Variety {2} x UV-C exposition {3}; $F = 5.20$; $p = 0.0004$).

Table 1. Results of the analysis of variance for repeated measurements: Impact of the year of testing, variety, and UV-C irradiation on potato tuber weight loss in individual storage periods.

Qualitive Predictor and Interaction	Sum of Square	Degrees of Freedom	Mean Square	Value of F-Snedecor test	Probability Test
Free Word	38815.53	1	38815.53	5391.355	0.0000
Year {1}	4.98	2	2.49	0.346	0.7078
Variety {2}	22.46	2	11.23	1.560	0.2104
UV-C exposition {3}	49.45	1	49.45	6.868	0.0089
{1}x{2}	55.41	4	13.85	1.924	0.1039
{1}x{3}	11.46	2	5.73	0.796	0.4513
{2}x{3}	12.18	2	6.09	0.846	0.4293
{1}x{2}x{3}	37.48	4	9.37	1.301	0.2673
Error	11533.74	1602	7.20		
Time {4}	8273.54	3	2757.85	4498.591	0.0000
{4}x{1}	0.80	6	0.13	0.217	0.9717
{4}x{2}	6.28	6	1.05	1.706	0.1152
{4}x{3}	23.42	3	7.81	12.732	0.0000
{4}x{1}x{2}	12.45	12	1.04	1.692	0.0618
{4}x{1}x{2}x{3}	1.83	6	0.31	0.498	0.8102
{4}x{2}x{3}	8.57	6	1.43	2.329	0.0301
{4}x{1}x{2}x{3}	13.60	12	1.13	1.849	0.0357
Error	2946.30	4806	0.61		

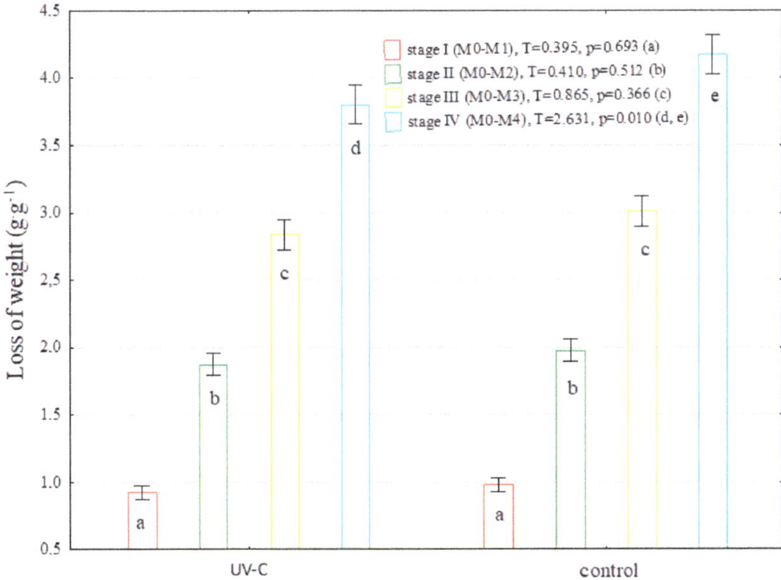

Figure 2. Effect of UV-C irradiation on weight loss (resulting from transpiration and respiration) of potato tubers in individual storage periods (error = mean value +/− 95% of confidence interval).

Table 2. Effect of variety and UV-C irradiation on weight loss (resulting from transpiration and respiration) of potato tubers in individual storage periods.

Variety	Exposition	Time	Loses (g·g⁻¹)	Homogeneous Groups					
				1	**2**	**3**	**4**	**5**	**6**
Fianna	UV-C	I	0.887	****					
Jelly	UV-C	I	0.927	****					
Fianna	control	I	0.948	****					
Syrena	UV-C	I	0.961	****					
Jelly	control	I	0.977	****					
Syrena	control	I	1.012	****					
Fianna	UV-C	II	1.785		****				
Jelly	UV-C	II	1.849		****				
Fianna	control	II	1.964		****				
Jelly	control	II	1.974		****				
Syrena	UV-C	II	1.991		****				
Syrena	control	II	1.992		****				
Fianna	UV-C	III	2.741			****			
Jelly	UV-C	III	2.804			****			
Fianna	control	III	2.863			****			
Syrena	UV-C	III	2.968			****			
Syrena	control	III	2.998			****			
Jelly	control	III	3.172			****			
Fianna	UV-C	IV	3.649				****		
Jelly	UV-C	IV	3.787				****	****	
Syrena	UV-C	IV	3.971				****	****	****
Fianna	control	IV	4.088				****	****	****
Syrena	control	IV	4.120					****	****
Jelly	control	IV	4.308						****

(****) Arrangement of homogeneous groups (Spjotvoll–Stoline test).

Table 3. Impact of the year of experiment, variety, and UV-C irradiation on weight and number of potato tuber sprouts after storage.

Qualitive Predictor and Interaction	Value of F-Snedecor Test	Probability Test
Free Word	10787.16	0.0000
Year {1}	0.22	0.9276
Variety {2}	12.58	0.0000
UV-C exposition {3}	10.11	0.0001
{1}×{2}	1.43	0.1767
{1}×{3}	1.72	0.1420
{2}×{3}	5.20	0.0004
{1}×{2}×{3}	1.82	0.0680

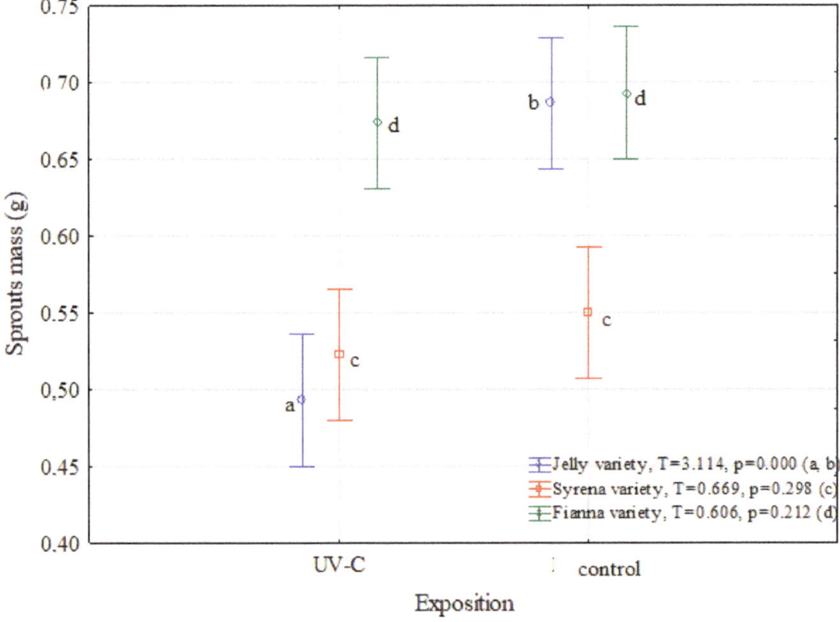

Figure 3. Effect of potato tuber UV-C irradiation on sprout weight after storage (statistically significant difference for the Jelly variety), (error = mean value +/− 95% of confidence interval).

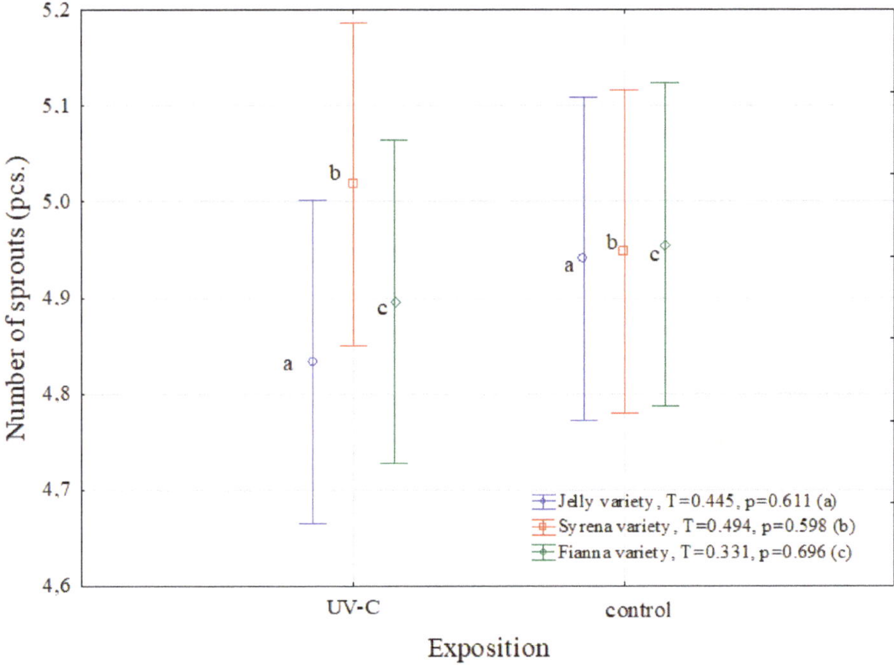

Figure 4. Effect of potato tuber UV-C irradiation on number of sprouts after storage (statistically insignificant differences), (error = mean value +/− 95% of confidence interval).

4. Discussion

The results of the conducted experiment, in which the reduction of natural losses measured after 6.5 months of storage is achieved by means of the UV-C irradiation of potato tubers before storage, were expected to be similar to the results obtained for early and medium-early varieties (Vineta and Ditta) and described by Jakubowski [55]. The tests carried out in this research and the results obtained allowed for a description of the phenomenon causing reduction of natural losses and a determination of the stage of potato storage in which the physical factor, in the form of UV-C radiation, significantly affects the tubers under exposure. The results of the experiment prove that the effects of UV-C on stored potato tubers occur in the final stage (IV) of their storage (Table 2), which corresponds to the phase of tubers awakening and beginning their sprouting. In stage IV of storage, all the tested varieties reacted to UV-C, demonstrating the reduction in natural losses in comparison to the control, and for the Jelly variety, these differences were statistically significant ($F = 2.329$; $p = 0.0301$). The reduction in the weight loss of the pre-treated tubers during storage could occur through changes in the epi- and cuticular wax morphology. Some information on this aspect of treatment are presented in other research, such as Charles et al. [56,57]. Sprouting of the tubers indicates the break of dormancy. Oxidative atmospheres, such as chlorine atmospheres, are known to control sprouting [58].

The analyzed potato tubers reacted to UV-C radiation demonstrating the reduction in sprout weight by 80 $g \cdot g^{-1}$, on average, for all the varieties, which amounts to approximately 14.2% compared to the control. For Jelly tubers, these differences were statistically significant (difference of 194 $g \cdot g^{-1}$, which is approximately 39% compared to the control) (Figure 3). The tubers of Jelly and Fiana varieties subjected to UV-C irradiation were also characterized by a lower number of sprouts compared to the tubers not being under exposure, by 0.11 and 0.05 pcs, respectively. A higher number of sprouts resulting from UV-C irradiation was noted for the tubers of Syrena variety, by 0.07 pcs compared to the control (Figure 4). The results of the experiment suggest that UV-C, in addition to neutralizing pests in areas of damaged potato periderm [55], may act as a sprouting inhibitor, which reduces respiration and, more particularly, transpiration. The obtained results also allow for the conclusion that UV-C does not interfere with the process of tuber transpiration and respiration at earlier stages (I-III) of storage. The inhibitory effect of ultraviolet in the C band may result from the fact that the effect of 253.7 nm wave on a biological object may lead to damaging its DNA chains, and at the same time UV is absorbed by DNA, RNA [59–63], protein, free purine, and pyrimidine bases, acting mutagenetically and inhibiting cell division in the irradiated organism [64–66]. The preliminary tests carried out by Pristijono et al. [41] on freshly harvested potatoes (Solanum tuberosum 'Innovator') that were exposed to UV-C light revealed that UV-C irradiation significantly affected the number of sprouts. UV-C irradiation also affected the sprout length since irradiated potatoes had significantly shorter sprouts than those of untreated potatoes. Despite the fact that storage conditions were different in this experiment (storage in air 20°C), the authors [41] conclude that these results indicate promise for UV-C as a potential postharvest treatment to reduce the incidence of sprouting in potato tubers.

5. Conclusions

1. A significant impact of potato tuber UV-C irradiation on the size of natural losses was observed.
2. A reduction in potato tuber weight loss caused by transpiration and respiration was shown in comparison to the control sample.
3. Jelly variety reacted to UV-C radiation, demonstrating the reduction in the sprout weight.
4. The result of the experiment indicates that the proposed physical UV-C method can be applied in practice and can be used as a way of reducing the natural defects of stored potato tubers.

6. Patents

Jakubowski T. Patent: The method and device for increasing the storage life of potato tubers with the participation of radiation UV-C (in Polish; Sposób i urządzenie do zwiększania trwałości przechowalniczej bulw ziemniaczanych przy udziale promieniowania UV-C: P.419392, data zgłoszenia 07-11-2016).

Jakubowski T., Sobol Z. Patent: The method for modifying the color of potato products and a device to modify the color of potato products (in Polish; Sposób modyfikowania barwy wyrobów z ziemniaków i urządzenie do modyfikowania barwy wyrobów z ziemniaków: P.425887, data zgłoszenia 11-06-2018).

Author Contributions: Conceptualization, T.J. and J.B.K.; methodology, T.J.; validation and formal analysis, T.J. and J.B.K.; investigation, resources, and data curation, T.J. and J.B.K.; writing—original draft preparation, writing—review and editing, and visualization, J.B.K. All authors have read and agreed to the published version of the manuscript.

Funding: This research received no external funding.

Conflicts of Interest: The authors declare no conflict of interest.

References

1. Turtoi, M. Ultraviolet light treatment of fresh fruits and vegetables surface. *J. Agroaliment. Process. Technol.* **2013**, *19*, 325–337.

2. Beuchat, L.R. *Surface Decontamination of Fruits and Vegetables Eaten Raw: A Review*; World Health Organization: Geneva, Switzerland, 1998.

3. Yaun, B.; Sumner, S.; Eifert, J.; Marcy, J. Inhibition of pathogens on fresh produce by ultraviolet energy. *Int. J. Food Microbiol.* **2004**, *90*, 1–8. [CrossRef]

4. Allende, A.; McEvoy, J.; Tao, Y.; Luo, Y. Antimicrobial effect of acidified sodium chlorite, sodium chlorite, sodium hypochlorite, and citric acid on *Escherichia coli* O157:H7 and natural microflora of fresh-cut cilantro. *Food Control* **2009**, *20*, 230–234. [CrossRef]

5. Oms-Oliu, G.; Rojas-Graü, A.; Gonzáles, L.A.; Varela, P.; Soliva-Fortuny, R.; Hernando Hernando, I.; Pérez Munuera, I.; Fiszman, S.; Martín-Belloso, O. Recent approaches using chemical treatments to preserve quality of freshcut fruit: A review. *Postharvest Biol. Technol.* **2010**, *57*, 139–148. [CrossRef]

6. Kader, A.A.; Zagory, D.; Kerbel, E.L. Modified atmosphere packaging of fruits and vegetables. *Crit. Rev. Food Sci. Nutr.* **1989**, *28*, 1–30. [CrossRef] [PubMed]

7. Soliva-Fortuny, R.C.; Elez-Martínez, P.; Martín-Belloso, O. Microbiological and biochemical stability of fresh-cut apples preserved by modified atmosphere packaging. *Innov. Food Sci. Emerg. Technol.* **2004**, *5*, 215–224. [CrossRef]

8. Saxena, A.; Singh Bawa, A.; Srinivas Raju, P. Use of modified atmosphere packaging to extentd shelflife of minimally processed jackfruit (*Artocarpus heterophyllus* L.) bulbs. *J. Food Eng.* **2008**, *87*, 455–466. [CrossRef]

9. Oliveira, M.; Usall, J.; Solsona, C.; Alegre, I.; Vinas, I.; Abadias, M. Effects of packaging type and storage temperature on the growth of foodborne pathogens on shreddred 'Romaine' lettuce. *Food Microbiol.* **2010**, *27*, 455–466. [CrossRef]

10. Sandhya. Modified atmosphere packaging of fresh produce: Current status and future needs. *LWT—Food Sci. Technol.* **2010**, *43*, 381–392. [CrossRef]

11. Abadias, M.; Alegre, I.; Oliveira, M.; Altisent, R.; Vinas, I. Growth potential of *Escherichia coli* O157:H7 on fresh-cut fruits (melon and pineapple) and vegetables (carrot and escarole) stored under different conditions. *Food Control* **2012**, *27*, 37–44. [CrossRef]

12. Harvey, J.M. Optimum environments for the transport of fresh fruits and vegetables. *Int. J. Refrig.* **1981**, *4*, 293–298. [CrossRef]

13. Tano, K.; Oule, M.K.; Doyon, G.; Lencki, R.W.; Arul, J. Comparative evaluation on the effect of storage temperature fluctuation on modified atmosphere packages of selected fruit and vegetables. *Postharvest Biol. Technol.* **2007**, *46*, 212–221. [CrossRef]

14. Zhang, R.; Beuchat, L.R.; Chinnan, M.S.; Shewflet, R.L.; Haung, Y.W. Inactivation of Salmonella Montevideo on tomatoes by applying cellulose-based edible Films. *J. Food Prot.* **1996**, *59*, 808–812. [CrossRef] [PubMed]

15. Vina, S.Z.; Mugridge, A.; Garcia, M.A.; Ferreyra, R.M.; Martino, M.N.; Chaves, A.R.; Zaritzky, N.E. Effects of polyvinylchloride films and edible starch coatings on quality aspects of refrigerated Brussels sprouts. *Food Chem.* **2007**, *103*, 701–709. [CrossRef]

16. Raybaudi-Massilia, R.M.; Mosqueda-Melgar, J.; Martin-Belloso, O. Edible alginate-based coating as carrier of antimicrobials to improve shelf-life and safety of fresh-cut melon. *Int. J. Food Microbiol.* **2008**, *121*, 313–327. [CrossRef]

17. Falguera, V.; Quintero, H.P.; Jimeez, A.; Munoz, J.A.; Ibarz, A. Edible films and coatings: Structures, active functions and trends in their use. *Trends Food Sci. Technol.* **2011**, *22*, 292–303. [CrossRef]

18. Gol, N.B.; Patel, P.R.; Ramana Rao, T.V. Improvement of quality and shelf-life of strawberries with edible coatings enriched with chitosan. *Postharvest Biol. Technol.* **2013**, *85*, 185–195. [CrossRef]

19. Shama, G. Ultraviolet light. In *Handbook of Food Science, Technology, and Engineering*; Hui, Y.H., Ed.; CRC Press: Boca Raton, FL, USA, 2005; pp. 122-1–122-14.

20. Gardner, D.W.; Shama, G. Modeling UV-induced inactivation of microorganisms on surfaces. *J. Food Prot.* **2000**, *63*, 63–70. [CrossRef]

21. Lu, J.Y.; Stevens, C.; Khan, V.A.; Kabwe, M.; Wilson, C.L. The effect of ultraviolet irradiation on shelf-life and ripening of peaches and apples. *J. Food Qual.* **1991**, *14*, 299–305. [CrossRef]

22. D'hallewin, G.; Schirra, M.; Manueddu, E.; Piga, A.; Ben-Yehoshua, S. Scoparone and scopoletin accumulation and ultraviolet-C induced resistance to postharvest decay in oranges as influenced by harvest date. *J. Am. Soc. Hortic. Sci.* **1999**, *124*, 702–707. [CrossRef]

23. Lamikanra, O.; Kueneman, D.; Ukuku, D.; Bett-Garber, K.L. Effect of Processing Under Ultraviolet Light on the Shelf Life of Fresh-Cut Cantaloupe Melon. *J. Food Sci.* **2005**, *70*, C534–C539. [CrossRef]

24. Darvishi, S.; Fatemi, A.; Davari, K. Keeping quality of use of fresh 'Kurdistan' strawberry by UV-C radiation. *World Appl. Sci. J.* **2012**, *17*, 826–831.

25. Ben-Yehoshua, S.; Rodov, V.; Kim, J.J.; Carmeli, S. Preformed and induced antifungal materials of citrus fruits in relation to the enhancement of decay resistance by heat and ultraviolet treatments. *J. Agric. Food Chem.* **1992**, *40*, 1217–1221. [CrossRef]

26. Rodov, V.; Ben-Yehoshua, S.; Kim, J.J.; Shapiro, B.; Ittah, Y. Ultraviolet illumination induces scoparone production in kumquat and orange fruit and improves decay resistance. *J. Am. Soc. Hortic. Sci.* **1992**, *117*, 788–792. [CrossRef]

27. Mercier, J.; Arul, J.; Julien, C. Effect of UV-C on phytoalexin accumulation and resistance to Botrytis cinerea in stored carrots. *Phytopathology* **1993**, *139*, 17–25. [CrossRef]

28. Mercier, J.; Roussel, D.; Charles, M.T.; Arul, J. Systemic and local responses associated with UV and pathogen-induced resistance to Botrytis cinereal in stored carrot. *Phytopathology* **2000**, *90*, 981–986. [CrossRef]

29. De Capdeville, G.; Wilson, C.L.; Beer, S.V.; Aist, J.R. Alternative disease control agents induce resistance to blue mold in harvested 'Red Delicious'apple fruit. *Phytopathology* **2002**, *92*, 900–908. [CrossRef]

30. Marquenie, D.; Michiels, C.; Geeraerd, A.; Schenk, A.; Soontjens, C.; Van Impe, J.; Nicolai, B. Using survival analysis to investigate the effect of UV–C and heat treatment on storage rot of strawberry and sweet cherry. *Int. J. Food Microbiol.* **2002**, *73*, 187–196. [CrossRef]

31. Marquenie, D.; Geeraerd, A.H.; Lammertyn, J.; Soontjens, C.; Van Impe, J.F.; Michiels, C.W.; Nicolai, B.M. Combinations of pulsed white light and UV-C or mild heat treatment to inactivate conidia of Botrytis cinerea and *Monilia fructigena*. *Int. J. Food Microbiol.* **2003**, *85*, 185–196. [CrossRef]

32. Marquenie, D.; Michiels, C.W.; Van Impe, J.F.; Schrevens, E.; Nicolai, B.M. Pulsed white light in combination with UV-C and heat to reduce storage rot of strawberry. *Postharvest Biol. Technol.* **2003**, *28*, 455–461. [CrossRef]

33. Lammertyn, J.; De Ketelaere, B.; Marquenie, D.; Molenberghs, G.; Nicolai, B.M. Mixed models for multicategorical repeated response: Modeling the time effect of physical treatments on strawberry sepal quality. *Postharvest Biol. Technol.* **2003**, *30*, 195–207. [CrossRef]

34. Kinay, P.; Yildiz, F.; Sen, F.; Yildiz, M.; Karacali, I. Integration of pre- and postharvest treatments to minimize Penicillium decay of *Satsuma mandarins*. *Postharvest Biol. Technol.* **2005**, *37*, 31–36. [CrossRef]

35. Artes, F.; Conesa, A.; Lopez-Rubira, V.; Artes-Hernandez, F. UV–C treatments for improving microbial quality in whole and minimally processed bell peppers. In *The Use of UV as a Postharvest Treatment: Status and Prospects, Proceedings of the International Conference on Quality Management of Fresh Cut Produce, Bangkok, Thailand, 6–8 August 2007*; Ben-Yehoshua, S., D'Hallewin, G., Erkan, M., Rodov, V., Lagunas, M., Eds.; ISHSS: Leuven, Belgium, 2006; pp. 12–17.

36. Gonzalez-Aguilar, G.A.; Wang, C.Y.; Buta, J.G.; Krizek, D.T. Use of UV-C irradiation to prevent decay and maintain postharvest quality of ripe 'Tommy Atkins' mangoes. *Int. J. Food Sci. Technol.* **2001**, *36*, 767–773. [CrossRef]

37. Gonzalez-Aguilar, G.A.; Zavaleta-Gatica, R.; Tiznado-Hernandez, M.E. Improving postharvest quality of mango 'Haden' by UV-C treatment. *Postharvest Biol. Technol.* **2007**, *45*, 108–116. [CrossRef]

38. Perkins-Veazie, P.; Collins, J.K.; Howard, L. Blueberry fruit response to postharvest application of ultraviolet radiation. *Postharvest Biol. Technol.* **2008**, *47*, 280–285. [CrossRef]

39. Romanazzi, G.; Mlikota Gabler, F.; Smilanick, J.L. Preharvest chitosan and postharvest UV irradiation treatments suppress gray mold of table grapes. *Plant Dis.* **2006**, *90*, 445–450. [CrossRef]

40. Khademi, O.; Zamani, Z.; Poor Ahmadi, E.; Kalantari, S. Effect of UV-C radiation on postharvest physiology of persimmon fruit (*Diospyros kaki* Thunb.) cv. 'Karaj' during storage at cold temperature. *Int. Food Res. J.* **2013**, *20*, 247–253.

41. Pristijono, P.; Bowyer, M.C.; Scarlett, C.J.; Vuong, Q.V.; Stathopoulos, C.E.; Golding, J.B. Effect of UV-C irradiation on sprouting of potatoes in storage. In Proceedings of the VIII International Postharvest Symposium: Enhancing Supply Chain and Consumer Benefits-Ethical and Technological Issues, Cartagena, Spain, 21–24 June 2016; pp. 475–478.

42. Rocha, A.B.; Honório, S.L.; Messias, C.L.; Otón, M.; Gómez, P.A. Effect of UV-C radiation and fluorescent light to control postharvest soft rot in potato seed tubers. *Sci. Hortic.* **2015**, *181*, 174–181. [CrossRef]

43. Stevens, C.; Khan, V.A.; Lu, J.Y.; Wilson, C.L.; Chalutz, E.; Droby, S.; Kabwe, M.K.; Haung, Z.; Adeyeye, O.; Pusey, L.P.; et al. Induced resistance of sweet potato to Fusarium root rot by UV-C hormesis. *Crop Prot.* **1999**, *18*, 463–470. [CrossRef]

44. Clasen, B.; Stoddard, T.; Luo, S.; Demorest, Z.; Li, J.; Cedrone, F. Improving cold storage and processing traits in potato through targeted gene knockout. *Plant. Biotechnol. J.* **2016**, *14*, 169–176. [CrossRef]

45. Elmore, E.; Briddon, A.; Dodson, T.; Muttucumaru, N.; Halford, G.; Mottram, S. Acrylamide in potato crisps prepared from 20 UK-grown varieties: Effects of variety and tuber storage time. *Food Chem.* **2016**, *182*, 1–8. [CrossRef] [PubMed]

46. Hardigan, D.; Hirsch, N.; Manrique-Carpintero, A. The contribution of the Solanaceae coordinated agricultural project to potato breeding. *Potato Res.* **2014**, *57*, 215–224. [CrossRef]

47. El-Awady Aml, A.; Moghazy, M.; Gouda, A.; Elshatoury, A. Inhibition of sprout growth and increase storability of processing potato by antisprouting agent. *Trends Hortic. Res.* **2014**, *4*, 31–40. [CrossRef]

48. Castronuovo, D.; Tataranni, G.; Lovelli, S.; Candido, V.; Sofo, A.; Scopa, A. UV-C irradiation effects on young tomato plants: Preliminary results. *Pak. J. Bot.* **2014**, *46*, 945–949.

49. Katerova, Z.; Ivanov, S.; Prinsen, E.; Van Onckelen, H.; Alexieva, V.; Azmi, A. Low doses of ultraviolet-B or ultraviolet-C radiation affect ACC, ABA and IAA levels in young pea plants. *Biol. Plant.* **2009**, *53*, 365–368. [CrossRef]

50. Hassan, H.; Abd El-Rahman, A.; Liela, A. Sprouting suppression and quality attributes of potato tubers as affected by post-harvest UV-C treatment under cold storage. *Int. J. Adv. Res.* **2016**, *4*, 241–253. [CrossRef]

51. Pietruszewski, S.; Martínez, E. Magnetic field as a method of improving the quality of sowing material. *Int. Agrophysics* **2015**, *29*, 377–389. [CrossRef]

52. Kasyanov, G.; Syazin, I.; Grachev, A.; Davidenko, T.; Vazhenin, E. Features of usage of electromagnetic field of extremely low frequency for the storage of agricultural products. *J. Electromagn. Anal. Appl.* **2013**, *5*, 236–241. [CrossRef]

53. Zhu, Y.; Koutchma, T.; Warriner, K.; Zhou, T. Reduction of Patulin in Apple Juice Products by UV Light of Different Wavelengths in the UVC Range. *J. Food Prot.* **2014**, *77*, 963–967. [CrossRef]

54. Falguera, V.; Pagán, J.; Ibarz, A. Effect of UV irradiation on enzymatic activities and physicochemical properties of apple juices from different varieties. *Food Sci. Technol.* **2011**, *44*, 115–119. [CrossRef]

55. Jakubowski, T. Use of UV-C radiation for reducing storage losses of potato tubers. *Bangladesh J. Bot.* **2018**, *47*, 533–537. [CrossRef]

56. Charles, M.T.; Mercier, J.; Makhlouf, J.; Arul, J. Physiological basis of UV-C-induced resistance to Botrytis cinerea in tomato fruit: I. Role of pre-and post-challenge accumulation of the phytoalexin-rishitin. *Postharvest Biol. Technol.* **2008**, *47*, 10–20. [CrossRef]

57. Charles, M.T.; Makhlouf, J.; Arul, J. Physiological basis of UV-C induced resistance to Botrytis cinerea in tomato fruit: II. Modification of fruit surface and changes in fungal colonization. *Postharvest Biol. Technol.* **2008**, *47*, 21–26. [CrossRef]

58. Tweddell, R.J.; Boulanger, R.; Arul, J. Effect of chlorine atmospheres on sprouting and development of dry rot, soft rot and silver scurf on potato tubers. *Postharvest Biol. Technol.* **2003**, *28*, 445–454. [CrossRef]

59. Onik, J.; Xie, Y.; Duan, Y.; Hu, X.; Wang, Z.; Lin, Q. UV-C treatment promotes quality of early ripening apple fruit by regulating malate metabolizing genes during postharvest storage. *PLoS ONE* **2019**, *14*, e0215472. [CrossRef]

60. Wu, X.; Guan, W.; Yan, R.; Lei, J.; Xu, L.; Wang, Z. Effects of UV-C on antioxidant activity, total phenolics and main phenolic compounds of the melanin biosynthesis pathway in different tissues of button mushroom. *Postharvest Biol. Technol.* **2016**, *118*, 51–58. [CrossRef]

61. Najeeb, U.; Xu, Z.; Ahmed, M.; Rasheed, G.; Jilani, M.; Naeem, W.; Shen, W. Ultraviolet-C mediated physiological and ultrastructural alterations in *Juncus effuses* L. shoots. *Acta Physiol. Plant.* **2011**, *33*, 481–488. [CrossRef]

62. Cools, K.; Alamar, M.; Terry, L. Controlling sprouting in potato tubers using ultraviolet-C irradiance. *Postharvest Biol. Technol.* **2014**, *98*, 106–114. [CrossRef]

63. Kowalski, W. *Ultraviolet Germicidal Radiation Handbook*; Springer: Berlin/Heidelberg, Germany, 2009; pp. 38–90. ISBN 978-3-642-01998-2.

64. Bhattacharjee, C.; Sharan, R. UV-C radiation induced conformational relaxation of pMTa4 DNA in *Escherichia coli* may be the cause of single strand breaks. *Int. J. Radiat. Biol.* **2005**, *81*, 919–927. [CrossRef]

65. Schreier, W.; Schrader, T.; Koller, F.; Gilch, P.; Crespo-Hernandez, C.; Swaminathan, V.; Carell, T.; Zinth, W.; Kohler, B. Thymine dimerization in DNA is an ultrafast photoreaction. *Science* **2007**, *315*, 625–629. [CrossRef]

66. Quek, P.; Hu, J. Indicators for photoreactivation and dark repair studies following ultraviolet disinfection. *J. Ind. Microbiol. Biotechnol.* **2008**, *35*, 533–541. [CrossRef] [PubMed]

 *sustainability*

Article

Health-Promoting Properties of Fresh and Processed Purple Cauliflower

Joanna Kapusta-Duch [1],*, Anna Szeląg-Sikora [2], Jakub Sikora [2], Marcin Niemiec [3], Zofia Gródek-Szostak [4], Maciej Kuboń [2], Teresa Leszczyńska [1] and Barbara Borczak [1]

[1] Department of Human Nutrition, Faculty of Food Technology, University of Agriculture in Krakow, 30-149 Krakow, Poland
[2] Institute of Agricultural Engineering and Informatics, University of Agriculture in Krakow, 30-149 Krakow, Poland
[3] Department of Agricultural and Environmental Chemistry, University of Agriculture in Krakow, 31-120 Krakow, Poland
[4] Department of Economics an Organization of Enterprises, Cracow University of Economics, 31-510 Krakow, Poland
* Correspondence: joannakapustaduch@interia.pl; Tel.: +48-12-662-48-16

Received: 21 June 2019; Accepted: 20 July 2019; Published: 24 July 2019

Abstract: Plant-based foods should be fresh, safe, and natural, with nutritional value and processed in sustainable ways. Among all consumed vegetables, Brassica vegetables are considered to be the most important ones. As they are eaten in large quantities and frequently, they may constitute an important source of nutrients and bioactive compounds in a daily diet. This work is aimed at assessing the effect of technological processing (blanching and traditional cooking in water and in a convection steam oven) as well as the method of frozen storage (in PE-LD zipper bags and vacuum packing) on the content of selected components in purple cauliflower. The material was examined for the content of dry matter, vitamin C, total polyphenols, anthocyanins, thiocyanates, nitrates, and nitrites, as well as antioxidant activity. All technological processes caused significant changes in the contents of examined nutritive and non-nutritive compounds as well as in antioxidant activity or the level of selected chemical pollutions. A trend was also observed towards lower constituents' losses as a result of convection steaming, compared to traditional cooking in water. Moreover, the reduction in the content of examined compounds was smaller in vacuum-packed and frozen-stored vegetables then in those stored in zipper PE-LD bags.

Keywords: nutritional value; brassica vegetables; antioxidative properties; quality of food; nitrates and nitrites; frozen storage; processing of vegetables

---

1. Introduction

Sustainable food consumption is a significant aspect of sustainable development. Throughout the last two decades, Brassica crops were of intense interest to many researchers due to their health benefits [1]. Numerous epidemiological and pharmacological works showed a significant role of a diet abundant in Brassica vegetables, which may protect against many chronic diseases, including type II diabetes, cardiovascular disease, age-related macular degeneration, dementia, immune dysfunction, obesity, and some kinds of cancers [2,3]. Cauliflower (*Brassica oleracea* var. *botrytis*) belongs to the very popular Brassica species and is broadly used as a dish or an ingredient of soups or salads. It is rich in vitamins B_1, B_2, B_3, B_5, B_6, folic acid and C, E, K, as well as omega-3 fatty acids, dietary fiber, potassium, phosphorus, magnesium manganese, and iron. Chemical components contents in Brassica vegetables vary among cultivars, which is especially relevant to cauliflower because, in addition to the large group of white-curded cultivars, breeding techniques have resulted in commercially available genotypes forming green, purple, and orange curds, with enhanced synthesis of chlorophylls, anthocyanins, and carotenoids, respectively. Cauliflower genotypes show differences in the content of bioactive compounds, as well as in the chemical composition [4–6]. In addition, this vegetable contains also a lot of valuable and healthy plant's metabolites, including flavonoids, terpenes, S-methylcysteine sulfoxide, sulfur-containing glucosinolates, coumarins, and other minor compounds. These compounds of cauliflower, and other Brassica vegetables, were found to be effective in the protection of some kinds of cancer as cancer-fighting components [4,7,8]. The number of studies has suggested protective effects of these compounds on human health. Brassica species are commonly present in a diet as additives to meat dishes and other products rich in fat, which constituents favour cell transformation and cancer growth [9]. Therefore, some of Brassica crops are classified as functional foods [7].

Thermal treatment of different types of food products leads to various physical, chemical and biological changes occurring in nutrients as well as non-nutrient compounds. In vegetable processing, cooking is the most commonly used technique; although, the application of high temperature may affect the basic chemical composition of as well as activity of bioactive compounds. Two contrary phenomena may be responsible for changes in the content of bioactive compounds during this process: denaturation of enzymes that are involved in degradation of nutrients and bioactive compounds and softening effect of cooking, which increases the extractability of bioactive compounds. As a result, their amount in cooked products is higher than in the raw material [10]. Therefore, in order to evaluate their accessibility with a diet, more knowledge should be gained about the role and the final concentration of bioactive compounds before and after food processing.

New thermal technologies with higher energy efficiency, less nutrient loss and less environmental impacts are being developed. It is believed that for example the steam blanching is relatively inexpensive and retains more water-soluble ingredients and minerals in comparison to water blanching.

The main drawback to vegetables is that they are perishable and available seasonally. Food technology is focused on the development of such methods of food processing, which will affect its chemical composition to the least extent. In view of this, freezing, which is a simple and fast method, is one the most universal and convenient ways of food preservation. Knowledge about optimization of freezing and storing processes is crucial to preserve beneficial and bioactive compounds of vegetables [6]. In this context, the importance of packaging and its role is huge, especially considering its fundamental role in the product protection against the external conditions as well as mechanical damages. In order to guarantee high product quality in terms of their sensory and nutritional features, selection of suitable packaging materials is required [11].

As for packaging properties, low-density polyethylene (PE-LD) is characterized by low permeability to water vapor and good permeability to gases, especially carbon dioxide [12]. Vacuum packaging, as a static form of hypobaric storage, is widely used in food industry. This method allows oxidative reactions to be reduced effectively in a product, at relatively low cost [13].

Until now, the majority of studies concerned other than examined here *Brassicas* cultivars. In consequence, less information is available about such Brassicas like, for example, colored varieties of cauliflower.

High quality foods obtained with sustainable practices during preharvest need proper postharvest practices, including thermal treatments and storage. The aim of the work was to assess the effect of technological processing (blanching, traditional cooking in water and in a convection steam oven) as well as the method of frozen storage (in PE-LD zipper bags and vacuum packing) on the content of selected components in purple cauliflower.

2. Materials and Methods

2.1. Material

The experimental material was purple cauliflower (*Graffiti* cv.) cultivated at the Producer Cooperative "Traf" in Tropiszów (Poland). The experimental field was located in the northern outskirts of the Krakow. The climate of the region is humid continental (Dfb) according to Köppen's classification. The purple cauliflower was grown in black soil on loess framework with neutral pH. Mineral fertilization was applied according to the fertility of soil and the nutritional requirements of the species and condition treatments (mechanical weed control, diseases, and pests) were carried out during the growing season (depending on soil and weather conditions). Mineral fertilization included 400 kg Polifoski PK (MgS) 15-24-(6-7), 100 kg of Saletrzak $NH_4NO_3 + CaCO_3$ and twice foliar application of Folicare NPK 18:18:18; 5:17:40.

Vegetable samples were prepared for analyses directly after harvest. Properly prepared representative medium samples of vegetables (fresh, blanched, cooked in water, convection steamed and frozen, and stored in different package types for 2 and 4 months), were examined for the content of dry matter, vitamin C, total polyphenols, anthocyanins, thiocyanates, nitrates and nitrites as well as antioxidant activity. In addition, the contents of protein, fat, carbohydrates, ash and fiber were determined in fresh material.

The first step of processing (before thermal treatment) included leaf removing, washing in running water, and dividing the heads into roses 4–6 cm in diameter and 5 cm in length. Then, the vegetables were mixed in order to obtain the representative average laboratory samples (a minimum of three for each analysis performed on the fresh material and the same procedure was on material after cooking). Analyses of fresh vegetables were carried out immediately after the pretreatment.

The blanching of material was carried out in the HENDI steam convection oven (model G715RXSD) for 5 min at 100 °C and then cooking to the consumer's softness in the same oven for 20 min at 100 °C. After blanching, the material was chilled and dried at room temperature for about 20 min. Another part of the fresh material was washed, dried using the filter paper, shredded mechanically, frozen at −22 °C, and, finally, freeze-dried in the Christ Alpha 1–4 apparatus. Afterwards, the material was comminuted in the Knifetec 1095 Sample Mill (Tecator) until reaching a homogenous sample with the possibly smallest diameter of particle. Adequately labelled samples were stored at −22 °C in plastic containers in a Liebherr GTS 3612 chamber freezer (Germany), until analyses of protein, fat, dietary fiber and ash.

Another part of the vegetables was cooked in a stainless steel pot using an electric cooking plate, in unsalted water, and in the initial phase of hydrothermal treatment—without a lid but in accordance with the principle "from farm to fork". The proportion of water to the raw material is 5:1 by weight. The cooking time applied was 15 min. The boiled vegetables were then prepared as described for fresh, blanched and steamed vegetables.

The material blanched in a convection steam oven was divided and packed in two types of packaging: the first batch in the conventional polyethylene (PE-LD) zipper bags (0.915–0.935 g/cm^3 in density and 230 × 320 mm in size); the remainder in the special vacuum bags adapted for this purpose, applying a RAMON vacuum packaging machine (60% vacuum; pressure: 0.10 MPa). Next, hermetically sealed samples were stored at −22 °C in a Liebherr GTS 3612 chamber freezer (Germany). Analyses were carried out on the raw material, the material blanched, cooked and the frozen product. Frozen samples were analyzed after 2 and 4 months of frozen storage. The experimental material, taken from every package (on average: 3 roses differing in diameter—from the smallest up to the largest), was collected and then homogenized using a homogenizer (CAT type X 120) to obtain a mean representative sample.

2.2. Analytical Methods

The experimental material was taken from every container (on average: 3 roses differing in diameter—from the smallest up to the largest) and homogenized using a homogenizer (CAT type X 120) in order to obtain a mean representative sample.

At the same time, 70% methanol extracts have been prepared to determine total polyphenols (calculated per chlorogenic acid) by means of the colorimetric measurement of colorful substances occurring as a result of the reaction between phenolic compounds and a Folin–Ciocalteu reagent (Sigma) [14]. The same extract was also used to measure antioxidant activity based on the ABTS$^+$

free radical scavenging ability by a colorimetric assessment of the amount of the ABTS$^+$ free radical solution, which remained unreduced by the antioxidant in the products [15]. In addition, 70% acidified methanolic extracts (5 g of raw vegetables in 80 mL of 70% acidified methanol solution) were prepared [16], which were then used to determine anthocyanin content.

The amount of total phenols in the extracts was measured by a spectrophotometric method at 760 nm on a RayLeigh UV-1800 spectrophotometer, according to the procedure described by Folin–Ciocalteu. The results were calculated using a chlorogenic acid equivalent (CGA) and expressed in milligrams per 100 g of fresh or dry mass, based on a standard curve.

Antioxidant activity was measured by means of colorimetric determination of the quantity of the colored solution of ABTS$^+$ free radical (2,2′-azinobis-(3-ethylbenzothiazoline-6-sulfonic acid)) because of its reduction performed by the antioxidants present in the product examined. Absorbance was measured at 734 nm on a RayLeigh UV-1800 spectrophotometer. The values obtained for every sample were compared to the concentration–response curve of the standard Trolox solution and expressed as micromoles of Trolox equivalent per gram of fresh and dry mass (TEAC).

Anthocyanin content, measured as anthocyanins' absorbance in visible light, was determined in buffer solutions at two pH values: 1 and 4.5. At pH 1 anthocyanins took the form of a red-colored flavic cation, while at pH 4.5 they had changed to the colorless pseudobase. Anthocyanin content was converted to mg of cyanidin-3-glucoside [17].

The dry matter content in the vegetable samples was found in agreement to PN-90/A-75101/03 [18]. The determination principle comprised determining the decrease in mass upon removal of water from the product during thermal drying at the temperature of 105 °C, under normal pressure conditions.

The contents of total ascorbic acid and dehydroascorbic acid were determined using 2,6-dichlorophenoloindophenol in accordance with PN-A-04019:1998 [19]. Extraction of the ascorbic acid was performed using oxalic acid solution. Vitamin C content was expressed as milligrams per 100 g of dry mass (mg/100 g d.m.).

Determination of thiocyanate was based on extraction of the sample with trichloroacetic acid (TCA) and the reaction with ferric ions. Under acidic conditions, blood-red coloration was created due to the formation of $Fe(SCN)^{2+}$ to $Fe(SCN)_6{}^{3-}$ complexes.

Determinations of nitrates and nitrites were carried out in accordance with the Polish standard PN-92/A-75112 [20]. The colorimetric method was used to determine these contaminants based on nitrite colored reaction with Griess I and II. Previously nitrates must be reduced to nitrites. Nitrate content was established using Griess I (sulfanilamide, Sigma-Aldrich) and Griess II (*n*-(1-Naphtyl)ethylene-diamine dihydrochloride, water solution, Sigma-Aldrich). The principle of this method is to induce in acidic conditions, a color reaction of nitrate(III) with *n*-(1-Naphtyl)ethylene-diamine dihydrochloride.

Raw, freshly prepared, lyophilized samples were also examined for protein content using a Tecator Kjeltec 2200; fat content, using a Tecator Soxtec Avanti 2050; ash content, by means of dry mineralisation in muflon oven at 525 °C in accordance with PN-A-79011-8:1998 [21]; and dietary fiber content, with the enzymatic-gravimetry method using the Tecator Fibertec System E.

Protein content was determined by mineralization of the product in concentrated sulfuric acid (IV) (aqueous mineralization), followed by alkalizing the solution, distillation of the ammonia released, and its qualitative determination [22]. As for fat content, at first fat extraction was performed from the dried material with an organic solvent (petroleum ether), then distilling off the solvent, drying the residue, and determining the mass of the extracted "crude fat" [23]. Analysis of dietary fiber content was conducted according to Polish standard [24] by means of enzymatic and gravimetric methods. Lyophilized samples were subjected to gelatinization with a thermally stable α-amylase, then digested by enzymes involving protease and amyloglucosidase to remove protein and starch present in the sample. Soluble dietary fiber was precipitated by adding ethanol. The sediment was then filtered off, washed in ethanol and acetone, and, after drying, weighed. Half the samples was analyzed for the presence of protein and the remainder incinerated. Total dietary fiber has been calculated as the mass of sediment minus the mass of protein and ash.

In addition, the percentage of carbohydrates has been calculated, as the difference between 100 g of fresh product and the sum of water (g), total fat (g), protein (g), and mineral compounds—ash (g).

2.3. Statistical Analysis

All analyses were conducted in three parallel replications and mean ± standard deviations (SD) were calculated for the values obtained. Significance of differences between mean values of raw, blanched, cooked and frozen stored material was checked by one-way analysis of variance (ANOVA). The significance of differences was estimated with the Duncan test at the critical significance level of $p \leq 0.05$. The Statistica 10.1 (Statistica, Tulsa, OK, USA) program was applied.

3. Results

3.1. Basic Composition and Dietary Fiber

Dry matter content in the vegetable varies depending on the process applied, therefore, to show only an effect of the applied process, all the results presented below have been calculated per the dry matter unit. Fresh vegetable contained 19.8 g dry matter, 2.36 g proteins, 0.14 g fat, 0.53 g ash, 16.3 g total carbohydrates, and 2.36 g dietary fiber per 100 g of fresh vegetable (Table 1).

Table 1. Selected antioxidative and bioactive compounds, antioxidant activity, basic composition, and nitrates and nitrites of raw purple cauliflower.

Component	Unit	Mean
Dry Mass	g/100 g	9.18 ± 0.04
Vitamin C	mg/100 g d.m.	689.54 ± 1.54
Total Polyphenols	mg CGA/100 g d.m.	1376.36 ± 3.85
Antioxidant Activity	μmol Trolox/g d.m.	79.85 ± 0.46
Thiocyanates	(SCN) mg/100 g d.m.	26.25 ± 1.69
Anthocyanins	μmol/g d.m.	78.21 ± 5.85
Total Protein	g/100 g d.m.	25.70 ± 1.56
Fat	g/100 g d.m.	1.55 ± 0.21
Ash	g/100 g d.m.	5.73 ± 0.29
Total Carbohydrates	g/100 g d.m.	66.72 ± 1.16
Dietary Fiber	g/100 g d.m.	25.67 ± 1.63
Nitrates	mg NaNO$_3$/kg d.m.	605.23 ± 23.72
Nitrites	mg NaNO$_2$/kg d.m.	17.97 ± 0.31

Technological treatments, both traditional cooking and cooking in the convection steam oven, resulted in a significant ($p \leq 0.05$) increase in dry matter content by 17.1 and 12.7% respectively, compared to fresh cauliflower. Frozen storage of the analyzed material for 2 and 4 months led to a substantial ($p \leq 0.05$) increase in the dry matter content respectively by 31.1 and 32.2% (conventional, in a PE-LD bag) and 30 and 32.2% (in vacuum) compared to the blanched product (Table 2).

Table 2. Content of selected compounds and antioxidant activity of purple cauliflower depending on the technological processing.

The Kind of Processing	Dry Mass g/100 g	Vitamin C mg/100 g d.m.	Total Polyphenols mg CGA/100 g d.m.	Antioxidant Activity μmol Trolox/g d.m.	Thiocyanates (SCN) mg/100 g d.m.	Anthocyanins μmol/g d.m.	Nitrates mg NaNO$_3$/kg d.m.	Nitrites mg NaNO$_2$/kg d.m.
fresh	9.18[a] ± 0.04	689.54[f] ± 1.54	1376.36[f] ± 3.85	79.85[d] ± 0.46	26.25[d] ± 1.69	78.21[c] ± 5.85	605.23[e] ± 23.72	17.97[c] ± 0.31
blanched	8.85[a] ± 0.07	683.62[f] ± 7.99	1358.76[f] ± 13.58	80.79[d] ± 0.80	26.89[d] ± 0.31	80.34[c] ± 8.15	580.79[e] ± 33.56	17.29[bc] ± 0.48
cooked	10.75[b] ± 0.35	305.30[d] ± 0.26	493.40[a] ± 1.84	47.53[a] ± 2.76	12.27[a] ± 1.18	28.47[a] ± 5.92	219.16[b] ± 7.10	11.26[a] ± 1.18
steamed	10.35[b] ± 0.21	362.42[e] ± 4.24	1143.77[e] ± 15.30	61.55[c] ± 1.78	19.13[bc] ± 0.96	67.73[bc] ± 8.20	470.53[d] ± 4.10	14.11[ab] ± 1.50
after 2 months of frozen storage (zipper bags)	11.60[c] ± 0.14	245.69[b] ± 19.51	968.97[c] ± 6.10	52.76[b] ± 0.98	19.39[bc] ± 1.10	52.76[b] ± 11.83	211.81[b] ± 2.80	13.62[a] ± 1.10
after 2 months of frozen storage (vacuum)	11.50[c] ± 0.14	280.87[c] ± 2.46	1013.04[d] ± 7.38	59.39[c] ± 0.37	20.09[c] ± 1.23	60.43[bc] ± 17.09	291.39[c] ± 7.26	14.61[abc] ± 0.86
after 4 months of frozen storage (zipper bags)	11.70[c] ± 0.14	197.43[a] ± 1.21	899.23[b] ± 15.59	44.87[a] ± 1.21	13.08[a] ± 1.33	43.68[ab] ± 8.70	148.03[a] ± 6.41	22.99[d] ± 1.45
after 4 months of frozen storage (vacuum)	11.70[c] ± 0.28	232.48[b] ± 8.46	976.32[c] ± 11.72	53.08[b] ± 2.05	16.50[b] ± 2.05	53.85[b] ± 8.46	234.02[b] ± 8.70	33.16[e] ± 2.78

Values are presented as mean value ± standard deviation and expressed in dry matter. Means in columns with different superscript letters in common (a, b, c, d, e) differ significantly ($p \leq 0.05$).

3.2. Vitamin C

Studies revealed that fresh product contained 63.2 mg/100 g (689.54 mg per 100 g dry matter) vitamin C (Table 2). Traditional cooking and convection steaming led to significant ($p \leq 0.05$) losses of ascorbic acid of 55.7 and 47.4% respectively, compared to the fresh vegetable. In the frozen stored material, all changes were significant in comparison with the blanched cauliflower ($p \leq 0.05$). Losses of vitamin C in products stored for 2 months were 64.1% and 58.9%, respectively, for the conventionally packed and vacuum packed ones. In the material stored for 4 months, vitamin content decreased by 71.1% (traditional packaging) and by 66.0% (vacuum packaging). At every stage of the study, losses were significantly ($p \leq 0.05$) lower in the vegetables stored in vacuum pouches compared to those kept in conventional l zipper bags.

3.3. Total Polyphenols

It has been revealed that 100 g of purple cauliflower had 126.35 mg of total polyphenols, expressed as chlorogenic acid (1376.36 mg/100 g dry matter) (Table 2). In comparison with the fresh vegetable, the applied hydrothermal treatment caused statistically significant ($p \leq 0.05$) losses of these components: by 64.2% in traditional cooking and by 16.9% in convection steaming. Losses of total polyphenols after 2 and 4-months of frozen storage in a traditional way and in vacuum, were respectively 28.7 and 33.8% as well as 25.4 and 28.1% and were statistically significant ($p \leq 0.05$) compared to the blanched material. At every stage of this experiment, they were significantly ($p \leq 0.05$) lower for steam-cooked vegetables and those frozen stored in vacuum.

3.4. Anthocyanins

Purple cauliflower contained 7.18 mg anthocyanins per 100 g fresh material (78.21 mg/100 g dry matter). Traditional cooking in water significantly reduced their content (63.6%), compared to fresh vegetables. However, the remaining heat treatments (blanching and convection steaming) had no significant effect on anthocyanin content in this cauliflower. After 2 and 4 months of frozen storage, anthocyanin content compared to blanched vegetables decreased significantly ($p \leq 0.05$) by respectively 34.3 and 45.6% (traditional packaging) and 24.8 and 33% (vacuum packaging). At every stage of the research, these losses were significantly ($p \leq 0.05$) lower in the vacuum packed vegetables than in those stored in PE-LD bags.

3.5. Antioxidant Activity

The study revealed that antioxidant activity in purple cauliflower was 7.33 μmol Trolox per 1 g fresh of vegetable matter (79.85 μmol Trolox/g dry vegetable matter). The applied technological treatments (except for blanching) caused significant ($p \leq 0.05$) reductions in antioxidant activity compared to not processed vegetables, by 40.5% for traditional cooking and by 22.9% for convection steaming. Frozen storage of vegetables in PE-LD bags and vacuum pouches for 2 and 4 months resulted in a significant fall in antioxidant activity, compared to blanched vegetables, of 34.6 and 44.5%, and 26.5 and 34.3%, respectively. At all stages of the experiment the reduction of this parameter was significantly ($p \leq 0.05$) lower in steam-boiled vegetables and in those stored in vacuum.

3.6. Thiocyanates

The content of thiocyanates (SCN) in purple cauliflower was 2.41 mg/100 g fresh mass (26.25 mg/100 g dry mass). As in the case of the majority of the discussed components as well as antioxidant activity, blanching did not change significantly ($p \leq 0.05$) the content of thiocyanates compared to the fresh material. However, losses were recorded after traditional cooking (53.3%) and convection steaming (27.1%). In turn, 2- and 4-month frozen storage of cauliflower reduced significantly ($p \leq 0.05$) the content of these components compared to blanched material by 27.9 and 51.3% (traditional storage in PE-LD bags) and by 25.3 and 38.6% (vacuum storage), respectively.

3.7. Nitrates and Nitrites

This study showed that purple cauliflower had 55.56 mg of nitrates, expressed as potassium nitrate (KNO_3), per 1000 g fresh mass (605.23 mg/1000 g dry mass). Traditional cooking and convectional steaming led to a significant ($p \leq 0.05$) decrease in these compounds compared to fresh vegetables, by 63.8 and 22.2% respectively. In addition, 2- and 4-month frozen storage in two package types—conventional PE-LD and vacuum packaging—caused a significant reduction in nitrates: by 63.5 and 74.5%, and 49.8 and 59.7%, respectively, compared to blanched vegetables. At every stage of this work, the reduction of this parameter was significantly ($p \leq 0.05$) lower in steam-cooked vegetables and in those frozen stored in vacuum.

The content of nitrites in 1000 g of fresh material was 1.65 mg (17.97 mg/1000 g dry matter). The applied hydrothermal treatment reduced significantly ($p \leq 0.05$) nitrites compared to the fresh material, by 37.3% (traditional cooking) and 21.5% (convection steaming). Two months' frozen storage of blanched cauliflower led to decreases in the content of nitrites in the examined material; however, only the difference recorded for conventional packaging (21.2%) was statistically significant ($p \leq 0.05$). In the case of 4-month stored cauliflower, significant ($p \leq 0.05$) increases in the content of nitrites were recorded in the products packed conventionally (by 33%) and in vacuum (by 91.8%).

4. Discussion

4.1. Basic Composition and Dietary Fiber

This study revealed that protein content (g per fresh and dry mass) in purple cauliflower was close to that found by Rumpel [25] and Kunachowicz et al. [26] and Kahlon et al. [27], according to whom, its values were 2.5 g and 2.4 g/100 g fresh weight of white cauliflower and 27.8 g/100 g dry matter, respectively. Fat content in white cauliflower, determined by the latter author, was 1.7 g/100 g dry matter, which is a higher value compared to that obtained in this work. Rumpel [25] reports that the content of assimilable carbohydrates in white cauliflower is 2.6 g/100 g of fresh vegetable, while Schonhof et al. [28] gives two values—2.27 and 2.56—depending on the cultivar, which differs from our findings. On the other hand, the content of dietary fiber found by Kahlon et al. [27] was similar (62.1 g/100 g dry mass) to our results. Both Rumpel [25] and Kunachowicz et al. [26] determined dietary fiber in white rose cauliflowers at the levels of 2.9 and 2.4 g/100 g fresh weight respectively, which is close to our findings. According to Puupponen-Pimiä et al. [29], the content of this constituent was 30.2 g/100 g dry matter unit. The values reported by Kahlon et al. [27] for ash content (10.1 g/100 g dry matter), considerably exceeding that obtained in this work (5.73 g/100 g dry matter). However, Ali [30] noted a similar content of micro- and macroelements, but in white cauliflower florets.

4.2. Dry Matter

According to the literature [31,32], dry matter content in the fresh white rose of Romanesco cauliflowers ranges from 6.96 to 13.95 g/100 g, which agrees with our results. Cooking in water as well as convection steaming increased its content compared to the fresh material, by 17.1% and 12.7% respectively. Similar increases, of 18.0% (white rose cauliflower) and 12.5% (green rose cauliflower), were also reported by Gębczyński and Kmiecik [33]. The reduced water content in the examined material was probably caused by the release of water from the tissues damaged during high-temperature processes [34].

This work showed that after 2-month frozen storage at −22 °C, the content of dry matter increased, which is congruent with the findings of Gębczyński and Kmiecik [33] and Cebula et al. [35], who noted mean increases of 22.3 and of 10.5% (white rose cultivar) and 14.5% (*Romanesco* cv.), respectively. An increase in dry matter in the frozen and then stored material may probably be explained by the phenomenon of denaturing cell walls at low temperature and resulting release of water [36].

4.3. Vitamin C

In 100 g of fresh cauliflower there was 63.20 mg of vitamin C, which is consistent with the findings of Kaulman et al. [37] and Volden et al. [38], who noted values of 63 mg and 64.2 mg (*Graffiti* cv.), respectively. The obtained results were in agreement also with those of Bhandari and Kwak [39] and Picchi et al. [40], who reported that vitamin C in cauliflower ranged from 396.7–649.7 and 346–638 mg of 100 g dry weight, respectively.

This study showed minimal losses of this component due to blanching, which was carried out using a convection steam oven. Cooking in water reduced vitamin C content by 55.7%, while convection steaming by 47.4%. Lower losses of vitamin C in this process were observed by Filipiak-Florkiewicz [31], Davey et al. [41], and Pellegrini et al. [42], and were 38.2% (white cauliflower) and 36.9% (*Romanesco* variety), 14.1–29.5%, and 50.9%, respectively.

Technological treatments for example, pre-processing or blanching may lead to considerable losses particularly in vitamin C. Their extent depends on the applied temperature, length of the exposure to this temperature, and a degree of the product comminuting [43]. Conventional frozen storage of cauliflower did result in a vitamin C decrease of 64.1 (after two months) and 71.1% (after 4 months). Volden et al. [38] noted losses in this constituent content after 3-month frozen storage, of 22.7% for the cultivar Graffiti and 24.1% for the green cauliflower cultivar *Celio*.

Incedayi and Suna [44] treated one group of cauliflower florets with 1,5% citric acid solution and the another group with 0.5% Ca-ascorbate plus citric acid 1% solution. Afterwards, florets were packed in two MAP gas mixtures and into biaxially oriented polypropylene film (BOPP). After 15 days of storage at 4 °C, vitamin C content decreased by 11.5%. Vacuum packaging led to a significantly lower ($p \leq 0.05$) reduction in the vitamin C content of 62.4%, on average, compared to conventional packing in zipper bags.

The losses in vegetable constituents are caused by both internal factors like, for example, respiration, production and the effect of ethylene, or changes in chemical composition, and external (environmental) ones such as relative humidity of the air, temperature, and gas composition of the surrounding atmosphere. Vacuum packaging reduces the access of oxygen to the material, which protects it against losses, for example, oxidation of vitamin C [45]. Quality of plant raw material is affected by factors connected with cultivars and by the broadly understood agricultural engineering factors [46].

4.4. Total Polyphenols

In most of the literature polyphenols are determined in various cauliflower varieties with white florets and calculated as total polyphenol content per 100 g fresh material. Volden et al. [38] and Pellegrini et al. [42] reported per 100 g of fresh vegetable 59.9 and 146 mg of total polyphenols, respectively. The results stated by Kaulman et al. [37], Bahorun et al. [47], and Kaur and Kapoor [48] are much lower: 63.8 mg, 27.8 mg, and 96 mg/100 g, respectively. Significant differences are most probably the result of the variety of the analyzed cultivars, growing conditions and the results' calculation methods.

Compared to fresh cauliflower, losses in total polyphenol content due to cooking in water were substantially larger than from blanching; slightly lower values were noted by Pellegrini et al. [42] and Filipiak-Florinkiewicz [31] for white varieties: 45.4 and 43.6%, respectively. Our results showed that losses resulting from traditional cooking were substantially larger than those found by Mazzeo et al. [49] and Picchi et al. [40] for varieties with white florets, being 30.8% and 18.9% respectively. As for cooking in a convection steam oven, the obtained results are almost identical with that reported by Pellegrini et al. [42] for white cauliflower (17.7%). Frozen storage of purple cauliflower led to a decline in the level of these compounds. Losses in the material stored for 4 months, noted by Gębczyński and Kmiecik [33], were much lower and were 11.6% (white cauliflower) and 20.4% (green cauliflower). This is consistent with the findings of Puupponen-Pimiä et al. [29], who observed a reduction in the content of these compounds by only 14.3% after 6 months of frozen storage.

Chassagne-Berces et al. [50] also noted that long-lasting storage of raw materials enhances the processes of enzymatic or chemical oxidation of polyphenolic compounds, the extent of which depends on the raw material or medium parameters such as temperature, pH, water activity, time, and oxygen content.

4.5. Anthocyanins

As for anthocyanin content, the results reported by Lo Scalzo et al. [51] and Li et al. [52] were lower (4.21 mg/100 g fresh weight) and higher (201.11. mg/100 g dry matter), respectively, compared to our findings. The same authors [51] noted that blanching reduced the content of these components by 64.6%, which markedly exceeds our results. Anthocyanins are known to be thermolabile compounds. However, due to the high temperature and short time of blanching, losses may be insignificant that explains such results [53]. Lo Scalzo et al. [51] and Palermo et al. [10] found larger losses of these compounds of 80.2 and 80%, respectively, after traditional cooking in water. According to these authors, there were no significant changes in the convection-steamed material, which is in line with our results. As opposed to Volden et al. [38], who did not observe significant changes in the material frozen stored for 3, 6, and 12 months, frozen storage of the blanched material caused significant losses in the content of these constituents.

4.6. Antioxidant Activity

Volden et al. [38] found that the cultivar Graffiti is characterized by strong antioxidant activity (25.7 μmol Trolox equivalent per 1 g of fresh vegetable). Antioxidant activity determined by Beecher et al. [53], Filipiak-Florkiewicz [31], Murcia et al. [54], and Mazzeo et al. [49] in other cauliflower varieties was similar or higher: 6.5, 6.15, 9.9, and 17.3 μmol Trolox equivalent per 1 g of fresh vegetable, respectively. Such differences may result from various processing conditions including, for example, methods applied for the measurement of antioxidant activity, different extractants, or various extraction times. Culinary treatments may contribute to the reduction of antioxidant activity due to the penetration of antioxidants into the solution or their degradation under elevated temperature [55]. In this case, however, there were no changes in antioxidant activity due to blanching. Losses observed by Volden et al. [38] and Filipiak-Florkiewicz [31] were larger, by 28.0% in the cultivars Graffiti and Celio, as well as by 10.4% and 25.4% in white and green cultivars, respectively. Cooking lowered antioxidant activity of purple cauliflower. Mazzeo et al. [49] and Filipiak-Florkiewicz [31] found reductions of 35.3 and 22.5%, respectively, compared to the fresh material (white cultivar) and of 34.7 (white cultivar) and 37.3% (green cultivar).

Similar decreases in antioxidant activity resulting from 3-month frozen storage, of 30.4% and 26.6%, were observed by Volden et al. [38] for the cultivars *Graffiti* and *Celio*. According to Murcia et al. [54], 24-h freezing caused a fall in antioxidant activity of 0.6%; whereas, Drużyńska et al. [56] noted an increase in this parameter after 48 h.

4.7. Thiocynates

Glucosinolates alone are not biologically active; only their enzymatic derivatives are. When the plant tissue is damaged, hydrolysis of glucosinolates takes place by the endogenous enzyme 'myrosinase' (thioglucoside glucohydrolase EC 3:2:3:1) with a release of a range of breakdown products, including isothiocyanates, indoles, nitriles, oxazolidines, and thiocyanates [7]. This study showed that purple cauliflower contained 2.41 mg/100 of thiocyanates (SCN) per g fresh weight. To the best of our knowledge, there are few studies concerning not only the amount of glucosinolates and their profile, but also their breakdown products. According to Kapusta-Duch et al. [7], cooking decreased significantly the concentration of isothiocynates in green and purple cauliflowers, by 11.0 and 42.4%, respectively, in comparison with raw vegetables. Similar results were found in this work.

4.8. Nitrates and Nitrites

Nitrites and indirectly nitrates, when consumed with food in too large quantities, can be harmful to health [57]. Among vegetables, Brassicas are characterized by medium or low degree of accumulation of these compounds. However, due to the consumed mass they may constitute significant proportion in the daily food ration [58]. In comparison with our results, most authors report several times higher or similar contents of nitrates in various cauliflower cultivars. For example, the values reported were 143–354 mg [59], 210.6 mg [60], 171.2 mg [61], or 61 mg/kg fresh mass [62]. As far as the content of nitrites is concerned, the results obtained by researchers, depending on the examined cultivar, were lower or markedly higher compared to our findings. For example, Leszczyńska et al. [62] found 3.49 mg and 47 mg nitrites in white and green cauliflower respectively, while Gajewska et al. [60] stated 0.8 mg of these constituents per kg of fresh vegetable. According to Leszczyńska et al. [62], the reductions in these compounds due to blanching were much greater than that in this work, amounting to 72.2% for white rose cauliflower and 37.9% for the cultivar Romanesco. Cooking was also responsible for the reduction of nitrates, which agrees with the findings of Shimada and Ko [63], who determined losses of these compounds within the range of 14 to 79%. Filipiak-Florkiewicz et al. [61] observed higher losses of nitrates, compared to those obtained in this work, of up to 72.2% in white rose cauliflower. In comparison with blanching, frozen storage lowered the content of nitrates, on average by 61.9%. Leszczyńska et al. [62] noted a decline in these constituents by 78.7% in white rose cauliflower.

Compared to fresh vegetables, no significant change in the content of nitrites was found due to blanching. However, losses reported by Leszczyńska et al. [62] due to this process were significant and amounted to 8.9% on average. Cooking purple cauliflower led to an average 29.4% decrease in nitrite content compared to fresh vegetables. Similar decreases, of 22.9 (white rose cauliflower) and 17% (*Romanesco* cv.) obtained Filipiak-Florkiewicz et al. [61] and Leszczyńska et al. [62].

Storage of the material contributes to an increase of nitrite content in cauliflower as a result of the reduction of nitrates which takes place in tissues due to enzyme or bacteria activity [64]. After 2-months frozen storage, the content of nitrites decreased, on average by 18.3%, while after a 4 months increased significantly, by 62.4% on average, compared to blanched vegetables. A similar upward trend (151%) for the level of nitrites noted Leszczyńska et al. [62] for green rose cauliflower, frozen and stored for 4 months. Filipiak-Florkiewicz et al. [61], who cooked a vegetable previously frozen stored for the identical period of time, also observed a 105.7% increase in the amount of these compounds.

The results obtained indicate that further studies are needed to explain changes in nutritive and non-nutritive components during selected treatments like cooking or freezing not only purple cauliflower, but other colored Brassica vegetables. The results presented by other authors and those obtained in this work are not clear-cut, so the research problem seems to be interesting. Therefore, the research goal should be to optimize hydrothermal processes and to select the best type of packaging in order to make the best possible use of pro-health substances occurring in Brassica vegetables in human nutrition in accordance with the idea of sustainable development.

5. Conclusions

Brassica are among the top 10 economic vegetables in the world and a source of valuable nutrients, but their content is strongly affected by the method of their preparation. This study clearly showed that technological treatments had a significant effect on nutrient and health-promoting compounds in purple cauliflower. The nutritional content in Brassica vegetables have been reported to vary during the growth period due to climatic and agronomical factors, but also the postharvest treatments are valid factors influenced on quality of these vegetables.

All technological processes, i.e., blanching, cooking, and frozen storage, led to significant changes in the content of the examined nutritive and non-nutritive compounds as well as antioxidant activity or the level of selected chemical pollutions. A trend was also observed towards lower losses of constituents due to convection steaming, compared to traditional cooking in water. Moreover, the reduction of examined compounds was smaller in vacuum-packed and frozen-stored vegetables then in those stored in zipper PE-LD bags.

Author Contributions: Conceptualization, J.K.D., T.L. and M.K.; Methodology, B.B.; Software, A.S.S. and Z.S.G.; Validation, A.S.S., J.S. and M.N.; Formal Analysis, J.S.; Data Curation, B.B.; Writing Original Draft Preparation, J.K.D., BB.; Writing Review & Editing, J.K.D., A.S.S., J.S. and M.N.; Visualization, Z.S.G.; Supervision, T.L. and M.K.

Funding: This research received no external funding.

Conflicts of Interest: The authors declare no conflict of interest.

References

1. Šamec, D.; Urlić, B.; Salopek-Sondi, B. Kale (*Brassica oleracea* var. *acephala*) as a superfood: Review of the scientific evidence behind the statement. *Crit. Rev. Food Sci. Nutr.* **2018**, 1–12. [CrossRef]
2. Dos Reis, L.C.; De Oliveira, V.R.; Hagen, M.E.; Jabloński, A.; Flôres, S.H.; De Oliveira, R.A. Effect of cooking on the concentration of bioactive compounds in broccoli (*Brassica oleracea* var. Avenger) and cauliflower (*Brassica oleracea* var. Alphina F_1) grown in an organic system. *Food Chem.* **2015**, *172*, 770–777. [CrossRef] [PubMed]
3. Shahidi, F.; Ambigaipalan, P. Phenolics and polyphenolics in foods, beverages and spices: Antioxidant activity and health effects—A review. *J. Funct. Foods.* **2015**, *18*, 820–897. [CrossRef]
4. Ahmed, F.A.; Ali, R.F. Bioactive compounds and antioxidant activity of fresh and processed white cauliflower. *Biomed. Res. Int.* **2013**, *2013*, 367819. [CrossRef] [PubMed]
5. Florkiewicz, A.; Filipiak-Florkiewicz, A.; Topolska, K.; Cieślik, E.; Kostogrys, R.B. The effect of technological processing on the chemical composition of cauliflower. *Ital. J. Food Sci.* **2014**, *26*, 275–281. Available online: https://www.researchgate.net/publication/286116360_The_effect_of_technological_processing_on_the_chemical_composition_of_cauliflower (accessed on 9 August 2018).
6. Kalisz, A.; Sękara, A.; Smoleń, S.; Grabowska, A.; Gil, J.; Komorowska, M.; Kunicki, E. Survey of 17 elements, including rare earth elements, in chilled and non-chilled cauliflower cultivars. *Sci. Rep.* **2019**, *9*, 5416. [CrossRef]
7. Kapusta-Duch, J.; Kusznierewicz, B.; Leszczyńska, T.; Borczak, B. Effect of cooking on the contents of glucosinolates and their degradation products in selected *Brassica* vegetables. *J. Funct. Foods* **2016**, *23*, 412–422. [CrossRef]
8. Manchali, A.; Murthy, K.N.Ch.; Patil, B.S. Crucial facts about health benefits of popular cruciferous vegetables. *J. Funct. Foods.* **2012**, *4*, 94–106. [CrossRef]
9. Ribeiro-Santos, R.; Andrade, M.; Sanches-Silva, A.; De Melo, N.R. Essential Oils for Food Application: Natural Substances with Established Biological Activities. *Food Bioproc. Tech.* **2017**, 1–29. [CrossRef]
10. Palermo, M.; Pellegrini, N.; Fogliano, V. The effect of cooking on the phytochemical content of vegetables. *J. Sci. Food Agric.* **2014**, *94*, 1057–1070. [CrossRef]
11. Ghaani, M.; Cozzolino, C.A.; Castelli, G.; Farris, S. An overview of the intelligent packaging technologies in the food sector. *Trends Food Sci. Technol.* **2016**, *51*, 1–11. [CrossRef]
12. Hussein, Z.; Caleb, O.J.; Opara, U.L. Perforation-mediated modified atmosphere packaging of fresh and minimally processed produce—A review. *Food Pack. Shelf Life.* **2015**, *6*, 7–20. [CrossRef]
13. Chowdhury, S.; Nath, S.; Biswas, S.; Dora, K.C. Effect of vacuum packaging on extension of shelf life of Asian Sea-Bass fillet at 5 ± 1 °C. *J. Exp. Zool. India* **2017**, *20*, 1241–1245.
14. Swain, T.; Hillis, W.E. The phenolic constituents of *Prunus domesticus* (L.). the quantity of analysis of phenolic constituents. *J. Sci. Food Agric.* **1959**, *10*, 63–68. [CrossRef]
15. Re, R.; Pellegrini, N.; Proteggente, A.; Pannala, A.; Yang, M.; Rice-Evans, C. Antioxidant activity applying an improved ABTS radical cation decolorization assay. *Free Radic. Biol. Medic.* **1999**, *26*, 1231–1237. [CrossRef]
16. Pellegrini, N.; Del Rio, D.; Colombi, B.; Bianchi, M.; Brighenti, F. Application of the 2′2-azobis (3-ethylbenzothiazoline-6-sulfonic acid) radical cation assay to flow injection system for the evaluation of antioxidant activity of some pure compounds and bevereges. *J. Agric. Food Chem.* **2003**, *51*, 164–260. [CrossRef] [PubMed]
17. Benvenuti, S.; Pellati, F.; Melegani, M.; Beertelli, D. Polyphenols, Anthocyanins, Ascorbic Acid and Radical Scavenging Activity of Rubus, Ribes and Aronia. *J. Food Sci.* **2004**, *69*, 164–169. [CrossRef]

18. Polish Standard. *PN-90/A-75101/03. Fruit and Vegetable Products. Preparation of Samples for Physico-Chemical Studies. Determination of Dry Matter Content by Gravimetric Method*; Polish Committee for Standardization: Warsaw, Poland, 1990. (In Polish)

19. Polish Standard. *PN-A-04019:1998. Food products—Determination of Vitamin C*; Polish Committee for Standardization: Warsaw, Poland, 1998. (In Polish)

20. Polish Standard. *PN-92/A-75112. Fruit and Vegetable Products. Determination of Nitrates and Nitrites*; Polish Committee for Standardization: Warsaw, Poland, 1992. (In Polish)

21. Polish Standard. *PN-A-79011-8:1998. Dry Food Mixes. Test Methods. Determination of Total Ash And Ash Insoluble In 10 Percent (m/m) Hydrochloric Acid*; Polish Committee for Standardization: Warsaw, Poland, 1998. (In Polish)

22. Polish Standard. *PN-EN ISO 8968-1:2004. Milk—Determination of Nitrogen Content—Part 1: Determination of Nitrogen by the Kjeldahl Method*; Polish Committee for Standardization: Warsaw, Poland, 2004. (In Polish)

23. Polish Standard. *PN-A-79011-4:1998. Dry Food Mixes. Test Methods. Determination of Fat Content*; Polish Committee for Standardization: Warsaw, Poland, 1998. (In Polish)

24. Polish Standard. *PN-A-79011-15:1998P. Dry Food Mixes. Test Methods. Determination of Dietary Fiber Contents*; Polish Committee for Standardization: Warsaw, Poland, 1998. (In Polish)

25. Rumpel, J. *Cultivation of Cauliflower*, 1st ed.; Publisher Hortpress: Warszawa, Poland, 2002; pp. 6–13. (In Polish)

26. Kunachowicz, H.; Nadolna, I.; Iwanow, K.; Przygoda, B. *The Nutritional Value of Selected Foods and Typical Dishes*, 6th ed.; PZWL: Warszawa, Poland, 2014. (In Polish)

27. Kahlon, T.; Chin, M.; Chapman, M. Steam cooking significantly improve in vitro bile acid binding of beets, eggplant, asparagus, carrots, green beans and cauliflower. *Nutr. Res.* **2007**, *27*, 750–755. [CrossRef]

28. Schonhof, I.; Krumbein, A.; Bruckner, B. Genotypic effects on glucosinolates and sensory properties of broccoli and cauliflower. *Nahr. Food* **2004**, *48*, 25–33. [CrossRef]

29. Puupponen-Pimiä, R.; Häkkinen, S.T.; Aarni, M.; Suortti, T.; Lampi, A.-M.; Eurola, M.; Piironen, V.; Nuutila, A.M.; Oksman-Caldentey, K.-M. Blanching and long-term freezing affect various bioactive compounds of vegetables in different ways. *J. Sci. Food Agric.* **2003**, *83*, 1389–1402. [CrossRef]

30. Ali, A.M. Effect of food processing methods on the bioactive compound of cauliflower. *Egypt. J. Agric. Res.* **2015**, *93*, 117–131. Available online: http://www.arc.sci.eg/ejar/UploadFiles/Publications/1043314%D8%A7%D9%84%D8%A8%D8%AD%D8%AB%20%D8%A7%D9%84%D8%AB%D8%A7%D9%86%D9%89%20%D8%AA%D9%83%D9%86%D9%88%D9%84%D9%88%D8%AC%D9%8A%D8%A7.pdf (accessed on 25 August 2018).

31. Filipiak-Florkiewicz, A. *Effect of Hydrothermal Treatment on Selected Health-Promoting Properties of Cauliflower (Brassica oleracea var. Botrytis L.)*; Scientific Papers; University of Agriculture in Krakow: Kraków, Poland, 2011; p. 347. (In Polish)

32. Florkiewicz, A. Sous-vide method as alternative to traditional cooking of cruciferous vegetables in the context of reducing losses of nutrients and dietary fibre. *Ż.N.T.J.* **2018**, *25*, 45–57. [CrossRef]

33. Gębczyński, P.; Kmiecik, W. Effects of traditional and modified technology, in the production of frozen cauliflower, on the contents of selected antioxidative compounds. *Food Chem.* **2007**, *1*, 229–235. [CrossRef]

34. Ramesh, M.N. The performance evaluation of a continuous vegetable cooker. *Int. J. Food Sci. Technol.* **2000**, *2*, 377–384. [CrossRef]

35. Cebula, S.; Kunicki, E.; Kalisz, A. Quality changes in curds of white, green an Romanesco cauliflower during storage. *Pol. J. Food Nutr. Sci.* **2006**, *56*, 155–160.

36. Evans, J. Emerging Refrigeration and Freezing Technologies for Food Preservation. In *Innovation and Future Trends in Food Manufacturing and Supply Chain Technologies*; Woodhead Publishing: Sawston, UK; Cambridge, UK, 2016; pp. 175–201. [CrossRef]

37. Kaulman, A.; Jonville, M.; Schneider, Y.; Hoffmann, L. Carotenoids, polyphenols and micronutrient profiles of *Brassica oleraceae* and plum varieties and their contribution to measures of total antioxidant capacity. *Food Chem.* **2014**, *155*, 240–250. [CrossRef] [PubMed]

38. Volden, J.; Bengtsson, B.G.; Wicklund, T. Glucosinolates, L-ascorbic acid, total phenols, anthocyanins, antioxidant capacities and colour in cauliflower (*Brassica oleracea* L. ssp. *Botrytis*); effect of long-term freezer storage. *Food Chem.* **2009**, *112*, 967–976. [CrossRef]

39. Bhandari, S.R.; Kwak, J.H. Chemical composition and antioxidant activity in different tissues of Brassica vegetables. *Molecules* **2015**, *20*, 1228–1243. [CrossRef]

40. Picchi, V.; Migliori, C.; Scalzo, R.L.; Campanelli, G.; Ferrari, V.; Di Cesare, L.F. Phytochemical content in organic and conventionally grown Italian cauliflower. *Food Chem.* **2012**, *130*, 501–509. [CrossRef]

41. Davey, M.; Montagu, M.; Inze, D.; Sanmartin, M.; Kanellis, A.; Smirnoff, M.; Benzie, J.; Strain, J.; Favell, D.; Fletcher, J. Review: Plant L-ascorbic acid: Chemistry, function, metabolism, bioavailability and effects of processing. *J. Sci. Food Agric.* **2000**, *80*, 825–860. [CrossRef]

42. Pellegrini, N.; Chiavaro, E.; Gardana, C.; Mazzeo, T.; Contino, D.; Gallo, M.; Riso, P.; Fogliano, V.; Porrini, M. Effect of different cooking methods on colour, phytochemical concentration and antioxidant capacity of raw and frozen *Brassica* vegetables. *J. Agric. Food Chem.* **2010**, *58*, 4310–4321. [CrossRef] [PubMed]

43. Abushita, A.A.; Daood, H.G.; Biacs, P.A. Change in carotenoids and antioxidant vitamins in tomato as a function of varietal and technologicals factors. *J. Agric. Food Chem.* **2000**, *48*, 2075–2081. [CrossRef] [PubMed]

44. İncedayi, B.; Suna, S. Effects of modified atmosphere packaging on the quality of minimally processed cauliflower. *Acta Aliment.* **2012**, *41*, 401–413. [CrossRef]

45. Van Ooijen, I.; Fransen, M.L.; Verlegh, P.W.J.; Smit, E.G. Atypical food packaging affects the persuasive impact of product claims. *Food Qual. Prefer.* **2016**, *48*, 33–40. [CrossRef]

46. Szeląg-Sikora, A.; Niemiec, M.; Sikora, J.; Chowaniak, M. Possibilities of designating swards of grasses and small-seed legumes from selected organic farms in Poland for feed. In Proceedings of the IX International Scientific Symposium "Farm Machinery and Processes Management in Sustainable Agriculture, Lublin, Poland, 22–24 November 2017; pp. 365–370. [CrossRef]

47. Bahorun, T.; Luximon-Ramma, A.; Crozier, A.; Aruoma, O. Total phenol, flavonoid, proanthocyanidin and vitamin C levels and antioxidant activities of Mauritian vegetables. *J. Sci. Food Agric.* **2004**, *4*, 1553–1561. [CrossRef]

48. Kaur, C.; Kapoor, H.C. Anti-oxidant activity and total phenolic content of some Asian vegetables. *Int. J. Food Sci. Tech.* **2002**, *37*, 153–161. [CrossRef]

49. Mazzeo, T.; N'Dri, D.; Chiavaro, E.; Visconti, A.; Fogliano, V.; Pellegrini, N. Effect of two cooking procedures on phytochemical compounds, total antioxidant, capacity and colour of selected frozen vegetables. *Food Chem.* **2011**, *28*, 617–633. [CrossRef]

50. Chassagne-Berces, S.; Fonseca, F.; Citeau, M.; Marin, M. Freezing protocol effect on quality properties of fruit tissue according to the fruit, the variety and the stage of maturity. *LWT Food Sci. Technol.* **2010**, *43*, 1441–1449. [CrossRef]

51. Lo Scalzo, R.; Genna, A.; Branca, F.; Chedin, M.; Chassaigne, H. Anthocyanin composition of cauliflower (*Brassica oleracea* L. var. *botrytis*) and cabbage (*B. oleracea* L. var. *capitata*) and its stability in relation to thermal treatments. *Food Chem.* **2008**, *107*, 136–144. [CrossRef]

52. Li, H.; Deng, Z.; Zhu, H.; Hu, C.; Liu, R.; Young, J.; Tsao, R. Highly pigmented vegetables: Anthocyanin composition and their role in antioxidant activities. *Food Res. Int.* **2012**, *46*, 250–259. [CrossRef]

53. Beecher, G.R.; Gebhardt, S.E.; Haytowitz, D.B.; Holden, J.M.; Wu, X.; Prior, R.L. Lipophilic and hydrophilic antioxidant capacities of common foods in the United States. *J. Agric. Food Chem.* **2004**, *52*, 4026–4037. [CrossRef]

54. Murcia, M.A.; Lopez-Ayerra, B.; Garcia-Carmona, F. Effect of processing and different blanching times on broccoli: Proximate composition and fatty acids. *LWT Food Sci. Technol.* **2009**, *32*, 238–243. [CrossRef]

55. Cartea, M.E.; Francisco, M.; Soengas, P.; Velasco, P. Phenolic Compounds in *Brassica* Vegetables. *Molecules* **2011**, *16*, 251–280. [CrossRef] [PubMed]

56. Drużyńska, B.; Stępień, K.; Piecyk, M. The influence of cooking and freezing on contents of bioactive components and their antioxidant activity in broccoli. *Bromat. Chemia Toksykol.* **2009**, *42*, 169–176. Available online: http://ptfarm.pl/pub/File/bromatologia_2009/bromatologia_2_2009/Bromat%202,2009%20s.%20169-176.pdf (accessed on 24 July 2010). (In Polish).

57. Sindelar, J.J.; Milkowski, A.L. Human safety controversies surrounding nitrate and nitrite in the diet. *Nitric Oxide* **2012**, *26*, 259–266. [CrossRef] [PubMed]

58. Bahadoran, Z.; Mirmiran, P.; Jeddi, S.; Azizi, F.; Ghasemi, A.; Hadaegh, F. Nitrate and nitrite content of vegetables, fruits, grains, legumes, dairy products, meats and processed meats. *J. Food Comp. Anal.* **2016**, *51*, 93–105. [CrossRef]

59. Samantaria, P. Nitrate in vegetables: Toxity, content, intake and EC regulation. *J. Sci. Food and Agric.* **2006**, *86*, 10–17. [CrossRef]

60. Gajewska, M.; Czajkowska, A.; Bartodziejska, B. The content of nitrates and nitrites in selected vegetables on detail sale in Lodz region. *Ochr Śr. Zasobów Nat.* **2009**, *40*, 388–395. (In Polish)

61. Filipiak-Florkiewicz, A.; Cieślik, E.; Florkiewicz, A. Effect of technological processing on contents of nitrates and nitrites in cauliflower. *Żyw. Człow. Metab.* **2007**, *34*, 1197–1201. Available online: http://agro.icm.edu.pl/agro/element/bwmeta1.element.agro-article-e3ebf1f1-dd46-4dad-8ed5-c9d0526fceac?q=bwmeta1.element.agro-number-fcf445e1-46f5-442d-a933-e0d38017adce;75&qt=CHILDREN-STATELESS (accessed on 18 April 2008). (In Polish).

62. Leszczyńska, T.; Filipiak-Florkiewicz, A.; Cieślik, E.; Sikora, E.; Pisulewski, P. Effects of some processing methods on nitrate and nitrite changes in cruciferous vegetables. *J. Food Comp. Anal.* **2009**, *22*, 315–321. [CrossRef]

63. Shimada, Y.; Ko, S. Nitrate in vegetables. *Chugoku Gakuen J.* **2004**, *3*, 7–10. Available online: https://pdfs.semanticscholar.org/5cbe/9ce8f74a9c1c334008cd295ccd5a1553b9da.pdf (accessed on 15 May 2005).

64. Ranasinghe, R.A.S.N.; Marapana, R.A.U.J. Nitrate and nitrite content of vegetables: A review. *J. Pharmacogn. Phytochem.* **2018**, *7*, 322–328. Available online: https://www.researchgate.net/publication/326533977_Nitrate_and_nitrite_content_of_vegetables_A_review_RASN_Ranasinghe_and_RAUJ_Marapana (accessed on 13 January 2019).

Article

Modeling and Prediction of the Uniformity of Spray Liquid Coverage from Flat Fan Spray Nozzles

Paweł A. Kluza [1], Izabela Kuna-Broniowska [1,*] and Stanisław Parafiniuk [2]

[1] Department of Applied Mathematics and Computer Science, University of Life Sciences in Lublin, 20-612 Lublin, Poland; pawel.kluza@up.lublin.pl

[2] Department of Machinery Exploitation and Management of Production Processes, University of Life Sciences in Lublin, 20-612 Lublin, Poland; stanislaw.parafiniuk@up.lublin.pl

* Correspondence: izabela.kuna@up.lublin.pl

Received: 31 October 2019; Accepted: 23 November 2019; Published: 27 November 2019

Abstract: The effectiveness and quality of agricultural spraying largely depends on the technical efficiency of the nozzles installed in agricultural sprayers. The uniform spraying of plants results in a decrease in the amount of pesticides used in agricultural production and affects environmental safety. Both newly developed sprayers and those currently in use need quality control as well as an assessment of the performance of the spraying process, especially its uniformity. However, the models applied presently do not ensure accurate estimates or predictions of the spray liquid coverage uniformity of the treated surface. Generally, the distribution of the atomized liquid quantity is symmetrical and leptokurtic, which means that it does not fit well to the commonly used standard distribution. Therefore, there is a need to develop and design new tools for the evaluation, modeling, and prediction of such a process. The research problem studied in the present work was to find a new model for the distribution of atomized liquid quantity that could provide capabilities better than have been available so far to assess and predict the spraying process results. The research problem was solved through the formulation of a new function for the probability density distribution of sprayed liquid accumulation on the surface of the preset dimension size. The development of the new model was based on the results from a series of water atomization tests with an appropriate measurement device design based on the widely applied flat fan nozzles (AZ-MM type).

Keywords: flat fan nozzle; liquid coverage; coefficient of variation (CV); crop yields

---

1. Introduction

Spraying liquids is a technically and technologically important process in all areas of economy and everyday life. Particular importance is ascribed to the correctness of the process and the application thereof in agricultural production and the food industry, taking into account the effect of the sprayed liquids on the environment. A fundamental issue in these processes is their quality and efficiency. They determine the requirements for the structure and the use of technical equipment and the parameters of spraying liquids in practical terms. Hence, the uniformity and predictability of the distribution of the sprayed liquid are the basic characteristics.

The quality of agricultural spraying largely depends on the uniformity of the distribution of the sprayed liquid over the spraying surface. In field crops, currently slot nozzles are the most commonly used, which wear out during exploitation. A worn nozzle causes the amount of liquid flowing out to be greater and, thus, the uniformity of liquid distribution is disturbed. The current Directive 2009/128EC of the European Parliament and of the Control concerning the sustainable use of pesticides in agriculture contains guidelines regarding, among others, the testing of spray nozzles installed in agricultural sprayers [1]. In field sprayers, the even parameter is the even distribution of sprayed liquid on the sprayed surface. Specialized devices with a grooved table are used to measure the uniformity of spray distribution [2,3].

The stream of the sprayed liquid has the shape of a cone with characteristics depending on the design of the nozzle and the geometry of its working slot, which is subject to wear and can therefore change its shape due to mechanical damage or limescale deposition. Another consequence of nozzle wear is also the change in flow rate.

Field spraying is an example of the use of the liquid spraying process. One of the key issues in this process is the uniformity of the liquid amount distribution achieved from the spray boom.

The coefficient of variation is a measure of the process uniformity expressed (usually in percentages) as the ratio of the standard deviation to the arithmetic mean of the sample. This parameter is used most commonly for the comparison of the variation of a trait in two different distributions.

In the case of field spraying, its value should not exceed 10%, as specified by the EN ISO 16119-2 part 2 standard [4].

When the level defined in the standard is not exceeded, the spray uniformity is regarded to be correct. Its level is mainly determined by the shape of the distribution of the amount of liquid sprayed from a single nozzle, based on which spray boom is constructed. Distribution data from a single nozzle is replicated to simulate a complete virtual boom.

Manufacturers strive to improve the sprayer structure to ensure the best uniformity of surface coverage.

The field sprayer boom that will be equipped in such nozzles should provide the greatest uniformity of the treated surface spray, i.e., it should be characterized by the minimum value of the coefficient of variation for the sprayed liquid. To date, there has been no model ensuring a relatively low value of the coefficient of variation of liquid coverage from a virtual field boom containing all the same new nozzles and, hence, a very high uniformity of spraying. Similarly, there is no model that would satisfactorily determine the amounts of accumulated liquid allowing the achievement of the permissible value of the coefficient of variation defined by the standard.

Therefore, there is a need to propose a new model that will substantially increase the level of liquid spray uniformity from the spray boom, which will result in the improvement of the spray quality.

An analysis of liquid sprayed from a single nozzle and a sprayer in terms of the distribution characteristics has been carried out [5].

To ensure the most optimal spray, in [6–8], the impact of the sprayer and nozzle parameters on the spray distribution and on the coefficient of variation, which is a measure of the uniformity of treated surface coverage, was analyzed. Various types of nozzles have been tested in laboratory conditions to analyze this parameter [9].

The simulation of a virtual field boom illustrating the work of the sprayer with the analysis of spray uniformity has been carried out [10].

A review of various distribution patterns and spray coverage achieved by some nozzles has been presented [11].

Mathematical models describing the physical motion of a sprayed liquid particle [12] and presenting the wear of flat fan nozzles [13] have been developed. In [14], as well as [15], the impact of an external air stream simulating the wind with varied velocity on the distribution of sprayed liquid has been analyzed. It has compared the patterns of liquid distributions and the values of the coefficient of variation after the spraying process in experiments conducted on tables with 5 cm and 10 cm groove spacing [16].

The scope of the research in the present work is to find a new model for the distribution of atomized liquid quantity thanks to which better capabilities than have been available so far to assess and predict the spraying process results could be provided. Achieving this aim will help to increase environmental efficiency in agriculture.

2. Materials and Methods

2.1. Description of the Experiment

To obtain data for the development and verification of the new model, the experiments were conducted in the Laboratory of Techniques for Application of Agrochemicals, Department of Machinery Exploitation and Management of Production Processes, University of Life Sciences in Lublin.

It was assumed that mounting new nozzles of good quality to the spray boom would ensure the most uniform spray distribution.

Typically, the working spray boom was located at a height of 0.5 m above the sprayed surface. The liquid outflow pressure was set at 0.3 MPa. The process time was preset to one minute. The spray angle of the flat fan nozzles was set at 110°. In this case, the width of the area sprayed with a single sprayer nozzle was approx. 1.5 m, and the surface coverage of all 25 nozzles mounted on the virtual boom was approx. 12.5 m.

The tests and measurements were carried out on grooved tables with a width corresponding to that of the virtual field boom. Two types of flat fan nozzles were included in the study. The results obtained from AZ-MM spray nozzles were used for the development of the model. The goodness of the fit of the model to the data was evaluated based on results obtained from RS-MM spray nozzles.

Fifty new nozzles were used in the experiment.

The tests were carried out using the blue nozzles, characterized by a liquid flow rate of 1.2 liters per minute.

Water was used as the test liquid, following the standard practice for this type of experiment conducted in laboratory conditions.

The experimental measurement stand shown in Figure 1 facilitated the analysis of the distribution of the atomized liquid sprayed from the nozzle. The results obtained in this design were used for the construction of a virtual field boom. The sprayed liquid reaches the table and falls along the grooves into 50 containers with the same volume arranged serially at equal distances from each other. Each measurement determines the amount of the liquid accumulated in the consecutive containers and is expressed in milliliters. The capacity of a single container is 250 ml. All experiments were carried out on the measurement stand with 5 cm groove spacing.

Figure 1. Measurement table (Department of Machinery Exploitation and Management of Production Processes, University of Life Sciences in Lublin).

With the parameters determined during the experimental process, the spray spectrum for a single nozzle covered approximately 30 containers located in the center of the measurement table. Hence, 0 mL value was in the 10 containers located on the left-hand side and 0 mL value was in the 10 containers located on the right side of the analyzed spectrum area.

The data used in the experiment were the results provided by all 50 available nozzles (25 of each type used for the development and verification of the model) included in the measurement stand. This yielded 50 different data sets. Each of these sprays duplicated 25 times simulated the data generated by the virtual spray boom as described below.

The width of the sprayed area was scaled to an appropriate number of the consecutively numbered containers located at an identical distance as on the measurement table. Distribution data from a single nozzle was replicated to simulate a complete virtual boom.

I.   The nozzle was located on the extreme left of the virtual field boom;

II.   Next, another identical nozzle was placed on the boom on the right side of the first one at a distance of 0.5 m (i.e., at a distance covered by 10 containers to the right);

III.   By mounting successive identical nozzles on the boom, the 25th container was reached (Figure 2). The total range of measurements comprised 270 containers. The amounts of the liquid collected in each container served for the calculation of the coefficient of variation.

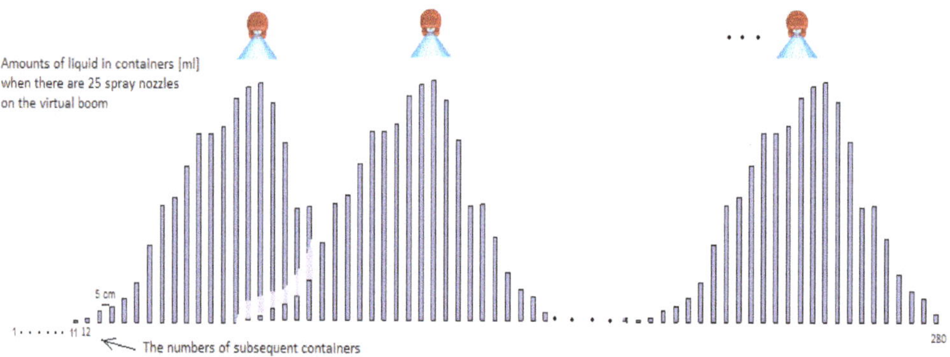

Figure 2. Data collection scheme (final stage 3).

In 50 measurement series performed in the experiment, data from 50 virtual field sprayer booms, created from single nozzle distribution, were collected. Each boom was assigned the same number as that of each consecutive nozzle with a specified degree of wear.

Within the adopted scope of work, we carried out a comprehensive statistical analysis of the distribution of the liquid amounts determined for each simulated virtual field boom and for the real data from each single nozzle.

2.2. Distribution of Droplets after the Spraying Process

The first attempts to match the distribution of droplets to the real sprays provided by flat fan nozzles were undertaken [17,18]. The researchers considered a triangular pattern. The effect of changes in the sprayer nozzle angle on the symmetry and shift of spray distribution was emphasized [17]. The density function of the triangular distribution is described by an equation that combines the nozzle height, the set angle, and the distance between water-sensitive papers:

$$f(x) = x\left(H - d_i ctg\frac{\propto}{2}\right) \tag{1}$$

where:

x is the distance between water-sensitive papers [m];
H is the distance between the spray nozzle and the sprayed surface [m];
d_i is the distance between the ith water-sensitive paper and the nozzle [m];

α is the preset angle of the nozzle spray [°].

Another probability distribution that was analyzed in the description of spray was beta distribution [18]. It is defined by two shape parameters denoted as α and β, which are the exponents of a random variable in the formula and determine the shape of the distribution. The density function of this distribution is as follows:

$$f(x) = \frac{\Gamma(\alpha + \beta)}{\Gamma(\alpha)\Gamma(\beta)} x^{\alpha-1}(1-x)^{\beta-1} \tag{2}$$

where:

x is a random variable specifying the distance between two water-sensitive papers [m];

$\alpha, \beta > 0$ are shape parameters estimated from the experimental data;

$\Gamma(z) = \int_0^{+\infty} t^{z-1}e^{-t}dt$ is the special gamma function for $z > 0$.

Joint research was conducted on this distribution and found that it was sometimes more suitable for the description of spray than the normal distribution [5].

This most popular distribution mentioned above was fitted and applied in comprehensive investigations of spray [19]. The issue was divided by the researchers into two stages. The first stage consisted of the selection of an appropriate model and density function, whereas the second stage was focused on the estimation of parameters determining the spray quality, i.e., pressure and nozzle height and size. Given the basic fact that the function of spray distribution density is a symmetrical curve with an approximate bell shape, the distribution of sprayed liquid was effectively fitted to the doubly truncated normal distribution [19]. This indicates that the values of the random variable had upper and lower limits determined by a certain value. The form of the probability density function $f(x)$ in this case is as follows:

$$f(x) = \frac{\frac{1}{\sigma\sqrt{2\pi}}\exp\left(-\frac{(x-\mu)^2}{2\sigma^2}\right)}{F\left(\frac{b-\mu}{\sigma}\right) - F\left(\frac{a-\mu}{\sigma}\right)} \tag{3}$$

where:

x is a random variable specifying the distance between two water-sensitive papers [m];

u is the expected value of the random variable;

σ is the standard deviation for normal distribution;

$F(x)$ is the cumulative distribution function that complies with $f(x)$;

a, b are the lower and upper cut-off points of normal distribution, respectively.

After the development of the model presented above, a relationship was discovered between its parameters and factors that influence spraying [19]. By the application of the multiple regression equation, the researchers were able to express the standard deviation and spray width with the coefficients of determination of 0.98 and 0.976, respectively:

$$\begin{aligned}
\sigma &= -78 + 27.1\ln(H) + 6.15\ln(P) + 1.72\ \ln(Q_2) \\
W &= -401 + 1.39\ln(H) + 36.5\ \ln(P) + 8.94\ \ln(Q_2)
\end{aligned} \tag{4}$$

where:

σ is the standard deviation [cm];

W is the spray width [cm];

H is the nozzle height [cm];

P is the spray pressure [105 Pa];

Q_2 is the standard flow rate at a pressure of 0.2 MPa [l/min].

Another model [20] is the so-called mean distribution model. It consists of the calculation of the mean values for the parameters of the distribution of liquid amounts collected from several single nozzles. This approach generates highly precise results, but at the cost of a substantially greater number of measurements. Additionally, since different types of nozzles are often used, this will very likely be a random model.

Figure 3 shows a comparison of the consistency of the data obtained from triangular, beta, and normal distributions with the measurement results generated by a Teejet 110 04 VS nozzle at a pressure of 0.2 MPa and a spray height of 0.5 m [21].

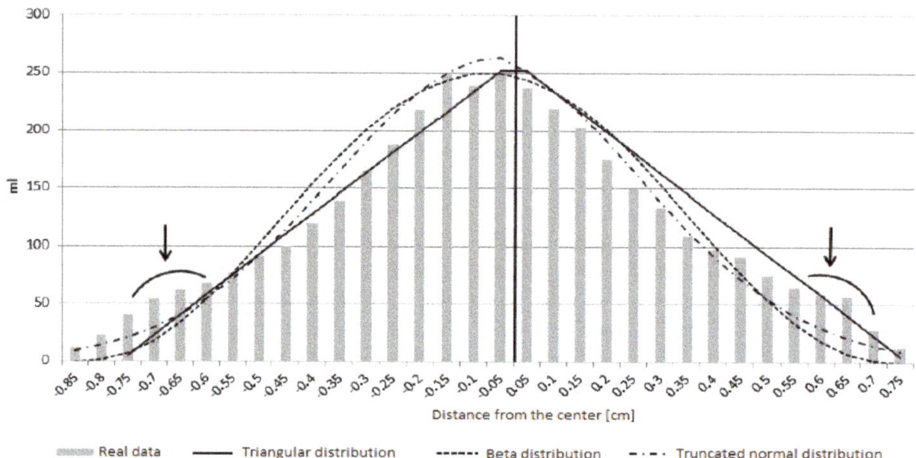

Figure 3. Example of the fit of results in selected models to real spray data [21].

An evident conclusion prompted by the analysis of this figure is the fact that none of the characterized models generate results that match real data adequately.

2.3. Development of a New Model of Liquid amount Distribution

Based on the determinants and data presented in the previous subsection, an original density function of a new distribution model describing the accumulation of liquid amounts after the spraying process was defined for a single nozzle as follows:

$$f(x) = \begin{cases} a(x-0.5)^p, & x \in (0.5; 0.9) \\ b(x-1.25)^q + c, & x \in < 0.9; 1.6) \\ (-1)^p a(x-2)^p, & x \in < 1.6; 2) \\ 0, & x \in (-\infty; 0.5 > \cup < 2; +\infty) \end{cases} \tag{5}$$

where:

x is the real variable with values from the spray range $(0.5; 2)$ [m];

$a > 0$ is a shape parameter;

$b < 0$ is a shape parameter;

$c > 0$ is a shift parameter;

$p, q \geq 2$ are shape parameters (p = natural number, q = even number)

By increasing the values of parameters a, p, or q, the shape of the density curve becomes leptokurtic (thin). Reversely, i.e., when the values of any of the three parameters decrease, the shape becomes more flattened.

The opposite is noted in the case of parameter b, i.e., the higher its value, the greater the flattening of the curve in an appropriate range.

The method for the construction of the density function presented above ensures symmetry of distribution. The new model is symmetrical and is a probability density function.

Using the optimization method, i.e., the Microsoft Excel Solvers tool, the shape of parameters p, q, and b was selected in a way providing a properly set value of the coefficient of variation for the liquid amount after the spraying process.

After the determination of equations required for the correct determination of the probability distribution for a single nozzle, an original model was developed for the determination of the liquid distribution after simultaneous spraying with 25 nozzles in order to reflect the real work of the sprayer.

Therefore, based on the developed form of the density function, a function reflecting the amount of the liquid at any point in the spray area was defined:

$$g(x) = \sum_{n=0}^{24} \left[a(x - 0.5 - 0.5n)^p + (-1)^p a(x - 2 - 0.5n)^p + b(x - 1.25 - 0.5n)^q + c \right] \tag{6}$$

The arguments of the function are points across the entire width of the sprayed area located at a distance of 0.05 m from each other.

The total width of the area sprayed by the boom is identical to that of all liquid-collecting containers aligned and assigned numbers from 1 to 290. According to ISO 16122-2: 2015 (E), the data range taken into account in the calculation of the coefficient of variation of the sprayed liquid distribution includes the amounts of liquid collected in containers located in the center of the spray area width for the second nozzle mounted on the boom to the center of the spray area width for the penultimate nozzle on the boom. Therefore, the sum range in Equation (6) corresponds to containers 36 to 256.

Based on the method of calculation of the coefficient of variation, the function dependent on all parameters a, b, c, p, and q is defined with the following formula:

$$V = \frac{s}{\overline{g}} \cdot 100\% \tag{7}$$

where the following equations are valid for i representing natural numbers and denoting the consecutive numbers of the liquid-accumulating containers after the spraying process:

$$\overline{g} = \frac{1}{221} \sum_{i=36}^{256} g(0.05i) \tag{8}$$

$$s^2 = \frac{1}{221} \sum_{i=36}^{256} (g(0.05i) - \overline{g})^2 \tag{9}$$

and $\overline{g}$ and s^2 are the mean and standard deviation, respectively, for the values of function $g(x)$.

Next, all parameters of the probability distribution were selected for function V to reach the minimum, i.e., to simulate the work of a field boom equipped with the new (model) nozzles, thus providing the most uniform spray of the treated surface, assuming values of 5%, 7%, and 10%, and simulating the work of a field boom with nozzles characterized with an adequate wear degree.

The flow rate in all nozzles was fixed and the extension of the spray swath from virtual boom sprayer could be unlimited, because any number of nozzles may be added.

If we consider the case that the spacing of the grooves on the measurement table is 5 cm, then in the model we have, $a = 15.352$; $b = -2.024$; $c = 1.23$ for $p = 3$, $q = 2$, and $V = 10\%$.

On the over hand, if we consider that the spacing of the grooves on the measurement table is 10 cm, then in the model we have, $a = 14.87$; $b = -2.55$; $c = 1.259$ for $p = 3$, $q = 2$, and $V = 10\%$.

3. Results

To check the goodness of the fit of the data obtained with the new model to real results, the RS-MM 110 03 nozzles were used in such a way that each of them multiplied 25 times formed a single virtual field boom. The STATISTICA 13 program, supported by the Statsoft company from Poland in 2013, was used to perform the linear regression analysis of data generated by the model in comparison with the measurements provided by the virtual field booms. Figure 4 shows an example of a correlation diagram of the amount of liquid after the spraying process from the boom equipped with the new nozzles (model generating V = 2.16%) and from a boom equipped with worn nozzles (generating V = 9.37% wear degree) with a fitted regression line.

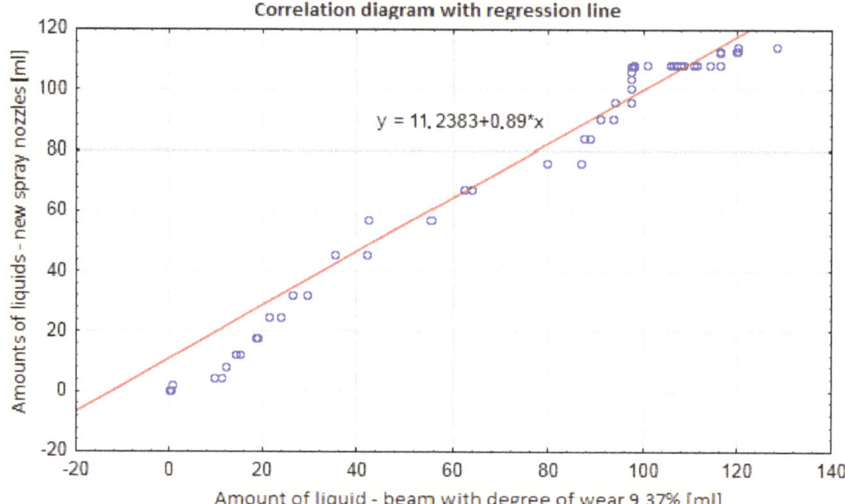

Figure 4. Correlation diagram of the amounts of liquid after the spray from a boom characterized by a 9.37% wear degree (axis OX) and from a boom equipped with the new spray nozzles (axis OY).

We analyzed the values of coefficients of correlation and coefficients of determination from each regression analysis of the data generated by the model and those obtained with the use of virtual field booms equipped with worn nozzles.

The coefficients of correlation have a value of approximately 0.95 and higher, which indicates a strong correlation (dependence) between the model and experimental data.

The coefficients of determination, i.e., the measure of the fit of the model to the experimental data, are higher than 0.9, which indicates that the model efficiently explains the distribution of the experimental data.

Therefore, the developed model is characterized by a very high goodness of fit to the experimental data.

The new model can be used for the description of the distribution of the liquid amounts after the spraying process, even with a very low value of the coefficient of variation equal to 2.16%.

This result should be regarded by manufacturers of sprayers as a guarantee of high uniformity of surface coverage.

Additionally, to test the compatibility of the data from the new distribution with the experimental data, the chi-square test for observed and expected values was applied, besides regression analysis.

A null hypothesis of the compatibility of the examined distributions with an alternative hypothesis of the absence of compliance was adopted. The critical value for the chi-square distribution with 28 degrees of freedom was 41.333 at the significance level α = 0.05. The values of the test statistics for the results of each nozzle RS-MM 110 03 duplicated on the boom in relation to the data obtained with the new model. Experimental values from each RS-MM 110 03 nozzle correspond to an individual deposition in each groove of the distribution test bench. The higher the value of the χ^2 test statistic is, the more the nozzle, which generates a virtual boom sprayer, is worn out and, in consequence, the higher the value of the coefficient of variation (CV) generated from the boom.

A critical area at the significance level $\alpha = 0.05$ is the range <41.337; $+\infty$). The analysis of all the results shown in Table 4 indicates that there is no ground for the rejection of the null hypothesis about the consistency of distributions at the significance level of 0.05. Hence, the tested distributions can be considered identical. For comparison, data generated by the model (for $V = 2.16\%$) and data from measurements of RS-MM 110 03 spray nozzles constituting the virtual field booms are presented in a graph (Figure 5). The Figure shows the difference between the level of uniformity of model data simulating the work of the new nozzles and of the model provided by the data from the worn nozzles.

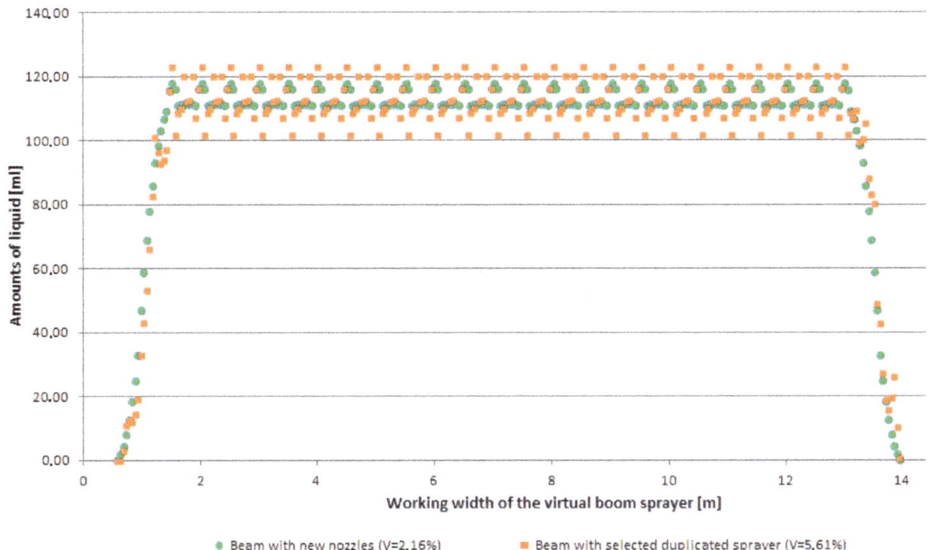

Figure 5. Liquid amounts from the boom equipped with the new nozzles and from the selected virtual boom.

Additionally, we considered the fact that the total amounts of liquid accumulated along the entire booms are the same in both distributions.

These calculations conducted for the measurement table with a 10 cm spacing of grooves, reflecting conditions of nozzle tests and analyses performed by producers, were analogous, although with one exception. Every second value of variable x, denoting the amount of liquid accumulated on the 55th centimeter, 65th centimeter, up to 13.95 meters of the length of the virtual field boom, was deleted from all values of the measurement table with grooves located 5 cm apart from each other.

Data provided by the proposed model generating the values of the coefficient of variation of $V = 2.16\%$ and $V = 10\%$, which simulated spraying performed with both a new nozzle and an adequately worn nozzle, were scaled to obtain the same total amounts of sprayed liquid. Figures 6 and 7 present the amounts of liquid obtained at different values of model parameters for two nozzles with normalized flow rates (the new nozzle and another one with a moderate or permissible degree of wear).

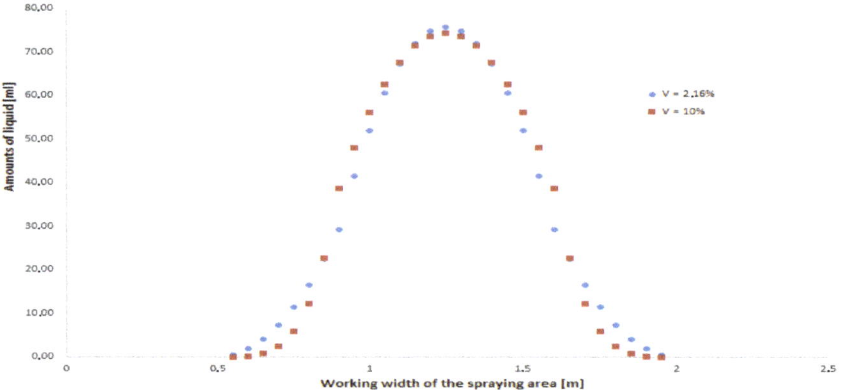

Figure 6. Amounts of liquid sprayed with one nozzle generating $V = 2.16\%$ (new) and $V = 10\%$ ($p = 4, q = 2$).

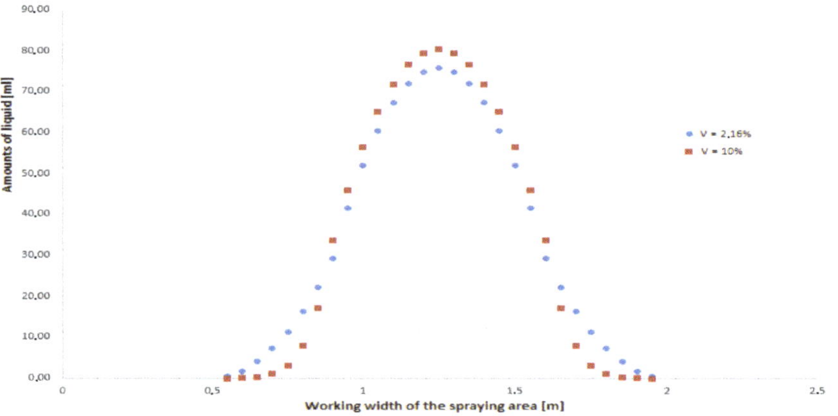

Figure 7. Amounts of liquid sprayed with one nozzle generating $V = 2.16\%$ (new) and $V = 10\%$ ($p = 3, q = 2$).

The analysis of the data presented in the graphs revealed the mean ranges of deviations of the amounts of sprayed liquid (Tables 1 and 2), which reflect the moderate or permissible wear of the new (model) nozzle.

Table 1. Changes in the amounts of liquid sprayed by the new nozzle yielding the relevant values of the coefficient of variation ($p = 4, q = 2$).

Range [m]	No. of Containers	Mean Change [mL]	Coefficient of Variation
<0.6; 0.85>	12–17	−4	
<0.9; 1.6>	18–32	+3.4	10%
<1.65; 1.9>	33–38	−4	

Table 2. Changes in the amounts of liquid sprayed by the new nozzle yielding the relevant values of the coefficient of variation ($p = 3, q = 2$).

Range [m]	No. of Containers	Mean Change [mL]	Coefficient of Variation
<0.6; 0.85>	12–17	−5.6	
<0.9; 1.6>	18–32	+4.7	10%
<1.65; 1.9>	33–38	−5.6	

In other words, the data in the tables facilitate observation and inference on the identification of the adequate nozzle wear.

The tables and graphs presented above can be used for the development of a diagnostic model showing ranges of deviations in the amounts of liquid sprayed by the new nozzle that causes permissible wear, i.e., at the coefficient of variation of 10% (Figures 8 and 9).

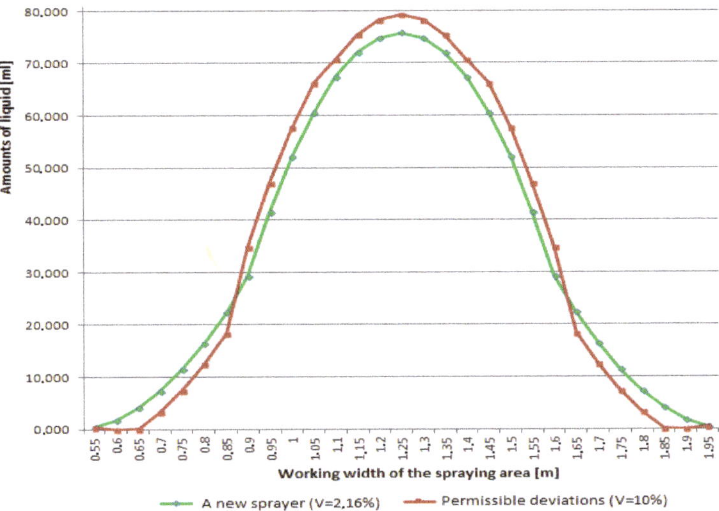

Figure 8. Permissible average ranges of deviations of the amounts of liquid sprayed by the new nozzle ($p = 4$, $q = 2$).

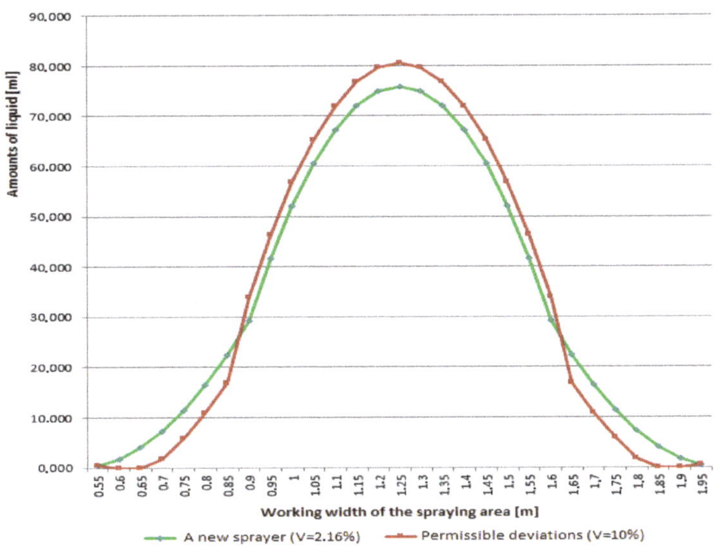

Figure 9. Permissible average ranges of deviations of the amounts of liquid sprayed by the new nozzle ($p = 3$, $q = 2$).

To present the conditions under which producers test new nozzles, a similar analysis of data was carried out for a model simulating measurements from the table with 10 cm groove spacing.

The simulation amounts of liquids from the new nozzle were compared to those from the nozzle with 10% wear (Figure 10).

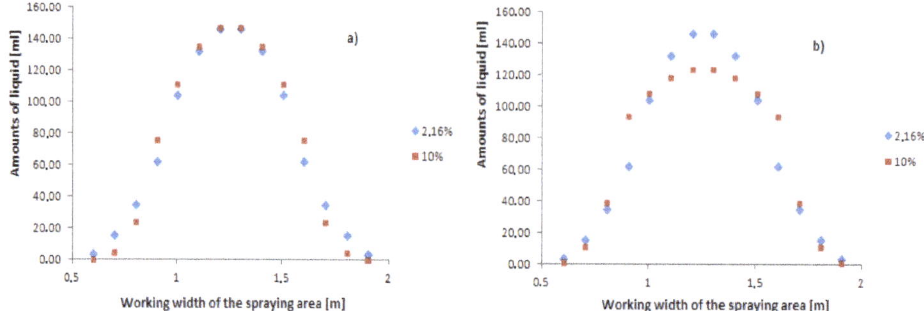

Figure 10. Amounts of liquid sprayed with one nozzle generating a value of the coefficient of variation of $V = 2.16\%$ and $V = 10\%$ on the boom: (**a**): $p = 4$, $q = 2$; (**b**): $p = 3$, $q = 2$.

Next, following the results presented above, as in the case of the measurement table with 5 cm groove spacing, Tables 3 and 4 were compiled to show the simulations of the moderate or permissible nozzle wear. Examples (Figure 11) were prepared to present the mean ranges of deviations indicating the wear of the analyzed nozzle in accordance with the adopted model.

Table 3. Changes in the amounts of liquid sprayed by the new nozzle yielding the relevant values of the coefficient of variation ($p = 4$, $q = 2$).

Range [m]	No. of Containers	Mean Change [mL]	Coefficient of Variation
<0.6; 0.9>	6–9	−9.8	
<1; 1.5>	10–15	+3.8	10%
<1.6; 1.9>	16–19	−9.8	

Table 4. Changes in the amounts of liquid sprayed by the new nozzle yielding the relevant values of the coefficient of variation ($p = 3$, $q = 2$).

Range [m]	No. of Containers	Mean Change [mL]	Coefficient of Variation
<0.6; 0.7>	6–8	−3.1	
<0.8; 1>	9–10	+13.9	
<1.1; 1.4>	11–14	−17.7	10%
<1.5; 1.7>	15–16	+13.9	
<1.8; 1.9>	17–19	−3.1	

The analysis of the results shown in Tables 1–4 and in Figures 5–11 allows the conclusion that the greater amounts of liquid in the central spray area and the lower amounts on its left and right side indicate the higher degree of nozzle wear. This facilitates the assessment of the degree of wear relative to the quality of the produced model.

Based directly on the data obtained without the application of the new model, the amounts of liquid from two adjacent containers were added, which yielded results that could be obtained at the 10 cm groove spacing. Hence, the spraying treatment exhibited greater uniformity, as the minimum value of the coefficient of variation was 1.85%, which was lower than in the initial case by ca. 0.3%.

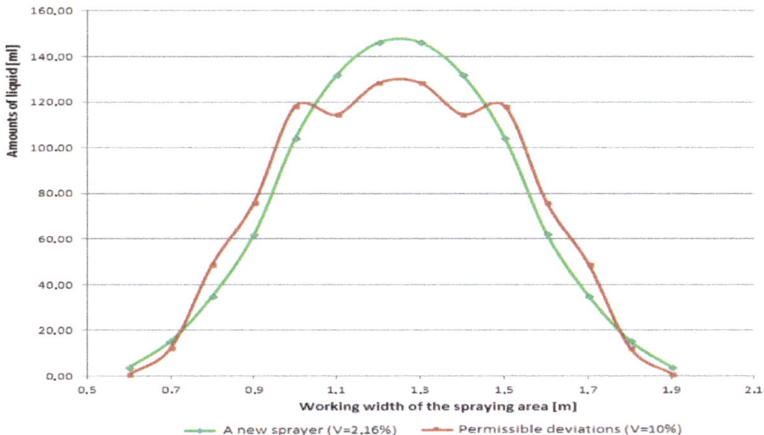

Figure 11. Permissible mean ranges of deviations for the amounts of liquid sprayed by the new nozzle ($p = 3, q = 2$).

Additionally, similarly lower values were obtained for each nozzle wear degree, which are summarized in Table 5.

Table 5. Comparison of the coefficients of variation for different nozzle wear degrees at the 5 cm and 10 cm groove spacing.

		Mean Wear		Permissible Wear	
p	2	3	4	3	4
q	2	2	2	2	2
V_{10cm}	1.85%	4.74%	4.61%	9.43%	9.52%
V_{5cm}	2.16%	5%	5%	10%	10%

These findings confirm the thesis that a greater distance between the grooves reduces the value of the coefficient of variability of data [22].

4. Discussion

The spraying process is applied in various areas, for example in fuel spraying [23], in the precipitation profiles of a fixed spray-plate sprinkler [24], in aerial spray application based on thermal imaging technology [25], and in evaluating irrigation system performance, which mainly depends on the uniformity of water application [26] or by using an unmanned aerial vehicle [27]. Another improvement introduced to examine the uniformity of spraying coverage was adding adjuvant to a liquid [28].

There are many factors that affect the uniformity of spray liquid coverage from spray nozzles, such as height of boom and nozzle pressure. The range of the coefficient of variation due to the height of the nozzle ranges from 8% to 17.6%, due to nozzle pressure from 7.6% to 20.3%, and for fixed height and nozzle pressure from 7.2% to 21.3% [29].

Coefficients of variation, in the case of repetitions performed with the same technique, were sometimes quite large, ranging from 7.5% to 24.0% [30], but we achieved with the same approach a value of this coefficient equal to 1.85%, as in Table 5.

When the technique of pulse width modulation is used, the coefficient of variation is around 10%. Increasing the signal means increasing the value of this factor [31]. In the same technique when for various kinds of nozzles the fixed pressure is equal to 207, 276, or 476 kPa, the values of the coefficient of variation range from 5.3% to 20.1% [32].

The best results of the uniformity were obtained in a wind tunnel and had values of CV from 0.5% to 7.6%, where 75% of all nozzle types tested had CV values below 4%. However, the use of water alone, like in our research, caused the largest differences in CV values [33]. In relation to the working width of the entire boom sprayer, the generated CV values were generally around 10% [34].

Considering the literature on the subject of the uniformity of field spraying and looking at the results achieved there, our model is the first one thanks to which we are able to forecast the permissible wear of the atomizer nozzle and then determine its usefulness.

5. Conclusions

The tools were designed and developed for the description and characterization of the distribution of the liquid amounts sprayed on an area with a specified size with the use of flat fan nozzles.

An original probability density function was proposed, which efficiently reflects the accumulation of the amounts of liquid sprayed from a single nozzle.

A new model of the distribution of the amounts of liquid after the spraying process was developed, yielding a minimum value of the coefficient of variation for the amount of liquid that ensures a very high uniformity of surface coverage.

The generation of the model of the distribution of sprayed liquid amounts simulating the operation of the new nozzle should help to maintain the accepted level of spraying uniformity as long as possible, which will allow the optimization of the process.

By the determination of the distribution of the spray ensured by the new nozzle with the use of the model, the level of nozzle wear can be assessed.

The application of the solution presented in this study will facilitate nozzle quality control consisting of the effective monitoring of wear degree and will contribute to the extension of nozzle service life, in compliance with accepted standards.

The results obtained from the model were standardized to the 10 cm spacing of the grooves on the measurement table, i.e., the distance for which manufacturers carry out tests and analyses of nozzles. This mode of presentation of the results will facilitate quality control aimed at the optimization of the field spraying process.

The solution of the research problem yielded a new diagnostic model as a tool for atomization and spray quality control and the assessment of nozzle wear. The model will have a positive effect on the quality, duration, and optimization of the field spraying process as well as the condition of the natural environment through the possibility of the application of an appropriate amount of sprayed agents.

Author Contributions: P.A.K. proposed the mathematical model and verified obtained data. I.K.-B. generally corrected the manuscript. S.P. provided data and formulated technical conclusions.

Funding: This research received no external funding.

Conflicts of Interest: The authors declare no conflict of interest.

References

1. The European Parliament and the Council of the European Union. Directive 2009/128EC establishing a framework for Community action to achieve the sustainable use of pesticide. *Off. J. EU* **2009**, *309*, 71–86.
2. Herbst, E.; Herbst, K. Ernst Herbst Pruftechnik e.K.-Plant protection equipment, test engineering, agricultural technology. *Julius-Kühn-Archiv* **2010**, *426*, 127.
3. Lodwik, D.; Pietrzyk, J. Automated Test Station for Transverse Spray Non-Uniformity. *J. Res. Appl. Agric. Eng.* **2013**, *58*, 103–106.
4. International Standarization Organization (ISO). ISO 16119. In *Agriculture and Forest Machinery–Environmental Requirements for Sprayers—Part 2*; ISO: Geneva, Switzerland, 2013.
5. Mawer, C.J.; Miller, P.C.H. Effect of roll angle and nozzle spray pattern on the uniformity of spray volume distribution below a boom. *Crop Prot.* **1989**, *8*, 217–222. [CrossRef]

6. Nowakowski, T. Interaction between selected spraying parameters on the variability coefficient of liquid transverse distribution. *Agric. Eng.* **2007**, *3*, 135–141.

7. Nowakowski, T.; Chlebowski, J. The impact of liquid pressure and design of fan atomizers on spraying angle. *Agric. Eng.* **2008**, *1*, 319–323.

8. Nuyttens, D.; Baetens, K.; De Schampheleire, M.; Sonck, B. Effect of nozzle type, size and pressure on spray droplet characteristics. *Biosyst. Eng.* **2007**, *97*, 333–345. [CrossRef]

9. Parafiniuk, S.; Sawa, J.; Huyghebaert, B. The evaluation of the technical condition of the field toolbar of the spraying machine with the use of the method of survey of selected sprayers. *Agric. Eng.* **2011**, *5*, 207–215.

10. Parafiniuk, S.; Tarasińska, J. Work simulation of the sprayer field boom with the use of R program. *J. Cent. Eur. Agric.* **2013**, *14*, 166–175. [CrossRef]

11. Rojek, G. Analysis Of The Spray Distribution And Coverage In Variable Working Conditions of Selected Nozzles. Ph.D. Thesis, University of Life Sciences in Wroclaw, Wroclaw, Poland, 2013.

12. Szewczyk, A.; Wilczok, G. Theoretical and actual liquid distribution for selected atomizer setting parameters. *Agric. Eng.* **2007**, *8*, 265–271.

13. Zhu, H.; Rowland, D.L.; Dorner, J.W.; Derksen, R.C.; Sorensen, R.B. Influence of Plant Structure, Orifice Size, and Nozzle Inclination on Spray Penetration into Peanut Canopy. *Trans. ASAE* **2002**, *45*, 1295–1301.

14. Szewczyk, A.; Wilczok, G. Theoretical description of sprayed liquid distribution in conditions of frontal air stream operation. *Agric. Eng.* **2008**, *5*, 292–299.

15. Wilczok, G. Analysis of Spray Liquid Distribution during Spraying in Variable Working Conditions of Sprayers. Ph.D. Thesis, University of Life Sciences in Wroclaw, Wroclaw, Poland, 2008.

16. Świechowski, W.; Hołownicki, R.; Doruchowski, G.; Godyń, A. Comparison of the methods evaluating flat spraying nozzles. *Probl. Agric. Eng.* **2006**, *4*, 5–12.

17. Nation, H.J. Spray nozzle performance and effects of boom height on distribution. In *Departmental Note n° DN/S/777/1925*; National Institute of Agricultural Engineering Silsoe: Wrest Park, Silsoe, Bedford, UK, 1976; unpublished.

18. Mawer, C.J. *The Effect of Nozzle Characteristics and Boom Attitude on the Volume Distribution below a Boom*; Div. Note DN 1462; AFRC Institute of Engineering Research: Silsoe, Bedford, UK, 1988.

19. Leunda, P.; Debouche, C.; Caussin, R. Predicting the transverse volume distribution under an agricultural spray boom. *Crop Prot.* **1990**, *9*, 111–114.

20. Sinfort, N.; Bellon, V.; Sevila, F. Image analysis for in-flow measurement of particle size. *Food Control* **1992**, *3*, 84–90. [CrossRef]

21. Huyghebaert, B. Verification of Measurement Methods of Flat Fan Nozzles Working Parameters Used in Agriculture. Ph.D. Thesis, University of Life Sciences in Lublin, Lublin, Poland, 2015.

22. Sawa, J.; Kubacki, K.; Huyghebaert, B. Equivalence of the criteria of assessing results of tests in legalizing crop sprayers. *Agric. Eng.* **2001**, *4*, 1.

23. Li, T.; Nishida, K.; Hiroyasu, H. Droplet size distribution and evaporation characteristics of fuel spray by a swirl type atomizer. *Fuel* **2011**, *90*, 2367–2376. [CrossRef]

24. Sayyadi, H.; Nazemi, A.H.; Sadraddini, A.A. Characterising droplets and precipitation profiles of a fixed spray-plate sprinkler. *Biosyst. Eng.* **2014**, *119*, 13–24. [CrossRef]

25. Jiao, L.; Dong, D.; Feng, H.; Zhao, X.; Chen, L. Monitoring spray drift in aerial spray application based on infrared thermal imaging technology. *Comput. Electron. Agric.* **2016**, *121*, 135–140. [CrossRef]

26. Irmak, S.; Odhiambo, L.O.; Kranz, W.L.; Eisenhauer, D.E. Irrigation Efficiency and Uniformity, and Crop Water Use Efficiency. *Biol. Syst. Eng. Pap. Publ.* **2011**, *451*, 1–8.

27. Xue, X.; Lan, Y.; Sun, Z.; Chang, C.; Hoffmann, W.C. Develop an unmanned aerial vehicle based automatic aerial spraying system. *Comput. Electron. Agric.* **2016**, *128*, 58–66. [CrossRef]

28. Griesang, F.; Decaro, R.A.; dos Santos, C.A.M.; Souza Santos, E.; de Lima Roque, N.H.; da Costa Ferreira, M. How Much Do Adjuvant and Nozzles Models Reduce the Spraying Drift? Drift in Agricultural Spraying. *Am. J. Plant Sci.* **2017**, *8*, 2785–2794. [CrossRef]

29. Sehsah, E.M.E.; Kleisinger, S. Study of some parameters affecting spray distribution uniformity pattern. *MJ Agric. Eng.* **2009**, *26*, 69–93.

30. Nuyttens, D.; Zwertvaegher, I.K.A.; Dekeyser, D. Spray drift assessment of different application techniques using a drift test bench and comparison with other assessment methods. *Biosyst. Eng.* **2017**, *154*, 14–24. [CrossRef]

31. Mangus, D.L.; Sharda, A.; Engelhardt, A.; Flippo, D.; Strasser, R.; Luck, J.D.; Griffin, T. Analyzing the nozzle spray fan pattern of an agriculture sprayer using puls width modulation technology to generate an on-ground coverage map. *Trans. ASABE* **2017**, *60*, 315–325. [CrossRef]

32. Butts, T.R.; Luck, J.D.; Fritz, B.K.; Hofmann, W.C.; Kruger, G.R. Evaluation of spray pattern uniformity using three unique analyses as impacted by nozzle, pressure, and pulse-width modulation duty cycle. *Pest. Manag. Sci.* **2019**, *75*, 1875–1886. [CrossRef] [PubMed]

33. Ferguson, J.C.; O'Donnel, C.C.; Chauhan, B.S.; Adkins, S.W.; Kruger, G.R.; Wang, R.; Ferreira, P.H.U.; Hewitt, A.J. Determining the uniformity and consistency of droplet size across spray drift reducing nozzles in a wind tunnel. *Crop Prot.* **2015**, *76*, 1–6. [CrossRef]

34. Balsari, P.; Gil, E.; Marucco, P.; van de Zande, J.C.; Nuyttens, D.; Herbst, A.; Gallart, M. Field-crop-sprayer potential drift measured using test bench: Effects of boom height and nozzle type. *Biosyst. Eng.* **2017**, *154*, 3–13. [CrossRef]

Review

The Review of Biomass Potential for Agricultural Biogas Production in Poland

Katarzyna Anna Koryś [1,2], **Agnieszka Ewa Latawiec** [1,2,3,4,*], **Katarzyna Grotkiewicz** [3] and **Maciej Kuboń** [3,5]

[1] International Institute for Sustainability, Estrada Dona Castorina 124, Rio de Janeiro 22460-320, Brazil; k.korys@iis-rio.org

[2] Rio Conservation and Sustainability Science Centre, Department of Geography and the Environment, Pontifícia Universidade Católica, Rio de Janeiro 22453900, Brazil

[3] Department of Production Engineering, Logistic and Applied Computer Sciences, University of Agriculture in Kraków, Balicka 116B, 30-149 Kraków, Poland; katarzyna.grotkiewicz@urk.edu.pl (K.G.); maciej.kubon@urk.edu.pl (M.K.)

[4] School of Environmental Science, University of East Anglia, Norwich NR4 7TJ, UK

[5] Institute of Technical Sciences, State Vocational East European Higher School in Przemyśl, Książąt Lubomirskich 6, 37-700 Przemyśl, Poland

[*] Correspondence: a.latawiec@iis-rio.org

Received: 8 October 2019; Accepted: 8 November 2019; Published: 19 November 2019

Abstract: Adequate management of biomass residues generated by agricultural and food industry can reduce their negative impacts on the environment. The alternative use for agricultural waste is production of biogas. Biomass feedstock intended as a substrate for the agricultural biogas plants may include energy crops, bio-waste, products of animal and plant origin and organic residues from food production. This study reviews the potential of selected biomass residues from the agri-food industry in terms of use for agricultural biogas production in Poland. The most common agri-food residues used as substrates for biogas plants in Poland are maize silage, slurry, and distillery waste. It is important that the input for the agricultural biogas installations can be based on local wastes and co-products that require appropriate disposal or storage conditions and might be burdensome for the environment. The study also discusses several limitations that might have an unfavourable impact regarding biogas plants development in Poland. Given the estimated biomass potential, the assumptions defining the scope of use of agricultural biogas and the undeniable benefits provided by biogas production, agricultural biogas plants should be considered as a promising branch of sustainable electricity and thermal energy production in Poland, especially in rural areas.

Keywords: biomass; agricultural biogas plants; agricultural waste; sustainable and renewable energy; organic residue management; Poland

---

1. Introduction

Global environmental change caused by excessive exploitation of natural resources and combustion of fossil fuels causes a range of negative impacts on human health and functioning of ecosystems humans ultimately depend on. Generation of solid waste is increasing rapidly as a result of industrialization, global urbanization and economic development [1,2]. Projections indicate that by the 2050 the word population will reach 9.7 billion [3]. There is therefore an increasing pressure on land and water resources to supply food and industrial products. Inability to effectively manage wastes as recently highlighted by scientists and decision-makers lead to serious environmental and socio-economic problems that need urgent and reliable solutions [4,5]. In response to this problem, a number of scientific initiatives were founded with the goal of creating new usages for organic residues [6]. One of the solutions of organic waste management can be their use for production of biogas, during anaerobic digestion (AD) [7].

Biomass [8] represents an important source of alternative energy and provides an opportunity to decrease environmental problems such as pollution and depletion of natural resources [6]. It is widely available and regenerates in a relatively short time [9]. There are many possibilities of biomass processing. For example, agricultural residues and municipal solid residues for biogas and biofuel production [10,11], the use of organic waste for vermicomposting [12], as a supplement during combustion with hard or brown coal or use of agricultural, urban, woody and industrial pyrolyzed residues to improve soil quality [13,14]. This paper discusses the potential of using residual biomass in Poland for the energy sector. It focuses on the assessment of the suitability of biomass from agricultural production and by-products or residues from the agri-food industry as a substrate for biogas production.

2. Methods

The first stage of the study was the literature review, which was performed in English and in Polish languages. The search was conducted on the online databases such as Google Scholar, Scopus, Web of Science, and Science Direct (without restriction to year). For this purpose, we provide an extensive combination of keywords in English and Polish (Table S1). The second stage was the selection of the scientific papers (articles, conference papers and PhD theses), based on title and abstract content. The selected documents were archived in the literature database Mendeley. Additionally, we reviewed the bibliography in the key documents with the aim of finding other relevant titles on the specific subject ("snowball" search method).

3. The Role of Biomass in Energy Production in Poland

The rational use of renewable energy sources is essential for sustainable development. Nowadays, Poland is the second largest coal consumer in the EU (after Germany), and the 10th largest coal (here referred to both: hard coal and lignite) consumer in the world, with consumption of 77 million tonnes of coal per year. In addition, the country is a leader in hard coal extraction and import. For example, in 2016 Poland produced 70.7 million tonnes of hard coal and imported 8.3 million tonnes [15]. Current data provided by the World Energy Council from 2017 showed that 92% of electricity and 89% of heat in Poland is generated from coal [16]. Biomass co-firing with fossil fuels might be a promising solution, allowing an increase in the share of renewable energy sources (RES) in the total energy in Poland [17]. Supporting energy production from renewable energy sources has become an essential objective of the European Union's policy as a strategy to pursue sustainable development [18,19]. In the case of Poland, the need for RES development results from the commitments of the "3 × 20" climate package imposed by the EU. By year 2020 Poland is obliged to obtain 15% of RES in gross final energy use and reduce emissions of air pollution. The use of biomass as the renewable energy source might be an essential aspect to achieve these obligations [20].

Biomass is one of the oldest as well as the most promising development of RES in Poland, with a great potential to be used for energy purposes [21]. It is mostly related to favourable geographical and climatic conditions for biomass production, wide range of its application, and its large resources [21–23]. In case of other RES, the limiting factors may result from unfavourable topography or insufficient resources (hydropower, wind energy), as well as the costs of production: for example, the high price of the solar cells for solar energy [21]. The energy obtained from biomass in Poland comprise approximately 80% of the entire energy pool, which is obtained from RES [20]. The technical potential of biomass in Poland is around 900 PJ/year [24]. The structure of primary energy production from renewable energy sources in Poland from 2016 is shown in Figure S1 [25]. Additionally, Poland is one of the largest exporters of biomass in Europe [21]. The annual biomass energy potential that can be managed is estimated as follows: over 20 million tons of waste straw, 4 million tons of wood waste (sawdust, tree cortex, sawdust, pellets), 6 million tons of sewage sludge from the paper industry and pulp, food and municipal waste [26].

From an environmental point of view, biomass can be considered better than coal. Previous studies demonstrated that combustion of biomass emits less SO_2 compared to coal and has zero-balance carbon dioxide emission [27,28]. Furthermore, the ash gained from the combustion of biomass is returned to the soil, where the plants used for thermal process were cultivated and collected, is consistent with the principles of sustainable development [21]. Despite the fact that electric energy in Poland is obtained mainly from coal [17], it has been demonstrated that using biomass from agriculture co-products for energy purposes is fully justified [23].

4. The Potential of Use of Agricultural Residues for Biogas Production in Poland

In Poland, the production of agricultural biogas has a great potential to grow given increasing demand for heat and electricity from renewables. Moreover, availability of feedstock (substrates) presents interesting opportunities for potential investors, farmers in particular [29]. Because the production of agricultural biogas requires a daily input of substrates [30], location of agricultural biogas installation and its capacity to process the biomass is determined by the constant supply of raw material [31]. This ensures high biogas yield, stability of the fermentation process and possibility of formed digestate utilization [32]. The composition and capacity of biogas obtained from biomass depends on many indicators, including moisture and physical state of the feedstock, technology used, temperature and pressure. As the input for agricultural power plant, both animal and plant origin substrates can be applied as well as waste from the agri-food industry [30,33].

4.1. Utilization of Biomass from Animal and Plant Production

Large breeding farms and agri-food processing generate a significant amount of organic waste [34]. In accordance with the requirements of environmental protection and waste management, if the breeding farm presents more than 40,000 places of poultry, or pig stock more than 2000, farmers are obliged to dispose at least 70% of animals manure on their own farmlands [35]. Semi-liquid and liquid slurry from animal farming are valuable fertilizer, however, their improper storage, application and spillage can lead to environmental pollution and cause odour problems [36]. Their utilization as the substrate for agricultural biogas plants could be one of the options. This solution might be practically attractive for farmers that produce large amounts of organic waste characterized by high energy value. In Poland, approximately 80,750 tons of manure and 35 million m^3 of slurry is provided as the organic waste of agricultural production. About one third of this amount could be processed into biogas for local use [34]. Banaszkiewicz and Wysmyk [37] estimated that the total technical potential to produce agricultural biogas from livestock excreta in Poland is 674 million m^3, i.e., 26.2 PJ. Furthermore, the anaerobic digestion conducted in agricultural biogas plants provides stabilization and deodorization of raw manure, also changing the category of fertilizer from "natural" to "organic" as the final product, so it can be disposed-of easier [36]. Nevertheless, digesting of raw manure as the only substrate under thermophilic conditions might be unprofitable due to some exploitation issues and inhibition of biogas production due to higher N amounts in animal excreta than in other organic waste [38,39]. A mixture of slurry with plant-biomass enhances C/N (carbon to nitrogen) ratio and nutrient balance, contributing to the improvement of biogas quality and lowering production costs [40,41].

In the case of farming, not only animal excrements can be used as a substrate for an agricultural biogas plant. Straw is commonly used in agriculture as feed for farm animals, however, its surplus is not suitable for agricultural purposes and may be burdensome for the environment. Currently, the surplus of straw is estimated to be 10–11 million tonnes per year. [42]. Nevertheless, straw has been recognized as a valuable biomass for the energy sector. This raw material can provide 934 TJ of energy. Assuming that the average calorific value of coal is 24 MJ kg^{-1}, the evaluated biomass could replace over 9.16 million tonnes of coal. However, it should be considered that straw combusting arouses controversy due to CO_2 emissions [43].

In Poland, approximately 25% of arable lands can be used for the growing of annual energy crops with the aim of agricultural biogas production [44]. Among them, maize, more precisely maize silage combined with slurry is used often for the co-fermentation process [33]. In comparison with other grain plants, maize shows higher biogas efficiency (Table S2), high yielding potential, harvesting and silage [45–47]. It is worth mentioning that silages of various plants, especially corn silage, are used for large-scale energy purposes. In Poland, maize growing is concentrated mainly in the western and northern part of the country [48]. The spatial analysis conducted by Jędrejek and Jarosz [31] demonstrated that the most suitable voivodeships for the construction of 50–100 kW micro-biogas plant using 100% maize silage are: Lower Silesian, Pomerania, Lubusz, Greater Poland, Warmian-Masurian and West Pomeranian. In the communities located in the voivodeships mentioned above, there is also a prospect of building micro-biogas plants obtaining biogas from a blend of maize silage (70%) and slurry (30%). Nonetheless, cultivation of monoculture—as in the case of maize intended for silage—might have an unfavourable impact for the environment, especially for soil [31]. Thus, due to the growing interest of advantages resulting from agricultural biogas production, the number of installations built in recent years and high costs of the feedstock, more attention is paid for alternative substrates for biogas plants.

4.2. Biomass from the Agri-Food Industry

The Polish agri-food industry generates large amounts of organic waste [49]. Replacing or co-firing of biomass obtained from agricultural crops with the raw material from agri-food production can be a promising solution for biogas production in Poland [50]. The usage applies to processed and unprocessed waste from the agri-food production, e.g., fruit processing [51], residues from diary industry, [52,53], distillery waste [54], meat processing [55] or fresh vegetables and fruits [56,57]. Theoretically, any biodegradable biomass that contains carbohydrates, proteins and fats can be used as the substrate for the agricultural biogas plant, however the prerequisite for the profitability of using raw material is the content of the organic dry matter amount above 30% [58,59]. It should be emphasized that regarding the methane fermentation process, it is important to use technologies based on the use of by-products from agriculture that do not compete with the production of food or forage [44].

4.2.1. Fruits Residues

During extraction of juice from fruits, a by-product called fruit pomace is separated. A certain amount of leftover goes to landfill, contributing to environmental pollution, while a significant part can be applicable as an energy source [50]. The use of fruit pomace as an input for a biogas facility has been well-described in literature. For example, Pilarska et al. [60] showed that the quantity of biogas and methane obtained from apple pomace was as follow: 203.64 m^3/ton of fresh substance and 101.36 m^3/ton of fresh substance. As a comparison, the results for activated sludge in the current study is: 4.38 m^3/ton of fresh substance for biogas and 2.21 m^3/ton of fresh substance for methane. Moreover, Prask et al. [61] demonstrated that grape pomace obtained during wine production might be successfully used in agricultural biogas plants, both as a substrate or co-substrate. It is important to highlight that fruit residues contain low concentrations of heavy metals [62]. Thus, as the result of processing the fruit residues during the methane fermentation, valuable organic fertilizer which is nutrient-rich and free from heavy-metals can be obtained [63].

4.2.2. Dairy Industry

The main by-product of the dairy industry is whey, whose annual production in Poland is 2–3 million m^3 [64]. Utilization of whey may be problematic, because it contains chemical substances, resistant for biodegradation in the conventional wastewater treatment [64]. Due to the content of lactose, which is the source of energy for many microbial groups like lactic acid bacteria, whey can be used in numbers of biotechnological processes, including methane fermentation, instead. Wesołowska-Trojanowska and Targoński [52] demonstrated that from one ton of substrate, up to 55 m^3 of biogas, containing about 78% of methane, can be obtained. In addition, the product does not contain sulphur compounds and can be directly used for combustion in steam boilers without prior desulphurisation. Nevertheless, it should be noticed that depending on the milk processing technology used in the dairy plant, acid, sweet and casein whey may be formed, differing mainly in pH. Whey is characterized by extremely high chemical oxygen demand (COD)—approximately 50,000 mg O_2/dm^3, and nitrogen (Kjeldahl nitrogen (Nog) 600 mg N/dm^3; nitrate nitrogen (N-NO_3) 2.5 mg/dm^3; N-NH_4 + 60 mg/dm^3). Therefore, despite the high biogas potential estimated, it is not recommended to use whey as the only substrate in the methane fermentation process, mainly due to the low C/N ratio required for the correct course of the process [53].

4.2.3. Meat Industry and Post-Slaughter Waste: Inconveniences Continuation

Meat industry, pork particularly, is one of the most important products of Polish agriculture [55]. However, the meat processing generates annually about 18 million tons of waste [65]. It poses a serious environmental and epidemiological threat and should be disposed of properly [44,65,66]. Due to the restrictive regulation regarding disposal of slaughterhouse waste, meat producers are struggling with the high costs of animal waste management [67]. Methane fermentation in purpose of their utilization seems to be an optimal solution [55]. The research conducted in the Institute of Biosystems Engineering (University of Life Sciences in Poznań) showed that the waste from the meat industry might be promising for biogas efficiency, due to the high content of protein and fats [68]. For example, the amount of biogas obtained from 1 ton of the content of the digestive tract was 275.77 m^3, including methane: 194.38 m^3/t fresh matter, which is 70.48% of gas. Also, it has been demonstrated that the mixture of slurry and digestive tract content of pigs is an energetically effective input in the methane fermentation process, producing high-energy biogas with a methane content exceeding 60% [55].

Nevertheless, the use of waste from slaughterhouse and meat processing in Poland as a substrate for biogas plants is associated with certain problems. First, this type of waste requires the prior sanitary treatment (thermal), in accordance with the Regulation of the European Parliament and Council Regulation (EC) No 1069/2009 of 21 October 2009. The exception applies to the content of the digestive tract [65]. Second, the technological issues related with processing of selected organs. For instance, brains and spinal cords that cannot be used as a substrate for a biogas plant, as they might be the potential source of pathogenic prions [69]. Third: regarding the UE legislation, post-slaughterhouse wastes are divided into three categories (Animal by-product, ABPs), based on the risk they pose: category I ABPs classed as "particular risk" (SRM—from polish language: "material szczególnego ryzyka"), category II ABPs—"high-risk", category III—"low-risk". Only waste classified to category II and III can be used to produce agricultural biogas, after earlier treatment [70].

An appropriate and efficient fermentation process depends on the quality and proper balance of the supplied substrate. In the case of meat waste, the other, considerable issue is the high amount of nitrogen which makes a lot of difficulties in running the anaerobic fermentation process properly. Balance of the C/N components ratio is an important element because in the fermentation process the organic nitrogen from the substrate is converted into ammonium nitrogen, which is partly used for the synthesis of protein of newly emerging bacterial cells. In addition, ammonia is formed with excess nitrogen, which inhibits bacterial growth at low concentrations, while higher carbon to nitrogen ratio causes a decrease in the amount of methane due to disruption of the carbon metabolism [63].

Despite of mentioned inconveniences, a number of studies showed that waste of meat processing has a higher biomethane yield, i.e., compared to maize silage, the basic biogas input used in Europe. Furthermore, the biomass fermentation can be an effective tool to reduce the unfavourable effects of improper waste management from the meat industry, and the electricity and heat obtained can be an additional source of income for meat processing plants or used for their technological purposes [55,69].

4.2.4. Distillery Waste

Distillery stillage is a main by-product obtained during ethanol production [71,72]. Poland is one of the largest spirits producers worldwide and the amount of distillery stillage exceeds up to 12 times of alcohol, resulting in an estimated several million tons per year [73]. It creates a great problem with disposal of its surplus. Distillers contain, in addition to organic carbon (Ct) compounds, mineral nutrients necessary for plants and can be used to fertilize or improve the soil quality. A common feature of decoctions is too low phosphorus content in relation to nitrogen and potassium and relatively high Ct content [74]. Grain stillage is characterized by high content of B vitamins, minerals, exogenous amino acids, dietary and lactogenic value and favourable protein–oat ratio; therefore, it is used in the forage industry [75]. In turn, molasses stillage is not suitable for the forage purpose, however, for the economic reasons, is often used by biofuel producers as a substrate to produce ethanol [76]. Though molasses stillage may be used as fertilizer, the application on the field might be problematic due to its polluting potential and would generate further costs related with the constructions of the appropriate tanks for its storage [71,76]. At this point, attention should be paid to the hazard associated with excess potassium accumulation in the case of fertilization with this waste of crops grown for feed purposes [75]. Moreover, the period of usefulness of stillage is relatively short because of the risk of microbiological development [73].

Nevertheless, a number of studies were conducted with the aim of finding an optimal solution, regarding the difficulties associated with the large loads of distillery waste. The anaerobic digestion with biogas production is one of the alternative methods proposed for the utilization of stillage [49,54,77,78]. Biogas efficiency from stillage has been estimated in the range between 430–725 m^3 Mg^1 of dry organic matter, with methane content of 55% and is comparable to other substrates from the agri-food industry [79]. From the other hand, the use of stillage as the input for biogas plants can be justified in the case of a stable situation of the domestic ethanol industry and when the installation is located in vicinity of a distillery that generates great quantities of this waste. This would minimize the costs related with transport which might contribute to unfavourable economic stability [80,81]. Additionally, the high amount of potassium in molasses and accordingly in distillery waste, not only contaminates fields when it is used as fertilizer but may also inhibit bacteria involved in the anaerobic digestion. Therefore, when supplying the fermentation chamber, special attention should be paid to quality as well as chemical and elemental composition of the substrate [74,82].

4.2.5. Fresh Fruits and Vegetables: The Controversial Case

Utilization of fresh agricultural products for energy purposes seems to be unjustified and controversial. On other hand, in certain cases might be necessarily. Fresh fruits and vegetables require appropriate storage conditions, which most of the farms are deprived of. The amount of unused fresh biomass might be problematic for many food producers, as the product that stays with the farmers, become a waste and need to be disposed. For instance, the embargo imposed by the Russian Federation in 2014, had significant influence on Poland´s economic situation, causing the saturation of local markets with fresh agricultural products [56,57]. Thus, the farmers were recommended to use the waste as the substrate for biogas plant, because it would be profitable from the economic point of view [83]. The study conducted by Smurzyńska et al. [57], with the aim to determine the biogas yield and dynamic of the fermentation process of surplus of fresh vegetables and fruits (biogas and methane productivity tests were carried out for the following vegetables and fruits: eggplants, pumpkins, cauliflower, cabbage, peppers, tomatoes, cucumbers) demonstrated that the process of methane production has not been impaired by any inhibitory factors and biogas yield obtained from individual vegetables and fruits tested was comparable. Furthermore, a large number of polysaccharides in substrates tested contributed to a short but intense process of biogas production. A comparable study (also with the fruits covered by embargo, such as: apples, pears, nectarines, peaches and plums) showed that the biogas yield obtained from the tested fruits were also similar (653 m^3/Mg peaches, 829.66 m^3/Mg apples on organic matter) [56]. Based on these studies it can be concluded that biomass obtained from fresh non-traded agricultural products are a suitable substrate for biogas production, however, should be used only in case of exceptional situations.

5. Limitation for Development of Agricultural Biogas Plants in Poland

Agricultural biogas plants have huge potential that can positively influence both the environmental aspect and socio-economic development of a given area [34,84]. Currently there are 96 agricultural biogas plants in Poland under operation [85]. In Germany, for example, the number of installations is over 9400 and the country has similar potential for biogas production, when comparing arable land [85]. This difference can be explained by limiting factors such as the location. Construction of the installation may cause social resistance of the local communities. For instance, the surveys conducted with the residents of rural commune Kamionka in Lublin province, Poland, demonstrated that most of the residents were concerned about the unpleasant odour (60% of respondents) [86]. Other fears of the respondents were related to pollution, noise and risk of explosion. On the other hand, over 80% of the respondents indicated the benefits from an agricultural biogas facility, such additional income for the farmers, provision of the cheap energy for the community and positive impact on local environment. Moreover, some local producers were willing to cooperate with the owners of biogas facilities, by buying the energy, providing the biomass and using the post-fermentation pulp (digestate) for fertilizing purposes. The inconvenience related with cost of transport or seasonality of agri-food waste availability, characteristic for the areas with small farms that may cause input instability can be clarified by cooperation between farmers, producing different biomass [87]. Therefore, the decision regarding the selection of the place for the biogas installation should not only consider technological, environmental and legal issues, but also the social aspect should be taken into account [86].

The utilization of digestate might be another limiting factor for the development of agricultural biogas plants in Poland. The operating of agricultural biogas installations is associated with the generation of a large amount of post-fermentation pulp. This product, resulted from the digestion process contains more inorganic nitrogen than non-digested organic fertilizers, and, in consequence, more nitrogen in a form available for plants [88]. Previous studies conducted in EU countries demonstrated the possibility of using the digestate as a replacement for the traditional fertilizer or soil amendment, with the benefits both for the farmers (impact on the crop yields) and soil properties [89–93]. Nevertheless, in some cases the digestate management can be problematic for the biogas producers. When using digestate as the organic fertilizer, it is necessary to comply with several legal requirements regarding both storage methods of biomass intended for methane fermentation as the input and for fermentation pulp on the premises [93,94]. The other important issue of utilizing digestate for fertilizing purposes concerns is its classification. According to the law, digestate may be considered waste or by-product. However, numerous legal regulations for the digestate to be classified as a by-product and therefore qualify it for use as an organic fertilizer (or soil amendment), may be inconvenient for the producers (e.g., farmers, owners of the installation, etc.). Furthermore, the bio-fertilizers based on digestate require accurate physicochemical and microbiological tests in specialized research institutions and need to meet the procedures set by decision-makers [95]. These procedures apply to the use of digestate in agriculture as well as in gardening and forestry. The most important legal acts regarding the utilization and the requirements for the classification of digestate as a by-product, as well as legal requirements regarding methods of storage of substrates and digestate on biogas plant areas were discussed by Czekała et al. [94] and Łagocka et al. [96]. Nonetheless, due to the growing interest in energy obtained from the agricultural biogas installations in Poland, and consequently, an increased amount of post-digestate pulp attention should be given to the alternative methods of its management. In Italy for instance, the usage of digestate became a key factor to maintain profitability of biogas plants and to promote bioeconomy [88,97].

For the further development of agricultural biogas plants in Poland it is crucial to show farmers and residents the benefits of this type of investment. Promotion of bioeconomy as an important element of environmental sustainability and usage of renewable biological resources [98] as well as creating favourable conditions for research on cost-effective and implementing practical solutions [99].

6. Final Considerations

Agriculture plays an important role in the Polish economy and Poland is considered a producer and exporter of good quality products. The country has also considerable potential for biomass processing using agricultural, forest and municipal waste. Biomass from the residues of agri-food production and agricultural, especially bovine slurry, maize silage and distilleries has a great energy potential and is a valuable substrate for agricultural biogas production. Simultaneously it would be essential to implement and develop available and cost-effective technologies that convert biomass of agricultural origin into energy, while not competing with the food and forage market. The use of controversial products, for example fresh fruits and vegetables as a substrate should be considered with caution. Regarding the use of biomass for energy purposes, factors such an economic aspect, substrate availability and substrate storage should be taken into account. Utilization of digestate as a bio-fertilizer or soil amendment and its effect on crop yields is a priority for farmers. Nevertheless, here we propose further research on the impact of digestate on soil carbon sequestration, greenhouse gas emissions and on the use of digestate in degraded areas in order to restore soil ecosystem services. Agricultural biogas installations have the potential to contribute to the greening of the Polish energy sector but unless certain restrictions are overcome, the share of biomass for energy production might be limited.

Supplementary Materials: The following are available online at http://www.mdpi.com/2071-1050/11/22/6515/s1, Table S1: List of keywords used for literature search, Table S2: Comparison of biomass from agriculture plants in terms of biogas yield, Figure S1: Structure of primary energy production from RES in Poland, 2016.

Author Contributions: Conceptualization, K.A.K. and M.K.; methodology, K.A.K. and A.E.L.; literature review, K.A.K. and K.G., database creation, K.A.K.; writing-original draft preparation, K.A.K.; writing-review end editing, A.E.L., K.G., and M.K.; funding acquisition, M.K.

Funding: The publication was financed by the National Centre for Research and Development, under the strategy program "Natural environment, agriculture and forestry" BIOSTRATEG III (BIOSTRATEG3/345940/7/NCBR/2017).

Acknowledgments: Anonymous reviewers are gratefully acknowledged for their constructive review that significantly improved this manuscript.

Conflicts of Interest: The authors declare no conflict of interest.

References

1. Suocheng, D.; Tong, K.W.; Yuping, W. Municipal solid waste management in China: Using commercial management to solve a growing problem. *Util. Policy* **2011**, *10*, 7–11. [CrossRef]
2. Rosik-Dulewska, C.; Karwaczyńska, U.; Ciesielczuk, T. Możliwości wykorzystania odpadów organicznych i mineralnych z uwzględnieniem zasad obowiązujących w ochronie środowiska (Possibilities of using organic and mineral waste, regarding the principles applicable in environmental protection). *Rocz. Ochr. Srodowiska* **2011**, *13*, 361–376. (In Polish)
3. Department of Economic and Social Affairs, Population Division. *World Population Prospects: The 2015 Revision, Key Findings and Advance Tables*; United Nations: New York, NY, USA, 2015; p. 241.
4. Kostecka, J.; Koc-Jurczyk, J.; Brudzisz, K. Waste management in Poland and European Union. *Arch. Waste Manag. Environ. Prot.* **2014**, *16*, 1–10.
5. Jambeck, J.R.; Geyer, R.; Wilcox, C.; Siegler, T.R.; Perryman, M.; Andrady, A.; Narayan, R.; Law, L.K. Plastic waste inputs from land into the ocean. *Science* **2015**, *347*, 768–771. [CrossRef] [PubMed]
6. Rodriguez, A.; Latawiec, A.E. Rethinking Organic Residues: The Potential of Biomass in Brazil. *Mod. Concepts Dev. Agron.* **2018**, *1*, 1–5.

7. Grando, R.L.; De Souza Antune, A.M.; Da Fonseca, F.V.; Sánchez, A.; Barrena, R.; Font, X. Technology overview of biogas production in anaerobic digestion plants: A European evaluation of research and development. *Renew. Sustain. Energy Rev.* **2017**, *80*, 44–53. [CrossRef]

8. Renewable Energy Law of Poland [Dz.U. 2012 poz. 1229]. Available online: http://prawo.sejm.gov.pl/isap.nsf/download.xsp/WDU20120001229/O/D20121229.pdf (accessed on 8 October 2019).

9. Nowacka-Blachowska, A.; Resak, M.; Rogosz, B.; Tomaszewska, H. Zrównoważone wykorzystanie biomasy na terenie Dolnego Śląska (Sustainable use of biomass in Lower Silesia). *Gor. Odkryw.* **2016**, *6*, 48–53. (In Polish)

10. Roszkowski, A. Biomasa i bioenergia—Bariery technologiczne i energetyczne (Biomass and bioenergy—Technological and energy limitations). *Probl. Inz. Rol.* **2012**, *3*, 79–100. (In Polish)

11. Kacprzak, A.; Krzystek, L.; Ledakowicz, L. Badania biochemicznego potencjału metanogennego wybranych roślin energetycznych (Analysis on the biochemical methanogenic potential of selected energy plants). *Inżynieria I Aparatura. Chemiczna* **2010**, *49*, 32–33. (In Polish)

12. Ansari, A.A. Effect of Vermicompost on the Productivity of Potato (*Solanum tuberosum*), Spinach (*Spinaciaoleracea*) and Turnip (*Brassica campestris*). *World J. Agric. Sci.* **2008**, *4*, 333–336.

13. Castro, A.; Da Silva Batista, N.; Latawiec, A.E.; Rodrigues, A.; Strassburg, B.B.N.; Silva, D.; Araujo, E.; De Moraes, L.F.D.; Guerra, J.G.; Galvão, G.; et al. The effects of *Gliricidia*-Derived Biochar on Sequential Maize and Bean Farming. *Sustainability* **2018**, *10*, 578. [CrossRef]

14. Yuan, H.; Lu, T.; Wang, Y.; Chen, Y.; Lei, T. Sewage sludge biochar: Nutrient composition and its effect on the leaching of soil nutrients. *Geoderma* **2016**, *267*, 17–23. [CrossRef]

15. Climate and Energy Policies in Poland. 2017. Available online: http://www.europarl.europa.eu/RegData/etudes/BRIE/2017/607335/IPOL_BRI(2017)607335_EN.pdf (accessed on 8 October 2019).

16. World Energy Council. Available online: https://www.worldenergy.org/data/resources/country/poland/coal/ (accessed on 8 October 2019).

17. Dzikuć, M.; Piwowar, A. Ecological and economic aspects of electric energy production using the biomass co-firing method: The case of Poland. *Renew. Sustain. Energy Rev.* **2016**, *55*, 856–862. [CrossRef]

18. Dec, B.; Krupa, J. Wykorzystanie odnawialnych źródeł energii w aspekcie ochrony środowiska (The use of renewable energy sources in the aspect of environmental protection). *Przegląd Nauk. Metod. Eduk. Bezpieczenstwa* **2007**, *3*, 722–757. (In Polish)

19. Zuwała, J. Life cycle approach for energy and environmental analysis of biomass and coal co-firing in CHP plant with backpressure turbine. *J. Clean. Prod.* **2012**, *35*, 164–175. [CrossRef]

20. Ciepielewska, M. Development of Renewable Energy in Poland in the Light of the European Union Climate-Energy Package and the Renewable Energy Sources Act. 2016. Available online: http://dx.doi.org/10.18778/1429-3730.43.01 (accessed on 20 September 2018).

21. Gołuchowska, B.; Sławiński, J.; Markowski, G. Biomass utilization as a renewable energy source in polish power industry-current status and perspectives. *J. Ecol. Eng.* **2015**, *16*, 143–154. [CrossRef]

22. Ross, A.B.; Jones, J.M.; Chaiklangmuang, S.; Pourkashanian, M.; Williams, A.; Kubica, K.; Andersson, J.T.; Kerst, M.; Danihelka, P.; Bartle, K.D. Measurement and prediction of the emission of pollutants from the combustion of coal and biomass in a fixed bed furnace. *Fuel* **2002**, *81*, 571–582. [CrossRef]

23. Jasiulewicz, M. Potencjał energetyczny biomasy rolniczej w aspekcie realizacji przez Polskę Narodowego Celu Wskaźnikowego OZE i dyrektyw w UE w 2020 (Energy potential of agricultural biomass in the aspect of Poland's implementation of the National RES Target and EU directives in 2020). *Rocz. Nauk. Stowarzyszenia Ekon. Rol. Agrobiz.* **2014**, *16*, 70–76. (In Polish)

24. Bartosiewicz-Burczy, H. Potencjał i energetyczne wykorzystanie biomasy w rajach Europy Srodkowej (Biomass potential and its energy utilization in the Central European countries). *Energetyka* **2012**, *7*, 860–866. (In Polish)

25. Energy 2018. Available online: https://stat.gov.pl/files/gfx/portalinformacyjny/pl/defaultaktualnosci/5485/1/6/1/energia_2018.pdf (accessed on 8 October 2019).

26. Kuziemska, B.; Trębicka, J.; Wieremej, W.; Klej, P.; Pieniak-Lendzion, K. Benefits and risks in the production of biogas. *Zesz. Nauk. Uniw. Przyr. Humanist. Siedlcach Ser. Adm. Zarządzanie* **2014**, *103*, 99–113. (In Polish)

27. Maj, G. Emission factors and energy properties of agro and forest biomass in aspect of sustainability of energy sector. *Energies* **2018**, *11*, 1516. [CrossRef]

28. Sikora, J. The research on efficiency of biogas production from organic fraction of municipal solid waste mixed with agricultural biomass. *Infrastruct. Ecol. Rural Areas* **2012**, *2*, 89–98.

29. Sikora, J.; Tomal, A. Determination of the energy potential of biogas in selected farm household. *Infrastruct. Ecol. Rural Areas* **2016**, *3*, 971–982.

30. Czekała, W. Agricultural Biogas Plants as a Chance for the Development of the Agri-Food Sector. *J. Ecol. Eng.* **2018**, *19*, 179–183. [CrossRef]

31. Jędrejek, A.; Jarosz, Z. Regionalne możliwości produkcji biogazu rolniczego (Regional possibilities of agricultural biogas production). *Rocz. Nauk. Stowarzyszenia Ekon. Rol. Agrobiz.* **2016**, *18*, 61–65. (In Polish)

32. Cukrowski, A.; Oniszk-Popławska, A.; Mroczkowski, P.; Zowsik, M.; Wiśniewski, G.; The Institute of Renewable Energy, Warszawa. A Guide for Investors Interested in Building the Agricultural Biogas Plants. 2011. Available online: http://www.mg.gov.pl/node/13229 (accessed on 28 September 2018).

33. Piwowar, A.; Dzikuć, A.; Adamczyk, J. Agricultural biogas plants in Poland—Selected technological, market and environmental aspects. *Renew. Sustain. Energy Rev.* **2016**, *58*, 69–74. [CrossRef]

34. Obrycka, E. Korzyści społeczne i ekonomiczne budowy biogazowni rolniczych (Social and economic benefits of agricultural biogas plants). *Zesz. Nauk. Szk. Gl. Gospod. Wiej. Ekon. Organ. Gospod. Zywnosciowej* **2014**, *107*, 163–176. (In Polish)

35. Act of 10 July 2007 on Fertilizers and Fertilization Law of Poland. Available online: http://prawo.sejm.gov.pl/isap.nsf/download.xsp/WDU20071471033/T/D20071033L.pdf (accessed on 12 November 2019).

36. Marszałek, M.; Banach, M.; Kowalski, Z. Utylizacja gnojowicy na drodze fermentacji metanowej i tlenowej—Produkcja biogazu i kompostu (Utilization of liquid manure by methane and oxygen fermentation—Biogas and compost production). *Czas. Tech.* **2011**, *10*, 143–158. (In Polish)

37. Banaszkiewicz, T.; Wysmyk, J. Ecological aspect of utilization of agricultural feedstock. *Eur. Reg.* **2015**, *23*, 21–34. (In Polish)

38. Pawlak, J. Biogaz z rolnictwa—Korzyści i bariery (Biogas from agriculture—Benefits and limitations). *Probl. Agric. Eng.* **2013**, *3*, 99–108. (In Polish)

39. Borowski, S.; Domański, J.; Weatherley, L. Anaerobic co-digestion of swine and poultry manure with municipal sewage sludge. *Waste Manag.* **2014**, *34*, 513–521. [CrossRef] [PubMed]

40. Bujoczek, G.; Oleszkiewicz, J.; Sparling, R.; Cenkowski, S. High Solid Anaerobic Digestion of Chicken Manure. *J. Agric. Eng. Res.* **2000**, *76*, 51–60. [CrossRef]

41. Hryniewicz, M.; Grzybek, A. Available straw surplus for use for energy purposes in 2016. *Probl. Inz. Rol.* **2017**, *25*, 15–31. (In Polish)

42. Jarosz, Z. Energy potential of agricultural crops biomass and their use for energy purposes. *Sci. Noteb. Wars. Univ. Life Sci. SGGW Wars. Probl. World Agric.* **2017**, *17*, 81–92. (In Polish)

43. Jasiulewicz, M. Possibility of Liquid Biofuels, Electric and Heat Energy Production from Biomass in Polish Agriculture. *Polish J. Environ. Stud.* **2010**, *19*, 479–483.

44. Michalski, T. Biogazownia w każdej gminie—Czy wystarczy surowca (Biogas plant in every municipality—Is there enough raw material). *Wieś Jutra* **2009**, *3*, 12–14. (In Polish)

45. Romaniuk, W.; Domasiewicz, T.; Borek, K.; Borusiewicz, A.; Marczuk, T. *Analiza Potrzeb Techniczno-Technologicznych oraz Propozycje Rozwiązań w Produkcji Biogazu w Gospodarstwach Rodzinnych i Farmerskich (The Analysis of Technical Demands and Possible Solutions for Biogas Production in Family Agriculture and on Farms)*; Wydawnictwo Wyższej Szkoły Agrobiznesu w Łomży: Lomży, Poland, 2015. (In Polish)

46. Kowalczyk-Juśko, A.; Kościk, B.; Jóźwiakowski, K.; Marczuk, A.; Zarajczyk, J.; Kowalczuk, J.; Szmigielski, M.; Sagan, A. Effects of biochemical and thermochemical conversion of sorghum biomass to usable energy. *Przem. Chem.* **2015**, *94*, 1838–1840.

47. Król, A. Kiszonki—Cenny substrat do produkcji biogazu (Silage—A valuable substrate for biogas production). *Autobusy Tech. Eksploat. Syst. Transp.* **2011**, *10*, 249–254. (In Polish)

48. Księżak, J. Produkcja kukurydzy w różnych rejonach Polski (Maize production in various regions of Poland). *Wieś Jutra* **2009**, *3*, 16. (In Polish)

49. Daniel, Z.; Juliszewski, T.; Kowalczyk, Z.; Malinowski, M.; Sobol, Z.; Wrona, P. The method of solid waste classification from the agriculture and food industry. *Infrastruct. Ecol. Rural Areas* **2012**, *2*, 141–152.

50. Czyżyk, F.; Strzelczyk, M. Rational utilization of production residues generated in agri-food. *Arch. Waste Manag. Environ. Prot.* **2015**, *17*, 99–106.

51. Kruczek, M.; Drygaś, B.; Habryka, C. Pomace in fruit industry and their contemporary potential application. *World Sci. News* **2016**, *48*, 259–265.

52. Wesołowska-Trojanowska, M.; Targoński, Z. The whey utilization in biotechnological processes. *Eng. Sci. Technol.* **2014**, *1*, 102–119.

53. Michalska, K.; Pazera, A.; Bizukojć, M.; Wolf, W.; Sibiński, M. Innovative dairies-energy independence and waste-free technologies as a consequence of biogas and photovoltaic investments. *Acta Innov.* **2013**, *9*, 5–16.

54. Adamski, M.; Pilarski, K.; Dach, J. Possibilities of Usage of the Distillery Residue as a Substrate for Agricultural Biogas Plant. *J. Res. Appl. Agric. Eng.* **2009**, *54*, 10–15. (In Polish)

55. Kozłowski, K.; Cieślik, M.; Smurzyńska, A.; Lewicki, A.; Jas, M. The usage of waste from meat processing for energetic purposes. *Eng. Sci. Technol.* **2015**, *1*, 36–46.

56. Czekała, W.; Smurzyńska, A.; Cieślik, M.; Boniecki, P.; Kozłowski, K. Biogas Efficiency of Selected Fresh Fruit Covered by the Russian Embargo. In Proceedings of the 16th International Multidisciplinary Scientific Conference SGEM 2016, Albena, Bulgaria, 30 June–6 July 2016; Volume 3, pp. 227–234.

57. Smurzyńska, A.; Czekała, W.; Lewicki, A.; Cieślik, M.; Kozłowski, K.; Janczak, D. The biogas output of vegetables utilized in the polish market due to introduction of the Russian embargo. *Tech. Rol. Ogrod. Leśna* **2016**, *6*, 24–27.

58. Weiland, P. Biogas production: Current state and perspectives. *Appl. Microbiol. Biotechnol.* **2010**, *85*, 849–860. [CrossRef]

59. Kupryś-Caruk, M. Agri-food industry as a source of substrates for biogas production. *Postępy Nauk. Technol. Przem. Rolno Spoz.* **2017**, *72*, 69–85. (In Polish)

60. Pilarska, A.; Pilarski, K.; Ryniecki, A. The use of methane fermentation in the development of selected waste products of food industry. *Nauk. Inz. Technol.* **2014**, *4*, 100–111. (In Polish)

61. Prask, H.; Fugol, M.; Szlachta, J. Biogaz z wytłoków z białych i czerwonych winogron. *Przem. Ferment. Owocowo Warzywny.* **2012**, *5*, 45–46. (In Polish)

62. Bożym, M.; Florczak, I.; Zdanowska, P.; Wojdalski, J.; Klimkiewicz, M. An analysis of metal concentrations in food wastes for biogas production. *Renew. Energy* **2015**, *77*, 467–472. [CrossRef]

63. Kwaśny, J.; Banach, M.; Kowalski, Z. Technologies of biogas production from different sources—A review. *Chem. Tech. Trans.* **2012**, *17*, 83–102. (In Polish)

64. Maślanka, S.; Siołek, M.; Hamryszak, Ł.; Łopot, D. Zastosowanie odpadów z przemysłu mleczarskiego do produkcji polimerów biodegradowalnych. *Chemik* **2014**, *68*, 703–709. (In Polish)

65. Janczukowicz, W.; Zieliński, M.; Dębowski, M. Biodegradability evaluation of dairy effluents originated in selected sections of dairy production. *Bioresour. Technol.* **2008**, *99*, 4199–4205. [CrossRef] [PubMed]

66. Sobczak, A.; Błyszczek, E. Ways of management of by-products from meat industry. *Czas. Tech. Chem.* **2009**, *106*, 141–151.

67. Adhikari, B.B.; Chae, M.; Bressler, D.C. Utilization of Slaughterhouse Waste in Value-Added Applications: Recent Advances in the Development of Wood Adhesives. *Polymers* **2018**, *10*, 176. [CrossRef]

68. Zakrzewski, P. Technologia utylizacji odpadów poubojowych w instalacjach biogazowych (Technology of post-slaughter waste utilization in biogas installations). *Czysta Ènerg.* **2009**, *10*, 40–41. (In Polish)

69. Dach, J.; Kozłowski, K.; Czekała, W. Odpady poubojowe na biogaz—Czy to sie opłaca? (Post-slaughter waste for biogas—Is it profitable). *Biomasa* **2017**, *8*, 38–40.

70. EUR-Lex. Available online: https://eur-lex.europa.eu/eli/reg/2009/1069/oj (accessed on 8 October 2019).

71. Fuess, L.T.; Garcia, M.L. Anaerobic digestion of stillage to produce bioenergy in the sugarcane-to-ethanol industry. *Environ. Technol.* **2014**, *35*, 333–339. [CrossRef]

72. Dubrovskis, V.; Plume, I. Methane production from stillage. In Proceedings of the 16th International Scientific Conference Engineering for Rural Development, Latvia University of Life Sciences and Technology, Jelgava, 24–26 May 2017. [CrossRef]

73. Czupryński, B.; Kotarska, K. Recyrkulacja i sposoby zagospodarowania wywaru gorzelniczego (Recirculation and methods of managing stillage). *Inz. Apar. Chem.* **2011**, *50*, 21–23. (In Polish)

74. Skowrońska, M.; Filipek, T. Nawozowe wykorzystanie wywaru gorzelnianego (The use of stillage for fertilizing purposes). *Proc. ECOpole* **2012**, *6*, 267–271. (In Polish) [CrossRef]

75. Kotarska, K.; Kłosowski, G.; Czupryński, B. Zagospodarowanie wywaru gorzelniczego na cele paszowe (Managing of stillage for fodder purposes). *Przem. Ferment. Owocowo Warzywny* **1996**, *40*, 27–30. (In Polish)

76. Kotarska, K.; Dziemianowicz, W. Effect of Different Conditions of Alcoholic Fermentation of Molasses on Its Intensification and Quality of Produced Spirit. *Zywnosc Nauka. Technologia. Jakosc* **2015**, *2*, 150–159. (In Polish) [CrossRef]

77. Owczuk, M.; Wardzińska, D.; Zamojska-Jaroszewicz, A.; Matuszewska, A. The use of biodegradable waste to produce biogas as an alternative source of renewable energy. *Studia Ecol. Bioethicae UKSW* **2013**, *11*, 133–144. (In Polish)

78. Jasiulewicz, M. Implementation of the innovative investment in food industry and anaerobic digestion in the field of bioenergy. *Rocz. Nauk. Stowarzyszenia Ekon. Rol. Agrobiz.* **2017**, *19*, 88–94. (In Polish)

79. Romaniuk, W.; Domasiewicz, T. Substraty dla biogazowni rolniczych (Substrates for agricultural biogas plants). *Agrotech. Porad. Rolnika* **2014**, *11*, 74–75. (In Polish)

80. Janczak, D.; Kozłowski, K.; Zbytek, Z.; Cieślik, M.; Bugała, A.; Czekała, W. Energetic Efficiency of the Vegetable Waste Used as Substrate for Biogas Production. *Environment &Chem.* **2016**, *64*, 06002.

81. Kozłowski, K.; Lewicki, A.; Cieślik, M.; Janczak, D.; Czekała, W.; Smurzyńska, A.; Brzoski, M. The possibility of improving the energy and economic balance of agricultural biogas plant. *Tech. Rol. Leśna* **2017**, *3*, 10–13.

82. Kasprzycka, A. Causes of interference methane fermentation. *Autobusy* **2011**, *10*, 224–228. (In Polish)

83. Szymanska, D.; Lewandowska, A. Biogas power plants in Poland—Structure, capacity, and special distribution. *Sustainability* **2015**, *7*, 16801–16819. [CrossRef]

84. Rzeznik, W.; Mielcarek, P. Agricultural biogas plants in Poland. *Eng. Rural Dev.* **2018**, *17*, 1760–1765.

85. Biomass Media Group. Available online: http://rynekbiogazu.pl/2018/03/21/potencjal-rozwoju-sektora-biogazu-w-polsce (accessed on 8 October 2019).

86. Kowalczyk-Juśko, A.; Listosz, A.; Flisiak, M. Spatial and social conditions for the location of agricultural biogas plants in Poland (case study). *E3S Web Conf.* **2019**, *86*, 00036. [CrossRef]

87. Caruso, M.C.; Braghieri, A.; Capece, A.; Napolitano, F.; Romano, P.; Galgano, F.; Altieri, G.; Genovese, F. Recent updates on the use of agro-food waste for biogas production. *Appl. Sci.* **2019**, *9*, 1217. [CrossRef]

88. Bartoli, A.; Hamelin, L.; Rozakis, S.; Borzęcka, M.; Brandão, M. Coupling economic and GHG emission accounting models to evaluate the sustainability of biogas policies. *Renew. Sustain. Energy Rev.* **2019**, *106*, 133–148. [CrossRef]

89. Koszel, M.; Lorencowicz, E. Agricultural use of biogas digestate as a replacement fertilizers. *Agric. Agric. Sci. Procedia* **2015**, *7*, 119–124. [CrossRef]

90. Losak, T.; Hlusek, J.; Zatloukalova, A.; Musilova, L.; Vitezova, M.; Skarpa, P.; Zlamalova, T.; Fryc, J.; Vitez, T.; Marecek, J.; et al. Digestate from biogas plants is an attractive alternative to mineral fertilisation of kohlrabi. *J. Sustain. Dev. Energy Water Environ. Syst.* **2014**, *2*, 309–318. [CrossRef]

91. Šimon, T.; Kunzová, E.; Friedlová, M. The effect of digestate, cattle slurry and mineral fertilization on the winter wheat yield and soil quality parameters. *Plant Soil Environ.* **2015**, *61*, 522–527. [CrossRef]

92. Koszel, M.; Kocira, A.; Lorencowicz, E. The evaluation of the use of biogas plant digestate as a fertilizer in alfalfa and spring wheat cultivation. *Fresenius Environ. Bull.* **2016**, *25*, 3258–3264.

93. Pilarska, A.A.; Piechota, T.; Szymańska, M.; Pilarski, K.; Wolna-Maruwka, A. Ocena wartości nawozowej pofermentów z biogazowni oraz wytworzonych z nich kompostów (Evaluation of fertilizer value of digestate and its composts obtained from biogas plant). *Nauka Przyr. Technol.* **2016**, *10*, 35. (In Polish) [CrossRef]

94. Czekała, W.; Pilarski, K.; Dach, J.; Janczak, D.; Szymańska, M. Analysis of management possibilities for digestate from biogas plant. *Tech. Rol. Ogrod. Leśna* **2012**, *4*, 11–13. (In Polish)

95. Mystkowski, E. Poferment dla rolnictwa (Digestate for agriculture). *Rol. ABC* **2015**, *9*, 1–5. (In Polish). Available online: https://studylibpl.com/doc/1424615/masa-pofermentacyjna-nawozem-dla-rolnictwa (accessed on 8 October 2019).

96. Łagocka, A.; Kamiński, M.; Cholewński, M.; Pospolita, W. Korzyści ekologiczne ze stosowania pofermentu z biogazowni rolniczych jako nawozu organicznego (Ecological benefits from the use of digestate from agricultural biogas plants as organic fertilizer). *Kosmos* **2016**, *65*, 601–607. (In Polish)

97. Pantaleo, A.; De Gennaro, B.; Shah, N. Assessment of optimal size of anaerobic co-digestion plants: An application to cattle farms in the province of Bari (Italy). *Renew. Sustain. Energy Rev.* **2013**, *20*, 57–70. [CrossRef]

98. Konstantinis, A.; Rozakis, S.; Maria, E.A.; Shu, K. A definition of bioeconomy through bibliometric networks of the scientific literature. *AgBioForum* **2018**, *21*, 64–85.

99. Krzywonos, M.; Marciszewska, A.; Domiter, M.; Borowiak, D. Bioeconomy -current status, trends and prospects. The challenge for universities, business and government. *Pol. J. Agron.* **2016**, *27*, 71–79. (In Polish)

 *sustainability*

Article

Lyophilized Protein Structures as an Alternative Biodegradable Material for Food Packaging

Katarzyna Kozłowicz [1], Sybilla Nazarewicz [1], Dariusz Góral [1,*], Anna Krawczuk [2] and Marek Domin [1]

[1] Department of Biological Bases of Food and Feed Technologies, University of Life Sciences in Lublin, Głęboka 28, 20-612 Lublin, Poland; katarzyna.kozlowicz@up.lublin.pl (K.K.); sybilla_klap.94@o2.pl (S.N.); marek.domin@up.lublin.pl (M.D.)

[2] Department of Machinery Exploitation and Management of Production Processes, University of Life Sciences in Lublin, Akademicka 13, 20-950 Lublin, Poland; anna.krawczuk@up.lublin.pl

* Correspondence: dariusz.goral@up.lublin.pl; Tel.: +48-81-531-97-38

Received: 12 November 2019; Accepted: 5 December 2019; Published: 8 December 2019

Abstract: Considering the need for sustainable development in packaging production and environmental protection, a material based on lyophilized protein structures intended for frozen food packaging was produced and its selected thermophysical properties were characterized. Analyses of density, thermal conductivity and thermal diffusivity were performed and strength tests were carried out for lyophilized protein structures with the addition of xanthan gum and carboxymethyl cellulose. Packagings were made of new materials for their comparative assessment. Then, the surface temperature distribution during thawing of the deep-frozen product inside the packaging was tested. In terms of thermal insulation capacity, the best properties were obtained for sample B4 with a thermal conductivity of $\lambda = 0.06$ W·(mK)$^{-1}$), thermal capacity $C = 0.29$ (MJ·(m^{3}K)$^{-1}$) and thermal diffusivity $a = 0.21$ (mm^2·s^{-1}). The density and hardness of the obtained lyophilized protein structures were significantly lower compared to foamed polystyrene used as a reference material. Thermal imaging analysis of the packaging showed the occurrence of local freezing. Lyophilized protein structures obtained from natural ingredients meet the needs of consumers and are environmentally friendly. These were made in accordance with the principles of sustainable development and can be an alternative material used for the production of frozen food packaging.

Keywords: packaging; biodegradable material; lyophilized protein structure

---

1. Introduction

Packaging is an important factor in maintaining food quality and commercial attractiveness, facilitating its transport at the same time. Nowadays, packaging, in addition to protecting, informing or marketing functions, should show additional functional properties and meet many different requirements. Innovations play a special role in modifying packaging functionality and improving its barrier, strength and aging resistance properties while avoiding the use of environmentally harmful materials. This contributes to the creation of a new generation of packagings that allow to maintain and even improve the quality of the packaged product, which is a very desirable feature, especially in the case of food packaging [1,2]. Legal regulations, legislative pressure from governments and non-governmental organizations dealing with environmental protection [3,4], as well as consumers themselves make food producers more and more interested in providing new pro-ecological and sustainable solutions in production. At the same time, packaging made in accordance with the principles of sustainable development must be safe for health, life and the environment [5–7].

One of the directions of development of food packaging technology that meets the requirements of sustainable development is obtaining fully composted and biodegradable materials with the addition of natural fillers [8]. An alternative that can reduce carbon footprint, pollution risks and greenhouse gas emissions caused by the use of conventional polymers is the use of biopolymers from agroindustrial sources that are renewable and low cost [9]. Recent studies have shown that starch can be used to obtain foams [10–12] retaining its biodegradable character when converted to a thermoplastic material. Research on the use of natural fillers such as: plantain flour, wood fiber, sugarcane bagasse, asparagus peel fiber [13,14], sunflower protein and cellulose fiber [15], plant protein, palm oil [16] and grape stalks [1] help to improve the physical, chemical, mechanical and technological properties of the material. For example, protein derivatives are the most attractive biopolymers for edible film formulations because they provide high nutritional value, superior mechanical properties and exhibit the oxygen barrier [17]. Carboxymethylcellulose based films are easily water soluble as it contains a hydrophobic polysaccharide backbone and many hydrophilic carboxyl groups [18]. Carboxymethylcellulose improves protein film mechanical properties by increasing thermal stability and the elasticity modulus [19]. Gelatin films blended with xanthan gum characterize a transparent film with excellent ultraviolet light resistance, low total soluble matter and moisture content, low water vapour permeability, improved mechanical properties and thermal stability [20]. Xanthan gum is cross-linker to be blended with various materials; it may dissolves directly in many highly acidic, alkaline, alcoholic systems containing different components. It is also compatible with commercially available thickeners such as sodium alginate, carboxymethylcellulose and starch [21].

Biodegradable packaging, due to the possibility of full compostability, does not pose a threat to the environment. On the contrary, it can enrich the soil with nutrients. Despite significant technological progress, there are no packaging that meets all the requirements. For each group of products, especially chilled and frozen ones, the most rational packaging is selected taking into account technical, economic and legal conditions [22–24].

This paper discusses the need for sustainable development in packaging production to protect the environment, and a material based on lyophilized protein structures for frozen food has been produced and its selected thermophysical properties have been characterized. Hence, the present work used xanthan gum and carboxymethylcellulose as a crosslinking agents to form a potentially new natural and biodegradable materials. For this purpose, prototype packagings were made from the obtained lyophilized structures. The thawing kinetics of food stored in these packagings and in packaging made of foamed polystyrene were compared.

2. Materials and Methods

2.1. Research Material

The research material was lyophilized protein structures prepared on the basis of foams obtained from powdered albumin with high foaming activity containing 84.3% protein (*Basso*), modified by addition of carboxymethylcellulose (*Agnex, Białystok*) and xantan gum (*Agnex, Białystok*) in various percentage. Preparation of foams consisted of whipping albumin in distilled water for 5 minutes. Powdered carboxymethylcellulose and xanthan gum were added to the resulting foams and mixed for another 2–3 min. Four different foam variants with different percentages of individual components were prepared: 1-distilled water 88.0%, albumin 10.0% and carboxymethyl cellulose 2.0% (B1); 2-distilled water 88.0%, albumin 10.0% and xanthan gum 2.0% (B2); 3-distilled water 88.0%, albumin 6.0% and xanthan gum 6.0% (B3); 4-distilled water 88.0%, albumin 8.0% and xanthan gum 4.0% (B4). The obtained protein foams were transferred into 0.125 × 0.125 m plastic moulds. The samples prepared in this way were frozen in an air blast freezer at −30.0 °C. Then, after freezing, the sample was lyophilized for 72 h at a pressure of 20 Pa and an ice condensation temperature of −64 °C (ALPHA 2–4LD Plus freeze-dryer, Christ, Osterode am Harz, Germany) [25]. The obtained lyophilized protein structures with a thickness of 0.027 m are shown in Figure 1.

Figure 1. Cross-section of lyophilized protein structures: (**a**) B1, (**b**) B2, (**c**) B3, (**d**) B4.

2.2. Determination of Protein Structure Density

The density of the obtained samples (ρ) was calculated as the ratio of the mass of the samples to their volume. The volume of samples was determined using the formula for the cuboid volume (multiplying the dimensions: length, height and width).

2.3. Determination of the Strength of Protein Structures

For determination of the strength (hardness) of lyophilized protein structures, a fracture test was performer using a symmetrical knife with dimensions: blade thickness 0.003 m, blade angle 30° (texture analyzer LFRA 4500, Brookfield, Middleboro, MA, USA). Parameters of the device operation: knife speed 0.5 mm·s^{-1} and measurement accuracy 0.02 N. The dimensions of the tested samples were 0.10 × 0.04 m. The test for each sample was performed in 3 replications [26].

2.4. Measurement of Thermophysical Properties of Protein Structures

Lyophilized samples were analyzed for thermophysical properties such as thermal conductivity, heat capacity and thermal diffusivity using a KD2 Pro meter (Decagon Devices, Pullman, WA, USA) with the SH-1 probe. The measurement was carried out in 8 replications under the same conditions [27].

2.5. Making a Packaging Prototype

The packaging was made of 6 even cuboid walls with dimensions of 0.115 × 0.115 m and thickness of 0.18 m. The walls were joined with a specialist adhesive approved for food contact (*LOCTITE 454*). Only samples B3 and B4 were used for packaging. The remaining samples were rejected due to the soft and fragile structure that made them impossible to test. The reference packaging was made of foamed polystyrene (XPS) (Figure 2).

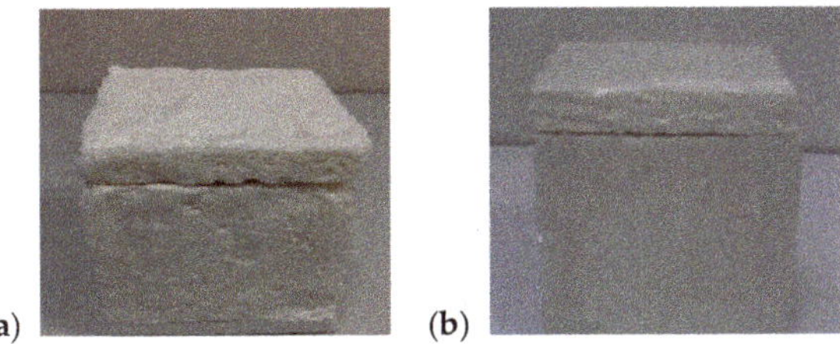

Figure 2. View of lyophilized protein structure of the sample B3 (**a**) and foamed polystyrene (**b**) packagings.

2.6. Thermovision Analysis of Lyophilized Protein Structures

For comparative assessment of the samples, the temperature fields recorded on their surface during thawing of the deep-frozen product (carrot with green peas) placed inside the packaging were tested. Thawing was carried out at an ambient temperature of 25 °C until the thawed product inside the packaging reached 23 °C. The temperature analyzer LB-515P cooperating with mini temperature and humidity data logger (type 23) was used to measure and record temperature. The mini data logger recorded temperature in a frozen product inside the packaging. The data obtained were used to compare thawing kinetics. For analysis of changes occurring during thawing of products on the surface of packaging walls, a Testo 882 thermal imaging camera (detector resolution 320×240 pixels, measuring accuracy ± 2 °C, thermal sensitivity < 50 mK), cooperating with the Testo IRSoft program Version 3.4., was used. Thermovision analysis of lyophilized structures was based on the detection of thermal bridges, i.e., points on the surface of the packaging wall with the highest temperature and histograms of temperature distribution along the line passing through the point with the lowest temperature (CS).

2.7. Statistical Analysis of Results

The results were evaluated with statistical analysis methods (determination of mean values, standard deviation) using the *Statistica* 13.1 program and analysis of variance (ANOVA) at a at the 95% confidence level. To verify the significance of differences between the average values, the Student's *t*-test was used, where the distribution of variables, followed a normal distribution.

3. Results and Discussion

3.1. Physical Properties of Obtained Lyophilized Protein Structures

Materials used for food packaging should have appropriate density. This parameter is closely related to the thermal conductivity coefficient. This relationship determines the decrease in the thermal conductivity coefficient as the material density increases. Material density is also important when mechanical properties are formed [28]. Table 1 shows the density and hardness of lyophilized protein structures modified by the addition of carboxymethylcellulose and xanthan gum in varying percentages. It was found that the density of lyophilized protein structures defined as the ratio of mass to volume increases with the increase of xanthan gum share in the samples B2, B3 and B4. The highest density (28.22 kg·m^{-3}), was noted for the sample B1 containing 10.0% albumin and 2.0% carboxymethyl cellulose, while the sample B2, which contained 10.0% albumin and 2.0% xanthan gum, had the lowest density (20.41 kg·m^{-3}). The percentage and type of modification used had a statistically significant impact on the density of lyophilized protein structures. The values of density of the obtained materials were significantly lower compared to foamed polystyrene which served as the reference sample. The density of the material is determined by the volume and dimensions of air bubbles which in turn depend on the viscosity of the environment and the conditions for the formation and dispersion of the gaseous phase [29].

Table 1. Density and bending force of lyophilized samples and foamed polystyrene.

Properties	B1	B2	B3	B4	XPS
Density (kg·m^{-3})	28.22 ± 0.28 [b]	20.41 ± 0.07 [c]	25.01 ± 0.11 [d]	24.51 ± 0.66 [d]	36.51 ± 1.17 [a]
p-value	0.01	0.001	0.003	0.002	
Bedning force (N)	0.68 ± 0.20 [b]	1.00 ± 0.21 [b]	4.49 ± 0.53 [c]	1.55 ± 0.11 [d]	20.84 ± 0.52 [a]
p-value	0.0001	0.0004	0.001	0.0004	

[a,b,c,d] Means in the same line indicated by different letters were significantly different (*p* value < 0.05). The results are expressed as mean ± SD ($n = 3$).

The conducted fracture test showed a much greater hardness of lyophilized protein structures with xanthan gum than those with carboxymethyl cellulose. The sample B3 containing 6.0% albumin and 6.0% xanthan gum had the highest hardness (4.49 N), whereas the lowest hardness (0.68 N) was noted for the sample B1, which contained 10.0% albumin and 2.0% carboxymethyl cellulose. The obtained lyophilized protein structures had a statistically significantly lower hardness compared to the reference material, which was foamed polystyrene (20.84 N). Hazirah et al. [30] showed that physical and mechanical properties of gelatin-carboxymethylcellulose films were best improved with 5% xanthan gum added. The use of xanthan gum a non-gelling nature, as an alternative crosslinking agent in gelatin/carboxymethylcellulose film blend have formed a blend of composite film and improved several physical and mechanical properties of gelatin/carboxymethylcellulose film blend alone. Results Lima et al. [31] demonstrated that films containing higher content of xanthan gum show the highest tensile strength and the lowest elongation. Xanthan gum addition did not affect the water vapor permeability, solubility, and moisture of films.

3.2. Characteristics of Thermophysical Properties of Obtained Protein Structures

Table 2 presents thermophysical properties (coefficient of thermal conductivity—λ, heat capacity—C, thermal diffusivity—a) of lyophilized protein structures modified with a different percentage of carboxymethyl cellulose and xanthan gum. The coefficient of thermal conductivity (λ) characterises the material's ability to conduct heat. It is defined as the heat flux per unit area of the material with a temperature gradient of $1\ \text{K}\cdot\text{m}^{-1}$ [28]. The study demonstrated that modification of material composition affected statistically significantly ($p < 0.05$) the value of thermal conductivity coefficient. The lowest value of this parameter ($0.06\ \text{W}\cdot(\text{mK})^{-1}$) was noted for protein structures obtained from 8.0% albumin and 4.0% xanthan gum (B4) and it was comparable to foamed polystyrene ($0.04\ \text{W}\cdot(\text{mK})^{-1}$), which is widely used as a material with insulating properties. Similar values of thermal conductivity were reported by Kozłowicz et al. [25]. The thermal conductivity of lyophilized gelatin structures ranged from 0.045 to $0.063\ \text{W}\cdot(\text{mK})^{-1}$. The values of thermal conductivity determined for lyophilized gelatin structures modified with hydrated paper pulp, ground extruded starch and hydrogel balls were in the range of $0.047–0.081\ \text{W}\cdot(\text{mK})^{-1}$ [32].

Table 2. Thermal properties of lyophilized samples and foamed polystyrene.

Properties	B1	B2	B3	B4	XPS
Thermal conductivity λ $(\text{W}\cdot(\text{mK})^{-1})$	0.12 ± 0.01 [b]	0.13 ± 0.01 [b]	0.12 ± 0.01 [b]	0.06 ± 0.00 [c]	0.04 ± 0.00 [a]
p-value	0.000	0.000	0.000	0.000	
Heat capacity C $(\text{MJ}\cdot(\text{m}^3\text{K})^{-1})$	0.40 ± 0.02 [b]	0.41 ± 0.03 [b]	0.41 ± 0.02 [b]	0.29 ± 0.02 [c]	0.26 ± 0.02 [a]
p-value	0.000	0.000	0.000	0.02	
Thermal diffusivity a $(\text{mm}^2\cdot\text{s}^{-1})$	0.30 ± 0.01 [b]	0.30 ± 0.01 [b]	0.28 ± 0.01 [b]	0.21 ± 0.01 [c]	0.18 ± 0.01 [a]
p-value	0.000	0.000	0.000	0.000	

[a,b,c] Means in the same line indicated by different letters were significantly different (p value < 0.05). The results are expressed as mean $\pm$ SD ($n = 3$).

Lyophilized protein structures modified with carboxymethylcellulose 2% (B1) and xanthan gum 2% (B2) and 6% (B3) had a significantly higher heat capacity C ($0.40\ \text{MJ}\cdot\text{m}^{-3}\cdot\text{K}^{-1}$ and $0.41\ \text{MJ}\cdot\text{m}^{-3}\cdot\text{K}^{-1}$) than lyophilized protein structures with addition of 4% xanthan gum (B4) ($0.29\ \text{MJ}\cdot\text{m}^{-3}\cdot\text{K}^{-1}$) and foamed polystyrene ($0.26\ \text{MJ}\cdot\text{m}^{-3}\cdot\text{K}^{-1}$). Heat capacity is defined as the ratio of the amount of heat absorbed by the material to the resulting increase in temperature. Materials with high heat capacity require more heat than those with low heat capacity to achieve the same effect of a temperature rise. In packaging, it is preferable to use materials with high heat capacity. Packagings made of such materials will maintain a constant temperature of the product inside.

Another value characterizing a material in terms of its thermal properties is thermal diffusivity a, which determines the ability of a given material to transfer heat within it and reduce temperature gradients. Thermal diffusivity in lyophilized protein structures, modified with carboxymethylcellulose and xanthan gum, ranged from $0.21–0.30\ \text{mm}^2\cdot\text{s}^{-1}$. The obtained thermal diffusivity values were significantly higher ($p < 0.05$) than those noted for foamed polystyrene ($0.18\ \text{mm}^2\cdot\text{s}^{-1}$). This value is determined by the structure, chemical composition and temperature describing the speed of heat conduction in the material [28].

3.3. Analysis of Thawing Kinetics

The thawing curve is a widely used data source for determining changes of the frozen product's temperature during thawing (Figure 3).

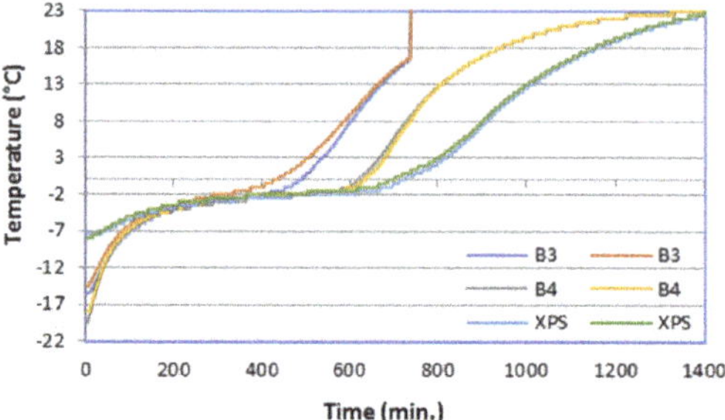

Figure 3. Thawing curves for carrots with green peas in packaging (B3—6.0% albumin, 6.0% xanthan gum, B4—8.0% albumin, 4.0% xanthan gum, XPS—foamed polystyrene).

Considering the temperature changes over time in the form of thawing curves summarized in diagrams (Figure 3), it was found that in a packaging made of lyophilized protein structures B3, the thawed product kept the temperature below 0 °C for the shortest time (temperature measured after 7.5 h). The temperature of 0 °C was reached by the thawed product in a packaging made of lyophilized protein structures B4 after 10 h. It was the time of thawing compared to that recorded for the product placed in the packaging made of foamed polystyrene (11.7 h).

3.4. Thermovision Analysis of Designed Packaging

The operation of the thermal imaging camera is based on the phenomenon of infrared radiation. The temperature distribution on the surface of the tested packaging is presented in the form of colored isotherms, where the individual color corresponds to points having the same temperature [33].

The packagings were subjected to thermovision analysis when testing changes in temperature during defrosting of the product. Figure 4 presents a thermovision image of packaging made of lyophilized protein structures of the B3 sample with characteristic parameters. The point CS1 specifying the lowest temperature occurring on the packaging wall (20.1 °C) and temperature profiles on the straight lines P1 and P2 are presented. A histogram was prepared for the thermovision image shown in Figure 5. The histogram is used to present the empirical distribution of features, so it is possible to use it to present the results obtained for certain quantitative variables. It was found that temperatures in the range of 21.4–22.5 °C had the largest share of the entire area, constituting about 65%. Analysis of the temperature distribution profile on the line P1 shows a minimum temperature of 20.1 °C (uneven material structure, uneven aeration) and a maximum temperature of 22.5 °C. The temperature distribution on the line P2 shows the minimum value of 21.8 °C and the maximum value of 24.6 °C (Figure 6).

Figure 7 presents a thermovision image of a packaging made of lyophilized protein structures (8.0% albumin, 4.0% xanthan gum) of the sample B4. At CS1, the lowest packaging surface temperature was 19.9 °C. The largest share in the whole area, constituting 82.0% of the total individual temperatures, was recorded for the range of temperature 23.3–25.0 °C (Figure 8). Analysis of the temperature distribution profile on the line P2 shows the temperature with minimum of 20.8 °C and maximum of 25.2 °C, while the temperature distribution on the line P3 shows the minimum value of 20.0 °C and the maximum value of 24.8 °C (Figure 9).

Figure 4. Infrared image of packaging surface with analysis of temperature profile and its extreme values (B3).

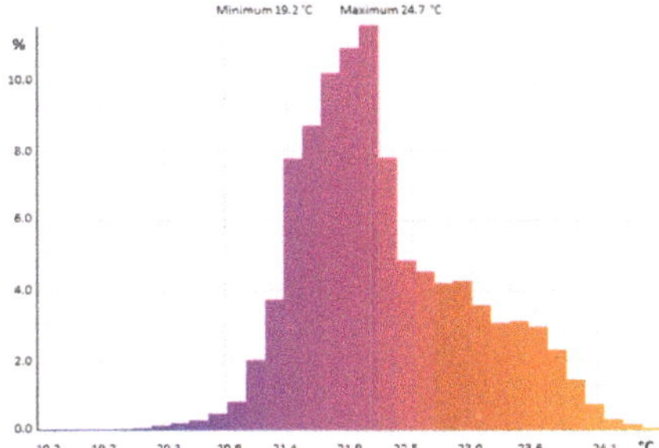

Figure 5. The histogram of temperature distribution on packaging surface (B3).

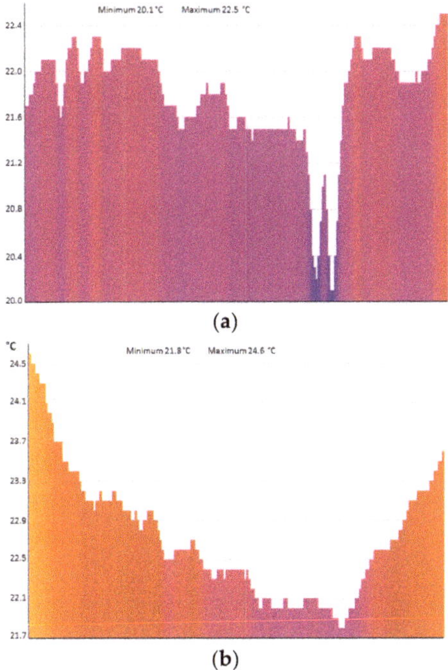

(**a**)

(**b**)

Figure 6. The temperature distribution profile of pack surface samples B3 in line (**a**) P1, (**b**) P2.

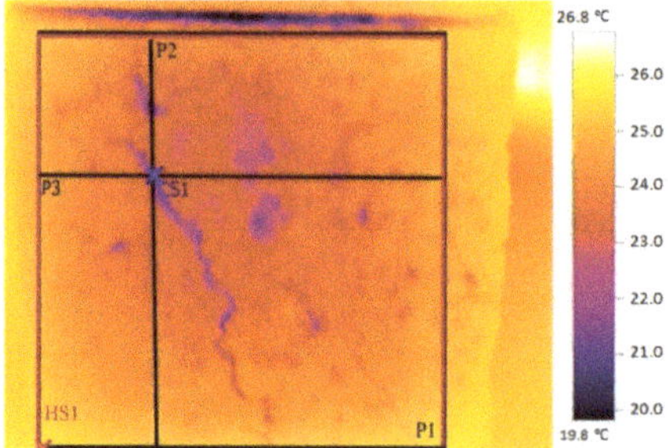

Figure 7. Infrared image of packaging surface with analysis of temperature profile and its extreme values (B4).

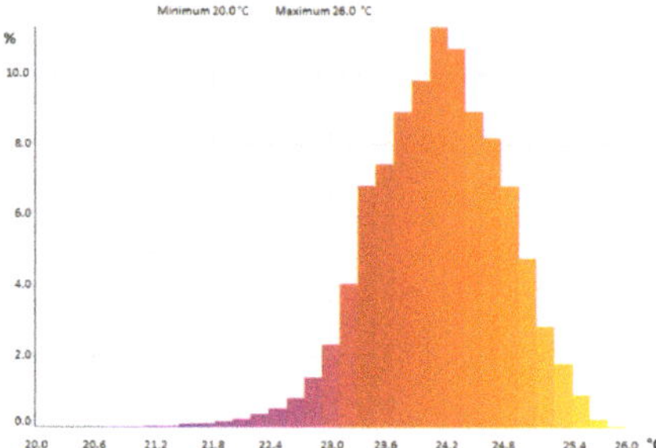

Figure 8. The histogram of temperature distribution on packaging surface (B4).

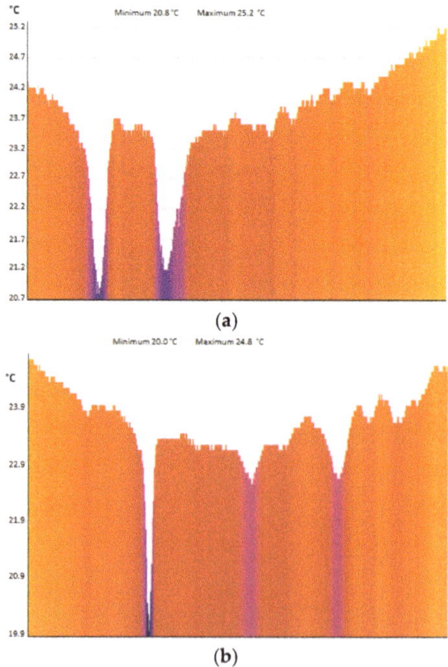

(a)

(b)

Figure 9. The temperature distribution profile of packaging surface samples B4 in line (a) P2, (b) P3.

For comparison purposes, packagings made of foamed polystyrene (XPS) were also subjected to the thermovision analysis. The image of the surface of packaging wall is shown in Figure 10 together with the profile analysis and indication of extreme values. The histogram of the temperature distribution for the entire surface area of the packaging (Figure 11) shows that about 40% of the total temperature is in the range 23.5–24.0 °C. Analysis of the temperature distribution profile on the line P1 shows the minimum temperature of 23.3 °C and maximum temperature of 24.8 °C. Whereas these values on the line P2 are, respectively, 24.7 °C and 25.7 °C (Figure 12).

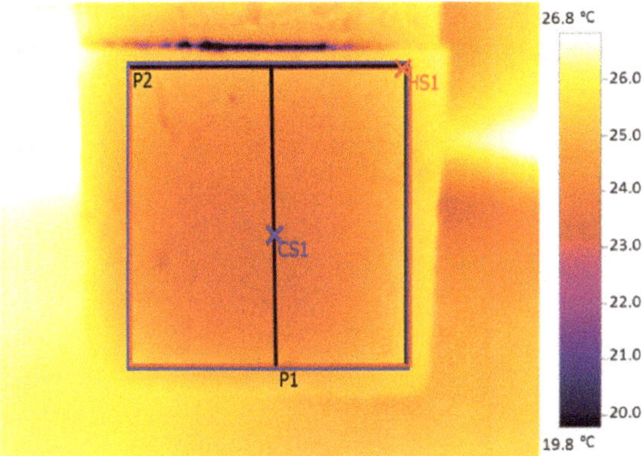

Figure 10. Infrared image of packaging surface with analysis of temperature profile and its extreme values (XPS).

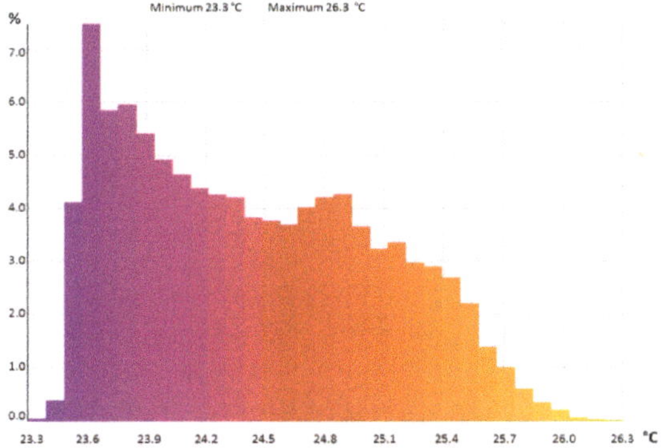

Figure 11. The histogram of temperature distribution on packaging surface (XPS).

The obtained results are consistent with the results of studies conducted previously by other authors who used lyophilized gelatin structures and lyophilized protein structures modified with agar and gelatin as packaging with good thermal insulation properties for frozen food [25,26]. The possible biodegradability of such material means that it can be a suitable replacement for foamed polystyrene used in the production of packaging.

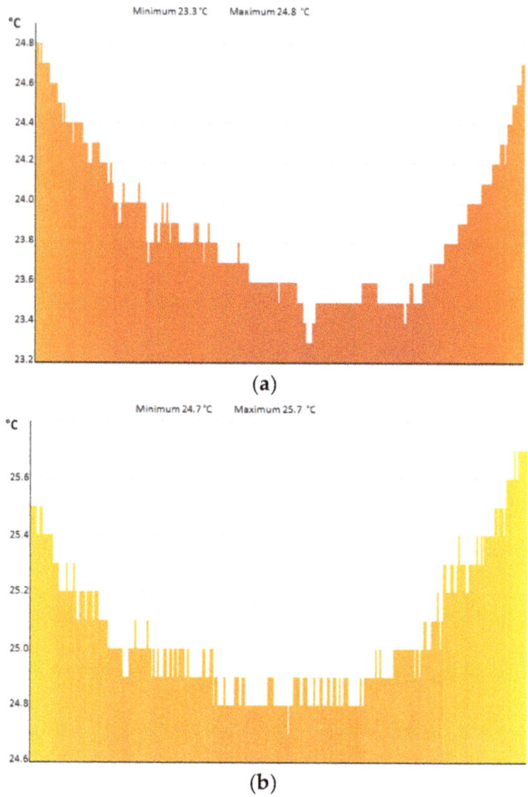

Figure 12. The temperature distribution profile of the foamed polystyrene packaging in line (**a**) P1, (**b**) P2.

The trend of using natural resources to produce packaging materials is now very common. Such materials include polysaccharides (corn and potato starch, cellulose, gums), animal proteins (casein, collagen, gelatin), vegetable (soy, gluten) and lipids (oils from fats) [34,35]. The use of proteins of animal and vegetable origin has proved to be promising in obtaining films and mixtures development of biodegradable packaging. The proteins hake films were more resistant and soluble in water, while gluten films showed greater elongation on cracking [36]. Microbial biopolymers such as gellan, bacterial cellulose, xanthan, pullulan, and curdlan are non-toxic, biocompatible, and biodegradable. They are used in the food industry as material for coating and packing purposes. The packagings prepared from these polymers are transparent and have good mechanical, moisture, and oxygen barrier properties, but weak water barrier properties. The blending of microbial gums with lipids, hydrocolloids or reinforcement agents improves the functional properties of materials [37]. However, it should continue to look for strategies that promote the improvement of the mechanical and barrier properties of these materials. Only in this way will it be possible to replace, partly or completely synthetic polymers with biodegradable polymers. These materials allow to maintain an appropriate quality of packaged raw materials. In addition, packaging made of natural materials does not require biodegradability testing.

4. Conclusions

The conducted research confirmed the possibility of using albumin foams in the form of lyophilized structures as an alternative material for frozen food packaging. In terms of thermal insulation, the proposed structures are also well suited to packaging material for frozen products as foamed polystyrene The best thermophysical properties were obtained for the sample B4 with thermal conductivity of $\lambda = 0.06$ W·(mK)$^{-1}$), thermal capacity $C = 0.29$ (MJ·(m^{3}K)$^{-1}$) and thermal diffusivity $a = 0.21$ (mm^2·s^{-1}). The density of the obtained lyophilizated protein structures was significantly lower compared to the reference material (foamed polystyrene). The obtained materials had statistically significantly lower hardness than foamed polystyrene. The analysis of thawing curves of frozen carrots with green peas showed that the product placed in a packaging made of foam polystyrene reached temperature 0 °C after the longest time (11.7 h). A slightly lower defrosting time was recorded in the product kept in packaging from material B4 (10 h). Thermovision analysis of the tested packagings showed the occurrence of local freezing. Lyophilized protein structures obtained from natural ingredients can be classified as materials made in accordance with the principles of sustainable development that meet the needs of consumers and which are environmentally friendly. These are completely biodegradable materials, suitable for composting and made from natural resources. In addition to biodegradability, lyophilized protein structures are distinguished by high insulation and low density, which provide adequate protection for stored frozen food. In addition, all the ingredients in the obtained lyophilized structures are 100% approved for contact with food, hence they can be a substitute for foamed polystyrene in the production of packaging.

Author Contributions: Conceptualization, K.K. and S.N.; methodology, K.K. and D.G.; prepared the materials, A.K.; analysed the data, M.D.; wrote the paper, K.K.; S.N., D.G.; visualization, D.G. All authors read and approved the final manuscript.

Funding: This research received no external funding.

Conflicts of Interest: The authors declare no conflict of interest.

References

1. Engel, J.B.; Ambrosi, A.; Tessaro, I.C. Development of biodegradable starch-based foams incorporated with grape stalks for food packaging. *Carbohydr. Polym.* **2019**, *225*, 115–234. [CrossRef] [PubMed]
2. Malherbi, M.; Schmitz, A.C.; Grando, R.C.; Bilck, A.P.; Yamashita, F.; Tormen, L.; Fakhouri, F.M.; Velasco, J.I. Corn starch and gelatine-based films added with guabiroba pulp for application in food packaging. *Food Packag. Shelf Life* **2019**, *19*, 140–146. [CrossRef]
3. Ustawa z dnia 13 Czerwca 2013 r., O Gospodarce Opakowaniami i Odpadami Opakowaniowymi (Dz.U. z 2019 r. poz. 542). Available online: http://prawo.sejm.gov.pl/isap.nsf/DocDetails.xsp?id=WDU20130000888 (accessed on 6 December 2019).
4. Strategia Wspólnoty w dziedzinie gospodarowania odpadami. *Off. J. Eur. Commun.* **1990**. No. C 122, 18.5.1990. Available online: https://eur-lex.europa.eu/legal-content/EN/TXT/?qid=1575734282491&uri=CELEX: 31990Y0518(01) (accessed on 6 December 2019).
5. Brody, A.L. Innovative food packaging solutions. *J. Food Sci.* **2008**, *73*, 9–15.
6. Wasilewska, A.; Pezała, A. Environmental aspects of sustainable packaging in FMCG industry. *Logistyka Odzysku* **2016**, *3*, 34–36.
7. Russell, D.A.M. Sustainable (food) packaging—An overview. *Food Addit. Contam.* **2014**, *31*, 396–401. [CrossRef]
8. Song, J.H.; Murphy, R.J.; Narayan, R.; Davies, G.B.H. Biodegradable and compostable alternatives to conventional plastics. *Philos. Trans. R. Soc. Lond. B. Biol. Sci.* **2009**, *364*, 2127–2139. [CrossRef]
9. Davis, G.; Song, J.H. Biodegradable packaging based on raw materials from crops and their impact on waste management. *Ind. Crops. Prod.* **2006**, *23*, 147–161. [CrossRef]
10. Chiarathanakrit, C.; Riyajan, S.A.; Kaewtatip, K. Transforming fish scale waste into an efficient filler for starch foam. *Carbohydr. Polym.* **2018**, *188*, 48–53. [CrossRef]
11. Machado, C.M.; Benelli, P.; Tessaro, I.C. Sesame cake incorporation on cassava starch foams for packaging use. *Ind. Crops. Prod.* **2017**, *102*, 115–121. [CrossRef]
12. Heydari, A.; Alemzadeh, I.; Vossoughi, M. Functional properties of biodegradable corn starch nanocomposites for food packaging applications. *Mater. Des.* **2013**, *50*, 954–961. [CrossRef]
13. Vargas-Torres, A.; Palma-Rodriguez, H.M.; Berrios, J.D.J.; Glenn, G.; Salgado-Delgado, R.; Olarte-Paredes, A.; Hernandez-Uribe, J.P. Biodegradable baked foam made with chayotextle starch mixed with plantain flour and wood fiber. *J. Appl. Polym. Sci.* **2017**, *134*, 455–465. [CrossRef]

14. Cruz-Tirado, J.P.; Siche, R.; Cabanillas, A.; Diaz-Sanchez, L.; Vejarano, R.; Tapia-Blacido, D.R. Properties of baked foams from oca (Oxalis tuberose) starch reinforced with sugarcane bagasse and asparagus peel fiber. *Procedia Eng.* **2017**, *200*, 178–185. [CrossRef]

15. Salgado, P.R.; Schmidt, V.C.; Ortiz, S.E.M.; Mauri, N.; Laurindo, J.B. Biodegradable foams based on cassava starch, sunflower proteins and cellulose fibers obtained by a baking process. *J. Food Eng.* **2008**, *85*, 435–443. [CrossRef]

16. Kaisangsri, N.; Kerdchoechuen, O.; Laohakunjit, N. Characterization of cassava starch based foam blended with plant proteins, kraft fiber and palm oil. *Carbohydr. Polym.* **2014**, *110*, 70–77. [CrossRef] [PubMed]

17. Ou, S.; Kwok, K.C.; Kang, Y. Changes in in vitro digestibility and available lysine of soy protein isolate after formation of film. *J. Food Eng.* **2004**, *64*, 301–305. [CrossRef]

18. Su, J.F.; Huang, Z.; Yuan, X.Y.; Wang, X.Y.; Li, M. Structure and properties of carboxymethyl cellulose/soy protein isolate blend edible films crosslinked by Maillard reactions. *Carbohydr. Polym.* **2010**, *79*, 145–153. [CrossRef]

19. Wiwatwongwana, F.; Pattana, S. Characterization on properties of modification gelatin films with carboxymethyl cellulose. In Proceedings of the 1st TSME International Conference on Mechanical Engineering, Ubon Ratchathani, Thailand, 20–22 October 2010; pp. 1–8.

20. Guo, J.; Ge, L.; Li, X.; Mu, C.; Li, D. Periodate oxidation of xanthan gum and its crosslinking effects on gelatin-based edible films. *Food Hydrocoll.* **2014**, *39*, 243–250. [CrossRef]

21. Sharma, B.R.; Naresh, L.; Dhuldhoya, N.C.; Merchant, S.U.; Merchant, U.C. Xanthan gum—A boon to food industry. *Food Promot. Chron.* **2006**, *1*, 27–30.

22. Sun, D. *Handbook of Frozen Food Processing and Packaging*, 2nd ed.; CRC Press: Boca Raton, FL, USA; New York, NY, USA, 2012; pp. 711–779.

23. Rozporządzenie (WE) nr 1935/2004 Parlamentu Europejskiego i Rady z dnia 27 Października 2004 r. w Sprawie Materiałów i Wyrobów Przeznaczonych do Kontaktu z Żywnością oraz Uchylające Dyrektywy 80/590/EWG i 89/109/EWG. Available online: https://eur-lex.europa.eu/legal-content/PL/TXT/?uri=celex%3A32004R1935 (accessed on 6 December 2019).

24. Grabowska, B. Frozen food packaging—Characteristics, review, norms and provisions. *Przem. Spożywczy* **2014**, *9*, 16–18.

25. Kozłowicz, K.; Góral, D.; Kluza, F.; Domin, M.; Kobus, Z.; Sagan, A.; Prazner, Ł. The porous gelatin structures as the material for packaging for frozen food. *Przem. Chem.* **2015**, *10*, 1742–1747.

26. Góral, D.; Kozłowicz, K.; Kluza, F.; Domin, M.; Blicharz-Kania, A.; Senetra, E.; Dziki, D.; Kocira, A.; Guz, T. Evaluation of thermophysical characteristics of freeze-dried protein foams as packaging material for frozen food. *Przem. Chem.* **2018**, *5*, 700–705.

27. Kozłowicz, K.; Góral, D.; Kluza, F.; Góral, M.; Andrejko, D. Experimental determination of thermophysical properties by line heat pulse method. *J. Food Meas. Charact.* **2018**, *12*, 2524–2534. [CrossRef]

28. Bornhorst, G.; Sarkar, A.; Singh, P.R. Terminal Properties of Frozen Foods. In *Engineering Properties of Foods*, 4th ed.; Rao, M.A., Rizvi, S.S.H., Datta, A.K., Ahmed, J., Eds.; CRC Press: Boca Raton, FL, USA; New York, NY, USA, 2014; pp. 247–280.

29. Marzec, A.; Jakubczak, E. Rheological properties of foams prepared for drying. *Acta Agrophys.* **2009**, *13*, 185–194.

30. Hazirah, M.A.S.P.; Isa, M.I.N.; Sarbon, N.M. Effect of xanthan gum on the physical and mechanical properties of gelatin-carboxymethyl cellulose film blends. *Food Pack. Shelf Life* **2016**, *9*, 55–63. [CrossRef]

31. Lima, M.; Carneiro, L.; Bianchini, D.; Dias, A.R.; Zavareze, R.; Prentice, C.; Moreira, A. Structure, thermal, physical, mechanical and barrier properties of chitosan films with the addition of xanthan gum. *J. Food Sci.* **2017**, *82*, 698–705. [CrossRef]

32. Kozłowicz, K.; Kluza, F.; Góral, D.; Nakonieczny, P.; Combrzyński, M. Modified gelatine structures as packaging material for frozen agricultural products. In Proceedings of the BIO Web Conference in Contemporary Research Trends in Agricultural Engineering, Kraków, Poland, 25–27 September 2017.

33. Jura, J.; Adamus, J. Thermography application for assessment of building thermal insulation. *Bud. O Zoptymalizowanym Potencjale Energetycznym* **2013**, *2*, 31–39.

34. Sanjay, M.R.; Arpitha, G.R.; Naik, L.L.; Gopalakrishna, K.; Yogesha, B. Applications of natural fibers and its composites: An overview. *Nat. Resour.* **2016**, *7*, 108–114. [CrossRef]

35. Folentarska, A.; Krystyjan, M.; Baranowska, N.M.; Ciesielski, W. Renewable raw materials as an alternative to receiving biodegradable materials. *J. Chem. Environ. Biotechnol.* **2016**, *19*, 121–124. [CrossRef]
36. Nogueira, D.; Martins, V.G. Use of different proteins to produce biodegradable films and blends. *J. Polym. Environ.* **2019**, *27*, 2027–2039. [CrossRef]
37. Alizadeh-Sani, M.; Ehsani, A.; Kia, E.M.; Khezerlou, A. Microbial gums: Introducing a novel functional component of edible coatings and packaging. *Appl. Microb. Biotechnol.* **2019**, *103*, 6853–6866. [CrossRef] [PubMed]

 *sustainability*

Article

Modelling Water Absorption in Micronized Lentil Seeds with the Use of Peleg's Equation

Izabela Kuna-Broniowska [1], Agata Blicharz-Kania [2,*], Dariusz Andrejko [2],
Agnieszka Kubik-Komar [1], Zbigniew Kobus [3], Anna Pecyna [3], Monika Stoma [4],
Beata Ślaska-Grzywna [2] and Leszek Rydzak [2]

[1] Department of Applied Mathematics and Computer Science, University of Life Sciences in Lublin,
20-612 Lublin, Poland; izabela.kuna@up.lublin.pl (I.K.-B.); agnieszka.kubik@up.lublin.pl (A.K.-K.)
[2] Department of Biological Bases of Food and Feed Technologies, University of Life Sciences in Lublin,
20-612 Lublin, Poland; dariusz.andrejko@up.lublin.pl (D.A.); beata.grzywna@up.lublin.pl (B.Ś.-G.);
leszek.rydzak@up.lublin.pl (L.R.)
[3] Department of Technology Fundamentals, University of Life Sciences in Lublin, 20-612 Lublin,
Poland; zbigniew.kobus@up.lublin.pl (Z.K.); anna.pecyna@up.lublin.pl (A.P.)
[4] Department of Power Engineering and Transportation, University of Life Sciences in Lublin, 20-612 Lublin,
Poland; monika.stoma@up.lublin.pl
* Correspondence: agata.kania@up.lublin.pl; Tel.: +48-81-531-9646

Received: 6 November 2019; Accepted: 25 December 2019; Published: 28 December 2019

Abstract: The aim of the paper was to investigate the effect of infrared pre-treatment on the process of water absorption by lentil seeds. The paper presents the effects of micronization on the process of water absorption by lentil seeds. As a source of infrared emission, 400-W ceramic infrared radiators ECS-1 were used. The seeds were soaked at three temperature values (in the range from 25 to 75 °C) for 8 h, that is, until the equilibrium moisture content was achieved. Peleg's equation was used to describe the kinetics of water absorption by lentil seeds. The results were compared with those obtained in the process of soaking crude seeds. On the basis of the conducted research, it was found that the infrared pre-treatment contributed to a substantial increase in the water absorption rate in the initial period of soaking lentil seeds (especially at 25 °C). Infrared irradiation can be an effective method for intensification of lentil seed hydration at an ambient temperature. It should be assumed that, in accordance with the principles of sustainable development, shortening the heating time will significantly reduce the energy consumption and cost of processing lentil seeds.

Keywords: legumes; infrared processing; acceleration of the process of hydration; Peleg's equation

---

1. Introduction

Water absorption by plant raw materials is influenced by both the process parameters (e.g., temperature, pressure) and the physical and chemical properties of the raw materials. The impact of the temperature of the soaking process on the rate and amount of absorbed water was investigated by, for example, Maskan [1] and Turhan et al. [2]. Their results clearly demonstrate the role of temperature (in the range from 6 to 100 °C) in the water absorption process. An elevated temperature was found to accelerate the water absorption rate in the case of peas, wheat, lupine, soybean, and faba bean.

Research is being carried out to develop a mathematical description of water absorption by granular plant materials. The process of water absorption in some cereal caryopses and legume seeds, that is, wheat, peas, or rice, has been investigated, and a model based mainly on the Fick diffusion equation has been developed [3–5]. However, prediction of the water absorption time based on Fick's law requires a very complex mathematical apparatus, which makes it very inconvenient for practical calculations in many cases. Hence, Peleg [6] proposed a two-parameter, non-exponential, empirical equation for modelling the water absorption process in raw materials and food products, as follows:

$$M_t = M_0 \pm \frac{t}{K_1 + K_2 \cdot t},$$ (1)

where

M_t—Water content after time t, % of d.w.;
M_0—Initial water content, % of d.w.;
K_1—Constant, h $\cdot$ %$^{-1}$;
K_2—Constant, %$^{-1}$.

Equation (1) contains the "±" sign. The "+" sign is used to model the water absorption process, and the "−" sign is used to model the drying process.

The main attribute of this equation, compared with other equations, is its simplicity [1]. This formula has been positively verified for several cereal and legume species [2,7–9].

The sorption rate R can be calculated from the first derivative of Peleg's Equation.

$$R = \frac{dM}{dt} = \pm \frac{K_1}{(K_1 + K_2 t)^2}$$ (2)

Peleg's constant K_1 refers to the initial sorption rate (R_0, i.e., the value of R at $t = t_0$).

$$R_0 = \frac{dM}{dt}\Big|_{t_0} = \pm \frac{1}{K_1}$$ (3)

At $t \rightarrow \infty$, Equation (4) describes the relationship between the equilibrium moisture content (M_e) and constant K_2.

$$M\big|_{t_\infty} = Me = M_0 \pm \frac{1}{K_2}$$ (4)

Linearization of Equation (1) gives the following formula:

$$\frac{t}{M_t - M_0} = K_1 + K_2 t.$$ (5)

Adjustment of the equation of such a line allows estimation of parameters K_1 and K_2.

The value of constant K_1 provides information about the mass transfer rate during the water absorption process, that is, the lower the K_1 value, the higher the water absorption rate [2,7,10, 11]. While the K_1 coefficient has been relatively precisely defined in many publications, there is a large discrepancy as far as the significance of constant K_2 is concerned. In investigations of water absorption by wheat-based products, Maskan [1] has found that constant K_2 refers to the maximum water absorption capacity, that is, the water absorption capacity increases at lower K_2 values. Similar conclusions were formulated by Abu-Ghannam and McKenna [7], as well as Sayar et al. [11]. However, other authors [10,12–14] did not confirm these findings.

Reduction of energy consumption in economy is the most efficient and best-recognized way of implementing the principle of sustainable development. In addition, from the consumer's point of view, the use of legume seeds is determined by the time of preparation of ready-to-eat products. Therefore, seed pre-treatment should ensure the shortest time of the subsequent hydrothermal treatment thereof (e.g., soaking, cooking). As suggested by Cenkowski and Sosulski [15], thermal infrared pre-treatment of seeds applied prior to soaking or cooking can shorten these treatments and provide instant-type products.

Infrared radiation is very widely used in various industries. In the food industry, it is used primarily for drying and pre-treatment of food. The micronization process allows, for example, reducing the content of anti-nutritional substances in legume seeds or increasing the efficiency of pressing oilseeds. Owing to the specific mechanism of heat transport during micronization, the heat and mass exchange conditions are definitely more favourable than when using traditional heat treatment methods. In practice, it results in more effective operation, and thus reduction of the duration of the process [16–18].

However, the available literature provides no reports on the possibility of using Peleg's equation to determine the rate of water absorption by micronized legumes.

Therefore, the paper presents an attempt to investigate the effect of infrared pre-treatment on the process of water absorption by lentil seeds.

2. Materials and Methods

2.1. Research Material

The lentil seeds (variety Anita) used in the experiment were purchased at a local supermarket in Lublin. Prior to the determinations, the raw material was purified by the removal of cracked and damaged seeds. It was stored in closed plastic containers at a temperature of 5 °C. Before further measurements, the lentil seeds were placed at room temperature (approximately 23 °C) for 3 h.

2.2. Measurement of Initial Moisture Content

The first stage of the research involved determination of the initial moisture content of the raw material (W) using the drying method. The material was dried for 3 h in a laboratory dryer SLN 15 STD at a temperature of 103 °C. The total protein content in the material was determined with the Kjeldahl method using an automated Foss Kjeltec 8400 distillation unit, and the total fat content was determined using a Soxtec 8000 apparatus with AN 310 software (Table 1).

Table 1. Effect of selected components in lentil seeds calculated per dry weight.

	Nutrients		Content [%]
	Protein		22.01 ± 0.53
	Fat		0.90 ± 0.02
Moisture		Crude seeds	8.40 ± 0.17
		Micronized seeds	6.00 ± 0.20

2.3. Micronization Process

A portion of seeds was subjected to the micronization process. As a source of infrared radiation emission, 400-W ceramic infrared radiators ECS-1 were used. These temperature radiators supplied by electricity (230 V) have a small fraction of visible radiation (dark radiators) in the spectrum and heat the entire plane uniformly (plane radiators) (Figure 1). The average temperature of the filament is approximately 500 °C and the emission wavelength is $\lambda = 2.5$–3.0 μm. The seeds were heated for 120 s at a temperature on the seed surface of 150 °C. Next, the seeds were left to dry in open containers at room temperature for 3 h.

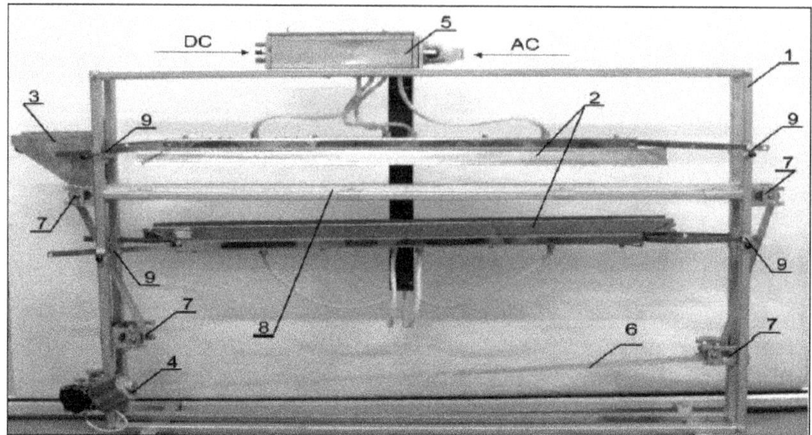

Figure 1. Laboratory device for infrared treatment of granular plant raw materials: DC - direct current, AC - alternating current, 1—frame, 2—infrared radiators, 3—feeding tank, 4—DC motor, 5—control unit, 6—conveyor belt, 7—rollers, 8—heating zone, 9—adjustment of the position of the heads.

2.4. Measurement of the Water Content

At 24 h after the micronization process, the moisture content in the micronized seeds was determined and the crude and micronized seeds were soaked. Samples of randomly selected seeds (excluding cracked and damaged seeds) weighing approximately 0.5 g (M_m) were placed in special baskets with a perforated wall and bottom. The material was soaked in a water bath (using distilled water) within a time range from 5 min to 8 h at a temperature of 25 °C, 50 °C, and 75 °C.

After the time intervals adopted in the experiment, the samples were removed, the excess water was dried on filter paper, and the water content in the seeds was determined.

The water content relative to dry weight was calculated using Equation (6):

$$M_t = \frac{M_1 + M_{H_2O}}{M_1} 100\%, \tag{6}$$

where

M_t—Water content after time t, % relative to dry weight [%];
M_{H2O}—Water weight after time t [g];
M_1—Dry weight [g].

The dry weight was calculated based on Equation (7):

$$M_1 = M_m \cdot (100\% - W), \tag{7}$$

where

M_m—Initial weight of the material [g];
W—Moisture [%].

The determinations were carried out in triplicate.

2.5. Statistical Analysis

The data were analysed statistically. A significance level of $\alpha = 0.05$ was assumed for inference. The analysis was carried out using analysis of variance (ANOVA) (StatSoft Polska, Poland, Cracow).

3. Results and Discussion

Table 1 shows the effect of selected components in lentil seeds calculated per dry weight. The lentil seeds were characterised by typical levels of essential nutrients, for example, protein and fat. For comparison, the investigations of the chemical composition of lentils conducted by Hefnawy [19] revealed a protein content of 26.6 ± 0.5 (g 100 g^{-1} dry weight basis) and fat content of 1.0 ± 0.08 (g 100 g^{-1} dry weight basis). Similarly, the water content in the crude seeds corresponded to the values demonstrated by other authors [19,20].

Figure 2 shows the effect of temperature and time on changes in the water content in the crude lentil seeds. Analysis of variance revealed a significant effect of both of these factors on the water content in the soaked seeds. The water absorption rate decreased with the time of soaking. The highest water content, that is, 142%, was observed after 5 h of soaking of lentil seeds at a temperature of 50 °C. Further, soaking did not change the water content in the treated lentil seeds significantly. The values obtained in this study are in agreement with the results reported by other authors. The mean water content in seeds was estimated at 129% by Chopra and Prasad [21], 140% by Pan and Tangratanavalee [22], and 147% by Quicazán et al. [8].

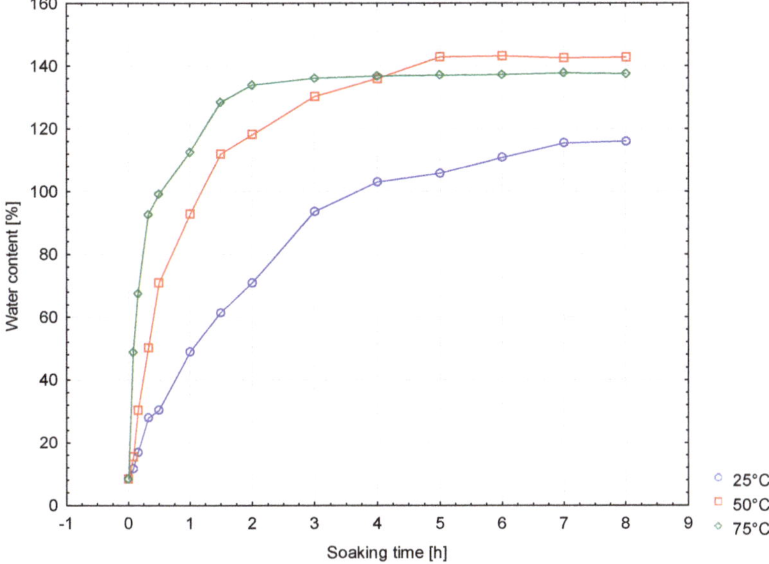

Figure 2. Effect of temperature and duration of soaking on the water content in crude lentil seeds.

In the initial soaking period, the lentil seeds exhibited the lowest water absorption rate at the temperature of 25 °C. An increase in the temperature to 50 °C and then 75 °C resulted in an increase in this parameter. After 3 h of soaking at 75 °C, the lentil seeds reached a maximum moisture level of 138%. Seeds soaked at the temperature of 50 °C absorbed water at a slower rate and reached the maximum level later, but this was by 4% higher than the value obtained at 75 °C. These results are in agreement with the findings presented by other authors. Quicazán et al. [8] reported a slightly higher level of water in seeds soaked at a temperature of 40 °C than the water content in seeds soaked at 80 °C. This phenomenon may be associated with the more intensive denaturation of proteins at a higher temperature, which in turn may reduce the final capacity of water absorption in lentil seeds.

Table 2 shows Peleg's coefficients and the goodness of fit of the model to the experimental data. It was shown that the K_2 parameter value changed slightly with the soaking temperature. This is in line with the results obtained by Sopade et al. [14], in experiments of cereal grain hydration, and Abu-Ghannam and McKenna [7], who investigated the process of hydration of bean seeds. The values of the K_1 parameter declined with an increase in the temperature of soaking. This trend is in good agreement with the results reported by other authors. In experiments involving soybean seed soaking, Quicazán et al. [8] observed a negative correlation between the soaking temperature and the value of the K_1 parameter. The dependence of K_1 on T (absolute temperature) is expressed by Equation (8).

$$K_1 = 0.0006T^2 - 0.4452T + 77.358 \tag{8}$$

Table 2. Parameters of Peleg's model for hydration of crude lentil seeds at three temperatures.

Parameter	T	Soaking Time	K_1	$1/K_1$	K_2	R^2	E
Unit	°C	h					%
Result	25	0–8	1.6845	0.59365	0.6893	0.9960	3.93
	50	0–8	0.5183	1.92938	0.6567	0.9957	6.08
	75	0–8	0.1543	6.48088	0.7461	0.9952	1.94

Notes: R^2—fit of the model; E—goodness of fit of the model.

Figure 3 shows the effect of the temperature and soaking time on changes in the water content in the micronized lentil seeds. The temperature and soaking time contributed to an increase in the water content in the micronized lentil seeds. In the initial soaking period, the lentil seeds exhibited the lowest rate of water absorption at a temperature of 25 °C. The increase in the temperature to 50 °C and then 75 °C increased the water absorption rate. After 4 h of soaking at both 50 °C and 75 °C, the lentil seeds achieved the maximum moisture level. The highest water content, that is, 138%, was observed at the temperature of 75 °C.

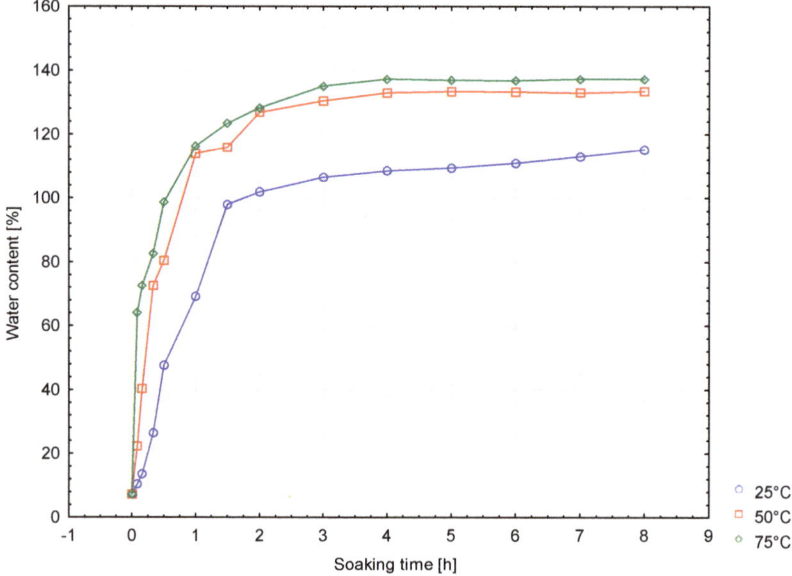

Figure 3. Effect of temperature and soaking time on changes in the water content in the micronized lentil seeds.

Table 3 shows Peleg's model coefficients and the goodness of fit of the model to the experimental data for the micronized seeds. The values of the K_1 parameter declined with the increase in the soaking temperature. In comparison with the crude lentil seeds, the values of the K_1 coefficients were significantly lower at all tested temperatures. This indicates a higher rate of water absorption in the micronized seeds. The dependence of K_1 on T (absolute temperature) is expressed by Equation (9):

$$K_1 = 0.0003T^2 - 0.1838T + 32.056 \tag{9}$$

Table 3. Parameters of Peleg's model for micronized lentil hydration at three temperatures.

Parameter	T	Soaking Time	K_1	$1/K_1$	K_2	R^2	E
Unit	°C	h					%
Result	25	0–8	0.7801	1.28189	0.7977	0.9698	17.38
	50	0–8	0.2920	3.42466	0.7299	0.9895	6.06
	75	0–8	0.1345	7.43494	0.7565	0.9791	4.18

Notes: R^2—fit of the model; E—goodness of fit of the model.

The values of the K_2 coefficient in the case of the micronized seeds changed as in the case of the crude seeds. The highest K_2 coefficient was noted at the temperature of 25 °C; next, it declined to the lowest value at 50 °C and reached an intermediate value at 75 °C. In general, all values of the K_2 coefficient for the micronized seeds were slightly lower than the K_2 values for the crude seeds. This was reflected in the final water content in the micronized seeds, which was slightly lower than that observed in the crude seeds.

4. Conclusions

The paper presents the effect of the infrared radiation treatment on the kinetics of water absorption by lentil seeds. The results were compared with the process of soaking crude seeds. In both cases, the changes in the water content were described using Peleg's model. An impact of the infrared pre-treatment on the water absorption rate and content in lentil seeds was demonstrated. The infrared pre-treatment largely contributes to an increase in the water absorption rate in the initial period of soaking lentil seeds. This is particularly evident at a temperature of 25 °C, at which micronization reduces the time required for achievement of 100% water content by more than half in comparison with crude seeds. At higher temperatures, these differences are smaller, that is, 40% and 17%, for the temperatures of 50 °C and 75 °C, respectively. The impact of micronization on the final water content in lentil seeds depends on the soaking temperature. A statistically significant effect of micronization on the water content was found at 25 °C (reduction of constant K_1 from 1.6845 to 0.7801). Slightly smaller changes in the rate of water absorption were observed at 50 °C and 75 °C.

Infrared treatment can be successfully used to accelerate the process of hydration of lentil seeds at an ambient temperature. As a consequence, limiting the damaging impact of production on the environment will ensure sustainable consumption and processing of legumes. The results obtained in the present study can be regarded as preliminary research on the possibility of using micronization to accelerate the absorption of water by seeds of other plants, including cereals.

Author Contributions: Conceptualization, I.K.-B., D.A. and M.S.; Data curation, A.P.; Formal analysis, I.K.-B. and Z.K.; Investigation, A.B.-K. and A.P.; Methodology, A.B.-K., D.A., A.K.-K. and B.Ś.-G.; Project administration, M.S.; Software, I.K.-B. and A.K.-K.; Supervision, I.K.-B. and D.A.; Validation, A.K.-K.; Visualization, A.B.-K., Z.K. and B.Ś.-G.; Writing–original draft, A.B.-K., Z.K. and A.P.; Writing–review & editing, D.A., A.K.-K. and L.R. All authors have read and agreed to the published version of the manuscript.

Funding: This research received no external funding.

Conflicts of Interest: The authors declare no conflict of interest.

Abbreviations

M_t	water content after time t, [% of dry weight];
M_o	initial water content, [% of dry weight];
K_1	constant, $[h \cdot \%^{-1}]$;
K_2	constant, $[\%^{-1}]$;
R	sorption rate, [-];
M_{H2O}	water weight after time t, [g];
M_1	dry weight, [g];
M_m	initial weight of the material, [g];
W	moisture, [%];
R^2	fit of the model, [-];
E	goodness of fit of the model, [%];

References

1. Maskan, M. Effect of processing on hydration kinetics of three wheat products of the same variety. *J. Food Eng.* **2002**, *52*, 337–341. [CrossRef]
2. Turhan, M.; Sayar, S.; Gunasekaran, S. Application of Peleg model to study water absorption in chickpea during soaking. *J. Food Eng.* **2002**, *53*, 153–159. [CrossRef]
3. Cunningham, S.E.; McMinn, W.A.M.; Magee, T.R.A.; Richardson, P.S. Modelling water absorption of pasta during soaking. *J. Food Eng.* **2007**, *82*, 600–607. [CrossRef]
4. Montanuci, F.D.; Perussello, C.A.; de Matos Jorge, L.M.; Jorge, R.M.M. Experimental analysis and finite element simulation of the hydration process of barley grains. *J. Food Eng.* **2014**, *131*, 44–49. [CrossRef]
5. Munson-McGee, S.H.; Mannarswamy, A.; Andersen, P.K. D-optimal designs for sorption kinetic experiments: Cylinders. *J. Food Eng.* **2011**, *104*, 202–207. [CrossRef]
6. Peleg, M. An empirical model for the description of moisture sorption curves. *J. Food Sci.* **1988**, *53*, 1216–1217, 1219. [CrossRef]
7. Abu-Ghannam, N.; McKenna, B. The application of Peleg's equation to model water absorption during the soaking of red kidney beans (*Phasedus vulgaris* L.). *J. Food Eng.* **1997**, *32*, 391–401. [CrossRef]
8. Quicazán, M.C.; Caicedo, L.A.; Cuenca, M. Applying Peleg's equation to modelling the kinetics of solid hydration and migration during soybean soaking. *Ing. Investig.* **2012**, *32*, 53–57.
9. Sopade, P.A.; Xun, P.Y.; Halley, P.J.; Hardin, M. Equivalence of the Peleg, Pilosof and Singh–Kulshrestha models for water absorption in food. *J. Food Eng.* **2007**, *78*, 730–734. [CrossRef]
10. Hung, T.V.; Liu, L.H.; Black, R.G.; Trewhella, M.A. Water absorption in chickpea (*C. arietinum*) and field pea (*P. sativum*) cultivars using the Peleg model. *J. Food Sci.* **1993**, *58*, 848–852. [CrossRef]
11. Sayar, S.; Turhan, M.; Gunasekaran, S. Analysis of chickpea soaking by simultaneous water transfer and water-starch reaction. *J. Food Eng.* **2001**, *50*, 91–98. [CrossRef]
12. Maharaj, V.; Sankat, C.K. Rehydration characteristics and quality of dehydrated dasheen leaves. *Can. Agric. Eng.* **2000**, *42*, 81–85.
13. Sopade, P.A.; Obekpa, J.A. Modeling water absorption in soybean, cowpea and peanuts at three temperatures using Peleg's equation. *J. Food Sci.* **1990**, *55*, 1084–1087. [CrossRef]
14. Sopade, P.A.; Ajisegiri, E.S.; Badau, M.H. The use of Peleg's equation to model water absorption in some cereal grains during soaking. *J. Food Eng.* **1992**, *15*, 269–283. [CrossRef]
15. Cenkowski, S.; Sosulski, F.W. Cooking characteristics of split peas treated with infrared heat. *Trans. ASAE* **1998**, *41*, 715–720. [CrossRef]
16. Khattab, R.Y.; Arntfield, S.D. Nutritional quality of legume seeds as affected by some physical treatments 2. Antinutritional factors. *Lebensm. Wiss. Technol.* **2009**, *42*, 1113–1118. [CrossRef]
17. Krajewska, M.; Ślaska-Grzywna, B.; Andrejko, D. Effect of Infrared Thermal Pre-Treatment of Sesame Seeds (*Sesamum Indicum* L.) on Oil Yield and Quality. *Ital. J. Food Sci.* **2018**, *30*, 487–496. [CrossRef]

18. Sobczak, P.; Zawiślak, K.; Panasiewicz, M.; Mazur, J.; Kocira, S.; Żukiewicz-Sobczak, W. Impact of Heat Treatment on the Hardness and Content of Anti-Nutritious Substances in Soybean Seeds. *Carpath. J. Food Sci. Technol.* **2018**, *10*, 46–52.

19. Hefnawy, T.H. Effect of processing methods on nutritional composition and anti-nutritional factors in lentils (*Lens culinaris*). *Ann. Agric. Sci.* **2011**, *56*, 57–61. [CrossRef]

20. Boye, J.I.; Aksay, S.; Roufik, S.; Ribéreau, S.; Mondor, M.; Farnworth, E.; Rajamohamed, S.H. Comparison of the functional properties of pea, chickpea and lentil protein concentrates processed using ultrafiltration and isoelectric precipitation techniques. *Food Res. Int.* **2010**, *43*, 537–546. [CrossRef]

21. Chopra, R.; Prasad, D.N. Standardization of soaking conditions for soybean seeds/cotyledons for improved quality of soymilk. *Indian J. Anim. Sci.* **1994**, *64*, 405–410.

22. Pan, Z.; Tangratanavalee, W. Characteristics of soybeans as affected by soaking conditions. *Lebensm. Wiss. Technol.* **2003**, *36*, 143–151. [CrossRef]

 *sustainability*

Article

Mechanical and Processing Properties of Rice Grains

Weronika Kruszelnicka [1,*], **Andrzej Marczuk** [2], **Robert Kasner** [1], **Patrycja Bałdowska-Witos** [1], **Katarzyna Piotrowska** [3], **Józef Flizikowski** [1] and **Andrzej Tomporowski** [1]

[1] Department of Manufacturing Techniques, Faculty of Mechanical Engineering, University of Science and Technology in Bydgoszcz, 85-796 Bydgoszcz, Poland; robert.kasner@gmail.com (R.K.); patrycja.baldowska-witos@utp.edu.pl (P.B.-W.); fliz@utp.edu.pl (J.F.); a.tomporowski@utp.edu.pl (A.T.)

[2] Faculty of Production Engineering, University of Life Sciences in Lublin, 20-612 Lublin, Poland; andrzej.marczuk@up.lublin.pl

[3] Faculty of Mechanical Engineering, Lublin University of Technology; 20-618 Lublin, Poland; k.piotrowska@pollub.pl

[*] Correspondence: weronika.kruszelnicka@utp.edu.pl

Received: 17 November 2019; Accepted: 9 January 2020; Published: 11 January 2020

Abstract: Strength properties of grains have a significant impact on the energy demand of grinding mills. This paper presents the results of tests of strength and energy needed the for destruction of rice grains. The research aim was to experimentally determine mechanical and processing properties of the rice grains. The research problem was formulated in the form of questions: (1) what force and energy are needed to induce a rupture of rice grain of the *Oryza sativa* L. of long-grain variety? (2) what is the relationship between grain size and strength parameters and the energy of grinding rice grain of the species *Oryza sativa* L. long-grain variety? In order to find the answer to the problems posed, a static compression test of rice grains was done. The results indicate that the average forces needed to crush rice grain are 174.99 kg m·s^{-2}, and the average energy is 28.03 mJ. There was no statistically significant relationship between the grain volume calculated based on the volumetric mass density V_ρ and the crushing energy, nor between the volume V_ρ and other strength properties of rice grains. In the case of Vs, a low negative correlation between strength σ_{min} and a low positive correlation between the power inducing the first crack were found for the grain size related volume. A low negative correlation between the grain thickness a_3, stresses σ_{min} and work W_{Fmax} was found as well as a low positive correlation between thickness a_3 and the force inducing the first crack F_{min}.

Keywords: rice; grinding; compressive strength; rupture energy

1. Introduction

The strength properties of grains have a significant impact on the energy demand of grinders [1–4]. During grinding of grains in a five-disc mill, a complex state of stress occurs in the material, with shear and compressive stresses prevailing [5,6]. Identification of the forces causing grain cracking (rupture) can be considered as the first step to determine the energy demand in the grain grinding process [7,8]. Two cases can be distinguished: static squeezing of grains and shearing of grains [9]. In the case of static compression, in order to determine the forces and stress and consequently, the work (energy) needed to crush one grain and then more than a dozen grains, a static compression test can be carried out. The ranges of probable forces destroying the grain may be determined in an experimental manner, and subsequently, the energy ranges of destroying its structure [10].

The subject of research on the physical-mechanical properties of rice grains has already been addressed by researchers, because the specificity of rice grain processing and the energy demand of processing lines, e.g., grinding, drying, pelleting, etc., depends largely on these properties [11–17]. In their research, Zeng, et al. [18] focused on modeling cracking of rice grains depending on the grain moisture and load speed, using the Discrete Element Method (DEM). The impact of humidity on the strength properties of white rice was also addressed in the work of Sadeghi et al. [19], and brown rice in the work by Cao et al. [20] and Chattopadhyay et al. [21]. Buggenhout et al. [22] studied the influence of physicochemical properties, in particular, the impact of grain husks and moisture on cracking phenomena during rice processing. Esehaghbeygi et al. [23] in turn, analyzed the effect of drying temperature and kinetic energy during the drying process on grain susceptibility to breaking. Similar research was conducted by Sarker et al. [24], Tajaddodi et al. [25], Nasirahmadi et al. [26] and Bonazzi and Courtois [27]. The influence of grain orientation on its mechanical properties under load was analyzed, among others, by Li et al., Shu et al., and Zareiforoush et al. [28–30]. They showed that rice grains are more flexible in horizontal orientation based on the results of static compression and three-point bending tests. Zareiforoush et al. [31], based on the conducted research, found that increasing the speed of load during the compression test results in lowering crushing forces and energy.

From the point of view of energy demand, processing of grainy biomaterials, particularly grinding, the strength of grains plays the key role. [32]. Considering the process of grinding, e.g., by means of grinding machines or roller mills, these are the compressive loads which prevail in the grinded material, whereas permanent deformation (fragmentation) occurs after exceeding the load value corresponding to the compressive strength limit. Strength is closely related to the power necessary to cause the strain and the grinded material cross-section field (hence being dependent on its geometric features). Thus, material fragmentation occurs upon application of appropriate forces, which, in the system of grinding machines, roller mills, is performed by rotary motion of rollers. In such a case, the force is a direct effect of torques, which, in turn, are related to the power of the devices affecting the energy demand of the grain processing. In general, the higher force to be applied the higher power, that is, the machine energy, is needed. The aspects concerning calculation of energy demand for grain grinding systems are described in detail in [33].

The strength of grains depends on the type of material, especially on its internal structure (porosity), moisture, components of the grain, and biological properties [34]. In the case of biomass grains, a significant diversification in terms of dimensions, physical and strength properties, can be observed even within one grain species which, apart from biological characteristics, is conditioned by the weather conditions and the cultivation method [35–37] Earlier research has shown that the energy needed to grind hard materials with higher strength is larger [38–40]. It was also observed that along with a moisture increase, the energy consumption increases as well [18–20,36]. The internal structure of the grain endosperm and tegument has an impact on the strength properties and the energy needed for grinding. The endosperms which are characterized by higher glassiness are usually harder; thus, for permanent deformation, it is necessary to use higher forces which, in turn, results in an increased energy demand as compared to materials whose endosperm is less glassy [36,41,42]. The glassiness of the endosperm also has an influence on the material fragmentation efficiency and the size of particles after division—the higher glassiness, the easier to separate the endosperm from bran and the grain disintegrates into smaller parts. [38,42]. Tests of physical properties and grinding energy, carried out for wheat grains have also revealed that the grinding energy is proportional to the grinded material mass [35]. Dziki and Laskowski [37] indicate, using the example of wheat, that the values of work and force to be applied for crushing the grain increase along with the grain thickness growth.

This study contains an analysis of strength properties of rice grains, mainly forces causing the first violation of the grain structure (first crack) and grain stiffness.

The aim of the research is to experimentally determine the mechanical and processing properties (strength and energy properties) of granular biomass (rice) accepted for research in the project "Intelligent monitoring of the grinding characteristics of grainy biomass". Determining the forces needed to break grains is of key importance when developing energy and environmental efficiency indicators for the grinding process and modeling grinding and crushing processes using the discrete element method DEM.

The research problem was formulated in the form of questions: (1) what strength and energy is needed to induce a rupture of rice grain of the species *Oryza sativa* L. long-grain variety? (2) what is the relationship between grain size and strength parameters and the energy of grinding rice grain of the species *Oryza sativa* L. long-grain variety?

2. Materials and Methods

2.1. Rice Grains Preparation

To determine the force needed to break the grain, a static compression test was carried out for 100 grains of rice of the species *Oryza sativa* L., a long-grained variety with a stabilized humidity equal to 13% ± 0.1%. *Oryza sativa* L., a long-grained variety of rice was accepted to be the research object due to its popularity, among others, in food industry [32]. Knowing the processing properties of this species, in particular, crushing energy, can significantly affect power demand of the processing devices, e.g., grinders. Samples of 100 individual rice grains were prepared and described by numbers (Figure 1). Then, three dimensions were measured with the vernier caliper: length, width and height of the grain.

Figure 1. Rice grains prepared for compression test.

2.2. Test Stand

A static compression test was carried out on an Instron 5966 testing machine (Figure 2). The samples were placed in the machine in a horizontal orientation, in such a way that the dimensions a_1 and a_2 were the large axis and small axis of cross-sectional areas of the grain subjected to the load (Figure 3). The load speed was 2 mm·min^{-1}.

Figure 2. The Instron 5966 testing machine.

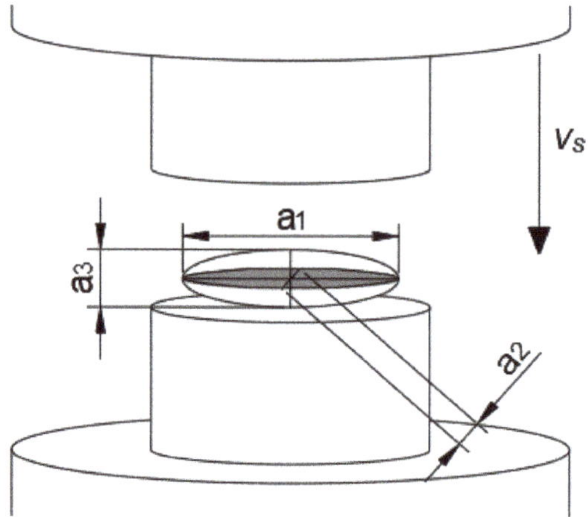

Figure 3. Position of the rice grain during the static compression test.

2.3. Research Methods

The analysis of the grain strength properties during a static compression test of a single grain was performed according to the plan shown in Figure 4. Three dimensions were measured with the vernier caliper: length, width and height of the grain.

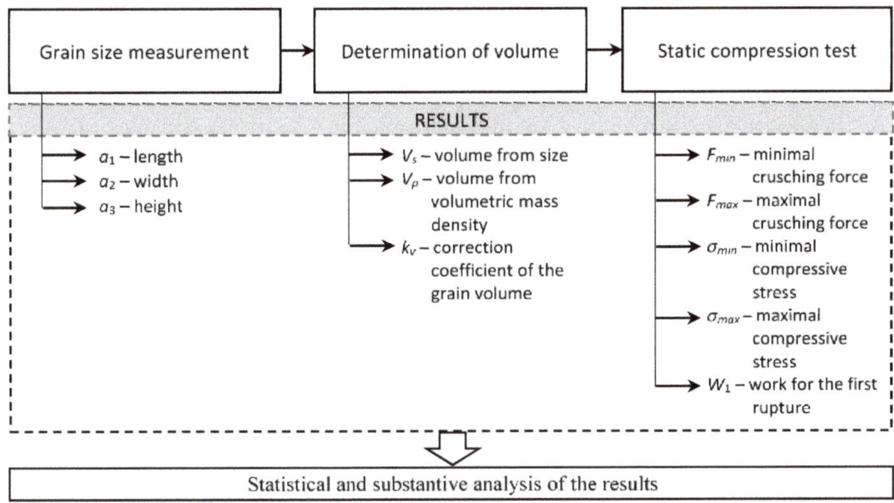

Figure 4. Plan and research program.

The volume of the studied seeds was determined in two ways: on the basis of the three measured dimensions a_1, a_2 and a_3 (V_s) and on the basis of the relationship of the grain mass and density (V_ρ).

The first method of determining the volume takes into consideration three basic grain sizes, i.e., the height and width length, and allows a simplified grain volume estimation with a certain error by aligning the grain shape to the cuboid with dimensions a_1, a_2, a_3 (Figure 5).

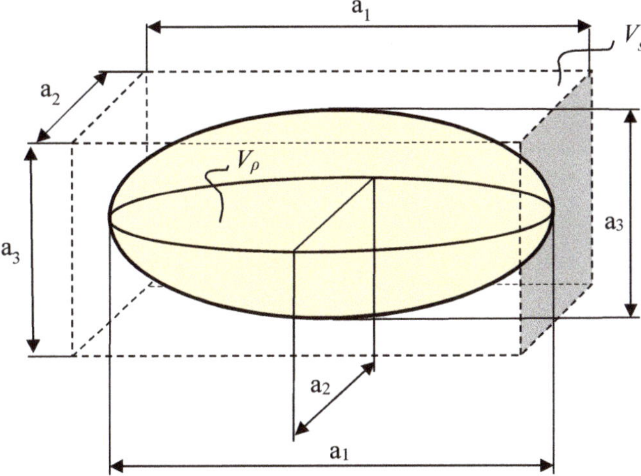

Figure 5. Graphical representation of grain volume determination based on the knowledge of three dimensions, a_1—length of the grain, mm, a_2—width of the grain, mm, a_3—height of the grain, mm, vs.—grain volume calculated based on three dimensions a_1, a_2, a_3 (volume of a cuboid with dimensions a_1, a_2, a_3), mm^3, V_ρ—grain volume calculated based on the volumetric mass density, mm^3.

For this interpretation, the formula for the estimated volume of grain will be:

$$V_s = a_1 \cdot a_2 \cdot a_3, \tag{1}$$

and equal to the volume of a cuboid circumscribed on the grain.

The second method of determining the volume is based on the knowledge of density ρ of the grain and its mass m. The grain density is determined from the dependence [43]:

$$\rho = m/V_\rho, \tag{2}$$

Hence, the volume determined on the basis of density V_ρ, will be [43]:

$$V_\rho = m/\rho, \tag{3}$$

The relationship between computational volume (V_s) and determined based on the grain density (V_ρ) can be determined by calculating the correction factor for the grain volume k_v taking into consideration the grain spherical and uneven shape. It can be determined experimentally and expressed by the dependence:

$$k_v = V_\rho/V_s, \tag{4}$$

hence:

$$V_\rho = k_v \cdot V_s, \tag{5}$$

The static compression test is mainly used for brittle materials, i.e., not showing significant plastic deformation. Rice can be considered as a fragile material with some (small) plastic deformability due to the internal structure that differentiates it from the cross-linked metal structure. A feature that characterizes fragile materials is compressive strength (R_c) [24]:

$$R_c = F_c/A_0, \tag{6}$$

where
F_c—the largest value of the compressive load at which the sample is crushed,
A_0—the initial cross-section of the sample.

If the compression diagram $l = f(F)$ has a part where shortening (Δl) is directly proportional to compressive force (F), then, on this basis, we determine Young's modulus (E) for this material. If this relationship is not directly proportional, then based on the first few results (where it is possible to assume that the material behaves linearly and elastically), we determine the mean value of Young's modulus (E). The value of the Young's modulus is determined by Hooke's law [24]:

$$E = F \cdot l/(\Delta l \cdot A_0), \tag{7}$$

where:
F—compressive force, $kg \cdot m \cdot s^{-2}$,
Δl—sample shortening corresponding to force (F), m,
l—the initial length of the sample, m,
A_0—area of the initial sample cross-section, m^2.

During the compression, some force F affects the grain and causes it to break (displacement s), so the elementary work done over the grain by force F causing the crack can be determined [24,44]:

$$dW = F \cdot ds. \tag{8}$$

During compression, we deal with a variable force F and displacement $s \rightarrow 0$, then, this characterized work can be determined by integrating both sides of the equation [24]:

$$W = \int_{s_1}^{s_2} F ds \tag{9}$$

Then, the work during compression of one grain is the area under graph $F = f(s)$ (Figure 6).

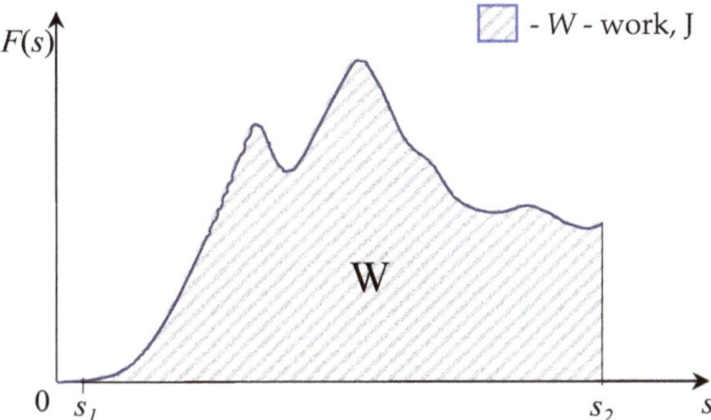

Figure 6. Graphic presentation of work during compression of grains, $F(s)$—compressive force, kg·m·s^{-2}, s—displacement, m, W—work, J, s_1—starting point of displacement, m, s_2—final displacement, m.

2.4. Analytical Methods

The statistical analysis tools available in MS Excel and Statistica were used in processing the results. Basic descriptive statistics of the examined physical and mechanical properties and rupture energy of rice grains were determined. The relationship between the grain volume and physical-mechanical properties and rupture energy was examined using Spearman's correlation analysis. A significance level $p < 0.05$ was adopted.

3. Results and Discussion

3.1. Results of Research on the Physical Properties of Rice Grains and Their Discussion

Firstly, the physical properties of rice grains were determined, such as grain length a_1, grain width a_2, grain height a_3, and grain volume determined based on the knowledge of grain mass and density V_ρ, volume determined on the basis of dimensions V_s, correction factor for volume k_v. The results of the tests after a basic statistical analysis are presented in Figures 7 and 8 in the form of a box plot.

Based on the analysis, it was found that the average grain length of rice was 6.38 mm, the average width was 1.91 mm, the average height was 1.51 mm. The average grain volume determined on the basis of the grain weight and density was equal to 14.82 mm^3, while the volume determined on the basis of grain size 18.44 mm^3. The correction factor for given volumes V_ρ and vs. assumed an average value of 0.82. The obtained results allowed to formulate the dependence of the real volume of the grain determined on the basis of density as a function of the volume determined on the basis of dimensions:

$$V_\rho = k_v \cdot V_s = 0.82 \cdot V_s \tag{10}$$

The results obtained for rice dimensions a_1, a_2, and a_3, are similar, although slightly smaller than those reported in the literature by other researchers, e.g., Sadeghi et al., Zareiforoush et al., and Zeng et al. [18,19,30]. Differences in dimensions may be caused primarily by the difference in the varieties and types of rice grains studied, the country of origin (the study of rice originated from Burma), grain humidity and growing conditions of grains (e.g., extensive, intensive cultivation, weather conditions that affect the grain size). The presented grain size results are an indispensable element of building models in computer simulations based on the DEM discrete element method [18], e.g., in the RockyDEM environment.

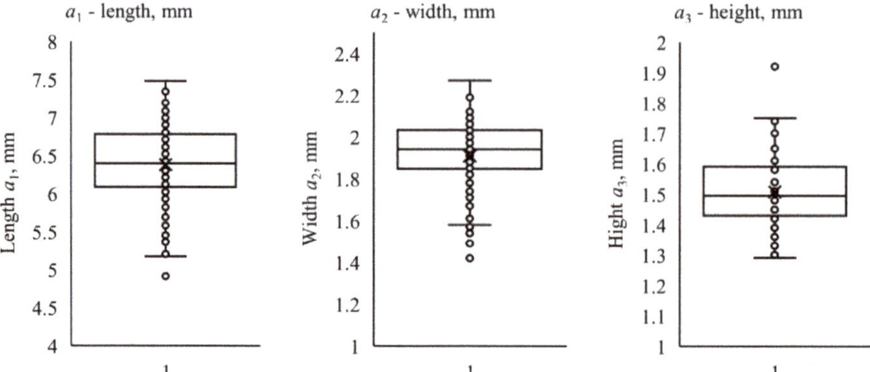

Figure 7. Results of the rice grains size determination.

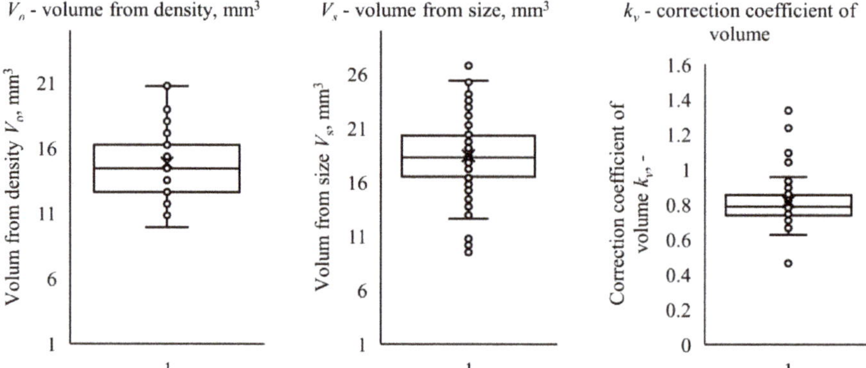

Figure 8. Results of rice grains volume determination.

3.2. The Results of Testing the Strength Properties of Rice Grains and Their Discussion

Figure 9 presents examples of graphs showing the characteristic of rice grain cracks during a compression test. Based on the presented curves, it can be stated that for each grain, the crack proceeded differently. This is caused by differences in internal structure between each grain, which is characteristic for biomaterials. However, noticeable are the characteristic points in the force-displacement graph marked as F_{min} and F_{max}. The point marked as F_{min} symbolizes the first crack of the grain, while the F_{max} point corresponds to the forces causing the breakdown of the grain into smaller fragments (Figure 10). Similar conclusions are presented in the work of Sadeghi et al. [19].

For rice grains, the forces inducing fracture F_{min} of the grain structure were within the range (35.86–198.71) kg m s^{-2}. The maximum forces F_{max} during the crack growth for rice were within the range (70.05–535.74) k g·m s^{-2}. Stresses σ_{min} for rice were in the range (2.21–17.38) MPa. Stresses σ_{max} for rice were in the range (5.00–29.61) MPa. Work W_{Fmin} for rice was in the range (1.88–56.55) mJ and W_{Famx} in the range (2.53–98.93) mJ (Figure 11).

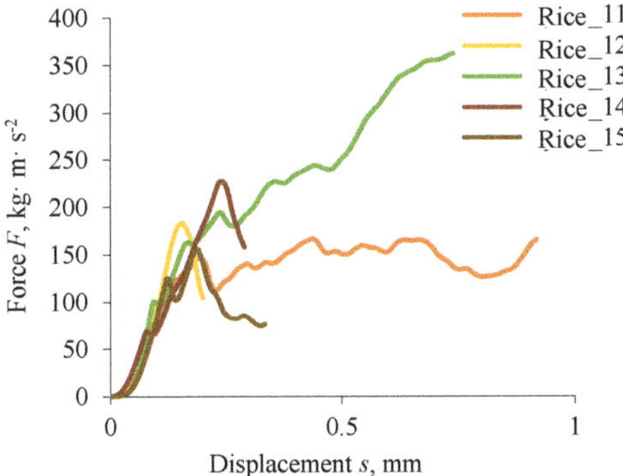

Figure 9. Exemplary curves for five rice grains from 100 tested illustrating the course of the rice grain compression process in the force-displacement coordinate.

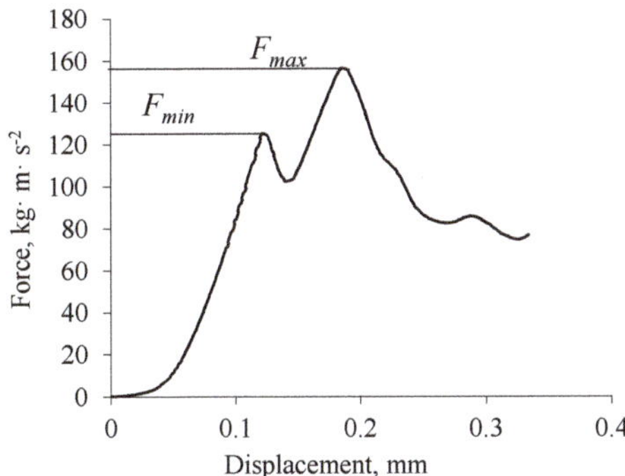

Figure 10. The cracking process for rice grains in function $F(s)$.

F_{min} – minimal crushing force, kg· m· s⁻² σ_{kmin} – minimal compressive stress, MPa W_{Fmin} – work for the first rupture, mJ

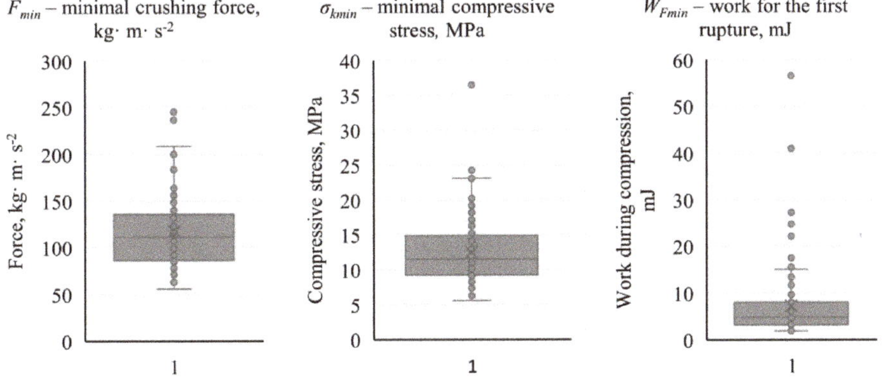

Figure 11. *Cont.*

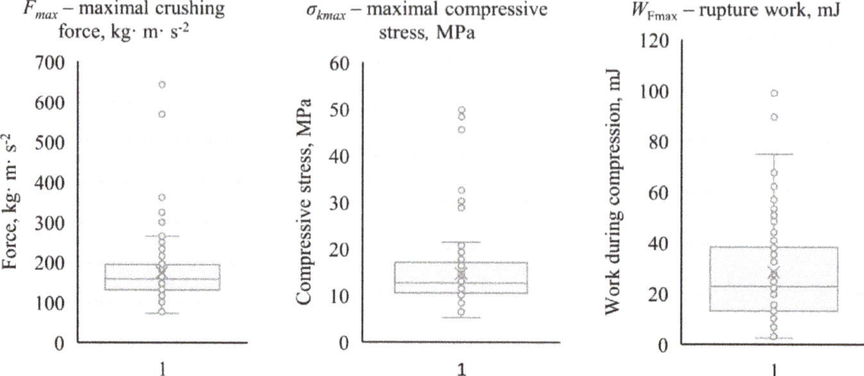

Figure 11. Results of statistical analysis of forces, stresses and energy during compression of rice grains.

Table 1 presents average values and basic results of a statistical analysis of the surveyed quantities. The obtained mean values of forces F_{max} are in accordance with the results presented by Lu and Siebenmorgen [45] obtained in the compression test at the load speed vs. 2 mm·min^{-1} (174.4–188.8 kg·m·s^{-2}), and higher than in the tests carried out by Sadeghi et al. [19] (169.06 kg·m s^{-2} for the Sorkheh type and 125.10 kg·m·s^{-2} for the Sazandegi type with v_s = 1,25 mm·min^{-1}), by Zareiforoush et al. [31] (125.69 kg·m·s^{-2} for the Alikazemi type and 109.96 kg·m·s^{-2} for the Hashemi type with v_s = 5 mm·min^{-1} and 117.38 kg m s^{-2} for the Alikazemi type and 88.33 kg m·s^{-2} for the Hashemi type with v_s = 10 mm·min^{-1}). Differences in the obtained values may result from different grain moistures, loading speeds of the samples and the type of rice used in the tests.

Table 1. Results of statistical analysis of examined mechanical properties of rice grains.

Parameter	Average	Standard Deviation	Median
Minimal crushing force F_{min}, kg m·s^{-2}	117.29	40.71	111.23
Maximal crushing force F_{max}, kg·m·s^{-2}	174.99	80.38	158.12
Minimal compressive stress σ_{kmin}, MPa	12.47	4.83	11.53
Maximal compressive stress σ_{kmax}, MPa	14.71	7.45	12.72
Stiffness C_k, N·mm^{-1}	1275.07	247.72	1322.11
Work for the first rupture W_{Fmin}, mJ	7.26	7.86	4.81
Rupture work W_{Fmax}, mJ	28.03	20.36	22.46

The average value of rupture work W_{Fmax} (28.03 mJ) is similar to the values presented, among others, in the work of Nasirahmadi et al. [26] (26.9 mJ for Fajr and 28.5 mJ for Tarom) and Sadeghi et al. [19] (24.45 mJ for Sazandegi) as well as in the work of Zareiforoush et al. [30,31], where the rupture energy assumed values of about 30 mJ.

Figures 12–18 summarize the results of mechanical properties of rice grains depending on its volume. The graphs show that mechanical properties of rice grains do not depend on its volume.

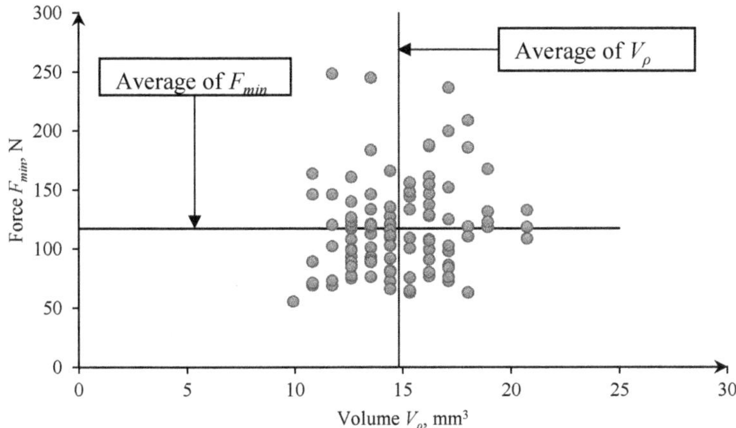

Figure 12. The scatter diagram of forces F_{min} for individual rice grains in relation to volume.

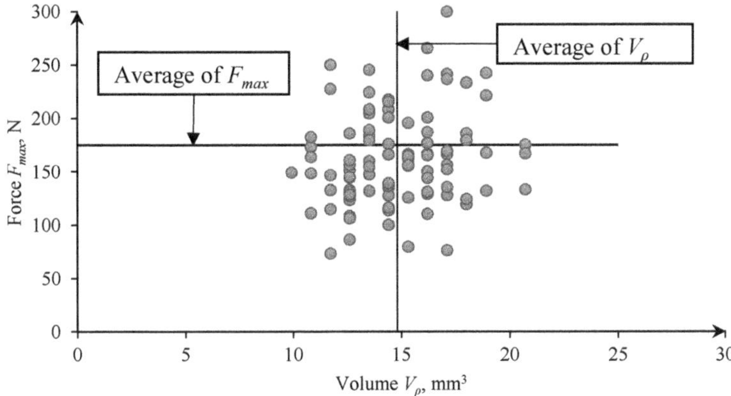

Figure 13. The scatter diagram of forces F_{max} for individual rice grains in relation to volume.

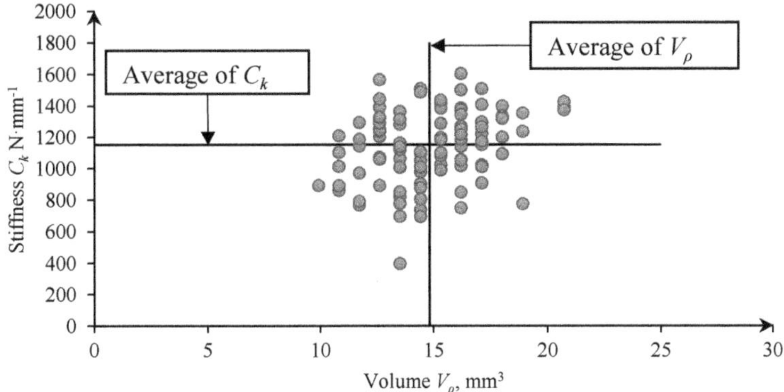

Figure 14. The scatter diagram of stiffnes for individual rice grains in relation to volume.

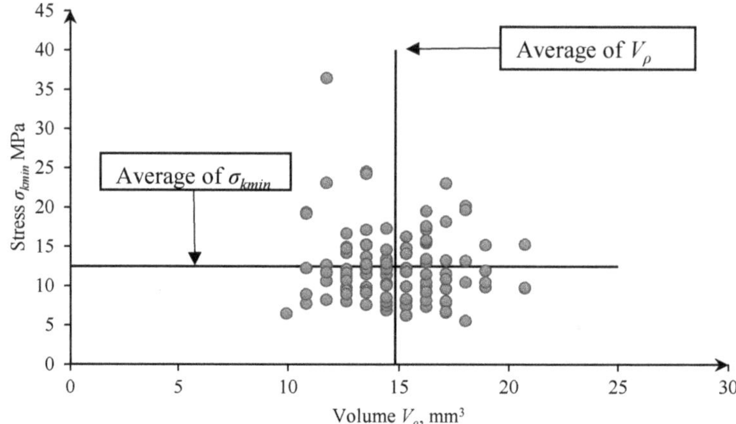

Figure 15. The scatter diagram of stress σ_{kmin} for individual rice grains in relation to volume.

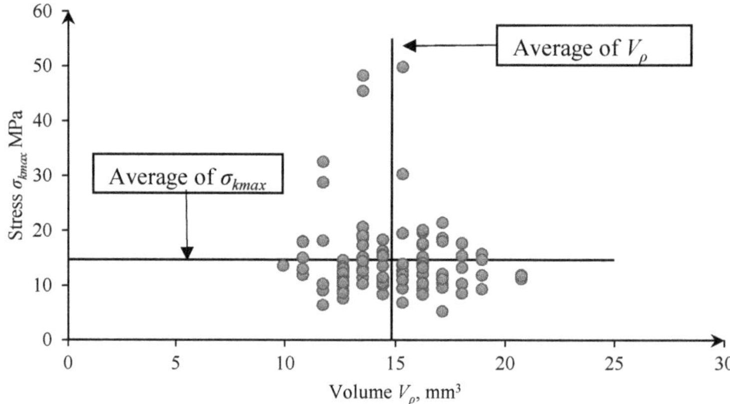

Figure 16. The scatter diagram of stress σ_{kmax} for individual rice grains in relation to volume.

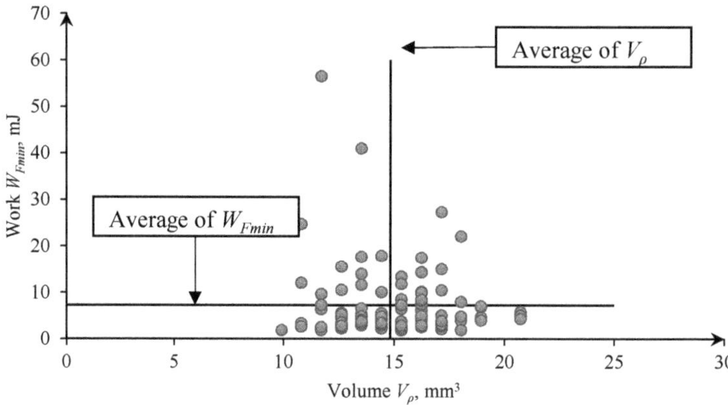

Figure 17. The scatter diagram of stress W_{Fmax} for individual rice grains in relation to volume.

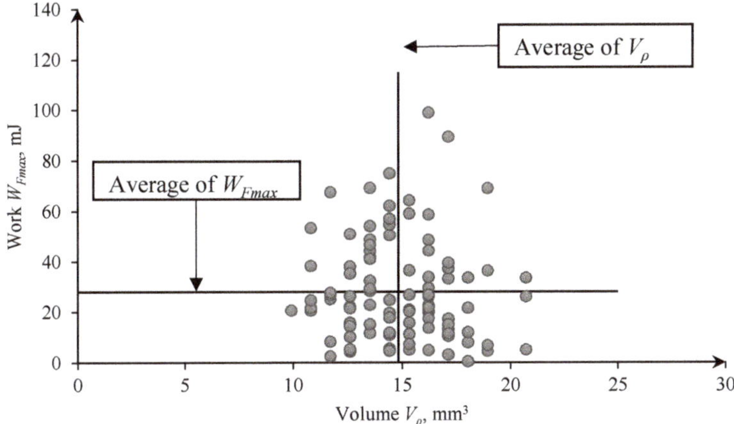

Figure 18. The scatter diagram of stress W_{Fmax} for individual rice grains in relation to volume.

The analysis of the Spearman correlation shows that there are no statistically significant relationships between the volume of the grain V_ρ and the tested strength properties, so there are no interdependencies between the variables (Tab 2). In the case of the grain dimension related volume vs. a low negative correlation between strength σ_{min} and low positive correlation between the force inducing the first crack were found (Table 2). An analysis of the correlation between the grain size related variables, that is, length a_1, width a_2, height a_3, mass m and volumes V_ρ and vs. showed that the volume of grains V_ρ is moderately positively correlated with length a_1 and height a_3 (Table S1). The surface of compression cross-section A_0 was in turn significantly correlated with the grain length a_1 and width a_2 (Table S1). However, no relevant correlations between the grain dimensions, volume and mass and its strength properties and energy needed to crush the grains, were found. Only low correlations occurred, including negative ones between the grain width a_2 and stresses σ_{min} and σ_{max} (Table S1). Contrarily to wheat, a dependence of grinding energy proportionality and its mass was not confirmed [35]. It was not possible to confirm distinct dependencies between the grain thickness (a_3) and the force value, either (for F_{min} positive correlation $R = 0.271$, statistically significant $p < 0.05$, for F_{max} statistically insignificant correlation (Table S1)) and work (W_{Fmin} and W_{Fmax} low correlations), which could be observed for wheat [37]. The volume is related to the grain dimensions, including compression cross-section A_0, whereas the cross section is related to compressive strength (according to dependence (6)). However, it was not possible to show significant dependencies between the grain cross section A_0 and values of compressive strength (low negative correlations between A_0 and σ_{min}, σ_{max} (Table S1)). The obtained results, including the results of dimension and volume scatter, confirm the significant variability and diversity of biological materials within one species. The diversity of the values confirms that each grain is characterized by a different internal structure. Such a diversification can indicate low quality of the grain and poor conditions of cultivation.

Table 2. Results of correlation analysis between the volume of grain and its strength properties.

		F_{min} [1]	F_{max} [2]	C_k [3]	σ_{min} [4]	σ_{max} [5]	W_{Fmin} [6]	W_{Fmax} [7]
	rhoSpearman coefficient	0.172	0.176	0.242	−0.018	−0.052	0.071	−0.057
V_ρ	Significance	0.088	0.08	0.015	0.859	0.604	0.482	0.573
	Number of samples n	100	100	100	100	100	100	100
	rhoSpearman coefficient	0.220	0.062	0.005	−0.280	−0.104	0.185	−0.138
V_s	Significance	0.028	0.543	0.958	0.005	0.305	0.066	0.174
	Number of samples n	100	100	100	100	100	100	100

[1] Minimal crushing force, [2] Maximal crushing force, [3] Minimal compressive stress, [4] Maximal compressive stress, [5] Stiffness, [6] Work for the first rupture, [7] Rupture work.

It is noteworthy that the grains were tested for only one orientation of grains in the strength testing machines. It is expected that the longitudinal and transverse orientation of grains would provide different values of destructive compression, forces and strength [28–30].

4. Conclusions

The main objectives of the study were achieved through the determination of energy (work) and compressive forces in the test of static compression and analysis of the dependence between the grain size and the grain strength parameters and grinding energy.

Based on the analyzes, it was found that the average grain length of rice was 6.38 mm, the average width was 1.91 mm, the average height was 1.51 mm. The average grain volume determined on the basis of the grain weight and density was equal to 14.82 mm^3, while the volume determined on the basis of grain size 18.44 mm^3. The correction factor for given volumes V_ρ and vs. assumed an average value of 0.82. The values of the grain dimension distribution can be a determinant in selection of structural features of the materials used, that is, performance parameters of roller mills: diameters of rollers and the size of the inter-roller gap; selection of the screen eye sieve

The average values of strength properties of rice grains were determined, such as $F_{min} = 117.9$ kg·m s^{-2}, $F_{max} = 174.99$ kg·m·s^{-2}, $\sigma_{min} = 9.80$ MPa, $\sigma_{max} = 14.71$ MPa, $C_k = 1150.26$ N·mm^{-1}, $W_{Fmin} = 7.26$ mJ, $W_{Fmax} = 28.03$ mJ, which coincided with the results of research carried out by other researchers. The determined ranges of forces, strength and compressive energy (work) are of applicable character and can be used in the design of machines dedicated to process rice. Knowing these values will allow, among others, the estimation of the power of devices, e.g., grinding machines, and roller mills, and in consequence, will minimize energy losses and energy demand for dedicated machines.

The analysis of the Spearman correlation showed that there are no statistically significant relationships between the volume V_ρ of the grain and the tested strength properties, so there are no interdependencies between the variables. In the case of the grain size volume vs. a low negative correlation between strength σ_{min} and low positive correlation between the force inducing the first crack (Table 2) were found. Dependence of grinding energy proportionality and the grain mass as well as clear, distinct dependencies between the grain thickness and the value of force and work (low negative correlation between thickness co a_3 and stresses σ_{min} and work W_{Fmax}, low positive correlation between thickness a_3 and force inducing the first crack F_{min}), that was proven for other biological grainy materials (wheat grains) could not be confirmed either. Based on these results, it was not possible to find significant dependencies between the grain cross-section A_0 and the values of compressive strength (only low negative correlations were found).

Based on the conducted tests, the crack was found to be different for each grain (Figure 9). The results, including the scatter of dimensions and volume, confirm the high variability and diversity of biological materials within one species. Diversification of the obtained values confirms that each grain is characterized by a different internal structure. Such a differentiation can indicate a poor quality of grain and weak cultivation conditions.

An analysis of the test results of rice grain strength properties provides the basis for determining the impact of biomass properties on the grinding process and, in subsequent stages, for the development of procedures for monitoring the grinding process using energy-environmental grinding efficiency models, the original CO_2 emission index for the intelligent monitoring system of usable characteristics of the grinding process.

The results can be used by other researchers to create models of materials (rice grains) for computer simulations of cracking, crushing, mixing using the discrete element method DEM.

Supplementary Materials: The following are available online at http://www.mdpi.com/2071-1050/12/2/552/s1, Table S1: Results of correlation analysis between the size of grain and its strength properties

Author Contributions: Conceptualization, W.K. and J.F.; methodology, W.K. and R.K.; software, W.K. and P.B.-W.; formal analysis, A.M., A.T., and J.F.; investigation, W.K., K.P. and P.B.-W.; resources, W.K.; data curation, A.M., K.P. and R.K.; writing—original draft preparation, W.K., P.B.-W., and A.T; writing—review and editing, W.K., R.K. and

J.F.; visualization, W.K.; supervision, A.T., J.F., A.M. and K.P.; project administration, W.K.; funding acquisition, W.K. All authors have read and agreed to the published version of the manuscript.

Funding: Scientific work financed by the budget resource for science in 2017–2021, as a research project under the "Diamentowy Grant" program. This research was funded by MINISTRY OF SCIENCE AND HIGHER EDUCATION OF POLAND, grant number DI2016 001646.

Conflicts of Interest: The authors declare no conflict of interest. The funders had no role in the design of the study; in the collection, analyses, or interpretation of data; in the writing of the manuscript, or in the decision to publish the results.

List of Symbols

a_1	length of the grain: mm
a_2	width of the grain, mm
a_3	height of the grain, mm
V_s	grain volume calculated based on three dimensions a_1, a_2, a_3, mm^3
V_ρ	grain volume calculated based on the volumetric mass density, mm^3
ρ	volumetric mass density, kg·m^{-3}
m	grain weight, g
k_v	correction coefficient of the grain volume
R_c	compressive strength, MPa
F_c	the largest value of the compressive load at which the sample is crushed, kg·m·s^{-2}
A_0	the initial cross-section of the sample, m^2
F	compressive force, kg m·s^{-2}
Δl	sample shortening corresponding to force (F), m
l	the initial length of the sample, m
A_0	area of the initial sample cross-section, m^2
s	displacement, m
W	work, J
dW	elementary work, J
ds	elementary displacement, m
F_{min}	minimal crushing force, kg·m s^{-2}
F_{max}	maximal crushing force, kg·m·s^{-2}
σ_{kmin}	minimal compressive stress, MPa
σ_{kmax}	maximal compressive stress, MPa
C_k	stiffness, N·mm^{-1}
W_{Fmin}	work for the first rupture, mJ
W_{Fmax}	rupture work, mJ

References

1. Tomporowski, A.; Flizikowski, J.; Kruszelnicka, W. A new concept of roller-plate mills. *Przem. Chem.* **2017**, *96*, 1750–1755.
2. Tomporowski, A.; Flizikowski, J.; Wełnowski, J.; Najzarek, Z.; Topoliński, T.; Kruszelnicka, W.; Piasecka, I.; Śmigiel, S. Regeneration of rubber waste using an intelligent grinding system. *Przem. Chem.* **2018**, *97*, 1659–1665.
3. Flizikowski, J.B.; Mrozinski, A.; Tomporowski, A. Active monitoring as cognitive control of grinders design. In *AIP Conference Proceedings*; AIP Publishing: Melville, NY, USA, 2017; Volume 1822, p. 020006.
4. Marczuk, A.; Caban, J.; Savinykh, P.; Turubanov, N.; Zyryanov, D. Maintenance research of a horizontal ribbon mixer. *Eksploat. Niezawodn.* **2017**, *19*, 121–125. [CrossRef]
5. Tomporowski, A.; Flizikowski, J. Motion characteristics of a multi-disc grinder of biomass grain. *Przem. Chem.* **2013**, *92*, 498–503.
6. Bochat, A.; Zastempowski, M. Kinematics and dynamics of the movement of the selected constructions of the disc cutting assemblies. In Proceedings of the Engineering Mechanics 2017, Brno University of Technology, Faculty of Mechanical Engineering, Institute of Solid Mechanics, Mechatronics and Biomechanics, Brno-Svratka, Czech Republic, 15–18 May 2017; Volume 23, pp. 170–173.

7. Zastempowski, M.; Bochat, A. Modeling of cutting process by the shear-finger cutting block. *Appl. Eng. Agric.* **2014**, *30*, 347–353.

8. Flizikowski, J.; Macko, M. Method of estimation of efficiency of quasi-cutting of recycled opto-telecommunication pipes. *Polimery* **2001**, *46*, 53–59. [CrossRef]

9. Kaczmarczyk, J.; Grajcar, A. Numerical simulation and experimental investigation of cold-rolled steel cutting. *Materials* **2018**, *11*, 1263. [CrossRef] [PubMed]

10. Kruszelnicka, W.; Shchur, T. Study of rice and maize grains grinding energy. *TEKA Comm. Mot. Energetics Agric.* **2018**, *18*, 71–74.

11. Szyszlak-Barglowicz, J.; Zajac, G. Distribution of heavy metals in waste streams during combustion of *Sida hermaphrodita* (L.) Rusby biomass. *Przem. Chem.* **2015**, *94*, 1723–1727.

12. Szyszlak-Barglowicz, J.; Zając, G.; Słowik, T. Hydrocarbon emissions during biomass combustion. *Polish J. Environ. Stud.* **2015**, *24*, 1349–1354. [CrossRef]

13. Kowalczyk-Jusko, A.; Kowalczuk, J.; Szmigielski, M.; Marczuk, A.; Jozwiakowski, K.; Zarajczyk, K.; Maslowski, A.; Slaska-Grzywna, B.; Sagan, A.; Zarajczyk, J. Quality of biomass pellets used as fuel or raw material for syngas production. *Przem. Chem.* **2015**, *94*, 1835–1837.

14. Rudnicki, J.; Zadrag, R. Technical state assessment of charge exchange system of self-ignition engine, based on the exhaust gas composition testing. *Polish Marit. Res.* **2017**, *24*, 203–212. [CrossRef]

15. Nizamuddin, S.; Qureshi, S.S.; Baloch, H.A.; Siddiqui, M.T.H.; Takkalkar, P.; Mubarak, N.M.; Dumbre, D.K.; Griffin, G.J.; Madapusi, S.; Tanksale, A. Microwave hydrothermal carbonization of rice straw: Optimization of process parameters and upgrading of chemical, fuel, structural and thermal properties. *Materials* **2019**, *12*, 403. [CrossRef] [PubMed]

16. Chang, K.-L.; Wang, X.-Q.; Han, Y.-J.; Deng, H.; Liu, J.; Lin, Y.-C. Enhanced enzymatic hydrolysis of rice straw pretreated by oxidants assisted with photocatalysis technology. *Materials* **2018**, *11*, 802. [CrossRef] [PubMed]

17. Mannheim, V. Examination of thermic treatment and biogas processes by Lca. *Ann. Fac. Eng. Hunedoara Int. J. Eng.* **2014**, *12*, 225–234.

18. Zeng, Y.; Jia, F.; Xiao, Y.; Han, Y.; Meng, X. Discrete element method modelling of impact breakage of ellipsoidal agglomerate. *Powder Technol.* **2019**, *346*, 57–69. [CrossRef]

19. Sadeghi, M.; Araghi, H.A.; Hemmat, A. Physico-mechanical properties of rough rice (*Oryza sativa* L.) Grain as affected by variety and moisture content. *Agric. Eng. Int. CIGR J.* **2010**, *12*, 129–136.

20. Cao, W.; Nishiyama, Y.; Koide, S. Physicochemical, mechanical and thermal properties of brown rice grain with various moisture contents. *Int. J. Food Sci. Technol.* **2004**, *39*, 899–906. [CrossRef]

21. Chattopadhyay, P.K.; Hamann, D.D.; Hammerle, J.R. Effect of deformation rate and moisture content on rice grain stiffness1. *J. Food Process Eng.* **1980**, *4*, 117–121. [CrossRef]

22. Buggenhout, J.; Brijs, K.; Celus, I.; Delcour, J.A. The breakage susceptibility of raw and parboiled rice: A review. *J. Food Eng.* **2013**, *117*, 304–315. [CrossRef]

23. Esehaghbeygi, A.; Daeijavad, M.; Afkarisayyah, A.H. Breakage susceptibility of rice grains by impact loading. *Appl. Eng. Agric.* **2009**, *25*, 943–946. [CrossRef]

24. Sarker, M.S.H.; Hasan, S.M.K.; Ibrahim, M.N.; Aziz, N.A.; Punan, M.S. Mechanical property and quality aspects of rice dried in industrial dryers. *J. Food Sci. Technol.* **2017**, *54*, 4129–4134. [CrossRef] [PubMed]

25. Talab, K.T.; Ibrahim, M.N.; Spotar, S.; Talib, R.A.; Muhammad, K. Glass transition temperature, mechanical properties of rice and their relationships with milling quality. *Int. J. Food Eng.* **2012**, *8*.

26. Nasirahmadi, A.; Abbaspour-Fard, M.H.; Emadi, B.; Khazaei, N.B. Modelling and analysis of compressive strength properties of parboiled paddy and milled rice. *Int. Agrophysics* **2014**, *28*, 73–83. [CrossRef]

27. Bonazzi, C.; Courtois, F. Impact of drying on the mechanical properties and crack formation in rice. In *Modern Drying Technology*; John Wiley & Sons, Ltd.: Hoboken, NJ, USA, 2011; pp. 21–49. ISBN 978-3-527-63166-7.

28. Li, Y.-N.; Li, K.; Ding, W.-M.; Chen, K.-J.; Ding, Q. Correlation between head rice yield and specific mechanical property differences between dorsal side and ventral side of rice kernels. *J. Food Eng.* **2014**, *123*, 60–66. [CrossRef]

29. Shu, Y.-J.; Liou, N.-S.; Moonpa, N.; Topaiboul, S. Investigating damage properties of rice grain under compression load. In Proceedings of the International Conference on Experimental Mechanics 2013 and Twelfth Asian Conference on Experimental Mechanics, Bangkok, Thailand, 25–27 November 2013; International Society for Optics and Photonics: Bangkok, Thailand, 2014; Volume 9234, p. 923402.

Sustainability **2020**, *12*, 552

30. Zareiforoush, H.; Komarizadeh, M.H.; Alizadeh, M.R.; Tavakoli, H.; Masoumi, M. Effects of moisture content, loading rate, and grain orientation on fracture resistance of paddy (*Oryza sativa* L.) grain. *Int. J. Food Prop.* **2012**, *15*, 89–98. [CrossRef]

31. Zareiforoush, H.; Komarizadeh, M.H.; Alizadeh, M.R. Mechanical properties of paddy grains under quasi-static compressive loading. *N. Y. Sci. J.* **2010**, *3*, 40–46.

32. Pandiselvam, R.; Thirupathi, V.; Mohan, S. Engineering properties of rice. *Agric. Eng.* **2015**, *XL*, 69–78.

33. Ligaj, B.; Szala, G. Obliczanie zapotrzebowania energii w procesach rozdrabniania materiałów ziarnistych na przykładzie ziaren zbóż. *Acta Mech. Autom.* **2009**, *3*, 97–99.

34. Tumuluru, J.S.; Tabil, L.G.; Song, Y.; Iroba, K.L.; Meda, V. Grinding energy and physical properties of chopped and hammer-milled barley, wheat, oat, and canola straws. *Biomass Bioenergy* **2014**, *60*, 58–67. [CrossRef]

35. Wiercioch, M.; Niemiec, A.; Roma, L. The impact of wheat seeds size on energy consumption of their grinding process. *Inzyneria Rol.* **2008**, *103*, 367–372.

36. Warechowska, M. Some physical properties of cereal grain and energy consumption of grinding. *Agric. Eng.* **2014**, *1*, 239–249.

37. Dziki, D.; Laskowski, J. Influence of wheat kernel geometrical properties on the mechanical properties and grinding ability. *Acta Agrophysica* **2003**, *2*, 735–742.

38. Warechowska, M.; Warechowski, J.; Skibniewska, K.A.; Siemianowska, E.; Tyburski, J.; Aljewicz, M.A. Environmental factors influence milling and physical properties and flour size distribution of organic spelt wheat. *Tech. Sci.* **2016**, *19*, 387–399.

39. Dziki, D. Ocena energochłonności rozdrabniania ziarna pszenicy poddanego uprzednio zgniataniu. *Inzyneria Rol.* **2007**, *11*, 51–58.

40. Dziki, D.; Cacak-Pietrzak, G.; Miś, A.; Jończyk, K.; Gawlik-Dziki, U. Influence of wheat kernel physical properties on the pulverizing process. *J. Food Sci. Technol.* **2014**, *51*, 2648–2655. [CrossRef] [PubMed]

41. Greffeuille, V.; Mabille, F.; Rousset, M.; Oury, F.-X.; Abecassis, J.; Lullien-Pellerin, V. Mechanical properties of outer layers from near-isogenic lines of common wheat differing in hardness. *J. Cereal Sci.* **2007**, *45*, 227–235. [CrossRef]

42. Greffeuille, V.; Abecassis, J.; Barouh, N.; Villeneuve, P.; Mabille, F.; Bar L'Helgouac'h, C.; Lullien-Pellerin, V. Analysis of the milling reduction of bread wheat farina: Physical and biochemical characterisation. *J. Cereal Sci.* **2007**, *45*, 97–105. [CrossRef]

43. Yenge, G.B.; Kad, V.P.; Nalawade, S.M. Physical properties of maize (*Zea mays* L.) grain. *J. Krishi Vigyan* **2018**, *7*, 125–128. [CrossRef]

44. Korczewski, Z.; Rudnicki, J. An energy approach to the fatigue life of ship propulsion systems marine 2015. In Proceedings of the VI International Conference on Computational Methods in Marine Engineering—The Conference Proceedings, Rome, Italy, 14–17 July 2015; Salvatore, F., Broglia, R., Muscari, R., Eds.; International Center Numerical Methods Engineering: Barcelona, Spain, 2015; pp. 490–501, ISBN 978-84-943928-6-3.

45. Lu, R.; Siebenmorgen, T.J. Correlation of head rice yield to selected physical and mechanical properties of rice kernels. *Trans. ASAE* **1995**, *38*, 889–894. [CrossRef]

 *sustainability*

Article

Optimization of Extraction Conditions for the Antioxidant Potential of Different Pumpkin Varieties (*Cucurbita maxima*)

Bartosz Kulczyński [1], Anna Gramza-Michałowska [1,*] and Jolanta B. Królczyk [2]

[1] Department of Gastronomy Sciences and Functional Foods, Faculty of Food Science and Nutrition, Poznań University of Life Sciences, 31 Wojska Polskiego Str., 60–624 Poznań, Poland; bartosz.kulczynski@up.poznan.pl

[2] Department of Manufacturing Engineering and Production Automation, Faculty of Mechanical Engineering, Opole University of Technology, ul. Prószkowska 76, 45–758 Opole, Poland; j.krolczyk@po.opole.pl

* Correspondence: anna.gramza@up.poznan.pl; +48-61-848-73-27

Received: 29 November 2019; Accepted: 7 February 2020; Published: 11 February 2020

Abstract: Antioxidants are a wide group of chemical compounds characterized by high bioactivity. They affect human health by inhibiting the activity of reactive oxygen species. Thus, they limit their harmful effect and reduce the risk of many diseases, including cardiovascular diseases, cancers, and neurodegenerative diseases. Antioxidants are also widely used in the food industry. They prevent the occurrence of unfavourable changes in food products during storage. They inhibit fat oxidation and limit the loss of colour. For this reason, they are often added to meat products. Many diet components exhibit an antioxidative activity. A high antioxidative capacity is attributed to fruit, vegetables, spices, herbs, tea, and red wine. So far, the antioxidative properties of various plant materials have been tested. However, the antioxidative activity of some products has not been thoroughly investigated yet. To date, there have been only a few studies on the antioxidative activity of the pumpkin, including pumpkin seeds, flowers, and leaves, but not the pulp. The main focus of our experiment was to optimize the extraction so as to increase the antioxidative activity of the pumpkin pulp. Variable extraction conditions were used for this purpose, i.e., the type and concentration of the solvent, as well as the time and temperature of the process. In addition, the experiment involved a comparative analysis of the antioxidative potential of 14 pumpkin cultivars of the *Cucurbita maxima* species. The study showed considerable diversification of the antioxidative activity of different pumpkin cultivars.

Keywords: pumpkin; *Cucurbita maxima*; antioxidative activity; Oxygen Radical Absorbance Capacity (ORAC); cluster analysis

---

1. Introduction

Food consumption is directly linked to maintaining health, well-being, and preventing hunger. In this context, proper nutrition is undoubtedly a key aspect of sustainable food, and thus also environmental sustainability. Plant products are the only alternative for the consumers to alter current meat products consumption and more sustainable way of living towards reducing negative impact on the environment. The application of plant matrices enriched with mineral components, instead of relevant animal products, allows the strive to develop practical alternatives that often change existing processes and products not corresponding with a sustainable production. Therefore, sustainable food production in an environmentally acceptable manner to meet the increasing demands of a growing population is an inevitable challenge for agricultural production. Antioxidants are natural molecules found in living organisms which prevent oxidative stress. These compounds are characterized by the ability to scavenge (neutralize) reactive oxygen species (ROS), including hydroperoxide radicals, superoxide anion radicals, singlet oxygen, hydrogen peroxide, and hydroxyl radicals [1,2]. The following compounds are usually listed as strong antioxidants: Carotenoids (e.g., beta-carotene, lycopene, astaxanthin), tocopherols (e.g., alpha-tocopherol, gamma-tocopherol), flavonoids (e.g., anthocyanins, flavanols, flavonones, isoflavones), phenolic acids (hydroxybenzoic acids, hydroxycinnamic acids), stilbenes, some vitamins (vitamin C), coenzyme Q10, sulphur compounds (e.g., allicin), mineral components (e.g., selenium, zinc) [1,3–5]. Products of plant origin (vegetables, fruits, mainly berries, herbs, spices, juices, wine, tea, some cereals, and grains) are rich sources of antioxidants. However, they can also be found in meat products [6]. The antioxidative activity of the compounds we consume every day is important for our health [1,7–10]. Antioxidants inhibit free radicals and thus they reduce the risk of various diseases of affluence, including cardiovascular diseases (atherosclerosis, hypertension, heart attack, stroke), diabetes, cancers, neurodegenerative diseases (e.g., Alzheimer's disease, Parkinson's disease), and osteoporosis [11–15]. Antioxidants may soothe inflammations and viral infections [16,17]. They are said to prevent age-related eye diseases [18,19]. Antioxidants are not only important to maintain good health, but they are also widely used in food technology. For example, these compounds inhibit oxidation and limit the degradation of phytosterols [20] and fats [21]. In this way, they may, for example, prevent the spoilage of meat products [22]. Antioxidants also preserve the right colour of food products [23,24]. It should be noted, however, that their activity depends on many factors, such as chemical structure and profile of antioxidants, various interactions (synergistic, antagonistic effect), and also the multidimensional characteristics of the food matrix [25].

The pumpkin (*Cucurbita* L.) is a plant material containing antioxidants, e.g., carotenoids, tocopherols, phenolic acids, and flavonols. It is commonly grown in Europe, Asia, South America, North America, and Africa [26]. It is estimated that the annual pumpkin production exceeds 20 million tonnes [27,28]. It is a valuable dietary component because its pulp is rich in carotenoids [29–31] and its seeds are a source of unsaturated fatty acids [32]. Pumpkin flowers and leaves are less popular, but they are also edible [33]. Pumpkin pulp can be consumed both raw and after being processed, e.g., cooked, or as compotes, jams, purees, and juices [34]. The most common pumpkin species are: *Cucurbita maxima*, *Cucurbita pepo*, and *Cucurbita moschata*. Each of these species has numerous varieties [34,35]. So far there have been few studies on the antioxidative activity of the pumpkin pulp. As a result, scientific publications do not provide a comparison of the antioxidative properties of the pulp of different pumpkin varieties. Most of the available studies only determine the antioxidative potential of pumpkin seeds and oils [36–38]. Therefore, it is difficult to point out the trends in the contribution of specific compounds and the impact of various factors to the total antioxidant potential.

Thus far, there have not been many studies comparing the cultivars-dependent antioxidant capacity of pumpkin pulp. Additionally, reference publications lack data on the correlation analysis between the antioxidant activity tests in *Cucurbita* maxima cultivars extracts and its cluster analysis. The gap in the current literature to which this study is addressed includes an assessment of the antioxidant capacity of selected pumpkin cultivars, determined by selected radicals scavenging assays. The aim of this study was to compare the antioxidative activity of the pulp of various pumpkin varieties of the *Cucurbita maxima* species. The study also assesses the influence of various pumpkin pulp extraction conditions on its antioxidative potential.

2. Materials and Methods

2.1. Chemicals and Reagents

Gallic acid, sodium carbonate, Folin and Ciocalteu's phenol reagent, (±)-6-hydroxyl-2,5,7,8-tetramethylchromane-2-carboxylic acid (Trolox), 2,2′-azino-bis(3-ethylbenzothiazoline-6-sulfonic acid) diammonium salt (ABTS), potassium persulphate, 2,2-diphenyl-1-picrylhydrazyl (DPPH), sodium acetate trihydrate, acetic acid, 2,4,6-tris(2-pyridyl)-s-triazine (TPTZ), iron(III) chloride hexahydrate, hydrochloric acid, iron (II) sulphate heptahydrate, iron (II) chloride tetrahydrate, 3-(2-pyridyl)-

5,6-diphenyl-1,2,4-triazine-4′,4″-disulfonic acid sodium salt (Ferrozine), ethylenediaminetetraacetic acid (EDTA), sodium phosphate monobasic dehydrate, sodium phosphate monobasic dehydrate, potassium phosphate dibasic, 2,2′-azobis(2-methylpropionamidine) dihydrochloride (AAPH), and fluorescein sodium salt were purchased from Sigma-Aldrich (Darmstadt, Germany).

2.2. Sample Collection

The pulp of 14 pumpkin cultivars of the *Cucurbita* maxima species ('Buttercup', 'Golden Hubbard', 'Galeux d'Eysines', 'Melonowa Żółta', 'Hokkaido', 'Jumbo Pink Banana', 'Marina di Chiggia', 'Flat White Boer Ford', 'Jarrahdale', 'Blue Kuri', 'Green Hubbard', 'Gomez', 'Shokichi Shiro', 'Porcelain Doll') was used as the research material. All the cultivars were purchased from 'Dolina Mogilnicy' Organic Farming Products Cooperative (Wolkowo, Poland). While the plants were being grown, they were irrigated, weeded, and the soil was loosened. The pumpkins were harvested in October 2018 and transported to a laboratory, where they were immediately cleaned. The experimental unit consisted of two randomly chosen pumpkins from each variety. All chemical analyses for each pumpkin were performed in triplicate. The edible pulp was cut into pieces, freeze-dried, and then subjected to analysis. The dried pulp was stored at room temperature, without access to oxygen and light.

2.3. Extract Preparation

Preliminary studies involved solvents most commonly used for the preparation of plant extracts (acetone, ethyl acetate, ethanol, methanol, water). The extraction was carried out for various concentrations of selected solvents (20%, 40%, 60%, 80%, 100%) and the extraction time (0.5, 1, 2, 3, 4 h). Tests were carried out in three temperature ranges, which are used in plant components extraction as the most effective (30, 50, and 70 °C). Variants with the highest DPPH radical scavenging values were selected for further stages of the study. As a result, the above factors were limited to the previously mentioned nine solvents, two times and three extraction temperatures. In order to optimize the extraction pumpkin extracts were prepared by weighing 5 g of freeze-dried pumpkin and dissolving them in 50 mL of a solvent (water, methanol, 80% water–methanol solution, ethanol, 80% water–ethanol solution, acetone, 80% water–acetone solution, ethyl acetate, 80% ethyl acetate, and water solution). Next, the whole was shaken in a water bath (SWB 22N) at 30, 50, or 70 °C for 1 or 2 h. The extracts were centrifuged (1500 rpm, 10 min) and filtered through paper. Next, their antioxidative activity was assessed. The extraction of all samples was triplicated. The most favourable extraction conditions were selected on the basis of the optimization results: 2 h and 70 °C, maintaining the ratio between the amount weighed and the volume of the solvent used (1:10). Extracts for the pumpkin cultivars were prepared according to these conditions to find differences in their antioxidative activity, which included determination of the ABTS and DPPH radical scavenging assay, oxygen radical absorbance capacity (ORAC), the ferric reducing antioxidant capacity assay (FRAP), and the iron chelating activity assay.

2.4. Share of Individual Plant Elements

The pumpkin was carefully peeled in order to determine the content of individual parts of the fruit. Then, the pulp was cleared of seeds. Each part of the pumpkin was weighed with an accuracy of three decimal digits. The results were used to calculate the percentage content of the skin, seeds, and pulp. Each sample was measured in triplicate.

2.5. Pumpkin Flesh Colour

The colour of the pumpkin flesh was determined with a colorimeter (Chroma Meter CR-410, Konica Minolta Sensing Inc., Osaka, Japan). The test consisted in measuring the colour of the sample in reflected light and calculating the L * a * b value (L—color brightness; a—color in the range from green to red; b—color from blue to yellow). Before the measurement a homogeneous pulp was prepared from the pumpkin flesh. The analysis was triplicated.

2.6. Moisture Content

The water content was measured by drying 1 ± 0.001 g of fresh pumpkin, weighed, and then the sample was dried at 105 °C for 3 h. Next, the samples were cooled to room temperature and their weight was checked. The samples were then placed in a dryer again for 30 min. The samples were cooled again and weighed. The procedure was repeated until the sample weight between two measurements differed by more than 0.004 g. Each sample was measured in triplicate.

2.7. Active Acidity

The potentiometric method was used to measure acidity (pH) with a pH meter (CP-401). The result of measurement of the potential difference between the indicator and comparative electrodes was recorded. Each sample was measured in triplicate.

2.8. DPPH Radical Scavenging Activity Assay

The analysis was made in accordance with the methodology invented by Brand-Williams et al. (1995) [39]. An amount of 0.01 g of 2,2-diphenyl-1-picrylhydrazyl (DPPH) was weighed and transferred into a 25-mL volumetric flask together with the solvent (80:20 methanol/water (*v/v*). Next, the flask was filled up to the marking level. A calibration curve for Trolox (Tx) was also prepared. Assay: An amount of 100 μL of the extract was collected and 2.0 mL of the solvent, as well as 250 μL of the DPPH reagent were added. The whole was shaken on a Vortex at ambient temperature and left in darkness for 20 min. Next, the absorbance was measured at a wavelength λ = 517 nm (Meterek SP 830). The extraction reagent + DPPH solution was used as a control sample. The assay was triplicated. The results were given as mg Trolox equivalents (Tx) per 100 g of dry mass using the calibration curve y = 81.2991x − 2.4922 with a confidence coefficient R^2 = 0.9963, and a relative standard deviation of residuals of 2.3%, slope 2.8450, and intercept 1.7469.

2.9. ABTS Radical Scavenging Assay

The analysis was conducted in accordance with the methodology invented by Re et al. (1999) [40]. An amount of 0.192 g of 2,2′-azino-bis(3-ethylbenzothiazoline-6-sulfonic acid) diammonium salt (ABTS) was weighed on an analytical balance with an accuracy of 0.001 g. Next, the weighed ABTS was transferred into a 50-mL volumetric flask. The rest was filled with deionized water up to the marking level. An amount of 0.0166 g of potassium persulphate ($K_2S_2O_8$) was weighed and transferred into a 25-mL volumetric flask. The rest was filled with deionized water up to the marking level. The ABTS and $K_2S_2O_8$ solution was mixed at a 1:0.5 ratio. The mixture was stored at room temperature in darkness for 16 h. Next, the mixture was diluted with a solvent to obtain an absorbance of 0.700 at a wavelength of λ = 734 nm. A calibration curve for Trolox was also prepared. Assay: An amount of 30 μL of the extract was collected into a tube and 3 mL of the ABTS reagent was added. The whole was mixed by shaking on a Vortex. After 6 min the absorbance was measured at a wavelength of λ = 734 nm (Meterek SP 830). The extraction reagent + ABTS solution was used as a control sample. The assay was triplicated. The results were given as mg Trolox equivalents (Tx) per 100 g of dry mass using the calibration curve y = 174.5970 - 0.2115 with a confidence coefficient R^2 = 0.9941, and a relative standard deviation of residuals of 2.5%, slope 9.5155, and intercept 2.6059.

2.10. Total Polyphenol Content

The total polyphenol content was measured with the method invented by Sanchez-Moreno et al. (1998) [41]. Preparation of saturated sodium carbonate solution: An amount of 10.6 g of sodium carbonate was weighed on an analytical balance and placed in a beaker. Then, 100 mL of deionized water was added and the whole was mixed on a magnetic stirrer. A calibration curve was prepared using gallic acid (GAE) as a standard. The Folin-Ciocalteau reagent (FCR) was diluted with deionized water at a 1:1 ratio. Assay: An amount of 125 μL of the FCR reagent, 2 mL of deionized water, 125 μL of the sample, and 250 µL of the saturated sodium carbonate solution were collected into a test tube. The whole was mixed thoroughly on a Vortex and left at ambient temperature for 25 min. Next, the absorbance was measured at λ = 725 nm (Meterek SP 830). The assay was triplicated. The results were given as mg Gallic Acid equivalents (GAE) per 100 g of dry mass using the calibration curve y = 0.0017x + 0.0071 with a confidence coefficient R^2 = 0.9943, and a relative standard deviation of residuals of 2.4%, slope 0.0001, and intercept 0.0124.

2.11. Ferric Reducing Antioxidant Power (FRAP) Assay

The assay consists of measuring the increase in absorbance of the FRAP reagent, which takes place after incubation with the active ingredients contained in the plant extract due to the reduction of Fe(III) ions. It can be monitored by measuring variation in absorbance at a wavelength of 593 nm. The assay was based on the methodology invented by Benzie and Strain (1996) [42]. Preparation of acetic buffer (300 mM; pH 3.6): 3.1 g of sodium acetate trihydrate was weighed and combined with 16 mL of glacial acetic acid. The whole was placed in a 1 L flask and the rest was filled with deionized water up to the marking level. TPTZ (2,4,6-tris(2-pyridyl)-s-triazine) (10 mM in 40 mM HCl): 1.46 mL of concentrated HCl was collected into a 1 L volumetric flask, which was filled with deionized water up to the marking level. Next, 0.031 g of TPTZ was weighed and dissolved in 10 mL of 40 mM HCl in a water bath at 50 °C. FeCl3 x $6H_2O$ (20 mM): 0.054 g of FeClx x $6H_2O$ was weighed and dissolved in 10 mL of deionized water. Before use the three reagents were mixed at a 10:1:1 ratio. Assay: An amount of 100 µL of the sample and 3 mL of the FRAP reagent (heated to 37 °C) was collected into a test tube. The whole was mixed on a Vortex and incubated at 37 °C for 4 min. Next, the absorbance was measured at a wavelength of λ = 593 nm (Meterek SP 830). A calibration curve was prepared using $FeSO_4$x $7H_2O$. The assay was triplicated. The results were given as mM FE(II) per 100 g of dry mass using the calibration curve y = 0.6884x - 0.0003 with a confidence coefficient R2 = 0.9977, and a relative standard deviation of residuals of 1.8%, slope 0.0166, and intercept 0.0100.

2.12. Iron Chelating Activity Assay

The method invented by Decker and Welch (1990) [43] was used to assay the ability of compounds to bind Fe(II) ions. Iron chloride tetrahydrate (2 mM): 0.0398 g of iron chloride tetrahydrate was weighed and transferred quantitatively into a 100-mL volumetric flask, which was filled with deionized water up to the marking level. Ferrozine (5 mM): 0.123 g of ferrozine was weighed and transferred quantitatively into a 50-mL volumetric flask, which was filled with deionized water up to the marking level. Assay: 1 mL of the extract was collected into a test tube. Next, 3.7 mL of deionized water was added. The whole was mixed on the Vortex. Next, 0.1 mL of iron chloride (2 mM) and 0.2 mL of ferrozine (5 mM) were added. The whole was mixed again and incubated at ambient temperature for 20 min. Next, the absorbance was measured at λ = 562 nm (Meterek SP 830). A calibration curve was also prepared using ethylenediaminetetraacetic acid (EDTA) disodium salt. The assay was triplicated. The results were given as ppm EDTA per 100 g of dry mass using the calibration curve y = 1.4053x − 2.8349 with a confidence coefficient R^2 = 0.9981, and a relative standard deviation of residuals of 1.2%, slope 0.0308, and intercept 1.4825.

2.13. Oxygen Radical Absorbance Capacity (ORAC) Assay

The ORAC method based on the methodology invented by Ou et al. (2001) [44] was used to determine the antioxidative capacity. A solution of 42 nM of fluorescein and 153 nM of AAPH as well as 0.075 M phosphate buffer (pH 7.4) were prepared. The extract was dissolved in 75 mM of phosphate buffer. The reaction mixture was prepared in a quartz cuvette to which 0.04 µM of disodium fluorescein in 0.075 M phosphate buffer was added. Next, the extract was added to the mixture, which was kept at 37 °C without access to light. Assay: The chemical reaction was initiated by adding 153 nM of the AAPH solution, which was a source of peroxide radicals. Fluorescence was measured with a Hitachi F-2700 spectrofluorometer at an excitation wavelength λ = 493 nm and emission wavelength

$\lambda = 515$ nm. The first measurement was made immediately after adding the AAPH solution. Then, the fluorescence of the samples was measured every 5 min (f1, f2 … …) for 45 min. A standard curve was prepared for the Trolox solution. The assay was triplicated. The results were given as μM of Trolox equivalent per 1 g of dry mass using the calibration curve y = 5.5280x - 3.5001 with a confidence coefficient $R^2 = 0.9943$, and a relative standard deviation of residuals of 2.9%, slope 0.2088, and intercept 2.2231.

2.14. Statistical Analysis

The results were analyzed statistically (STATISTICA 13.1 software, StatSoft Inc., Kraków, Poland). An analysis of variance was made to detect statistically significant differences. A multiple comparison analysis was made using post-hoc LSD tests. The significance level was assumed at $p = 0.05$. Pearson's linear correlation coefficients ($p = 0.05$; $p = 0.01$; $p = 0.001$) were calculated for the antioxidative activity of the samples obtained in various tests. The Ward method was used for a hierarchical cluster analysis, by means of which the pumpkin cultivars were grouped according to their antioxidative activity.

3. Results

3.1. General Characteristics of Pumpkin Cultivars

The pumpkin cultivars were characterized in general (Table 1). The visual assessment showed that the pulp of all the pumpkin cultivars was yellow or orange. The colorimetric measurement method showed that the brightness of the pumpkin pulp ranged from 48.38 to 64.43 (parameter L). There were differences in the pumpkin juice pH. The 'Gomez' (pH = 5.09), 'Buttercup' (pH = 5.16), and 'Golden Hubbard' (pH = 5.16) cultivars were characterized by the highest acidity. The lowest acidity was noted in the 'Hokkaido' (pH = 6.13) and 'Marina di Chiggia' (pH = 6.10) cultivars. The cultivars differed in the water content in the pulp. The highest content was found in the 'Porcelain Doll' (94.85%), 'Galeux d'Eysines' (94.78%), and 'Melonowa Żółta' (93.25%) cultivars, whereas the lowest content was found in the 'Blue Kuri' (78.76%) and 'Marina di Chiggia' (81.17%) cultivars. The cultivars differed in the content of skin, seeds, and pulp. The following cultivars had the highest percentage content of pulp: 'Jumbo Pink Banana' (87.11%), 'Galeux d'Eysines' (85.09%), 'Buttercup' (84.03%), 'Jarrahdale' (83.94%), 'Flat White Boer' (83.29%), and 'Melonowa Żółta' (83.21%). The smallest content of the pulp was found in the following varieties: 'Golden Hubbard' (63.29%), 'Green Hubbard' (63.47%), 'Porcelain Doll' (69.24%), 'Marina di Chiggia' (70.09%), and 'Hokkaido' (70.59%).

3.2. Optimization of Extraction of Pumpkin Antioxidative Components

3.2.1. DPPH Radical Scavenging Activity

The methanol–aqueous (80%) and aqueous extracts exhibited the highest antioxidative activity against the DPPH radical regardless of the extraction time and temperature (Table 2). The extracts prepared with acetone, acetone and water (80%), ethyl acetate and methanol exhibited the lowest ability to inactivate the DPPH radical. The analysis of the influence of the extraction time on the antioxidative activity of the samples showed that the samples extracted for 2 h exhibited a higher antioxidative activity only for the ethyl acetate (30 and 50 °C), methanol (30 °C), and 80% ethyl acetate (50 and 70 °C) extracts. The test also showed that the temperature of the process affected the DPPH radical scavenging ability. As the temperature increased, so did the antioxidative activity. The exceptions were the 80% ethanol extract (the activity decreased at 1 h and 50 °C, but it increased after 2 h) and the methanol extracts, whose activity decreased at 70 °C.

Table 1. General characteristics of used pumpkin cultivars.

Pumpkin Cultivars	Shape	Flesh Colour				pH of Pulp Juice	Moisture Content in the Pulp (%)	Share of Individual Pumpkin Plant Elements		
		Description	L	a	b			Skin Content (%)	Seeds Content (%)	Pulp Content (%)
Hokkaido	round, slightly elongated at the tail	orange	53.59 ± 1.97b	28.17 ± 0.74g	47.57 ± 1.23ab	6.13 ± 0.02h	88.47 ± 0.59e	17.82 ± 2.06de	11.46 ± 2.38e	70.59 ± 3.97b
Blue Kuri	round	yellow	48.38 ± 2.93a	21.29 ± 1.40de	43.23 ± 1.25a	5.94 ± 0.02f	78.76 ± 1.90a	18.14 ± 2.08e	8.08 ± 1.78cd	75.63 ± 2.46bc
Buttercup	round, slightly flattened	yellow	56.58 ± 4.13bc	27.76 ± 1.96g	48.62 ± 1.91bc	5.16 ± 0.03b	83.93 ± 0.84c	14.08 ± 1.89c	3.44 ± 0.86b	84.03 ± 2.99cd
Gomez	round	orange	56.43 ± 3.07bc	18.82 ± 1.78cd	50.76 ± 0.88c	5.09 ± 0.02a	91.33 ± 0.72f	18.73 ± 2.41e	5.82 ± 1.40c	79 ± 3.71c
Shokichi Shiro	round	orange	53.56 ± 2.59b	13.21 ± 0.87b	45.25 ± 1.80ab	5.72 ± 0.04e	88.79 ± 1.49e	20.7 ± 2.97ef	6.67 ± 1.76c	72.74 ± 2.15b
Jumbo pink banana	elongated	orange	55.78 ± 2.05b	25.81 ± 1.09f	49.96 ± 2.04c	5.49 ± 0.02d	90.61 ± 1.33f	10.39 ± 2.54a	4.2 ± 1.58bc	87.11 ± 3.13d
Golden hubbard	round, slightly elongated at the tail	orange	61.96 ± 2.65de	25.14 ± 1.95f	56.03 ± 2.79d	5.16 ± 0.02b	90.85 ± 1.03f	32.41 ± 3.45h	5.06 ± 1.02c	63.39 ± 3.10a
Flat White Boer Ford	round, clearly flattened	orange	55.01 ± 2.11b	25.25 ± 1.65f	45.89 ± 0.85ab	5.22 ± 0.03c	90.1 ± 0.83f	15.52 ± 2.04cd	3.72 ± 0.50b	83.29 ± 3.39cd
Jarrahdale	round, flattened, irregular	orange	63.59 ± 2.85f	20.02 ± 0.80d	58.13 ± 1.7de	6.06 ± 0.03g	85.78 ± 0.69d	12.53 ± 2.43ab	5.82 ± 0.96c	83.94 ± 4.27cd
Porcelain Doll	round, slightly flattened	orange	58.28 ± 1.55c	17.78 ± 1.80c	55.29 ± 2.49d	5.51 ± 0.03d	94.85 ± 0.78h	22.78 ± 2.84f	9.7 ± 1.31d	69.24 ± 3.61b
Galeux d Eysines	flattened	orange	55.47 ± 2.08bc	22.73 ± 1.01e	49.67 ± 1.21c	5.95 ± 0.04f	94.78 ± 0.45h	11.44 ± 2.13a	5.93 ± 1.07c	85.09 ± 2.71cd
Green hubbard	elongated	orange	59.38 ± 2.54cd	19.11 ± 0.57cd	60.94 ± 1.81e	5.69 ± 0.06e	88.26 ± 1.92e	32.54 ± 4.75h	2.79 ± 0.92a	63.47 ± 3.82a
Marina di Chiggia	round	yellow	64.43 ± 2.55f	8.36 ± 0.45a	60.07 ± 1.45e	6.10 ± 0.02h	81.17 ± 1.29b	26.7 ± 2.43fg	4.23 ± 0.79bc	70.09 ± 4.21b
Melonowa Żółta	round	orange	60.48 ± 1.97	20.2 ± 2.48d	55.97 ± 1.84d	5.93 ± 0.02f	93.25 ± 1.32g	14.37 ± 2.83c	2.45 ± 0.48a	83.21 ± 3.54e

L: Color brightness; a: Color in the range from green to red (positive values indicate the proportion of red, and negative values—green); b: Color from blue to yellow (positive values indicate the proportion of yellow, and negative values—blue); A–h: Means in the same column followed by the same letters shown in superscript do not significantly differ ($p < 0.05$) in terms of analyzed variables.

Table 2. Effect of extraction parameters on the antioxidants properties of pumpkin extracts determined by DPPH assay (mg Tx/100 g dm).

Solvent	Extraction Time and Temperature					
	1 h			2 h		
	30 °C	50 °C	70 °C	30 °C	50 °C	70 °C
Acetone	74.61 ± 1.25bA	90.94 ± 0.81bB	n.a.	82.37 ± 0.98bA	91.29 ± 0.96bB	n.a.
Acetone–water (80%)	88.81 ± 0.99cA	88.42 ± 1.08bA	n.a.	97.46 ± 1.24cA	111.51 ± 1.31bB	n.a.
Ethyl acetate	63.58 ± 123aA	89.98 ± 1.21bB	116.15 ± 1.54bC	59.57 ± 0.65aA	88.25 ± 1.1aB	133.59 ± 1.76bC
Ethyl acetate–water (80%)	106.93 ± 1.32dA	112.3 ± 1.46cB	161.67 ± 1.76cC	111.15 ± 1.18dA	111 ± 0.97bA	161.67 ± 1.76cB
Ethanol	127.78 ± 2.13fA	133.36 ± 1.13dB	177.16 ± 2.07dC	143.8 ± 1.54eA	156.56 ± 1.29dB	182.51 ± 1.86dC
Ethanol–water (80%)	115.37 ± 1.33eA	79.36 ± 1.61aB	114.93 ± 2.09bA	119.05 ± 1.04dA	132.28 ± 1.17cB	157.52 ± 2.06cC
Methanol	88.44 ± 0.46cA	88.58 ± 1.16bA	60.46 ± 1.60aB	82.29 ± 1.25bA	91.59 ± 1.47bB	84.33 ± 1.90aC
Methanol–water (80%)	136.31 ± 1.70gA	151.26 ± 1.98eB	198.95 ± 2.39fC	165.87 ± 1.38fA	180.41 ± 1.8eB	210.97 ± 2.13eC
Water	151.49 ± 1.33hA	158.37 ± 2.46eA	180.70 ± 3.43eB	168.27 ± 1.20fA	174.45 ± 2.76eB	183.40 ± 3.08dC

A–h: Different letters represent statistically significant differences ($p < 0.05$) between solvents (in column) in the variables under analysis; A–C: Different letters represent statistically significant differences ($p < 0.05$) between temperatures of extraction (separately for 1 and 2 h) in the variables under analysis; n.a.: Not analyzed.

3.2.2. ABTS Radical Scavenging

The aqueous and methanol–aqueous (80%) extracts exhibited the highest ability to scavenge ABTS cation radicals (Table 3). The acetone extract was characterized by the lowest antioxidative activity. The antioxidative activity increased along with the extraction time in most of the samples. The exceptions were: Acetone and 80% ethyl acetate (50 °C). The comparison of the results showed that the extraction temperature affected the efficiency of the extracts in sweeping the ABTS cation radical. As the temperature increased, the antioxidative activity of the samples increased, too. The antioxidative properties decreased only in the 80% acetone (1 h), 80% ethyl acetate (1 h, 70 °C), and 80% ethanol (1 h, 50 °C) extracts.

Table 3. Effect of extraction parameters on the antioxidants properties of pumpkin extracts determined by the ABTS assay (mg Tx/100 g dm).

Solvent	Extraction Time and Temperature					
	1 h			2 h		
	30 °C	50 °C	70 °C	30 °C	50 °C	70 °C
Acetone	24.62 ± 0.32aA	31.51 ± 0.97aB	n.a.	19.06 ± 0.19aA	27.5 ± 0.44aB	n.a.
Acetone–water (80%)	45.9 ± 1.04dA	45.18 ± 0.98cB	n.a.	55.30 ± 1.00dA	62.11 ± 0.35eB	n.a.
Ethyl acetate	24.9 ± 0.98aA	30.83 ± 0.90aB	41.23 ± 1.27bC	33.63 ± 1.02bA	35.76 ± 0.26bB	52.25 ± 1.11aC
Ethyl acetate–water (80%)	35.53 ± 1.23cA	44.57 ± 0.38cB	36.78 ± 0.55aA	40.48 ± 0.66cA	42.33 ± 0.42cB	50.05 ± 0.95aC
Ethanol	28.43 ± 0.50bA	38.21 ± 1.09bB	52.68 ± 1.01cC	37.89 ± 0.99bA	54.8 ± 1.06dB	60.69 ± 1.27bC
Ethanol–water (80%)	62.08 ± 0.48eB	57.24 ± 2.08dA	66.68 ± 0.52dC	58.89 ± 1.07dA	71.81 ± 1.90fB	73.80 ± 1.17cB
Methanol	35.47 ± 0.99cA	49.23 ± 0.71cB	62.26 ± 0.38dC	40.07 ± 0.62cA	58.86 ± 0.73dB	70.65 ± 0.44cC
Methanol–water (80%)	74.07 ± 1.11eA	84.44 ± 0.82eB	104.09 ± 1.02eC	82.27 ± 1.03eA	95.31 ± 0.32gB	110.84 ± 0.57dC
Water	79.30 ± 0.39eA	108.5 ± 1.53fB	108.16 ± 0.65eB	85.16 ± 0.86eA	127.31 ± 1.39hB	124.70 ± 0.59eC

A–i: Different letters represent statistically significant differences ($p < 0.05$) between solvents (in column) in the variables under analysis; A–C: Different letters represent statistically significant differences ($p < 0.05$) between temperatures of extraction (separately for 1 and 2 h) in the variables under analysis; n.a.: Not analyzed.

3.2.3. Total Phenolic Content (TPC)

The highest total phenolic content was found in the aqueous and methanol–aqueous (80%) extracts (Table 4). The lowest concentration of polyphenolic compounds was found in the ethyl acetate, ethyl acetate–aqueous (80%), and acetone extracts. The analysis of the influence of the extraction time on the content of polyphenolic compounds showed that it was higher in the samples extracted for 2 h than in the ones extracted for 1 h, except the acetone (30 and 50 °C), acetone–aqueous (30 °C), and ethyl acetate (70 °C) extracts, where the samples extracted for 1 h had higher content of polyphenols. The extraction temperature also affected the results. The total polyphenolic content in the extracts prepared at 70 °C was greater than in the samples extracted at 30 and 50 °C. The exceptions were 80% acetone (1 h), ethanol–aqueous (80%) (1 and 2 h), and methanol–aqueous (80%) extracts (1 and 2 h), where the content of polyphenolic compounds was lower at 50 °C. However, the highest content of polyphenols in the 80% ethanol–aqueous and 80% methanol–aqueous extracts was found at 70 °C.

3.2.4. Ferric Reducing Antioxidant Power (FRAP)

The methanol–aqueous (80%) and ethanol–aqueous (80%) extracts exhibited the highest ferric ion reducing capacity (Table 5). The analysis of variation in the ferric ion reducing capacity over time showed that the samples extracted for 2 h exhibited greater activity than the ones extracted for 1 h. Only the ethyl acetate, ethyl acetate (70 °C), ethanol–aqueous (80%) (30 °C), methanol (70 °C), and aqueous extracts were characterized by an inverse dependence. In most cases, the higher extraction temperature increased the ferric ion reducing capacity of the extracts. The higher extraction temperature caused a slight decrease in the antioxidative activity of the acetone extract only.

Table 4. Effect of extraction parameters on the total phenolic content (TPC) of pumpkin extracts (mg GAE/100 g dm).

Solvent.	Extraction Time and Temperature					
	1 h			2 h		
	30 °C	50 °C	70 °C	30 °C	50 °C	70 °C
Acetone	12.29 ± 0.22bA	24.10 ± 0.17cB	n.a.	7.49 ± 0.10aA	11.546 ± 0.10bB	n.a.
Acetone–water (80%)	89.51 ± 1.50dA	64.19 ± 0.71dB	n.a.	67.05 ± 0.96dA	74.51 ± 0.44dB	n.a.
Ethyl acetate	3.09 ± 0.06aA	8.09 ± 0.08aB	22.88 ± 0.11bC	5.2 ± 0.05aA	7.82 ± 0.03aB	18.54 ± 0.15aC
Ethyl acetate–water (80%)	12.82 ± 0.21bA	14.32 ± 0.16bB	18.49 ± 0.29aC	16.84 ± 0.45bA	24.95 ± 0.24cB	30.23 ± 0.21bC
Ethanol	46.27 ± 0.48cA	75.95 ± 1.01eB	130.59 ± 1.53cC	62.00 ± 0.52cA	102.35 ± 0.89fB	164.47 ± 1.88dC
Ethanol–water (80%)	106.52 ± 0.77fA	93.25 ± 0.44fB	132.86 ± 1.61cC	131.63 ± 1.22fA	128.11 ± 1.13gB	186.75 ± 2.05eC
Methanol	49.41 ± 0.46cA	92.88 ± 0.77fB	147.06 ± 1.28dC	58.51 ± 1.10cA	88.52 ± 0.61eB	136.67 ± 1.26cC
Methanol–water (80%)	128.62 ± 0.76gA	101.69 ± 1.12gB	162.51 ± 1.49eC	166.23 ± 0.97gA	160.12 ± 1.20hB	206.47 ± 2.44fC
Water	93.41 ± 1.46eA	155.83 ± 1.31hB	186.47 ± 2.93fC	108.2 ± 1.15eA	171.97 ± 2.15iB	237.94 ± 2.42gC

A–h: Different letters represent statistically significant differences ($p < 0.05$) between solvents (in column) in the variables under analysis; A–C: Different letters represent statistically significant differences ($p < 0.05$) between temperatures of extraction (separately for 1 and 2 h) in the variables under analysis; n.a.: Not analyzed.

Table 5. Effect of extraction parameters on ferric ion reducing antioxidant power (FRAP) of pumpkin extracts (mM Fe(II)/100 g dm).

Solvent	Extraction Time and Temperature					
	1 h			2 h		
	30 °C	50 °C	70 °C	30 °C	50 °C	70 °C
Acetone	71.32 ± 0.98dA	118.96 ± 1.38cB	n.a.	88.21 ± 0.99dA	127.91 ± 2.46dB	n.a.
Acetone–water (80%)	189.12 ± 1.65gA	180.17 ± 3.14eA	n.a.	249.59 ± 2.40iA	240.89 ± 2.28gA	n.a.
Ethyl acetate	28.56 ± 0.33bA	56.69 ± 0.63bB	182.27 ± 1.83cC	23.32 ± 0.23bA	40.82 ± 0.74bB	158.37 ± 1.50cC
Ethyl acetate–water (80%)	36.05 ± 0.67cA	55.57 ± 0.62bB	131.42 ± 2.57bC	43.08 ± 0.41cA	57.71 ± 0.64cB	119.87 ± 2.48bC
Ethanol	97.67 ± 1.52eA	172.86 ± 1.62dB	344.25 ± 4.09eC	128.13 ± 1.09fA	226.36 ± 2.40fB	351.06 ± 2.53eC
Ethanol–water (80%)	221.36 ± 2.03hA	217.73 ± 1.86fA	309.05 ± 2.01dB	171.18 ± 1.12gA	263.49 ± 3.09hB	346.79 ± 4.40eC
Methanol	106.1 ± 1.66fA	181.93 ± 1.81eB	360.74 ± 3.50fC	117.75 ± 2.22eA	222.13 ± 2.13fB	349.59 ± 3.99eC
Methanol–water (80%)	230.73 ± 3.54iA	231.35 ± 2.60gA	320.91 ± 4.65dB	199.39 ± 3.60hA	213.35 ± 3.21eB	290.15 ± 3.48dC
Water	13.15 ± 0.14aA	33.76 ± 0.25aB	85.06 ± 1.08aC	8.67 ± 0.25aA	22.51 ± 0.38aB	48.57 ± 1.33aC

A–i: Different letters represent statistically significant differences ($p < 0.05$) between solvents (in column) in the variables under analysis; A–C: Different letters represent statistically significant differences ($p < 0.05$) between temperatures of extraction (separately for 1 and 2 h) in the variables under analysis; n.a.: Not analyzed.

3.2.5. Iron Chelating Activity

The methanol–aqueous (80%) extracts exhibited the highest Fe(II) ion chelating activity (Table 6). The lowest Fe(II) chelating activity was characteristic of the ethyl acetate–aqueous (80%) and ethyl acetate extracts. As the extraction time increased, so did the Fe(II) ion chelating activity of all the samples except the ethyl acetate–aqueous (80%) extract (50 and 70 °C), and ethanol extract (50 °C). Apart from that, the assay showed that the Fe(II) ion chelating activity increased along with temperature. Only the chelating activity of the ethanol (30 vs. 50 °C), ethanol–aqueous (80%) (30 vs. 50 °C), and ymethanol–aqueous (80%) extracts (30 vs. 50 °C) decreased. However, the chelating activity of these extracts increased again at 70 °C.

Table 6. Effect of extraction parameters on iron chelating activity of pumpkin extracts (ppm EDTA/100 g dm).

Solvent	Extraction Time and Temperature					
	1 h			2 h		
	30 °C	50 °C	70 °C	30 °C	50 °C	70 °C
Acetone	1338.66 ± 23.26dA	1618.62 ± 17.42dB	n.a.	1435.13 ± 23.41dA	1781.61 ± 16.62dB	n.a.
Acetone–water (80%)	1116.18 ± 13.55cA	1452.60 ± 24.50cB	n.a.	1343.64 ± 12.54cA	1527.80 ± 12.67cB	n.a.
Ethyl acetate	385.00 ± 3.93bA	454.31 ± 3.17bB	1563.68 ± 22.42bC	589.78 ± 5.48bA	888.62 ± 6.08bB	1563.68 ± 25.42bC
Ethyl acetate–water (80%)	102.37 ± 5.13aA	174.30 ± 3.18aB	1276.61 ± 18.34aC	438.09 ± 4.22aA	723.49 ± 4.24aB	1236.57 ± 14.16aC
Ethanol	2542.92 ± 28.88fA	1809.22 ± 16.08eB	3023.88 ± 25.61dC	3146.40 ± 31.62fA	2787.12 ± 13.93eB	3486.03 ± 36.51dC
Ethanol–water (80%)	3641.36 ± 32.92hA	2715.41 ± 12.23gB	3864.89 ± 34.29eC	3905.50 ± 35.72hA	3219.77 ± 18.90fB	4144.63 ± 45.86fC
Methanol	3015.23 ± 33.00gA	3279.077.37.34hB	3854.02 ± 46.36eC	3497.55 ± 28.65gA	3579.78 ± 39.10gB	3974.26 ± 27.17eC
Methanol–water (80%)	3897.77 ± 44.08hA	3553.34 ± 28.12iB	4060.22 ± 28.37fC	4202.27 ± 37.15iA	4071.18 ± 39.20hB	4381.1 ± 52.25gC
Water	1796.45 ± 12.16eA	2108.03 ± 16.61fB	2159.47 ± 16.74cC	2009.48 ± 16.53eA	2566.80 ± 24.05eB	2224.65 ± 26.57cC

A–i: Different letters represent statistically significant differences ($p < 0.05$) between solvents (in column) in the variables under analysis; A–C: Different letters represent statistically significant differences ($p < 0.05$) between temperatures of extraction (separately for 1 and 2 h) in the variables under analysis; n.a.: Not analyzed.

3.3. Comparison of the Antioxidative Activity of Selected Extracts of Various Pumpkin Cultivars

3.3.1. DPPH Radical Scavenging Activity

The research showed that the 'Melonowa Żółta' (245.98 and 222.23 mg Tx/100 g dm), 'Hokkaido' (210.97 and 183.40 mg Tx/100 g dm), 'Galeux d'Eysiness' (206.99 mg Tx/100 g dm), and 'Buttercup' cultivars (185.19 mg/Tx/100 g dm) exhibited the highest antioxidative activity against the DPPH radical (Table 7). On the other hand, the aqueous extracts of the following pumpkin cultivars were characterized by the lowest antioxidative potential: 'Gomez' (34.11 mg Tx/100 g dm), 'Shokichi Shiro' (42.22 mg Tx/100 g dm), 'Golden Hubbard' (52.5 mg Tx/100 g dm), and 'Green Hubbard' (54.39 mg Tx/100 g dm). The methanol–aqueous (80%) extracts exhibited greater DPPH radical scavenging activity than the aqueous extracts in all the pumpkin cultivars. The pumpkin pulp exhibited low DPPH radical inhibiting activity (5.8 µmol Tx/g dm). The same experiment showed that other raw materials exhibited much higher antioxidative activity, e.g., artichokes (70.1 µmol Tx/g dm), lettuce (77.2 µmol Tx/g dm), spinach (50.9 µmol Tx/g dm), turmeric (57.6 µmol Tx/g dm). The following raw materials were characterized by lower DPPH radical inactivating ability: Leek (3.2 µmol Tx/g dm), cucumber (2.3 µmol Tx/g dm), celery (3.8 µmol Tx/g dm), carrots (3.5 µmol Tx/g dm), beans (3.8 µmol Tx/g dm) (44). The antioxidative activity of pumpkin seeds and skin (*Cucurbita pepo*) was confirmed in a test with DPPH radical. The test showed that the aqueous extracts (72.36% inhibition) and 70% ethanol extracts (71.0% inhibition) exhibited the highest DPPH radical scavenging activity in the pumpkin skin, whereas the 70% ethanol extract (20.5% inhibition) and 70% methanol extract (18.9% inhibition) exhibited the highest antioxidative activity in the pumpkin seeds. The radicals inhibition of the aqueous extract amounted to 4.12% [37]. Valenzuela et al. researched various pumpkin species seeds and found the highest total polyphenolic content in the *Cucurbita mixta* Pangalo species (275 µmol GAE/g of the extract). There were lower total polyphenolic content levels in the seeds of *Cucurbita maxima* Duchense (212.87 µmol GAE/g of the extract) and *Cucurbita moschata* (Duchense ex Lam.) species (118.79 µmol GAE/g of the extract) [45]. The above research showed that the methanol–aqueous and methanol extracts of pumpkin seeds were characterized by high DPPH radical scavenging activity, and amounted to 69.18% and 86.85%, respectively [46].

Table 7. Antioxidant activity of pumpkin extracts determined by ABTS and DPPH assays.

Pumpkin Cultivars	ABTS (mg Tx/100 g dm)		DPPH (mg Tx/100 g dm)	
	Aqueous–Methanol Extract	Aqueous Extract	Aqueous–Methanol Extract	Aqueous Extract
Hokkaido	110.84 ± 0.57cA	124.70 ± 0.59dB	210.97 ± 2.13hA	183.40 ± 3.08iB
Blue Kuri	85.12 ± 1.73aA	104.78 ± 1.45bB	145.44 ± 1.80fA	67.43 ± 0.88eB
Buttercup	116.37 ± 1.64dA	114.04 ± 1.00cA	185.19 ± 2.01gA	76.58 ± 1.13fB
Gomez	127.36 ± 1.66eA	107.19 ± 1.5bB	86.12 ± 1.90bA	34.11 ± 0.86aB
Shokichi Shiro	99.79 ± 1.71bA	123.75 ± 0.96dB	57.54 ± 0.88aA	42.22 ± 1.15bB
Jumbo pink banana	146.91 ± 0.68fA	173.11 ± 1.37gB	86.72 ± 0.96bA	72.9 ± 0.9fB
Golden hubbard	117.24 ± 1.20dA	152.01 ± 1.46gB	127.5 ± 1.64dA	52.5 ± 0.67dB
Flat White Boer Ford	113.07 ± 1.24cdA	95.92 ± 1.39aB	204.12 ± 2.28gA	96.49 ± 1.54gB
Jarahdale	95.90 ± 1.44bA	104.25 ± 1.28bB	136.90 ± 2.62eA	70.23 ± 1.28fB
Porcelain doll	109.74 ± 1.24cA	138.04 ± 1.47eB	82.42 ± 0.83bA	49.99 ± 1.21cB
Galeux d' Eysines	122.32 ± 1.19eA	110.07 ± 1.34bB	206.99 ± 2.24hA	127.85 ± 1.03hB
Green hubbard	103.87 ± 1.54cA	136.11 ± 1.13eB	101.98 ± 1.21cA	54.39 ± 0.75dB
Marina di Chiggia	118.28 ± 1.12dA	143.52 ± 2.43fB	105.63 ± 1.07cA	73.86 ± 0.85fB
Melonowa Żółta	152.86 ± 1.64fA	187.17 ± 2.55hB	245.98 ± 3.10iA	222.23 ± 3.87jB

A–i: Different letters represent statistically significant differences ($p < 0.05$) between antioxidative activity of pumpkin cultivars; A,B: Different letters represent statistically significant differences ($p < 0.05$) between antioxidative activity of extracts (aqueous–methanol vs. aqueous).

3.3.2. ABTS Radical Scavenging

The ABTS cation radical test showed that the 'Melonowa Żółta' pumpkin cultivar was characterized by the highest antioxidative potential (Table 7). The antioxidative activity of the aqueous extracts (187.17 mg Tx/100 g dm) was higher than that of the methanol–aqueous (80%) extracts (152.86 mg Tx/100 g dm). The lowest ABTS cation radical scavenging activity was noted in the 'Blue Kuri' (methanol–aqueous (80%) extract—85.12 mg Tx/100 g dm), 'Jarahdale' (methanol–aqueous (80%) extract—95.90 mg Tx/100 g dm), and 'Flat White Boer Ford' cultivars (aqueous extract—95.92 mg Tx/100 g dm). In most cases, the aqueous extracts exhibited greater antioxidative activity than the methanol–aqueous (80%) extracts. The research showed that the pumpkin pulp had relatively low ABTS cation radical scavenging ability (11.0 μM Tx/g dm), as compared with other vegetables: Artichoke (39.9 μM Tx/g dm), asparagus (37.5 μM Tx/g dm), broccoli (43.0 μM Tx/g dm), lettuce (85.8 μM Tx/g dm), radishes (61.7 μM Tx/g dm), and turmeric (118.6 μM Tx/g dm) (44). Sing et al. noted that pumpkin pulp (*Cucurbita maxima*) extracted with ethanol and water (50%) exhibited the highest ABTS radical scavenging activity, i.e., 2.04 μM Tx/g. The ABTS radical scavenging ability of the pumpkin extract was slightly lower than that of watermelon (2.24 μM Tx/g) and melon (2.78 μM Tx/g). The authors of the study also found that pumpkin skin extracts were characterized by higher antioxidative potential [47].

3.3.3. Total Phenolic Content (TPC)

The total phenolic content in the pumpkin cultivars was analyzed with the Folin-Ciocalteu method (Table 8). The analysis showed that the content of phenolic compounds in the pumpkin pulp varied depending on the cultivar and the extraction solvent. The highest concentration of total polyphenols was found in the following cultivars: 'Melonowa Żółta' (232.5 and 255.69 mg GAE/100 g dm), 'Hokkaido' (206.47 and 237.94 mg GAE/100 g dm), and 'Gomez' (172.63 and 188.22 mg GAE/100 g dm). The lowest content of phenolic compounds in the methanol–aqueous extracts was found in the 'Blue Kuri' (49.78 mg GAE/100 g dm) and 'Marina di Chiggia' cultivars (49.99 mg GAE/100 g dm). On the other hand, among the aqueous extracts, the lowest total polyphenolic level was found in the 'Blue Kuri' (65.66 mg GAE/100 g dm), 'Shokichi Shiro' (66.64 mg GAE/100 g dm), 'Flat White Boer Ford' (66.01 mg GAE/100 g dm), and 'Jumbo Pink Banana' cultivars (66.40 mg GAE/100 g dm). Apart from that, it is noteworthy that when water was used as the solvent, there was higher content of polyphenols in all the cultivars except for the 'Flat White Boer Ford'. Saavedra et al. conducted a study in which they measured the total polyphenolic content in fresh pumpkin skins and seeds. The highest content was found in aqueous extracts. The total polyphenolic content in the pumpkin skin was 741–1069 mg GAE/100 g dm, whereas in the seeds it was 234–239 mg GAE/100 g dm. The results showed that the polyphenol content in the pumpkin skin and seeds was higher than in its pulp. Only in the 'Melonowa Żółta' cultivar the polyphenol content (255.69 mg/100 g dm) was slightly higher than in the seeds. The Hokkaido cultivar had a similar content of polyphenols (237.94 mg/100 g dm) [37]. Kiat et al. also observed the presence of polyphenols in pumpkin seeds. They noted that the concentration of polyphenols in the methanol–aqueous extract (80%) (72 mg/100 g dm) was higher than in the methanol extract (44 mg/100 g dm) [48]. Bayili et al. showed that the total polyphenolic content in the pumpkin of the *Cucurbita pepo* species amounted to 100.2 mg/100 g fresh weight. The authors observed that the pumpkin contained more polyphenols than some other vegetables, e.g., tomatoes, eggplant, cucumber, and vegetable cabbage. The content of polyphenols was about two times greater in some cultivars, e.g.,: 'Hokkaido', 'Gomez', 'Porcelain Doll', 'Melonowa Żółta', but these results were calculated per dry mass [46]. Singh et al. observed that the content of polyphenolic compounds in pumpkin pulp (extracted with water) amounted to 13.92 mg GAE/100 g fresh weight. There was a higher content of polyphenolic compounds in the methanol–aqueous (19.36–30.69 mg/100 g dm) and ethanol–aqueous solvents (21.45–33.48 mg/100 g dm) [47]. Dar et al. found polyphenolic compounds in various extracts of pumpkin leaves (*Cucurbita pepo*). They observed that the polyphenolic content varied depending on the type of extract. The concentration of polyphenols in the ethanol extract was 40.37 mg/g GAE, in the aqueous extract—40.12 mg/g GAE, butanol extract—92.62 mg/g GAE, ethyl acetate extract—85.12 mg/g

GAE, chloroform extract—21.25 mg/g GAE, and n-hexane extract—12.50 mg/g GAE [49]. Oloyede et al. conducted research on the content of total polyphenols in pumpkin pulp (*Cucurbita pepo* Linn.). They found that the concentration of polyphenolic compounds in pumpkin depended on the degree of ripening. The content of polyphenolic compounds in ripe fruit was 33.5 mg/100 g dm, whereas in unripe fruit it was 10.3 mg/100 g dm. This content was similar to the results observed in methanol–aqueous extracts in the 'Blue Kuri', 'Marina di Chiggia', and 'Jumbo Pink Banana' cultivars [50].

Table 8. Total polyphenols content and oxygen radical absorbance capacity (ORAC) of pumpkin extracts.

Pumpkin Cultivars	Total Phenolic Content (mg GAE/100 g dm)		ORAC (µM Tx/g dm)
	Aqueous–Methanol Extract	Aqueous Extract	
Hokkaido	206.47 ± 2.44hA	237.94 ± 2.42iB	89.97 ± 2.07d
Blue Kuri	49.78 ± 0.76aA	65.66 ± 0.71aB	58.47 ± 0.70b
Buttercup	87.99 ± 1.52eA	115.24 ± 1.85fB	102.08 ± 1.84f
Gomez	172.63 ± 1.06gdA	188.22 ± 2.73hB	99.35 ± 3.05e
Shokichi Shiro	58.01 ± 1.08cA	66.64 ± 1.90aB	108.47 ± 2.32g
Jumbo pink banana	50.23 ± 1.61bA	66.40 ± 0.83aB	86.60 ± 1.19d
Golden hubbard	91.46 ± 1.16eA	101.05 ± 1.13eB	116.38 ± 1.91h
Flat White Boer Ford	70.99 ± 1.03dA	66.01 ± 1.33aB	43.04 ± 1.72a
Jarahdale	56.35 ± 0.90cA	72.61 ± 1.44bB	65.34 ± 1.83c
Porcelain doll	174.53 ± 1.37gA	189.40 ± 0.78hB	98.35 ± 2.36e
Galeux d' Eysines	113.40 ± 1.42fA	95.56 ± 1.36dB	94.87 ± 0.53e
Green hubbard	58.62 ± 1.36cA	126.08 ± 1.72gB	102.90 ± 1.64f
Marina di Chiggia	49.99 ± 1.49aA	77.71 ± 1.35cB	104.76 ± 1.84f
Melonowa Żółta	232.5 ± 2.63iA	255.69 ± 4.29jB	122.73 ± 3.39i

A–j: Different letters represent statistically significant differences ($p < 0.05$) between antioxidative activity (ORAC) and total polyphenolic content of pumpkin cultivars. A, B: Different letters represent statistically significant differences between antioxidative activity (ORAC) and total polyphenolic content ($p < 0.05$) of extracts (aqueous–methanol vs. aqueous).

3.3.4. Ferric Reducing Antioxidant Power (FRAP)

In order to determine the antioxidative potential of the pumpkin cultivars the ability of the aqueous and methanol–aqueous extracts to reduce ferric ions was also investigated. The following cultivars exhibited showed the highest antioxidative activity: 'Buttercup' (592.78 mM Fe(II)/100 g dm), 'Melonowa Żółta' (555.63 mM Fe(II)/100 g dm), 'Galeux d'Eysines' (524.90 mM Fe(II)/100 g dm), 'Flat White Boer Ford' (509.28 mM Fe(II)/100 g dm), and 'Porcelain Doll' (501.19 mM Fe(II)/100 g dm) (Table 9). The lowest ability to reduce the degree of ferric ion oxidation was found in the 'Hokkaido' (48.57 mM Fe(II)/100 g dm), 'Marina di Chiggia' (58.19 mM Fe(II)/100 g dm), and 'Green Hubbard' cultivars (66.34 mM Fe(II)/100 g dm). When 80% methanol was used as the solvent, all the cultivars exhibited higher antioxidative activity. Tiveron et al. observed that pumpkin pulp (*Cucurbita maxima*) was capable of reducing ferric ions at an amount of 19.5 µM Fe2+/g dm. This activity was greater than that of other vegetables, such as: Celery, carrot, cucumber, and leek. On the other hand, it was about 10–15 times lower than that of chicory, broccoli, spinach, and watercress [51]. Fidrianny et al. observed that the pumpkin leaf ethanol extract (*Cucurbita moschata*) exhibited relatively low ferric ion reducing activity, i.e., 1.37%, as compared with 7.39% for ascorbic acid. Ethyl acetate (1.37%) and hexane (0.28%) extracts exhibited lower ferric ion reducing ability [52]. One study showed that the compounds contained in pumpkin pulp (*Cucurbita maxima*) were capable of reducing ferric ions. The highest reducing activity was observed for the ethanol aqueous extract (50%) (3.23 µM Fe(II)/g) and ethanol aqueous extract (50%) (2.66 µM Fe(II)/g). The aqueous extracts exhibited the lowest activity (1.83 µM Fe(II)/g), which was consistent with the results of this study [46].

Table 9. Iron chelating activity and ferric ion reducing antioxidant power (FRAP) of pumpkin extracts.

Pumpkin Cultivars	Iron Chelating Activity (ppm EDTA/100 g dm)		FRAP (mM Fe(II)/100 g dm)	
	Aqueous–Methanol Extract	Aqueous Extract	Aqueous–Methanol Extract	Aqueous Extract
Hokkaido	4381.1 ± 52.25aA	2224.65 ± 26.57aB	290.15 ± 3.48bA	48.57 ± 1.33aB
Blue Kuri	11385.16 ± 56.32gA	7561.89 ± 46.15jB	337.39 ± 1.63cA	123.14 ± 1.73gB
Buttercup	5702.74 ± 59.19bA	2938.91 ± 30.21bB	592.78 ± 2.24jA	283.34 ± 2.43lB
Gomez	12804.62 ± 90.97gA	8125.81 ± 31.32kB	407.05 ± 1.72eA	111.54 ± 1.66fB
Shokichi Shiro	11350.7 ± 100.45gA	5029.95 ± 25.77fB	324.31 ± 1.60cA	84.50 ± 1.84dB
Jumbo pink banana	7499.11 ± 83.49dA	5194.75 ± 45.07fB	386.83 ± 2.43dA	140.94 ± 1.76hB
Golden hubbard	5635.76 ± 78.83bA	3255.99 ± 38.26cB	405.43 ± 2.81eA	95.31 ± 1.12eB
Flat White Boer Ford	7421.45 ± 78.67dA	6131.97 ± 24.33iB	509.28 ± 3.62gA	169.41 ± 1.16iB
Jarahdale	6197.18 ± 50.75cA	5314 ± 26.52gB	478.86 ± 4.04fA	110.17 ± 1.83fB
Porcelain doll	9321.91 ± 81.67fA	5782.54 ± 56.89hB	501.19 ± 5.79gA	202.2 ± 2.13jB
Galeux d' Eysines	6119.04 ± 84.52cA	4302.19 ± 65.47eB	524.90 ± 4.25hA	213.30 ± 2.47kB
Green hubbard	10760.30 ± 99.76gA	7349.57 ± 81.84jB	293.67 ± 2.28bA	66.34 ± 1.27cB
Marina di Chiggia	7753.32 ± 25.21eA	5715.51 ± 69.28hB	266.24 ± 2.97aA	58.19 ± 1.24bB
Melonowa Żółta	5615.31 ± 49.74bA	4095.36 ± 47.03dB	555.63 ± 6.94iA	364.90 ± 6.70mB

A–m: Different letters represent statistically significant differences ($p < 0.05$) between antioxidative activity (chelating properties and FRAP) of pumpkin cultivars. A,B: Different letters represent statistically significant differences ($p < 0.05$) between antioxidative activity of extracts (aqueous–methanol vs. aqueous).

3.3.5. Iron Chelating Activity

The analysis of metal ion binding properties is a method of determining the antioxidative potential of food. For this purpose, the ability to chelate Fe(II) ions by compounds contained in the extracts of various pumpkin cultivars was assayed (Table 9). The test revealed that the methanol–aqueous (80%) extracts of the following cultivars were characterized by the highest antioxidative activity: 'Gomez' (12,804.62 ppm EDTA/100 dm), 'Blue Kuri' (11,385.16 ppm EDTA/100 g dm), 'Shokichi Shiro' (11,350.7 ppm EDTA/100 g dm). By contrast, the aqueous extracts of the following cultivars exhibited the lowest ferric ion chelating ability: 'Hokkaido' (2,224.65 ppm EDTA/100 g dm), 'Buttercup' (2,938.91 ppm EDTA/100 g dm), 'Golden Hubbard' (3,225.99 ppm EDTA/100 g dm), 'Melonowa Żółta' (4,095.36 ppm EDTA/100 g dm), and 'Galeux d'Eysines' (4,302.19 ppm EDTA/100 g dm). As in the FRAP test, the methanol–aqueous (80%) extracts exhibited the highest antioxidative activity.

3.3.6. Oxygen Radical Absorbance Capacity (ORAC)

The antioxidative activity of the pumpkin cultivars was also measured with the oxygen radical absorbance capacity test (ORAC). This is a very high sensitivity method, which is mainly used to measure the activity of hydrophilic antioxidants [53]. It is thought to be one of the most preferable tests for measuring the antioxidative activity [54]. The ORAC test has been frequently used to prepare rankings of food products according to their antioxidative capacity [55]. The method is recommended to quantify the peroxide radical scavenging capacity. The highest antioxidative capacity was found in the following cultivars: 'Melonowa Żółta' (122.73 μM Tx/g dm), 'Jumbo Pink Banana' (117.72 μM Tx/g dm), and 'Golden Hubbard' (108.47 μM Tx/g dm) (Table 8). The following cultivars exhibited the lowest reactive oxygen species scavenging capacity: 'Flat White Boer Ford' (42.38 μM Tx/g dm), 'Blue Kuri' (58.47 μM Tx/g dm), 'Jarahdale' (65.00 μM Tx/g dm), 'Shokichi Shiro' (87.60 μM Tx/g dm), and 'Hokkaido' (89.97 μM Tx/g dm). So far, the ability of pumpkin pulp to absorb oxygen free radicals has not been analyzed. Parry et al. investigated the antioxidative activity (ORAC) of the oil extract from roasted pumpkin seeds. They noted that the product had very low antioxidative potential (1.1 μM Tx/g fat), as compared with parsley seed extract (1097.5 μM Tx/g fat), cardamom extract (941.5 μM Tx/g fat), milk thistle extract (125.2 μM Tx/g fat), and onion extract (17.5 μM Tx/g fat) [56].

3.3.7. Correlation between Antioxidative Properties of *Cucurbita maxima* Cultivars Observed in Different Tests

The results of the antioxidative activity assays were used to analyze correlations between the antioxidative activity exhibited both by the aqueous and aqueous–methanol extracts. As far as the aqueous extracts are concerned, there was a strong positive correlation between the results obtained in the ABTS and oxygen radical absorbance capacity (ORAC) tests ($\varrho = 0.64$; $p < 0.001$), as well as between the total polyphenolic content and the ORAC ($\varrho = 0.59$; $p < 0.001$) and ABTS tests ($\varrho = 0.41$; $p < 0.01$) (Table 10). The results showed a strong negative correlation between the ORAC and FRAP tests ($\varrho = -0.47$; $p < 0.01$), as well as between total polyphenolic content and the chelating activity ($\varrho = -0.47$; $p < 0.01$). Likewise, the analysis of the results for the aqueous–methanol extracts revealed a positive correlation between the total polyphenolic content and the antioxidative activity assayed in the ABTS test ($\varrho = 0.37$; $p < 0.05$) (Table 11). In addition, there was a positive correlation between both FRAP and DPPH tests ($\varrho = 0.45$; $p < 0.05$) and between the FRAP test and the total polyphenolic content ($\varrho = 0.39$; $p < 0.05$). On the other hand, there were negative correlations between the chelation activity and the total polyphenolic content ($\varrho = -0.63$; $p < 0.001$) and the DPPH radical scavenging ability ($\varrho = -0.53$; $p < 0.001$).

Table 10. Results of the correlation analysis between the antioxidant activity tests in *Cucurbita maxima* cultivars aqueous extracts.

	ABTS	DPPH	Total Phenolic Content	Iron Chelating Activity	FRAP	ORAC
ABTS	x	−0.01	0.41 **	−0.37 *	−0.38 *	0.64 ***
DPPH	−0.01	x	0.14	−0.26	0.32	−0.07
Total polyphenols content	0.41 **	0.14	x	−0.47 **	−0.24	0.59 ***
Iron chelating activity	−0.37 *	−0.26	−0.47 **	x	−0.01	−0.35 *
FRAP	−0.38 *	0.32	−0.24	−0.01	x	−0.47 **
ORAC	0.64 ***	−0.07	0.59 ***	−0.35 *	−0.47 **	x

$p < 0.05$ *; $p < 0.01$ **; $p < 0.001$ ***: Determination of statistically significant correlations between tested variables.

Table 11. Results of the correlation analysis between the antioxidant activity tests in *Cucurbita maxima* cultivars aqueous–methanol extracts.

	ABTS	DPPH	Total Phenolic Content	Iron Chelating Activity	FRAP
ABTS	x	−0.04	0.37 *	−0.39 *	−0.06
DPPH	−0.04	x	0.38	−0.53 ***	0.45*
Total Polyphenols Content	0.37 *	0.38	x	−0.63***	0.39*
Iron Chelating Activity	−0.39 *	−0.53 ***	−0.63 ***	x	−0.16
FRAP	−0.06	0.45 *	0.39 *	−0.16	x

$p < 0.05$ *; $p < 0.01$ **; $p < 0.001$ ***: Determination of statistically significant correlations between tested variables.

3.3.8. Cluster Analysis

A cluster analysis was made to isolate groups of the pumpkin cultivars according to their antioxidative activity (Figure 1). The assumption was that the selected groups of cultivars should differ from each other in terms of the most determining variables. The Ward method was used for a hierarchical cluster analysis. Figure 1 shows into which clusters the pumpkin cultivars were categorized.

Two groups of cultivars were distinguished on the basis of the analysis: Cluster 1 ('Buttercup', 'Golden Hubbard', 'Galeux d'Eysines', 'Melonowa Żółta', 'Hokkaido', 'Jumbo Pink Banana', 'Marina Di Chiggia', 'Flat White Boer Ford', 'Jarrahdale'), and cluster 2 ('Blue Kuri', 'Green Hubbard', 'Gomez', 'Shokichi Shiro', 'Porcelain Doll') (Table 12). The Student's t-test for independent samples was applied to check whether there were intergroup differences. The analysis showed that cluster 1 exhibited significantly greater ability to scavenge ABTS (120.49 vs. 104.84 mg Tx/100 g dm) and DPPH radicals

(190.21 vs. 95.24 mg Tx/100 g dm) (in the aqueous–methanol extracts) and DPPH radicals (91.74 vs. 49.49 mg Tx/100 g dm) (in the aqueous extracts). Apart from that, cluster 1 had higher total polyphenolic content (84.46 vs. 62.78 mg GAE/100 g dm). On the other hand, cluster 2 was characterized by greater ferric ion chelating capacity in the aqueous–methanol extracts (6210.87 vs. 11061.61 ppm EDTA/100 g dm) and aqueous extracts (4346.98 vs. 6745.29 ppm EDTA/100 g dm). There were no statistically significant differences between the groups in the oxygen radicals absorbance capacity (ORAC).

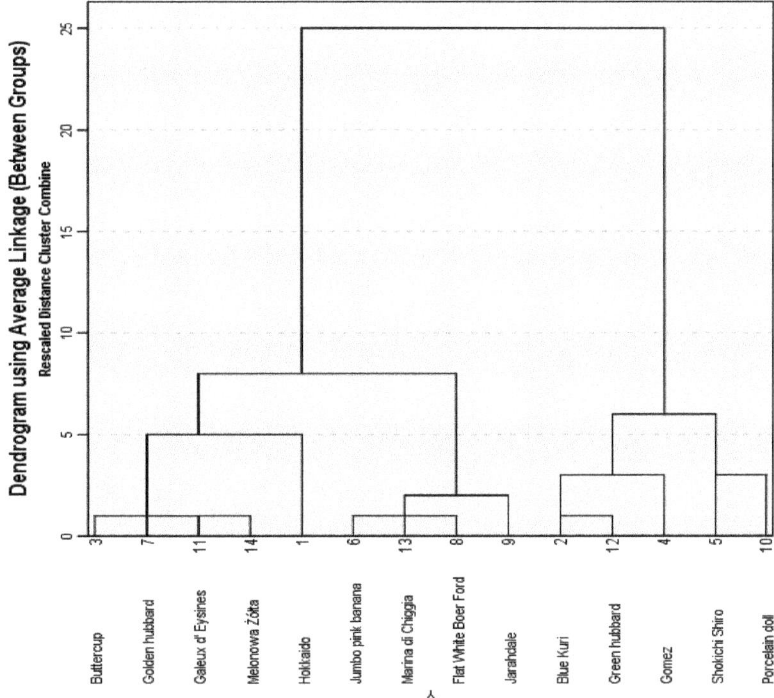

Figure 1. Dendrogram of studied pumpkin variety divisions.

Table 12. Variable levels divided into groups of pumpkin varieties.

		Cluster 1	Cluster 2	p Value
Aqueous–methanol extract	ABTS	120.49 ± 13.72	104.84 ± 14.01	$p < 0.05$
	DPPH	190.21 ± 99.12	95.24 ± 30.17	$p < 0.05$
	Total polyphenols content	84.46 ± 31.73	62.78 ± 9.72	$p < 0.05$
	Iron chelating activity	6210.87 ± 1188.94	11061.61 ± 1127.68	$p < 0.05$
	FRAP	415.28 ± 118.85	372.52 ± 76.37	NS.
Aqueous extract	ABTS	130.6 ± 25.69	122.04 ± 14.5	NS.
	DPPH	91.74 ± 40.25	49.49 ± 11.66	$p < 0.05$
	Total polyphenols content	95.54 ± 29.02	87.29 ± 22.81	NS.
	Iron chelating activity	4346.98 ± 1301.87	6745.29 ± 1191.65	$p < 0.05$
	FRAP	135.33 ± 71.86	117.54 ± 47.87	NS.
	ORAC	91.75 ± 24.88	93.51 ± 19.98	NS.

$p < 0.05$: Determination of statistically significant differences between the groups of varieties; NS: Not significant; dm: Dry mass; the results are expressed as the mean values ± standard deviation of the triplicate samples; the results are expressed in the following units: ABTS, DPPH (mg Tx/100 g dm), iron chelating activity (ppm EDTA/100 g dm), FRAP (mM Fe(II)/100 g dm), total phenolic content (mg GAE/100 g dm), ORAC (μM Tx/g dm).

4. Conclusions

The extraction methods were optimized to test the antioxidative activity of pumpkin pulp. The research showed that the aqueous and aqueous–methanol (80%) extracts exhibited the highest antioxidative potential. The best effects were achieved when the extraction was conducted at 70 °C for 2 h. The acetone and ethyl acetate extracts exhibited low antioxidative activity. The extraction optimization results were used for comparative analysis of the antioxidative potential of 14 pumpkin cultivars of the *Cucurbita maxima* species. To date there has not been such an extensive analysis of the antioxidative properties of the pumpkin pulp of various cultivars. The following tests were conducted to measure the antioxidative activity: ABTS, DPPH, FRAP, ORAC, chelating activity. The total polyphenolic content was also analyzed. The research clearly showed that the pumpkin cultivars under analysis were significantly diversified in their ability to scavenge free radicals and to reduce and chelate ferric ions. They also differed in the total polyphenolic content. The antioxidant activity of pumpkin extracts is generally affected by different variables such as the chemical structure and profile of antioxidants, e.g., carotenoids, tocopherols or phenolic compounds. However, this is the topic of the upcoming publication in which the profile of antioxidant compounds will be characterized. Antioxidants arise as easily extractable compounds, soluble, and as the residue of the extract. That is why it is difficult to identify and categorize the key trends in the contribution of various compounds and the concomitant influence of different factors to the total antioxidant capacity. Hence, as indicated in other studies on the antioxidant activity of the food matrix, consideration should also be given to substances bound in the extract residues. The following pumpkin cultivars exhibited high antioxidative activity: 'Melonowa Żółta', 'Hokkaido', 'Porcelain Doll', and 'Gomez'. The research showed that pumpkin pulp exhibited strong antioxidative properties, which might be significant for human health. The inclusion of pumpkin or pumpkin pulp-based food products into a diet may help protect the human body from the harmful effect of free radicals. It is important to reduce the risk of diseases of affluence. Properly selected powdered pumpkin pulp can be used in the food industry, e.g., as an additive increasing the stability of fat in meat.

Author Contributions: B.K., conceptualization, formal analysis, investigation, writing, review, and editing; A.G.-M., conceptualization, investigation, writing, editing, supervision, and funding acquisition; J.B.K., investigation, editing, and funding acquisition. All authors have read and agreed to the published version of the manuscript.

Funding: This research was funded by the National Science Centre, Poland, grant number 2018/29/B/NZ9/00461.

Conflicts of Interest: The authors declare no conflict of interest.

References

1. Oroian, M.; Escriche, I. Antioxidants: Characterization, natural sources, extraction and analysis. *Food Res. Int.* **2015**, *74*, 10–36. [CrossRef]
2. Nathan, C.; Cunningham-Bussel, A. Beyond oxidative stress: An immunologist's guide to reactive oxygen species. *Nat. Rev. Immunol.* **2013**, *13*, 349–361. [CrossRef]
3. Evans, J.R.; Lawrenson, J.G. Antioxidant vitamin and mineral supplements for preventing age-related macular degeneration. *Cochrane Database Syst. Rev.* **2017**, *2017*, CD000253. [CrossRef] [PubMed]
4. Prasad, K.; Laxdal, V.A.; Raney, B.L. Antioxidant activity of allicin, an active principle in garlic. *Mol. Cell. Biochem.* **1995**, *148*, 183–189. [CrossRef] [PubMed]
5. Pala, R.; Beyaz, F.; Tuzcu, M.; Er, B.; Sahin, N.; Cinar, V.; Sahin, K. The effects of coenzyme Q10 on oxidative stress and heat shock proteins in rats subjected to acute and chronic exercise. *J. Exerc. Nutr. Biochem.* **2018**, *22*, 14–20. [CrossRef] [PubMed]
6. Kulczyński, B.; Gramza-Michałowska, A. Characteristics of Selected Antioxidative and Bioactive Compounds in Meat and Animal Origin Products. *Antioxidants* **2019**, *8*, 335. [CrossRef]
7. Kulczyński, B.; Gramza-Michałowska, A. The importance of selected spices in cardiovascular diseases. *Postepy Hig. Med. Dosw.* **2016**, *70*, 1131–1141. [CrossRef]

8. Gramza-Michałowska, A.; Sidor, A.; Reguła, J.; Kulczyński, B. PCL assay application in superoxide anion-radical scavenging capacity of tea Camellia sinensis extracts. *Acta Sci. Pol. Technol. Aliment.* **2015**, *14*, 331–341. [CrossRef]

9. Salvador, A.C.; Król, E.; Lemos, V.C.; Santos, S.A.; Bento, F.P.; Costa, C.P.; Almeida, A.; Szczepankiewicz, D.; Kulczyński, B.; Krejpcio, Z.; et al. Effect of Elderberry (*Sambucus nigra* L.) Extract Supplementation in STZ-Induced Diabetic Rats Fed with a High-Fat Diet. *Int. J. Mol. Sci.* **2017**, *18*, 13. [CrossRef]

10. Gramza-Michałowska, A.; Kulczyński, B.; Xindi, Y.; Gumienna, M. Research on the effect of culture time on the kombucha tea beverage's antiradical capacity and sensory value. *Acta Sci. Pol. Technol. Aliment.* **2016**, *15*, 447–457. [CrossRef]

11. Kulczyński, B.; Gramza-Michałowska, A.; Kobus-Cisowska, J.; Kmiecik, D. The role of carotenoids in the prevention and treatment of cardiovascular disease–Current state of knowledge. *J. Funct. Foods* **2017**, *38*, 45–65. [CrossRef]

12. Sidor, A.; Drożdżyńska, A.; Gramza-Michałowska, A. *Black chokeberry (Aronia melanocarpa)* and its products as potential health-promoting factors—An overview. *Trends Food Sci. Technol.* **2019**, *89*, 45–60. [CrossRef]

13. Renaud, J.; Martinoli, M.-G. Considerations for the Use of Polyphenols as Therapies in Neurodegenerative Diseases. *Int. J. Mol. Sci.* **2019**, *20*, 1883. [CrossRef] [PubMed]

14. Shahidi, F.; De Camargo, A.C. Tocopherols and Tocotrienols in Common and Emerging Dietary Sources: Occurrence, Applications, and Health Benefits. *Int. J. Mol. Sci.* **2016**, *17*, 1745. [CrossRef] [PubMed]

15. Thayagarajan, A.; Sahu, R.P. Potential Contributions of Antioxidants to Cancer Therapy: Immunomodulation and Radiosensitization. *Integr. Cancer Ther.* **2018**, *17*, 210–216. [CrossRef] [PubMed]

16. Gramza-Michałowska, A.; Sidor, A.; Kulczyński, B. Berries as a potential anti-influenza factor–A review. *J. Funct. Foods* **2017**, *37*, 116–137. [CrossRef]

17. Arulselvan, P.; Fard, M.T.; Tan, W.S.; Gothai, S.; Fakurazi, S.; Norhaizan, M.E.; Kumar, S.S. Role of Antioxidants and Natural Products in Inflammation. *Oxid. Med. Cell. Longev.* **2016**, *2016*, 1–15. [CrossRef]

18. Buscemi, S.; Corleo, D.; Di Pace, F.; Petroni, M.L.; Satriano, A.; Marchesini, G. The Effect of Lutein on Eye and Extra-Eye Health. *Nutrients* **2018**, *10*, 1321. [CrossRef]

19. Braakhuis, A.; Raman, R.; Vaghefi, E. The Association between Dietary Intake of Antioxidants and Ocular Disease. *Diseases* **2017**, *5*, 3. [CrossRef]

20. Kmiecik, D.; Korczak, J.; Rudzińska, M.; Gramza-Michałowska, A.; Hes, M.; Kobus-Cisowska, J. Stabilisation of phytosterols by natural and synthetic antioxidants in high temperature conditions. *Food Chem.* **2015**, *173*, 966–971. [CrossRef]

21. Gramza-Michalowska, A.; Sidor, A.; Hes, M. Herb extract influence on the oxidative stability of selected lipids. *J. Food Biochem.* **2011**, *35*, 1723–1736. [CrossRef]

22. Hęś, M.; Gramza-Michałowska, A. Effect of Plant Extracts on Lipid Oxidation and Changes in Nutritive Value of Protein in Frozen? Stored Meat Products. *J. Food Process. Pres.* **2017**, *41*, e12989. [CrossRef]

23. Ismail, H.; Lee, E.; Ko, K.; Paik, H.; Ahn, D. Effect of Antioxidant Application Methods on the Color, Lipid Oxidation, and Volatiles of Irradiated Ground Beef. *J. Food Sci.* **2009**, *74*, C25–C32. [CrossRef] [PubMed]

24. Liu, F.; Xu, Q.; Dai, R.; Ni, Y. Effects of natural antioxidants on colour stability, lipid oxidation and metmyoglobin reducing activity in raw beef patties. *Acta Sci. Pol. Technol. Aliment.* **2015**, *14*, 37–44. [CrossRef]

25. Durazzo, A.; Lucarini, M. A current shot and re-thinking of antioxidant research strategy. *Br. J. Anal. Chem.* **2018**, *5*, 9–11. [CrossRef]

26. Tanaka, R.; Kikuchi, T.; Nakasuji, S.; Ue, Y.; Shuto, D.; Igarashi, K.; Okada, R.; Yamada, T. A Novel 3a-p-Nitrobenzoylmultiflora-7:9(11)-diene-29-benzoate and Two New Triterpenoids from the Seeds of Zucchini (*Cucurbita pepo* L.). *Molecules* **2013**, *18*, 7448–7459. [CrossRef]

27. Zhou, C.-L.; Mi, L.; Hu, X.-Y.; Zhu, B.-H. Evaluation of three pumpkin species: Correlation with physicochemical, antioxidant properties and classification using SPME-GC–MS and E-nose methods. *J. Food Sci. Technol.* **2017**, *54*, 3118–3131. [CrossRef]

28. Nishimura, M.; Ohkawara, T.; Sato, H.; Takeda, H.; Nishihira, J. Pumpkin Seed Oil Extracted from Cucurbita maxima Improves Urinary Disorder in Human Overactive Bladder. *J. Tradit. Complement. Med.* **2014**, *4*, 72–74. [CrossRef]

29. Kulczyński, B.; Gramza-Michałowska, A. The Profile of Carotenoids and Other Bioactive Molecules in Various Pumpkin Fruits (*Cucurbita maxima* Duchesne) Cultivars. *Molecules* **2019**, *24*, 3212. [CrossRef]

30. Kulczyński, B.; Gramza-Michałowska, A. The Profile of Secondary Metabolites and Other Bioactive Compounds in *Cucurbita pepo* L. and Cucurbita moschata Pumpkin Cultivars. *Molecules* **2019**, *24*, 2945. [CrossRef]

31. Zaccari, F.; Galietta, G. α-Carotene and β-Carotene Content in Raw and Cooked Pulp of Three Mature Stage Winter Squash "Type Butternut". *Foods* **2015**, *4*, 477–486. [CrossRef] [PubMed]

32. Montesano, D.; Blasi, F.; Simonetti, M.S.; Santini, A.; Cossignani, L. Chemical and Nutritional Characterization of Seed Oil from *Cucurbita maxima* L. (var. Berrettina) Pumpkin. *Foods* **2018**, *7*, 30. [CrossRef] [PubMed]

33. Kim, M.Y.; Kim, E.J.; Kim, Y.N.; Choi, C.; Lee, B.H. Comparison of the chemical compositions and nutritive values of various pumpkin (*Cucurbitaceae*) species and parts. *Nutr. Res. Pract.* **2012**, *6*, 21–27. [CrossRef] [PubMed]

34. Paris, H.S.; Daunay, M.-C.; Pitrat, M.; Janick, J. First Known Image of Cucurbita in Europe, 1503–1508. *Ann. Bot.* **2006**, *98*, 41–47. [CrossRef]

35. Lust, T.A.; Paris, H.S. Italian horticultural and culinary records of summer squash (*Cucurbita pepo, Cucurbitaceae*) and emergence of the zucchini in 19th-century Milan. *Ann. Bot.* **2016**, *118*, 53–69. [CrossRef]

36. Nawirska-Olszańska, A.; Kita, A.; Biesiada, A.; Sokół-Łętowska, A.; Kucharska, A.Z. Characteristics of antioxidant activity and composition of pumpkin seed oils in 12 cultivars. *Food Chem.* **2013**, *139*, 155–161. [CrossRef]

37. Saavedra, M.J.; Aires, A.; Dias, C.; Almeida, J.A.; De Vasconcelos, M.C.; Santos, P.; Rosa, E.A. Evaluation of the potential of squash pumpkin by-products (seeds and shell) as sources of antioxidant and bioactive compounds. *J. Food Sci. Technol.* **2015**, *52*, 1008–1015. [CrossRef]

38. Can-Cauich, C.A.; Sauri-Duch, E.; Moo-Huchin, V.M.; Betancur-Ancona, D.; Cuevas-Glory, L.F. Effect of extraction method and specie on the content of bioactive compounds and antioxidant activity of pumpkin oil from Yucatan, Mexico. *LWT-Food Sci. Technol.* **2019**, *285*, 186–193. [CrossRef]

39. Brand-Williams, W.; Cuvelier, M.; Berset, C. Use of a free radical method to evaluate antioxidant activity. *LWT-Food Sci. Technol.* **1995**, *28*, 25–30. [CrossRef]

40. Re, R.; Pellegrini, N.; Proteggente, A.; Pannala, A.; Yang, M.; Rice-Evans, C. Antioxidant activity applying an improved ABTS radical cation decolorization assay. *Free. Radic. Boil. Med.* **1999**, *26*, 1231–1237. [CrossRef]

41. Sánchez-Moreno, J.; Fulgencio, S. A procedure to measure the antiradical efficiency of polyphenols. *J. Sci. Food Agric.* **1998**, *76*, 270–276. [CrossRef]

42. Benzie, I.F.; Strain, J. Ferric reducing/antioxidant power assay: Direct measure of total antioxidant activity of biological fluids and modified version for simultaneous measurement of total antioxidant power and ascorbic acid concentration. *Methods Enzymol.* **1999**, *299*, 15–27. [PubMed]

43. Decker, E.A.; Welch, B. Role of ferritin as a lipid oxidation catalyst in muscle food. *J. Agric. Food Chem.* **1990**, *38*, 674–677. [CrossRef]

44. Ou, B.; Huang, D.; Hampsch-Woodill, M.; Flanagan, J.A.; Deemer, E.K. Analysis of Antioxidant Activities of Common Vegetables Employing Oxygen Radical Absorbance Capacity (ORAC) and Ferric Reducing Antioxidant Power (FRAP) Assays: A Comparative Study. *J. Agric. Food Chem.* **2002**, *50*, 3122–3128. [CrossRef] [PubMed]

45. Kiat, V.V.; Siang, W.K.; Madhavan, P.; Jin, C.J.; Ahmad, M.; Akowuah, G. FT-IR profile and antiradical activity of dehulled kernels of apricot, almond and pumpkin. *Res. J. Pharm. Biol. Chem. Sci.* **2014**, *5*, 112–120.

46. Tiveron, A.P.; Melo, P.S.; Bergamaschi, K.B.; Vieira, T.M.F.S.; Regitano-D'Arce, M.A.B.; Alencar, S.M. Antioxidant Activity of Brazilian Vegetables and Its Relation with Phenolic Composition. *Int. J. Mol. Sci.* **2012**, *13*, 8943–8957. [CrossRef] [PubMed]

47. Singh, J.; Shukla, S.; Singh, V.; Rai, A.K. Phenolic Content and Antioxidant Capacity of Selected Cucurbit Fruits Extracted with Different Solvents. *J. Nutr. Food Sci.* **2016**, *6*, 6. [CrossRef]

48. Bayili, R.G.; Abdoul-Latif, F.; Kone, O.; Diao, M.; Bassole, I.; Dicko, M. Phenolic compounds and antioxidant activities in some fruits and vegetables from Burkina Faso. *Afr. J. Biotechnol.* **2011**, *10*, 62.

49. Valenzuela, G.M.; Soro, A.S.; Tauguinas, A.L.; Gruszycki, M.R.; Cravzov, A.L.; Giménez, M.C.; Wirth, A. Evaluation Polyphenol Content and Antioxidant Activity in Extracts of *Cucurbita* spp. *OALib* **2014**, *1*, 1–6. [CrossRef]

50. Dar, P.; Farman, M.; Dar, A.; Khan, Z.; Munir, R.; Rasheed, A.; Waqas, U. Evaluation of Antioxidant potential and comparative analysis of Antimicrobial activity of Various Extracts of *Cucurbita pepo* L. Leaves. *J. Agric. Sci. Food Technol.* **2017**, *3*, 103–109.

51. Oloyede, F.; Agbaje, G.; Obuotor, E.; Obisesan, I. Nutritional and antioxidant profiles of pumpkin (*Cucurbita pepo* Linn.) immature and mature fruits as influenced by NPK fertilizer. *Food Chem.* **2012**, *135*, 460–463. [CrossRef] [PubMed]

52. Fidrianny, I.; Darmawati, A.; Sukrasno, S. Antioxidant capacities from different polarities extracts of cucurbitaceae leaves using FRAP, DPPH assays and correlation with phenolic, flavonoid, carotenoid content. *Int. J. Pharm. Pharm. Sci.* **2014**, *6*, 858–862.

53. Chensom, S.; Okumura, H.; Mishima, T. Primary Screening of Antioxidant Activity, Total Polyphenol Content, Carotenoid Content, and Nutritional Composition of 13 Edible Flowers from Japan. *Prev. Nutr. Food Sci.* **2019**, *24*, 171–178. [CrossRef] [PubMed]

54. Garrett, A.R.; Murray, B.K.; Robinson, R.A.; O'Neill, K.L. Measuring antioxidant capacity using the ORAC and TOSC assays. *Methods Mol. Biol.* **2010**, *594*, 251–262.

55. Haytowitz, D.B.; Bhagwat, S. USDA Database for the Oxygen Radical Absorbance Capacity (ORAC) of Selected Foods, Release 2. Available online: https://naldc.nal.usda.gov/download/43336/PDF (accessed on 2 May 2010).

56. Parry, J.; Hao, Z.; Luther, M.; Su, L.; Zhou, K.; Yu, L.L. Characterization of cold-pressed onion, parsley, cardamom, mullein, roasted pumpkin, and milk thistle seed oils. *J. Am. Oil Chem. Soc.* **2006**, *83*, 847–854. [CrossRef]

 *sustainability*

Article

Studies of a Rotary–Centrifugal Grain Grinder Using a Multifactorial Experimental Design Method

Andrzej Marczuk [1], Agata Blicharz-Kania [2,*], Petr A. Savinykh [3], Alexey Y. Isupov [3], Andrey V. Palichyn [4] and Ilya I. Ivanov [4]

[1] Department of Agricultural, Forestry and Transport Machines, University of Life Sciences in Lublin, Głęboka 28, 20-612 Lublin, Poland; andrzej.marczuk@up.lublin.pl

[2] Department of Biological Bases of Food and Feed Technologies, University of Life Sciences in Lublin, Głęboka 28, 20-612 Lublin, Poland

[3] Federal Agricultural Research Centre of the North-East Named after N.V. Rudnitskiy, Kirov 610007, Russia; peter.savinykh@mail.ru (P.A.S.); isupoff.aleks@yandex.ru (A.Y.I.)

[4] FSBEI HE Vologda State Dairy Farming Academy (DSFA) named after N.V. Vereshchagin, Vologda—Molochnoye 160555, Russia; zmij.hh@yandex.ru (A.V.P.); kadyichna@mail.ru (I.I.I.)

* Correspondence: agata.kania@up.lublin.pl; Tel.: +48-81-531-9646

Received: 26 August 2019; Accepted: 23 September 2019; Published: 27 September 2019

Abstract: A scientific and technical literature review on machines designed to grind fodder grain revealed that the existing designs of grinding machines—those based on destruction by impact, cutting, or chipping—have various drawbacks. Some disadvantages include high metal and energy intensity, an uneven particle size distribution of the ground (crushed) product, a high percentage of dust fraction, the rapid wear of work tools (units), and heating of the product. To eliminate most of the identified shortcomings, the design of a rotary–centrifugal grain grinder is proposed in this paper. The optimization of the grinder's working process was carried out using experimental design methodology. The following factors were studied: the grain material feed, rotor speed (rpm), opening of the separating surface, number of knives (blades) on the inner and outer rings, technical conditions of the knives (sharpened or unsharpened), and the presence of a special insert that is installed in the radial grooves of the distribution bowl. The optimization criteria were based on the amount of electricity consumed by and the performance of the rotary–centrifugal grain grinder. The quality of performance was quantified by the finished product, based on the percentage of particles larger than 3 mm in size. An analysis of the results of the multifactorial experiment allowed us to establish a relationship (interaction) between the factors and their influence on the optimization criteria, as well as to determine the most significant factors and to define further directions for the research of a centrifugal–rotary grain grinder. From our experimental results, we found that the grinder is underutilized in the selected range of factor variation. Furthermore, the number of knives installed at the second stage of the grinder, the gap (clearance) of the separating surface, and the technical condition of the knives are among the most important factors influencing the power consumption and the quality of the resulting product. A reduction in the number of knives at the first stage has a positive effect on all the selected optimization criteria; and by varying the factors in the selected range, it is possible to obtain a product corresponding to medium and coarse grinding.

Keywords: grain grinding; rotary–centrifugal grinder; construction optimization criteria

---

1. Introduction

In the cost structure of feed production, grinding represents a very labor- and energy-intensive process, constituting, according to various datasets, at least half of the total costs associated with the conservation, storage, and preparation of feed mixtures. Hence, implementing a sustainable design of tools and processes is becoming increasingly important for manufacturers [1]. There is an increasing use of concentrated feed in the composition of mixed fodder [2–4]. An important factor for improving the digestibility and ensuring the most complete extraction of the potential energy of the feed is the method of its grinding. To date, the most common method in agricultural production is grinding with hammer mills (disintegrators) [5–7]. However, modern requirements for the quality of the feed obtained in the grinding process require the minimization of metal and energy intensity.

The main approach for solving the problems discussed above is to improve existing machines and devices (units) by optimizing structural elements and the subsequent control of technological operating modes. Furthermore, one can develop and implement technical solutions based on new physical principles or by combining several previously known approaches. Thus, recently, the design of grinders has become of great practical interest in the field of grinding grain. It is possible to significantly reduce the specific energy consumption for the production of concentrated feed and obtain a product with a high uniformity of particle size by combining the grinding process with shearing and chipping [5,8–15].

Pushkarev A.S. and Fomin V.V. [3,16], in their research on the parameters of the work tools (units) of a centrifugal–rotary grain grinder, found that by modifying the curvilinear shape of the cutting pair, the specific energy intensity of the grinding process of grain is reduced; increasing the grinder's performance without changing its energy consumption. Further they found, the moisture content of the material being crushed has a significant effect on the specific energy intensity of the grinding process; an increase in the moisture content of grain, increases the energy consumption. However, the trend towards a decrease in energy intensity remains when curvilinear work tools are used. Finally, by taking into account the change in the optimal cutting angle of the material being processed as it moves in the channel of the work tool of a small-size centrifugal–rotary grinder, the grain size distribution (granulometric composition) is equalized, the dust fraction decreases, and there are no whole grains in the finished product [3,16].

Based on the analysis of Druzhynin R.A. and Ivanov V.V. [2,4] on the research of theorists and practitioners, it is recommended to use two-stage grinding for the manufacture of coarse and medium grinding products. Additionally, multi-stage crushers and grinders are more effective to obtain a fine grinding product. At the same time, the disadvantages of knife-type grinders compared to impact crushers are also noted [17]: intensive wear of the working parts (knives), a sharp decline in the quality of grinding as a result of wear of the working bodies (knives), and a higher specific energy consumption than crushers, with greater productivity.

Considering the existing industrial designs of grinders that have knives (blades) as work tools, it can be seen that they are used mainly in personal subsidiary farms, where high performance and drive power are not required, but a low price and simplicity of design are needed. Such crushers are widely represented on the market by the following manufacturers: Electromash, Greentechs, Bison, Kolos, Yarmash, Cyclone, Niva, Fermer, ZDN, etc. [17].The analysis of existing models of grinders on the market and scientific research on grinding and crushing suggests that the search continues for a method of grinding feed which improves the economic efficiency of the process, while simultaneously ensuring a high quality of the resulting product of concentrated feed.

The purpose of this study is to assess the structural elements and various operating modes of a rotary–centrifugal grinder according to key performance indicators. The task of the study is to determine the significance of factors (x_1–x_7) according to certain optimization criteria (y_1–y_3), allowing for more specific implementation in further research of rotary–centrifugal grain grinders.

2. Materials and Methods

2.1. Design of a Rotary–Centrifugal Device for Grinding Grain

Based on the analysis of the designs of crushers and grinders, we have developed and proposed the design of a rotary–centrifugal device for grinding grain [18]. The proposed device (Figure 1) consists of a stationary case or housing (1) with loading (input, 2) and output (3) nozzles. Two adjacent disks are coaxially mounted inside the housing (1): the upper (4) (Figure 2a) and the lower movable disks (5) (Figure 2b). On the work surface of the lower disk (5) there are annular protrusions (6), and on the work surface of the upper disk (4) there are installed knives (7), which are diamond-shaped with small cutting angles with respect to the large diagonals. The outer row of knives (8) forms a separating surface; thus, changing the angle of the knives makes it possible to continuously adjust the degree of grinding for the material. The lower disk (5) has grooves (9) in the radial direction which are made opposite in terms of angle to the direction of rotation of this disk. The lower disk (5) is mounted on the flange of the drive shaft (10) and is rotated by means of a pulley (11) mounted on it. The upper disk (4) is rigidly fixed to the stationary case (1). In the upper part of the stationary case, a receiving chamber (12) is installed, which is formed by vertical walls. The receiving chamber (12) communicates in its upper part with the loading nozzle (2) and is connected with the working chamber (14), which is the space between disks (4) and (5), through the radial windows (13).

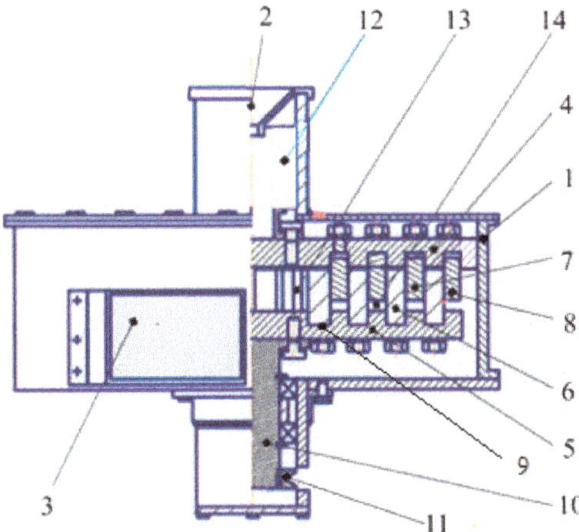

Figure 1. Design overview of a rotary–centrifugal device for grinding bulk materials. Parts include: 1—case; 2—loading (input); 3—output nozzles; 4—upper disk; 5—lower movable disk; 6—annular protrusions; 7—knives; 8—outer row of knives; 9—groove; 10—drive shaft; 11—pulley; 12—receiving chamber; 13—radial windows; and 14—working chamber.

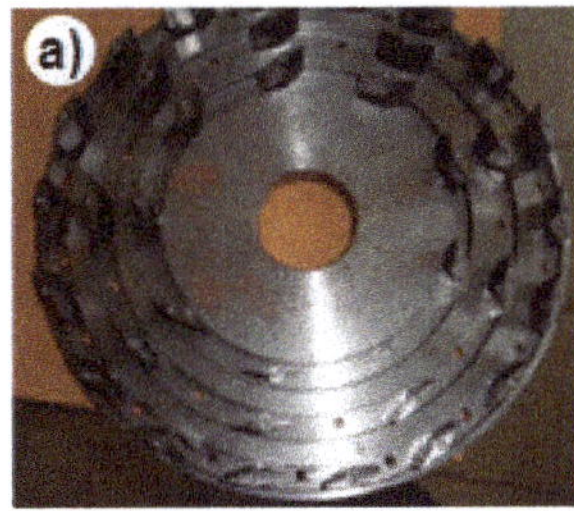

Figure 2. Photographs of the (**a**) upper disk and (**b**) lower disk.

The working process of grinding grain in the rotary–centrifugal device is carried out as follows (Figure 1). The incoming grain is subjected to the mechanical action of the first cutting pair; then, under the action of centrifugal forces, the pre-ground material moves along the grooves to the next pair. Then, the ground grains (groats or stock feed), having reached the outer row of the knives (8) which form the separating surface, pass into the gap between the knives (8) and, under the influence of the air flow created by the rotating lower disk (5), leave the housing (1) through the outlet nozzle (3).

The studies were carried out on an experimental installation of a rotary–centrifugal grinder (Figure 3) made on-site at the Federal State Budgetary Educational Institution of Higher Professional Education Vologda State Dairy Farming Academy (DSFA) named after N.V. Vereshchagin.

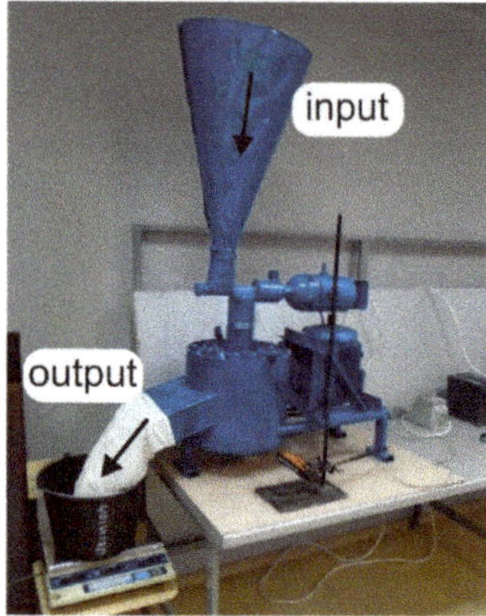

Figure 3. Overview of the experimental installation.

2.2. Description of a Multifactorial Experimental Design

To achieve the objectives of the study, the method of a multifactorial experimental design was applied. The analysis of the design of the grinder and a review of scientific and technical literature on grain grinding made it possible to identify a number of factors (Table 1) which allow us to most fully describe the process of grinding in the proposed centrifugal–rotary grinder. Below is a brief description of the selected factors (x_1–x_7) and how they vary:

- Grain was fed to the cumulative bunker of the experimental rotary–centrifugal installation with an auger conveyor and a frequency-controlled drive x_1, with a power inverter-controlled electric motor, Hyundai N700-220HF (Seul, South Korea);
- The rotation frequency of the lower disk (Figure 2b) x_2 was varied using a frequency-controlled drive with an electric motor controlled by an another Hyundai N700-220HF power inverter;
- The opening of the separating surface x_3 was adjusted by setting an appropriate size between the parallel planes of two adjacent knives, h_1 (Figure 5) on the outer row of knives (Figure 2a);
- The knives at the first x_4 and second x_5 stages of the upper disk (Figure 2a) were installed evenly, depending on the required quantity;
- To assess the impact of knife sharpening loss during operation, a factor of the technical condition of the knives x_6 was introduced as an experimental variable; i.e., "new" knives with a given angle of sharpening $\chi = 24°$ and "old" knives that have a much larger angle of sharpening (unsharpened knives); i.e., imitating their bluntness;
- The presence of inserts x_7 (Figure 4), installed in the slot of the distribution bowl of the lower disk (Figure 2b), was also considered. Eight inserts were installed in order to change the trajectory and rate of the material feed to the cutting pairs.

Table 1. Grinder experimental factors and the levels of their variations.

Factors	x_1 (kg·s^{-1})	x_2 (min^{-1})	x_3 (mm)	x_4 (pcs)	x_5 (pcs)	x_6	x_7
−1	0.023	800	2.5	9	18	"Old" ones	Present
+1	0.038	1200	3.2	3	9	"New" ones	Absent

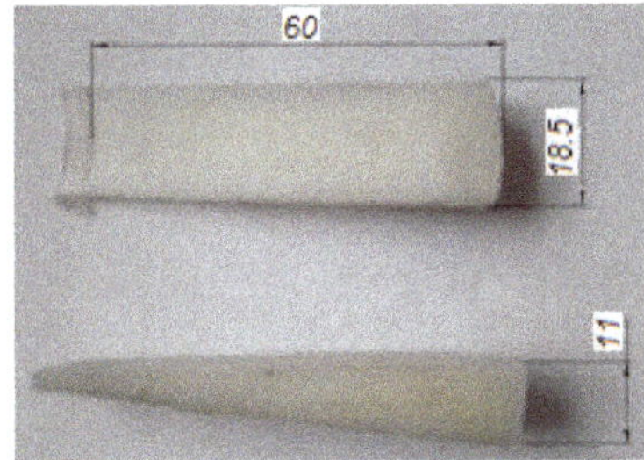

Figure 4. Geometry of the inserts with appropriate dimensions labeled.

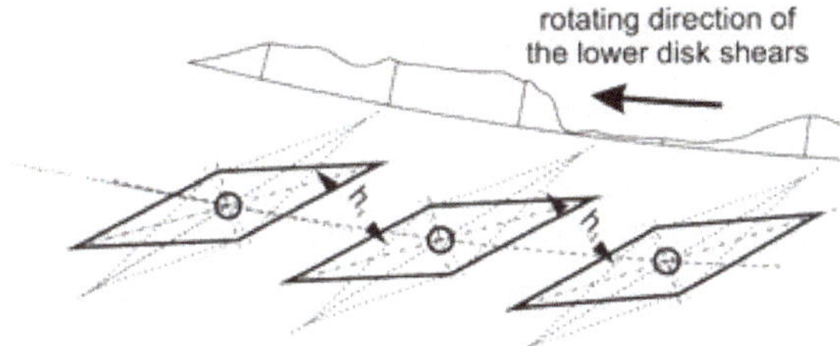

Figure 5. Geometry of the opening of the separating surface. h_1—size between the parallel planes of two adjacent knives.

Barley with 14% moisture content was used as the ground material. To reduce the amount of research, a type 2^{7-2} matrix of the fractional factor experiment was used.

The optimization criteria included the power consumption y_1 (kW) and the grinder performance (capacity) y_2 (kg/s). The measurement of power consumption was carried out using a Mercury 221 electrical energy meter connected to a PC via USB in CAN/RS-232/RS485 (Figure 6). The indicators were monitored instantly using a K-505 measurement kit.

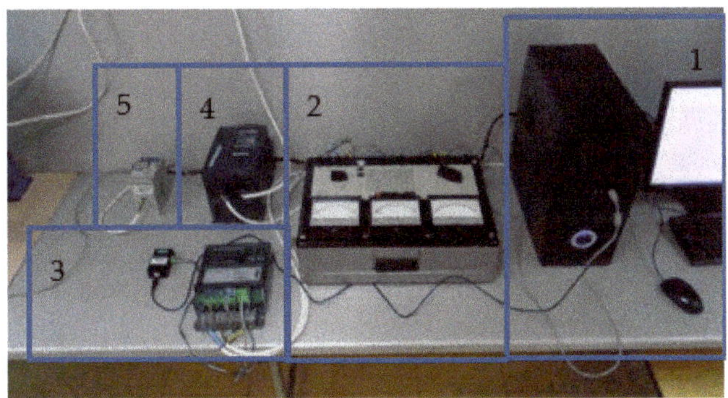

Figure 6. A set of measuring equipment used for the research. 1—personal computer; 2—K-505; 3—Mercury 221 electric energy meter with USB-CAN/RS-232/RS485 adapter; 4—frequency converter Hyundai N700-220HF; 5—circuit breaker

The performance (capacity) of the experimental installation y_2 was determined by monitoring the mass of the ground material per unit of time under a steady operating mode.

Zoo-technical requirements for mixed fodder concentrate according to GOST 9268-2015, GOST R 51550-2000, and other guidelines highlight a number of criteria for assessing the quality of the resulting product. However, the criteria directly dependent on the design and operating mode of the grain grinder. Some criteria are the percentage of particles greater than 3 mm y_3 after grinding, the grinding coarseness y_4, and the presence of whole grains in the grinding results y_5.

3. Results and Discussion

The results of the sieve analysis reveal that the most critical parameter for the compliance of the finished product with the zoo-technical requirements is the percentage of particles greater than 3 mm, y_3, which in our results is up to 60%. Screen sizing also shows the complete absence of whole grains y_5 in all samples. Calculations of the grain size (grinding coarseness) y_4 shows that grains are within the range of 1.5–3.2 mm. However, the use of this indicator as an optimization criterion at this stage is not informative, since the grain size depends largely on the percentage of particles greater than 3 mm. Thus, in this study, the most significant criterion for assessing the quality of the product is the percentage of particles greater than 3 mm, y_3. The critical particle size in different experiments depends on many factors (type of material, intended use, and subsequent processing) [19,20].

The processing of the results of the multifactorial experiment was carried out using the StatGraphics software package. A multivariate analysis of variance ANOVA (Table 2) and the regression results in Equations (1)–(3) show that the models obtained are statistically significant and describe the current processes with a reliability of at least 95%.

$$y_1 = 3.89 + 0.12x_1 + 0.2x_2 - 0.15x_3 + 0.3x_5 - 1.3x_6 + 0.28x_1x_3 + 0.12x_1x_5 - 0.28x_2x_4 + \\ +0.14x_3x_6 - 0.18x_3x_7 + 0.12x_4x_7 \tag{1}$$

$$y_2 = 0.021 + 0.004x_1 + 0.003x_2 - 0.001x_3 + 0.002x_6 + 0.002x_1x_2 - 0.001x_1x_3 + \\ +0.001x_2x_3 + 0.001x_2x_4 \tag{2}$$

$$y_3 = 13.1 + 1.71x_1 - 3.38x_2 - 5.6x_3 + 2.4x_4 + 5.75x_5 + 8.88x_6 - 2.6x_7 + 3.01x_1x_4 - \\ -1.68x_2x_6 - 8.16x_3x_6 + 3.2x_3x_7 \tag{3}$$

Table 2. Results of the model's analysis of variance (ANOVA).

	y_1	y_2	y_3
p-value	<0.0001	<0.0001	<0.0001
Error d.f.	20	18	20
Standard Error	0.464976	0.001784	4.21375
R-squared	0.9426	0.9564	0.9589

The search for a compromise solution aimed at reducing the amount of energy consumed y_1 and the percentage of particles exceeding 3 mm y_3, while increasing the performance (capacity) y_2 of the centrifugal–rotary grain grinder, was carried out using the StatGraphics software package (Table 3). The optimal values of the factors, obtained for the compromise solution in the selected range of factor variation, are presented in Table 4.

Table 3. Optimization criteria for the three factors, calculated using the StatGraphics software package.

Optimization Factor	Goal	Sensitivity	Lower Level	Upper Level	Average Predicted Value	Lower 95.0% Limit	Upper 95.0% Limit	Goal Achieved
y_1 (kW)	Minimize	Medium	-	-	2.59	2.00	3.18	0.76
y_2 (kg·s^{-1})	Maximize	Medium	-	-	0.032	0.0303	0.034	0.86
y_3 (%)	Minimize	Medium	0.0	10.0	−0.000022	−4.91	4.91	1.0

Table 4. Optimal values of the factors obtained using the StatGraphics software package.

Factor	x_1	x_2	x_3	x_4	x_5	x_6	x_7
Specified value	1	1	1	0.057	−1	1	1
Actual value	0.038 kg·s^{-1}	1200 min^{-1}	3.2 mm	6 pcs	9 pcs	"new" knives	no insert

The conducted analysis of the influence of factors on electricity consumption (1), and the performance of the centrifugal–rotary grinder (2), as well as the content of particles larger than 3 mm after grinding (3) reveals the following:

One of the most significant factors influencing the performance of the grinder y_2 is the feed of grain x_1 (Figure 7d). At the same time, an increase in the grain feed x_1 leads to an increase in all indicators y_1, y_2, and y_3. This is a consequence of an increased volume of material being transported by the lower disk of the grinder and an increased speed of transportation through the centrifugal–rotary grinder.

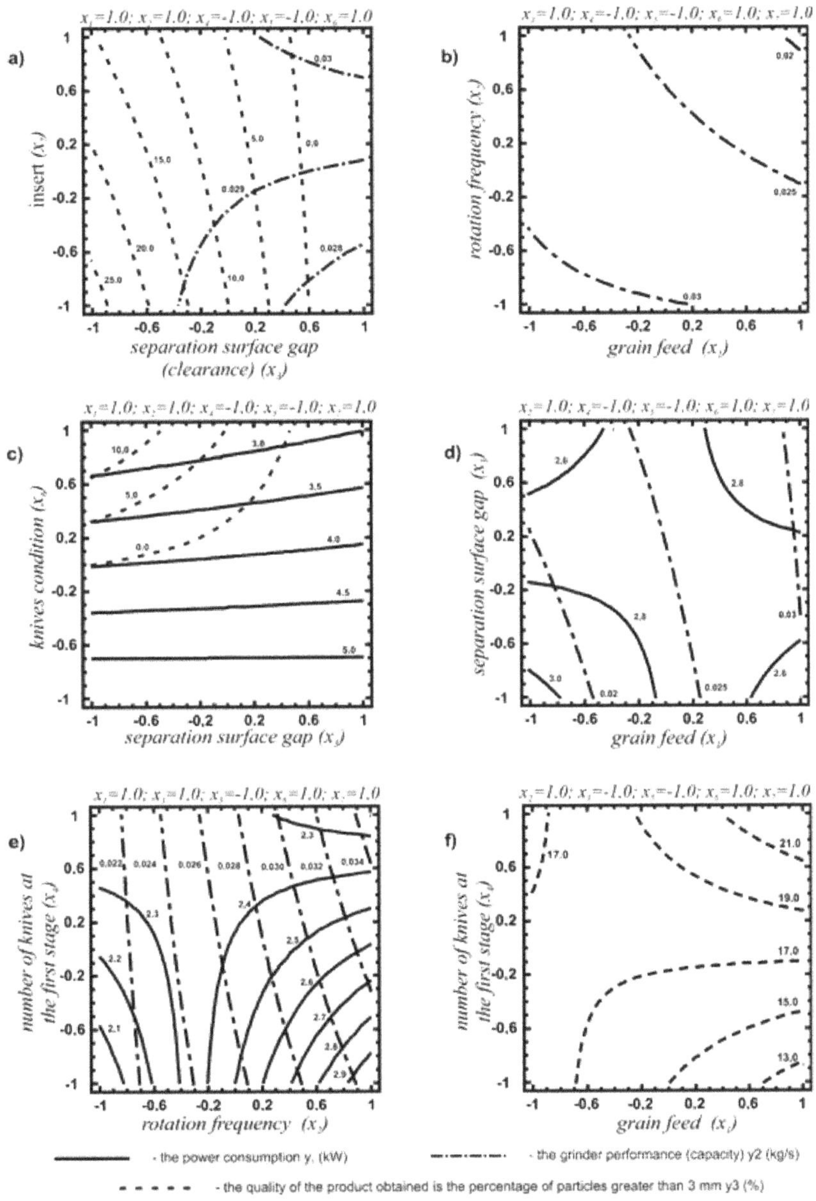

Figure 7. Two-dimensional sections of the response surface.

The increased rotation frequency of the lower disk x_2 leads to a directly proportional increase in the linear velocity of the work units, $v = 2\pi nR$ (n—rotation frequency, R—disc radius), at all stages, and an increase in the centrifugal inertial force. $F = 8m\pi^2 n^2 R$ (m—mass, n—rotation frequency, R—disc radius). As a result, the number of contacts of the grain material with the work units of the grinder is increased, while the time of its exposure in the work area is decreased. Thus, with increased values of the factor x_2, a decrease in the percentage of particles larger than 3 mm is observed with a simultaneous increase in the performance of the centrifugal–rotary grinder y_2. However, this leads to increased work for grinding and transporting the material, which is reflected in the total power consumption y_1. Bitra [21] notes that increasing speed affects the effective specific energy of hammers in different ways, depending on the type of raw material. In our experiments, the effective specific energy increases at certain rate and then decreases. Similar dependencies were shown by Moiceanu et al. [6].

Reducing the gap (clearance) of the separating surface x_3 naturally reduces the grinder performance y_2 and increases the power consumption y_1. As a result, there is an increase in the time required for the removal of the ground material from the work unit, as well as an increase in the mass of the transported ground material along the separating surface of the grinder. At the same time, reducing the gap of the separating surface x_3 does not lead to a logical decrease in the percentage of particles larger than 3 mm y_3. This phenomenon is explained by the fact that when there is a small gap of the separating surface, the particles of the material are mostly reflected from this "smooth" surface and then move along it until they pass through it. Likewise, a larger gap of the separating surface h (Figure 5) has an effect on the third grinding stage, since the knives of the separating surface have a larger approach angle and, as a result, have a lower reflectivity.

Changing the number of knives at the first grinding stage x_4 does not have a significant effect on the power consumption y_1 and grinder capacity y_2. An analysis of Equation (3), characterizing the quality of the product obtained, shows that reducing the number of knives from 9 to 3 at the first stage x_4 allows for a reduction of the percentage of particles over 3 mm in the finished product y_3 to 4.8%.

Reducing the number of knives from 18 to 9 at the second stage x_5 contributes to a reduction of particles over 3 mm y_3 in grinding by more than 11% and a reduction of the power consumption y_1 by 0.6 kW. This is understood as a result of an increase in the speed of transportation of the grain material through the grinder and a decrease in the mass accelerated by the lower disk.

Replacing the "old" knives x_6 with "new" ones is the main factor affecting energy consumption, and saves at least 2.6 kW of electricity. This can be explained by the fact that with the use of "old" knives, grinding takes place mainly by impact rather than cutting, resulting in a decrease in the number of particles larger than 3 mm y_3. At the same time, the performance (capacity) y_2 is reduced and the power consumption of the bulk grinder y_1 is increased. Ková č et al. [22] noticed that the basic factors affecting the properties of the ground material and the unit energy consumption in milling process are, among others, the knife angle and number of knives.

The presence of a special insert x_7 installed in the distribution bowls of the lower disk of the centrifugal–rotary grinder does not have any significant effect on the power consumption y_1 and the performance y_2. However, its installation increases the content of particles larger than 3 mm y_3 by 5.2%.

An analysis of the interactions of factors in the regressions, Equations (1)–(3), was carried out using (among others) two-dimensional sections (Figure 7), and shows the following results:

In general, installing a special insert x_7 into the distribution bowl of the grinder, in conjunction with the opening of the separating surface x_3, has negative effects on the percentage of particles larger than 3 mm y_3, and on the performance of the centrifugal–rotary grinder y_2. Thus, at a minimum gap (clearance) of the separating surface $x_3 = -1$, when the insert is installed, the percentage of particles larger than 3 mm is more than 25% and without the insert this value is 15%. However, when the gap is $x_3 = 0.6$ ($h = 30.5$ mm), particles larger than 3 mm are not observed in any case. Also, an increase in the gap (clearance) of the separating surface x_3 with the insert x_7 installed in the distribution bowl leads to a drop in the performance of the centrifugal–rotary grinder by 0.025 kg·s^{-1}, whereas without this insert the capacity increases by 0.008 kg^{-1} (Figure 7a);

The interaction of the factors x_1 and x_2 is most significant for regulating the performance of the grinder (y_2); reducing the feed and rotation frequency of the lower disk of the grinder leads to increased performance (Figure 7b);

The interaction of the factors of the condition of knives x_6 and the gap (clearance) of the separating surface x_3 is most significant for the percentage of particles larger than 3 mm (y_3). At the same time, regardless of the angle of sharpening of the knives, the number of particles larger than 3 mm y_3 decreases with an increasing gap of the separating surface. The deterioration of the sharpening of the knives leads to a twofold increase in power consumption y_1, which is more than 5 kW (Figure 7c). Branco et al. [23] suggested that knives should be replaced or sharpened periodically to ensure high efficiency in grinding.

Some of the determining parameters that affects the power consumption y_1 during grinding are the interactions of factors x_1 and x_3, as well as x_2 and x_4. When solving a compromise problem of increasing the grinder performance y_2 and reducing its power consumption y_1, it is necessary to increase the values of the factors x_3, x_2, and x_4, while decreasing the value of x_1 (Figure 7d,e).

With an increase in the feed x_1 and an increase in the number of knives at the first grinding stage x_4, the percentage of particles larger than 3 mm y_3 increases. As a result, the number of knives should be minimized (Figure 7f).

The search for a compromise solution was made using the StatGraphics software package, with equal significance of optimization criteria with the desired result shown in Table 4. The results of the optimization are shown in Table 4. These results suggest the compromised optimum is achieved by selecting the maximum values of the grain feed x_1 (in the selected range of factor variation, Table 1), the rotation frequency x_2, and opening of the separating surface x_3. Furthermore, "new" knives x_6 should be used without inserts in the distribution bowl x_7, where nine knives at the second stage x_5, and six at the first stage x_4 (Table 4) should be employed. Under these optimum conditions, it is possible to achieve a power consumption y_1 of 2.59 kW, a grinder capacity y_2 of 0.032 kg^{-1}, and the complete absence of particles larger than 3 mm y_3 after grinding. At the same time, the obtained values (Table 4) correspond in general to only 87.8% of the desired results, which indicates the insufficiency of the selected ranges for the factors to achieve the most optimal values of the power consumption y_1, performance (capacity) y_2, and quality of the resulting product, in terms of the content of particles exceeding 3 mm in size y_3.

4. Conclusions

The analysis of the results of this study (using the method of a multifactorial experimental design) into the operation of a centrifugal–rotary grinder suggests that the grinder is underutilized in the selected range of factor variation. The installation of special inserts in the distribution bowl of the lower disk (x_7) generally has a negative impact on the quality of the resulting product in terms of the content of particles larger than 3 mm. The number of knives installed at the second stage of the grinder (x_5), the gap (clearance) of the separating surface (x_3), and the technical condition of the knives (x_6) are among the most important factors influencing the power consumption and the quality of the resulting product. A reduction in the number of knives at the first stage (x_4) has a positive effect on all the selected optimization criteria. Finally, by varying the factors in the selected range, it is possible to obtain a product corresponding to medium and coarse grinding.

Summarizing the results of the study, we can conclude that, in further research the material feed (x_1) and rotor speed (x_2) should be increased, and the range of variation of the opening of the separating surface (x_3) should be extended. Further, the installation of a special insert (x_7) in the distribution bowl of the lower disk should be abandoned, or additional research related to changes in its shape and size should be carried out. Finally, the number of knives at the first stage (x_4) and at the second stage (x_5) should be reduced, and "new" knives (x_6) should be used in all cases.

Sustainability **2019**, *11*, 5362

Author Contributions: Conceptualization, A.M.; Formal analysis, A.M. and A.B.-K.; Funding acquisition, A.M.; Investigation, P.A.S., A.Y.I., A.V.P. and I.I.I.; Methodology, P.A.S., A.Y.I., A.V.P. and I.I.I.; Writing—original draft, P.A.S., A.Y.I., A.V.P. and I.I.I.; Writing—review & editing, A.M. and A.B.-K.

Funding: This research received no external funding.

Conflicts of Interest: The authors declare no conflict of interest.

Nomenclature

h_1—size between the parallel planes of two adjacent knives (mm); x_1—grain feed (kg·s^{-1}); x_2—rotation frequency of the lower disk (min^{-1}); x_3—separation surface gap (mm); x_4—number of knives at the first stages (pcs); x_5—number of knives at the second stages (pcs); x_6—knives condition ("Old" ones/"New" ones); x_7—presence of inserts (present/absent); y_1—power consumption (kW); y_2—grinder performance (capacity) (kg·s^{-1}); y_3—particles greater than 3 mm (%); v—linear velocity (m·s^{-1}); n—rotation frequency (s^{-1}); R—disc radius (m); F—centrifugal inertial force (N); m—mass (kg).

References

1. Linke, B.S.; Dornfeld, D.A. Application of axiomatic design principles to identify more sustainable strategies for grinding. *J. Manuf. Syst.* **2012**, *31*, 412–419. [CrossRef]
2. Druzhynin, R.A. Improving the Working Process of an Impact Centrifugal Grinding Machine. Ph.D. Thesis in Engineering Science, Voronezh State University, Voronezh, Russia, 2014; p. 169.
3. Fomin, V.V. Reducing the Energy Consumption and Improving the Uniformity of Grain Grinding in A Small-Size Centrifugal-Rotary Grinder. Abstract of a Ph.D. Dissertation, National Agricultural University FSEI HPE, Novosibirsk, Russia, 2010; p. 23.
4. Ivanov, V.V. Improving the Operating Modes of a Disk Grinder of Feed Grains. Abstract of a Ph.D. Dissertation, FSBEI HPE "Don State University", Rostov-on-Don, Russia, 2014; p. 132.
5. Ghorbani, Z.; Masoumi, A.; Hemmat, A. Specific energy consumption for reducing the size of alfalfa chops using a hammer mill. *Biosyst. Eng.* **2010**, *105*, 34–40. [CrossRef]
6. Moiceanu, G.; Paraschiv, G.; Voicu, G.; Dinca, M.; Negoita, O.; Chitoiu, M.; Tudor, P. Energy Consumption at Size Reduction of Lignocellulose Biomass for Bioenergy. *Sustainability* **2019**, *11*, 2477. [CrossRef]
7. Svihus, B.; Kløvstad, K.H.; Perez, V.; Zimonja, O.; Sahlström, S.; Schüller, R.B.; Prestløkken, E. Physical and nutritional effects of pelleting of broiler chicken diets made from wheat ground to different coarsenesses by the use of roller mill and hammer mill. *Anim. Feed Sci. Technol.* **2004**, *117*, 281–293. [CrossRef]
8. Ahmad, F.; Weimin, D.; Qishou, D.; Rehim, A.; Jabran, K. Comparative Performance of Various Disc-Type Furrow Openers in No-Till Paddy Field Conditions. *Sustainability* **2017**, *9*, 1143. [CrossRef]
9. Bulatov, S.Y.; Nechayev, V.N.; Savinykh, P.A. The development of a grain crusher for farm households, and the results of research on the optimization of its structural and technological pa-rameters. In *Theory, Development, Methods, Experiment, Analysis, Monograph*; Nizhniy Novgorod State Engineering and Economic Institute: Knyaginino, Russia, 2014; p. 156.
10. Marczuk, A.; Misztal, W.; Savinykh, P.; Turbanov, N.; Isupov, A.; Zyryanov, D. Improving efficiency of horizontal ribbon mixer by optimizing its constructional and operational parameters. *Ekspolatacja I Niezawodn. Maint. Reliab.* **2019**, *21*, 220–225. [CrossRef]
11. Sukhlyayev, V.A.; Molin, A.A.; Mezlyakov, I.N. *Device for Grinding Bulk Materials*; No 146644 RF 20-10-2014. IPC B02C 13/00; Bulletin No 29; Russian Federation: Moscow, Russia, 2014.
12. Sysuev, W.A.; Aleškin, A.V.; Savinyh, P.A.; Marczuk, A.; Wrotkowski, K.; Misztal, W. *Badania Rozwojowo-optymalizacyjne Urządzeń Do Obróbki Ziarna Zbóż I Pasz Objętościowych O Podwyższonej Wilgotności*; Towarzystwo Wydawnictw Naukowych Libropolis: Lublin, Poland, 2017; p. 156.
13. Sysuev, V.A.; Aleškin, A.V.; Savinyh, P.A.; Marczuk, A.; Wrotkowski, K.; Misztal, W. *Badanie Mobilnych Rozdrabniaczy Oraz Rozdrabniaczy-Mieszarek Pasz*; Towarzystwo Wydawnictw Naukowych Libropolis: Lublin, Poland, 2016; p. 103.
14. Sysuyev, V.A.; Aleshkin, A.V.; Savinykh, P.A. Feed-processing machines. In *Theory, Development, Experiment*; North-Eastern Zonal Agricultural Research & Development Institute: Kirov, Russia, 2008; Volume I, p. 640.
15. Yancey, N.; Wright, C.T.; Westover, T.L. Optimizing hammer mill performance through screen selection and hammer design. *Biofuels* **2013**, *4*, 85–94. [CrossRef]

16. Pushkarev, A.S. Using Work Units with Curvilinear-Shaped Cutting Elements. Abstract of a Ph.D. Dissertation, FSBEI HE Altai State Technical University named after I.I. Polzunova, Barnaul, Altai Krai, Russia, 2018; p. 22.

17. Mironov, K.Y. Improving the Efficiency of Grain Grinding Process with the Justification of the Parameters of Work Units of an Impeller Impact Crusher. Abstract of a Ph.D. Dissertation, GBOU VO "Nizhny Novgorod State Engineering and Economic University", Nizhny Novgorod, Russia, 2018; p. 142.

18. Savinykh, P.A. Device for grinding bulk materials. No 2656619 Russia, 06-06-2018. Solntsev, R.V. Centrifugal grain grinder. *Bull. Altai Agrar. Uni-Versity* **2010**, *4*, 76–80.

19. Murphy, A.; Collins, C.; Phillpotts, A.; Bunyan, A.; Henman, D. Influence of hammermill screen size and grain source (wheat or sorghum) on the growth performance of male grower pigs. In *Report Prepared for the Co-Operative Research Centre for An Internationally Competitive Pork Industry*; Project IB-107; Pork CRC: Willaston, SA, Australia, 2009.

20. Vidal, B.C.; Dien, B.S.; Ting, K.C.; Singh, V. Influence of Feedstock Particle Size on Lignocellulose Conversion—A Review. *Appl. Biochem. Biotechnol.* **2011**, *164*, 1405–1421. [CrossRef] [PubMed]

21. Bitra, V.S.; Womac, A.R.; Chevanan, N.; Miu, P.I.; Igathinathane, C.; Sokhansanj, S.; Smith, D.R.; Cannayen, I. Direct mechanical energy measures of hammer mill comminution of switchgrass, wheat straw, and corn stover and analysis of their particle size distributions. *Powder Technol.* **2009**, *193*, 32–45. [CrossRef]

22. Kovač, J.; Krilek, J.; Mikleš, M. Energy consumption of chipper coupled to a universal wheel skidder in the process of chipping wood. *J. For. Sci.* **2011**, *57*, 34–40. [CrossRef]

23. Branco, F.P.; Naka, M.H.; Cereda, M.P. Granulometry and Energy consumption as indicators of disintegration efficiency in a hammer mill adapted to extracting arrowroot starch (*Maranta arundinacea*) in comparison to starch extraction from cassava. *Eng. Agrícola* **2019**, *39*, 341–349. [CrossRef]

 *sustainability*

Article

Modeling and Simulation of Particle Motion in the Operation Area of a Centrifugal Rotary Chopper Machine

Andrzej Marczuk [1], **Jacek Caban** [1,*], **Alexey V. Aleshkin** [2], **Petr A. Savinykh** [2], **Alexey Y. Isupov** [2] **and Ilya I. Ivanov** [3]

[1]   Department of Agricultural, Forestry and Transport Machines, University of Life Sciences in Lublin, 20-612 Lublin, Poland
[2]   N.V. Rudnitsky North-East Agricultural Research Institute, Kirov 610007, Russia
[3]   FSBEI HE Vologda State Dairy Farming Academy (DSFA) named after N.V. Vereshchagin, Vologda—Molochnoye 160555, Russia
*   Correspondence: jacek.caban@up.lublin.pl; Tel.: +48-81-531-9612

Received: 11 August 2019; Accepted: 30 August 2019; Published: 6 September 2019

Abstract: The article presents approaches to the formation of a general computational scheme for modeling (simulating) the particle motion on an axisymmetric rotating curved surface with a vertical axis of rotation. To describe the complex particle motion over a given surface, the fundamental equation of particle dynamics in a non-inertial reference frame was used, and by projecting it onto the axes of cylindrical coordinates, the Lagrange's differential equations of the first kind were obtained. According to the proposed algorithm in C#, an application was developed that enables graphical and numerical control of the calculation results. The program interface contains six screen forms with tabular baseline data (input) and a table of a step-by-step calculation of results (output); particle displacement, velocity, and acceleration diagrams constructed along the axes of the system of cylindrical coordinates ρ and z; graphical presentation of the generate of the surface of revolution and the trajectory of the absolute motion of a particle over the axisymmetric rotating surface developed in polar coordinates. Examples of the calculation of the particle motion are presented. The obtained results can be used for the study and design of machines, for example, centrifugal rotary chopper machines.

Keywords: accelerator; axisymmetric surface; general equation of dynamics; non-inertial reference frame

1. Introduction

The manufacturing sector has one of the largest impacts on worldwide energy use and natural resource consumption. Traditionally, research on manufacturing processes was mainly conducted to improve efficiency and accuracy and to lower costs [1]. One of the modern challenges in the field of designing manufacturing systems is to determine the optimal level of their flexibility from the point of view of the production tasks being performed. Whether or not a production process to be executed is capable of achieving the assumed performance parameters depends, among others, on the reliability of the machines and technological devices that make up the system under design [2]. As commonly known, the oldest manufacturing techniques used by humans are the grinding, chopper, and milling processes. Research on grinding tries to enhance economic and ecological properties and performance to extend grinding applications in the overall process chain—on the one hand, in the direction of increased material removal rates, avoiding turning and milling, and on the other hand, in the direction of fine finishing, thus making further abrasive finishing processes, such as lapping and polishing, obsolete [3].

Many factors influence the sustainability of a manufacturing process, and a combination of qualitative and quantitative methods are needed to discover them [4]. To resolve these issues, namely, the identification of the type of underlying impact factors, the uncertainty of the influencing factors, and the incompleteness of the evaluation information, the researchers used the evaluation method with analytical hierarchy process [5,6] and simulations [2,7–9] in the manufacturing sector. Sustainability encompasses the three pillars of economic, environmental, and social sustainability [10–12]. In the investigation of Ahmad [10], generally, there were approximately 44% (25/57) of highly applicable indicators. Among them, there were 28% (7/25) environmental, 28% (7/25) economic, and 44% (11/25) social indicators [10]. We can, therefore, see that economic and environmental factors are treated on an equal level. Social indicators in sustainable development are highly rated. Nevertheless, in the case of the two previous ones, they are more adequate for production processes, which have a more indirect nature on social impact. Beekaroo et al. [13] revealed an index with nine environmental, four economic, and two social indicators which were pertinent in sustainability measurement.

One of the most commonly used methods focusing on environmental sustainability is the product life cycle assessment (LCA). This method is used in many sectors of the economy [14–16] such as energy sector [17–19], transport sector [18,19], and food and agricultural industry [20,21]. The limitation of using this method is the need for specific and very detailed databases covering a lot of quantitative data.

Another method uses sustainability indicators (SI). In this method, the indicator can be defined "as a measure or aggregation of measures from which one can infer about the phenomenon of interest" [11]. This method can capture all three dimensions of sustainable development and help with the evaluation on many levels (e.g., enterprises, objects, processes, and products). In particular, for perpetrators with limited means and resources, SI provide a good method of analyzing sustainable development. Enterprises can assess their real situation using indicators, raise awareness, and set targets. The results of a literature review show that energy costs and GHG (Greenhouse Gas) indicators [10,11,13,19,22–25] are becoming the most commonly used indicators in sustainable planning. On this basis [11,26], four main directions of future research on sustainability indicators can also be mentioned: implementation of new optimization methods; adding further sustainable development indicators; extension of the model on a larger scale of the production system; and loosening certain assumptions. Finding optimal indicators in production processes is not an easy task; generally, we focus on minimizing energy consumption, waste production, improving the efficiency of machinery and equipment, as well as production and logistic processes in a given company. Sustainability indicators are based on measured and/or estimated data that have to be normalized, scaled, and aggregated consistently [27].

The evolution of design criteria for grinding and chopper machines is driven by functional requirements, general trends in machine tools, and cost [28]. The primary functional requirements, as named by Möhring et al. [29], are similar for all machine tools: high static and dynamic stiffness, fatigue strength, damping, thermal and long-term stability, and low weight of moving parts. The grinding and chopper process is carried out in many areas of food manufacturing [30–32] and agriculture [33–35], as well as in the industrial sector [36–38]. The mixing process is highly complicated, with a number of affecting parameters, such as the particle properties, the structure and performance of the mixer, the mixing process parameters, and the particle feeding order [38]. The crushing method and the machinery used for this purpose should be adapted to the type of material being ground and, in particular, to its mechanical properties [33,39].

Laboratories and modern industry require fast and effective grinding in small volumes [40]. In the field of agricultural mechanization, centrifugal chopper machines with a horizontal rotor have recently attracted increasing interest. Their principle of action lies in the acceleration of grains due to centrifugal inertia forces with their subsequent grinding (fragmentation) by impact or cutting [41,42]. This is due to the lower energy consumption for the grinding process in comparison with other choppers (crushers), which, in addition to the direct costs of grain destruction, have the energy costs for drifting whole and crushed grain using air. In centrifugal rotary chopper machines, the material can be supplied to crusher/shredder hammers or to a cutting pair by centrifugal inertia forces created by a horizontally rotating rotor with working elements. In this, one of the key roles of a centrifugal rotary material grinding machine is the distribution bowl or accelerator, which functions to provide a uniform and stable supply of material to the chopper bodies and impart to the material (particles) located on its surface the necessary linear velocity and trajectory of motion. Despite the rotary blenders relying upon the action of gravity to cause the powder to cascade and mix within a rotating vessel, the convective blenders employ a paddle, impeller, blade, or screw which stirs the powder inside a static vessel [43].

Most of the centrifugal rotary grinding machines in their design have a central feed of the material into the accelerator made in the form of a flat disc. However, with this material feed, the acceleration of a particle that is closest to the axis of rotation will be difficult because of the lack of initial velocity and centrifugal forces to overcome the friction forces, which provokes the appearance of a stagnant zone and an increase in the cost of overcoming them. To solve this problem, two approaches can be used: feeding the material with an offset from the axis of rotation, leading to design complexity, or using the reflective surface of the splitter, which is a straight conical surface located coaxially with the axis of rotation, ensuring the separation of the particles to be ground from the axis of rotation and giving them the initial velocity. In centrifugal impact grinders using the "stone-crushing-stone" principle with self-lining, the trajectory of the particles to be ground (milled) is equally important.

Figure 1 shows the rotary chopper machine analyzed in this study.

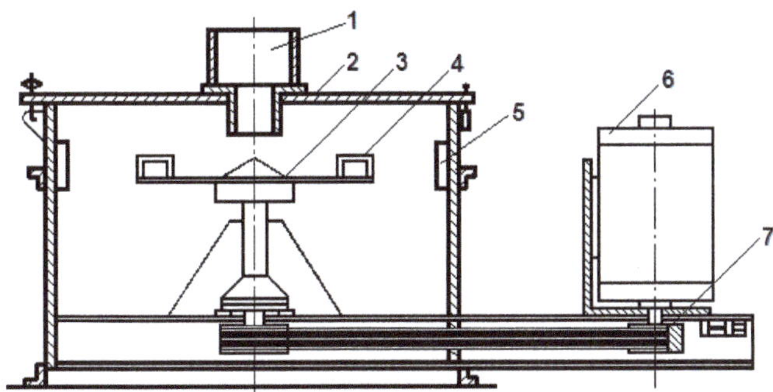

Figure 1. Tornado chopper machine: 1—feed-in tube; 2—case cover; 3—accelerator; 4—accelerating blade; 5—shredding element; 6—electric motor; 7—V-belt drive.

Figure 2 shows a scheme of self-lining formation in the accelerator.

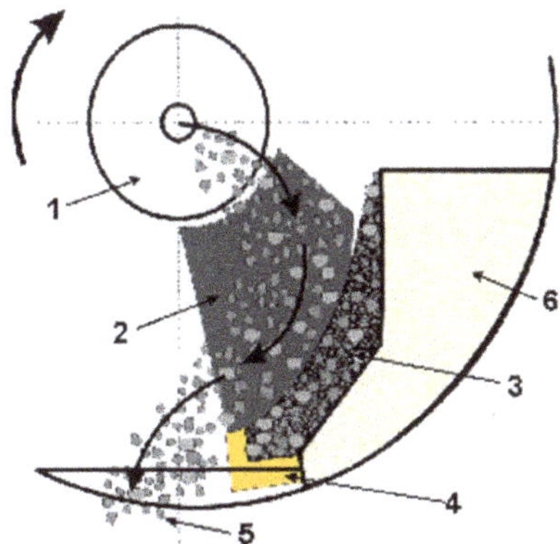

Figure 2. Scheme of self-lining formation in the accelerator: 1—a splitter; 2—wear plate; 3—self-lining pocket; 4—carbide blade; 5—descent of the material from the accelerator; 6—accelerator case.

The main contribution of this study is the modeling (simulating) of the particle motion on an axisymmetric rotating curved surface with a vertical axis of rotation, whose calculation can be used for the study and design of machines, for example, centrifugal rotary chopper machines. A mathematical model of motion was developed on the basis of step-by-step numerical integration of the obtained closed system of differential equations with an equation of constraint describing the surface of revolution. An application based on the algorithm in C# under MS Visual Studio (Microsoft, Redmond, Washington, USA) was developed that enables graphical and numerical control of the calculation results. Moreover, the results of the calculations of the particle motion trajectory and kinematic indicators are presented. The conducted simulation tests show possibilities of improving the process efficiency and shortening the operation times, which results in economic benefits during sustainable manufacturing.

2. Materials and Methods

Hypothesis. One of the methods to achieve this goal takes into account the form of the accelerator, allowing us to set the initial kinematic parameters and particle trajectory, which in turn will affect the geometric dimensions, metal consumption, and technological modes of operation of a centrifugal rotary chopper or its nodes, for example, rotor (crusher) speed and material feed rate, which would then reduce the specific energy consumption per unit of output.

Methods. To solve this problem, we modeled the motion of a particle of the material on a rotating curved surface. To do this, we used the basic law of particle dynamics (dynamics of a material point) and considered the motion of a particle on the surface of an axisymmetric spinning bowl. We used step-by-step numerical integration, which allowed us to obtain data for analyzing the particle.

3. Results and Discussion

3.1. Modeling Results

The curved surface rotates around the vertical axis z (Figure 3). The equation of the surface is described in cylindrical coordinates by the equation

$$\eta(\rho, \varphi, z) = 0 \tag{1}$$

where ρ is the cylindrical radius, φ is the polar (vectorial) angle, and z is the application (z-axis).

The radius vector of a material point moving on a curved surface in the coordinate axes, rotating with it, is a function of these three coordinates, which can transform over time without breaking Equation (1):

$$\vec{r} = \vec{r}(\rho, \varphi, z). \tag{2}$$

The differential equation of the relative motion of a point on a rotating surface is written in vector form [40]

$$m\frac{d^2\vec{r}}{dt^2} = m\vec{g} + \vec{N} + \vec{F_{TR}} + \vec{\Phi_e} + \vec{\Phi_c}, \tag{3}$$

where m is the mass of a particle, $\frac{d^2\vec{r}}{dt^2}$ is its acceleration, $m\vec{g}$ is the gravity force, $\vec{N}$ is the normal reaction of a bowl surface, $\vec{F_{TR}}$ is the friction force from the surface directed opposite to the relative velocity of a particle, $\vec{\Phi_e}$ is the centrifugal inertial force, and $\vec{\Phi_c}$ is the Coriolis inertial force.

Forces and accelerations on the direction of the cylindrical axes of coordinates are projected. The gravity force is opposite to the z axis,

$$m\vec{g} = m\vec{g}(0,\ 0,\ -mg), \tag{4}$$

where $\vec{g}$ is free-fall acceleration. The normal reaction of a rotating curved surface is

$$\vec{N} = \lambda \cdot grad(\eta),\tag{5}$$

where $\lambda = \lambda(t)$ is Lagrange's indeterminate multiplier [44], $grad(\eta)$ is the vector gradient to the equation of surface (1), which has projections on the axial cylindrical coordinate system with the φ axis directed perpendicular to the ρ, z, axes and passing through the moving point, so that the axes ρ, φ, z form the right-hand system of vectors:

$$\begin{cases} (grad(\eta))_\rho = \frac{\partial \eta}{\partial \rho} \\ (grad(\eta))_\varphi = \frac{\partial \eta}{\partial \varphi} \cdot \frac{1}{\rho}. \\ (grad(\eta))_z = \frac{\partial \eta}{\partial z} \end{cases}\tag{6}$$

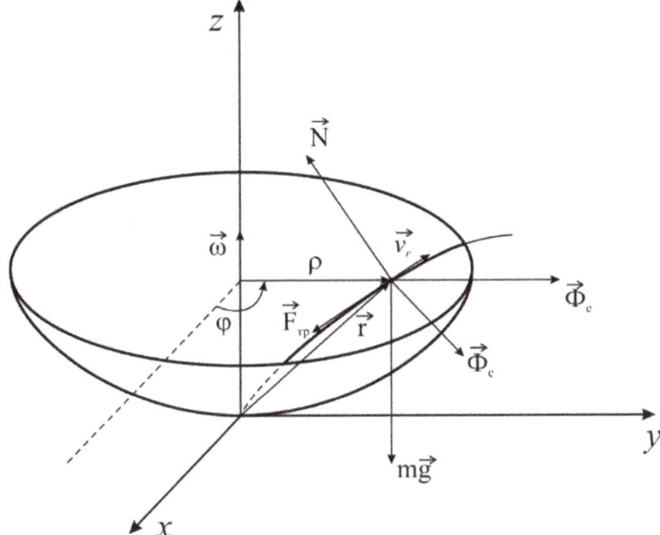

Figure 3. Design diagram of the motion of a particle on a rotating curved surface.

The projections of the normal reaction $\vec{N}$ are:

$$\begin{cases} N_\rho = \lambda \frac{\partial \eta}{\partial \rho} \\ N_\varphi = \lambda \frac{\partial \eta}{\partial \varphi} \cdot \frac{1}{\rho}. \\ N_z = \lambda \frac{\partial \eta}{\partial z} \end{cases}\tag{7}$$

The modulus of the normal reaction is found by

$$N = \left|\vec{N}\right| = |\lambda| \sqrt{\left(\frac{\partial \eta}{\partial \rho}\right)^2 + \left(\frac{\partial \eta}{\partial \varphi}\frac{1}{\rho}\right)^2 + \left(\frac{\partial \eta}{\partial z}\right)^2}.\tag{8}$$

At a constant angular velocity of rotation (spin rate) of the curved surface $\vec{\omega}$, the centrifugal inertial force $\vec{\Phi}_e$ is directed along the radius ρ in accordance with the result of vector products in its definition

$$\vec{\Phi}_e = -m\vec{\omega} \times \left(\vec{\omega} \times \vec{r}\right),\tag{9}$$

where $\vec{r}$ is determined by Equation (2), then

$$\begin{cases} \Phi_{e\rho} = m\omega^2\rho \\ \Phi_{e\varphi} = 0 \\ \Phi_{ez} = 0 \end{cases}.$$

(10)

Coriolis inertial force $\vec{\Phi}_c$ is by definition equal to

$$\vec{\Phi}_c = -2m\vec{\omega} \times \vec{v},$$

(11)

where $\vec{v}$ is the relative velocity of a material point in the moving axes of coordinates, and the modulus of relative velocity is associated with the cylindrical coordinates by

$$v = \sqrt{\dot{\rho}^2 + \left(\rho\dot{\varphi}\right)^2 + \dot{z}^2}.$$

(12)

Then, the projections of the Coriolis force equal

$$\begin{cases} \Phi_{c\rho} = 2m\omega\dot{\varphi}\rho \\ \Phi_{c\varphi} = -2m\dot{\rho}\omega \\ \Phi_{cz} = 0 \end{cases}.$$

(13)

The friction force of a particle on the bowl surface $\vec{F}_{TR}$ is determined by Coulomb's law through a normal reaction and it is opposite in direction to the relative velocity:

$$\vec{F}_{TR} = -\left|\vec{N}\right|f\frac{\vec{v}}{v}$$

(14)

$$\begin{cases} F_{TR\rho} = -\left|\vec{N}\right|f\frac{\dot{\rho}}{v} \\ F_{TR\varphi} = -\left|\vec{N}\right|f\frac{\dot{\varphi}\rho}{v} \\ F_{TRz} = -\left|\vec{N}\right|f\frac{\dot{z}}{v} \end{cases}.$$

(15)

Let us project Equation (3) on the axis of the moving system of cylindrical coordinates rotating together with the curved surface by using the found projections of all forces:

$$\begin{cases} m\left(\ddot{\rho} - \rho\dot{\varphi}^2\right) = \lambda\frac{\partial\eta}{\partial\rho} - Nf\frac{\dot{\rho}}{v} + m\omega^2\rho + 2m\omega\dot{\varphi}\rho \\ m\frac{1}{\rho}\frac{d}{dt}\left(\rho^2\dot{\varphi}\right) = -Nf\frac{\dot{\varphi}\rho}{v} - 2m\dot{\rho}\omega \\ m\ddot{z} = -mg + \lambda\frac{\partial\eta}{\partial z} - Nf\frac{\dot{z}}{v} \end{cases}.$$

(16)

Let us transform the second equation of the system, taking into account the projection of acceleration on the φ axis:

$$\frac{1}{\rho}\frac{d}{dt}\left(\rho^2\dot{\varphi}\right) = 2\dot{\rho}\dot{\varphi} + \rho\ddot{\varphi}.$$

(17)

By bringing all forces to a unit mass of a particle $m = 1$, we obtain

$$\begin{cases} \ddot{\rho} = \rho\dot{\varphi}^2 + \lambda\frac{\partial\eta}{\partial\rho} - Nf\frac{\dot{\rho}}{v} + \omega^2\rho + 2\omega\dot{\varphi}\rho \\ \ddot{\varphi} = \frac{1}{\rho}\left(-2\dot{\rho}\dot{\varphi} - Nf\frac{\dot{\varphi}\rho}{v} - 2\dot{\rho}\omega\right) \\ \ddot{z} = -g + \lambda\frac{\partial\eta}{\partial z} - Nf\frac{\dot{z}}{v} \end{cases}.$$

(18)

To obtain a closed system of equations, we supplement the system (18) with a coupling equation (of constraints) (1) (see Supplementary Material). Then, four unknown functions $\rho(t)$, $\varphi(t)$, $z(t)$, $\lambda(t)$ can be found by solving a system of three differential (18) and one algebraic (1) equations.

In the equation, various forms of execution of the surface of revolution are possible (1). Let us define the coupling equation in the form of an axisymmetric surface described by a power function passing through the given starting and final points of motion, written as

$$\frac{z}{z_K} = a_0 + \left(\frac{\rho}{\rho_K}\right)^n,$$

(19)

where a_0 is a summand determined from the condition of belonging to a given surface of revolution of the initial point of the trajectory with the coordinates

$$\rho(0) = \rho_0, \quad \varphi(0) = 0, \quad z(0) = 0$$

(20)

ρ_K, z_K are coordinates of the final point of the trajectory, where the surface described by Equation (19) ends for all values of the polar (vectorial) angle φ; n is an exponent in the equation of surface, which changes the degree of the bowl concavity and which can be varied to achieve the required parameters of the particle motion.

Thus, the origin of coordinates of the moving frame of reference, in which the particle motion is described, always corresponds along the z axis to the starting point of the motion trajectory, and the parabola vertex in the axial section of the surface of revolution lies on this axis below zero by the value a_0:

$$a_0 = -\left(\frac{\rho_0}{\rho_K}\right)^n.$$

(21)

Let us reduce Equation (19) to the form (1)

$$\eta(\rho, \varphi, z) = \frac{z}{z_K} - a_0 - \left(\frac{\rho}{\rho_K}\right)^n = 0,$$

(22)

and calculate the partial derivatives in the gradient expression to the constraint surface

$$\begin{cases} \dfrac{\partial \eta}{\partial \rho} = -n\left(\dfrac{\rho}{\rho_K}\right)^{n-1}\dfrac{1}{\rho_K} \\ \dfrac{\partial \eta}{\partial \varphi} = 0 \\ \dfrac{\partial \eta}{\partial z} = \dfrac{1}{z_K} \end{cases}.$$

(23)

These expressions in Equation (23) should be substituted in the projection of force $\vec{N}$, a normal reaction to the constraint surface in Equation (18).

From the last equation of the system (18) we find the Lagrange multiplier

$$\lambda = \frac{1}{\left(\frac{\partial \eta}{\partial z}\right)}\left(\ddot{z} + g + Nf\frac{\dot{z}}{v}\right).$$

(24)

Since, when moving along the surface, only two cylindrical coordinates are independent, then for step-by-step numerical integration, we take the first two differential Equations (18) as the basis and calculate the coordinate z and its time derivatives through coupling Equation (22):

$$z = z_K a_0 + z_K\left(\frac{\rho}{\rho_K}\right)^n$$

(25)

$$\dot{z} = nz_K \frac{1}{\rho_K}\left(\frac{\rho}{\rho_K}\right)^{n-1}\dot{\rho} \tag{26}$$

$$\ddot{z} = nz_K \frac{1}{\rho_K}(n-1)\left(\frac{\rho}{\rho_K}\right)^{n-2}\frac{\dot{\rho}^2}{\rho_K} + nz_K \frac{1}{\rho_K}\left(\frac{\rho}{\rho_K}\right)^{n-1}\ddot{\rho}. \tag{27}$$

In Equations (24)–(27), the coordinate values ρ and φ, as well as their time derivatives $\dot{\rho}$, $\dot{\varphi}$ at the next integration step, are assumed to be equal to their value at the end of the previous integration step. The procedure of numerical integration of the system of Equation (18) was carried out by the method of averaged acceleration [45], using the original algorithmic program for "Windows" coded in C# under MS Visual Studio.

3.2. Results and Discussion

The results of the analysis of the kinematic parameters of the motion and trajectory of the particle, which initiate motion, with no initial linear velocity v_0 and with an angular velocity ω_0 equal in magnitude to the angular velocity of rotation ω_e, for different curvilinear surfaces (Figure 4) of the accelerator, are presented in Figure 5. It shows that the use of a concave curvilinear surface ($n_a = 2$) allowed to double the value of the acquired velocity v_{abs} in comparison with the flat surface ($n_a = 0$) and to increase it 1.7 times compared to a conical surface ($n_a = -0.1$) at a descent from a rotating surface, which is a consequence of the time spent by the particle on the surface, for example, when $n_a = 2$, the time was $t = 2.05$ s, when $n_a = 0$, the time was $t = 0.84$ s, and when $n_a = -0.1$, the time was $t = 0.38$ s. When the shape of the rotating surface changed from concave $n_a = 2$ to convex $n_a = -0.1$, the total length of the path traveled by the particle—that is, its trajectory—decreased.

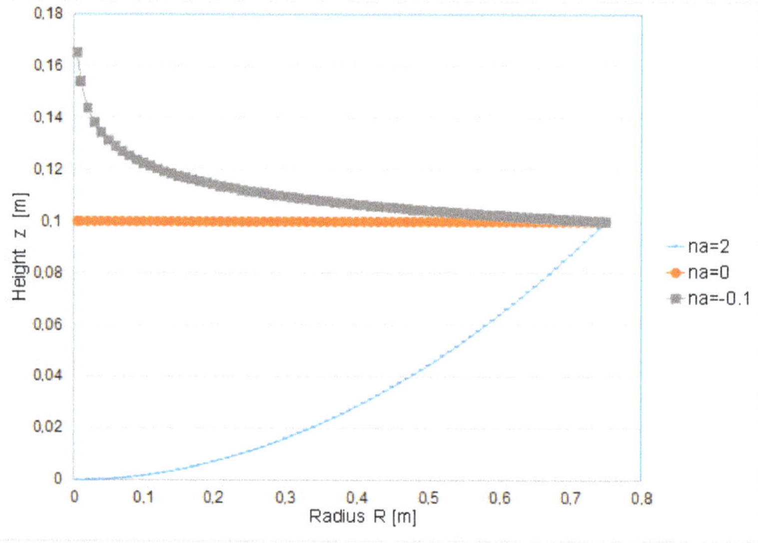

Figure 4. Variants of generates of the studied curvilinear surfaces of the accelerator.

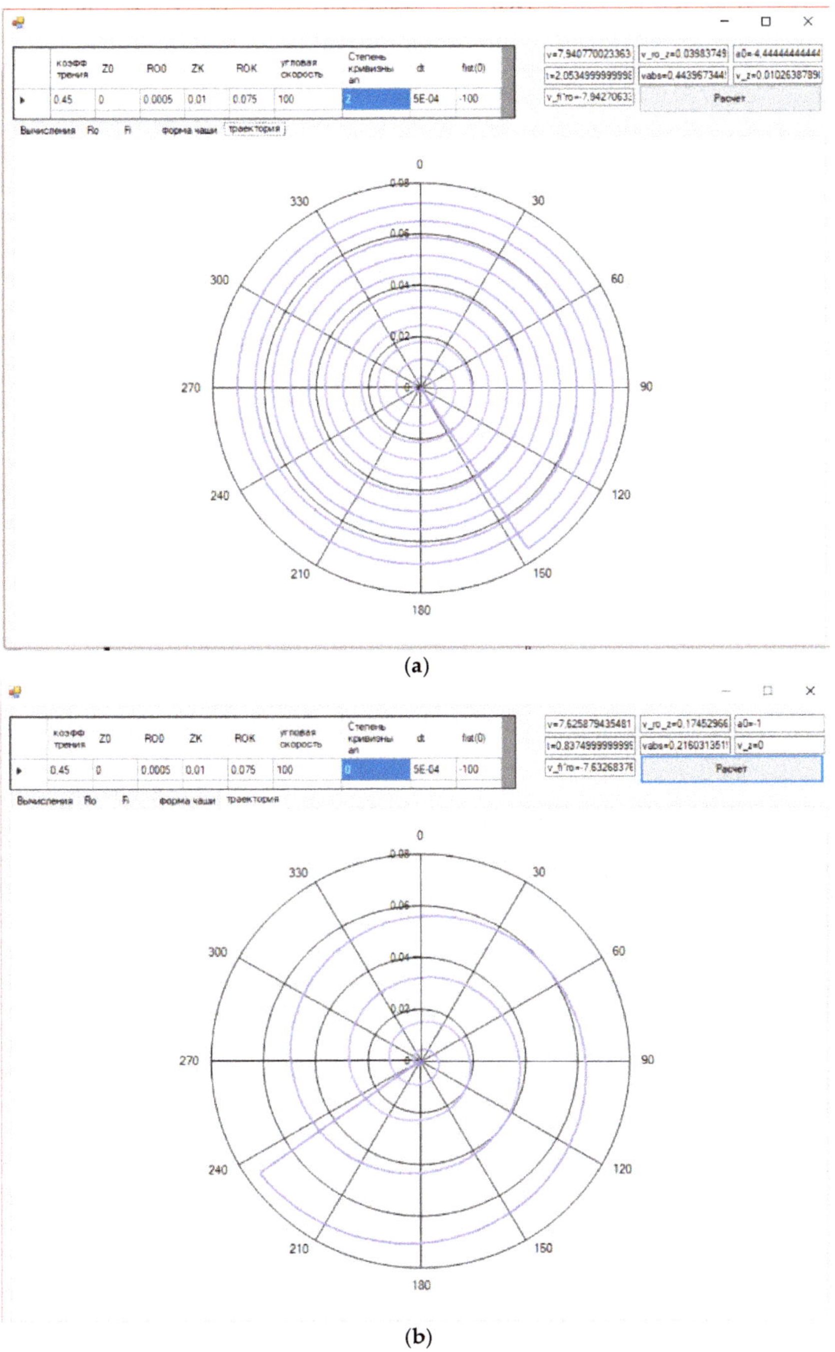

(**a**)

(**b**)

Figure 5. *Cont.*

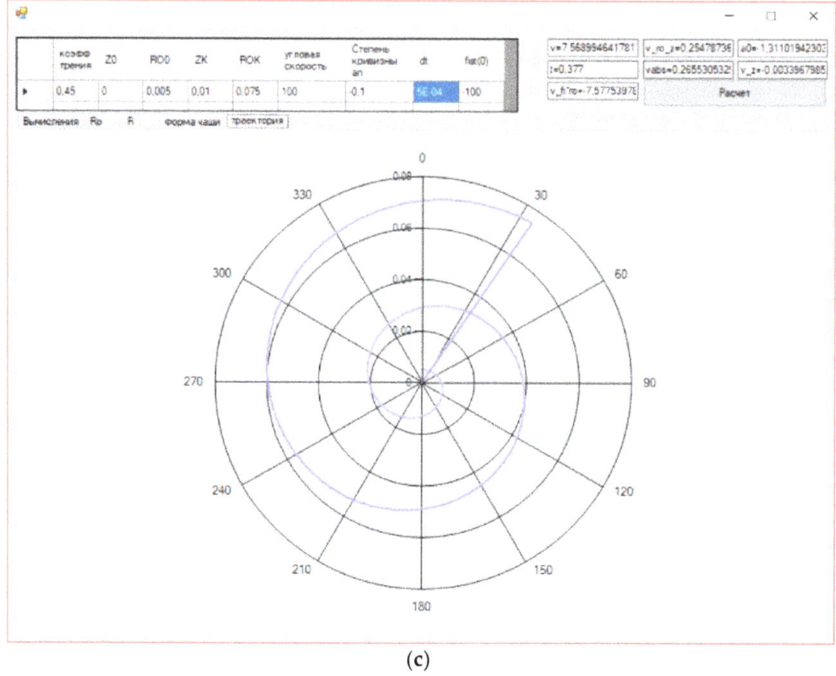

(**c**)

Figure 5. Data for analyzing the velocity and trajectory of the particle motion, which is introduced with no initial linear velocity and with angular velocity equal in magnitude to the angular velocity of rotation for different curvilinear surfaces of the accelerator: (**a**) $n_a = 2$ (concave); (**b**) $n_a = 0$ (straight line); (**c**) $n_a = -0.1$ (conical curvilinear).

The calculations were carried out for different variable characteristics of the surface, initial indicators of kinematic parameters, and friction coefficient. Figure 6a presents an example of tabular data with the results of calculations and diagrams of changes in the motion trajectory, velocity, and acceleration of a particle along the cylindrical axes ρ and z.

Analysis of the diagrams presented in Figure 6b, c made it possible to note that the particle reached its maximum velocity in 0.313 s, being at a distance of $\rho = 0.0092$ m from the axis of rotation. At the same time, the subsequent changes in velocity, down to its descent from the concave surface, had a damped harmonic nature.

Thus, curvilinear surfaces of concave nature combining conical and concave surfaces are of great interest for further research. The trajectories of the particles and their kinematic indicators can be used to determine the relationship between the design factors and the process parameters of centrifugal choppers, such as the time when the particle is on a dispensing bowl, the descent velocity of the particle from the bowl, and the shape of accelerating blades.

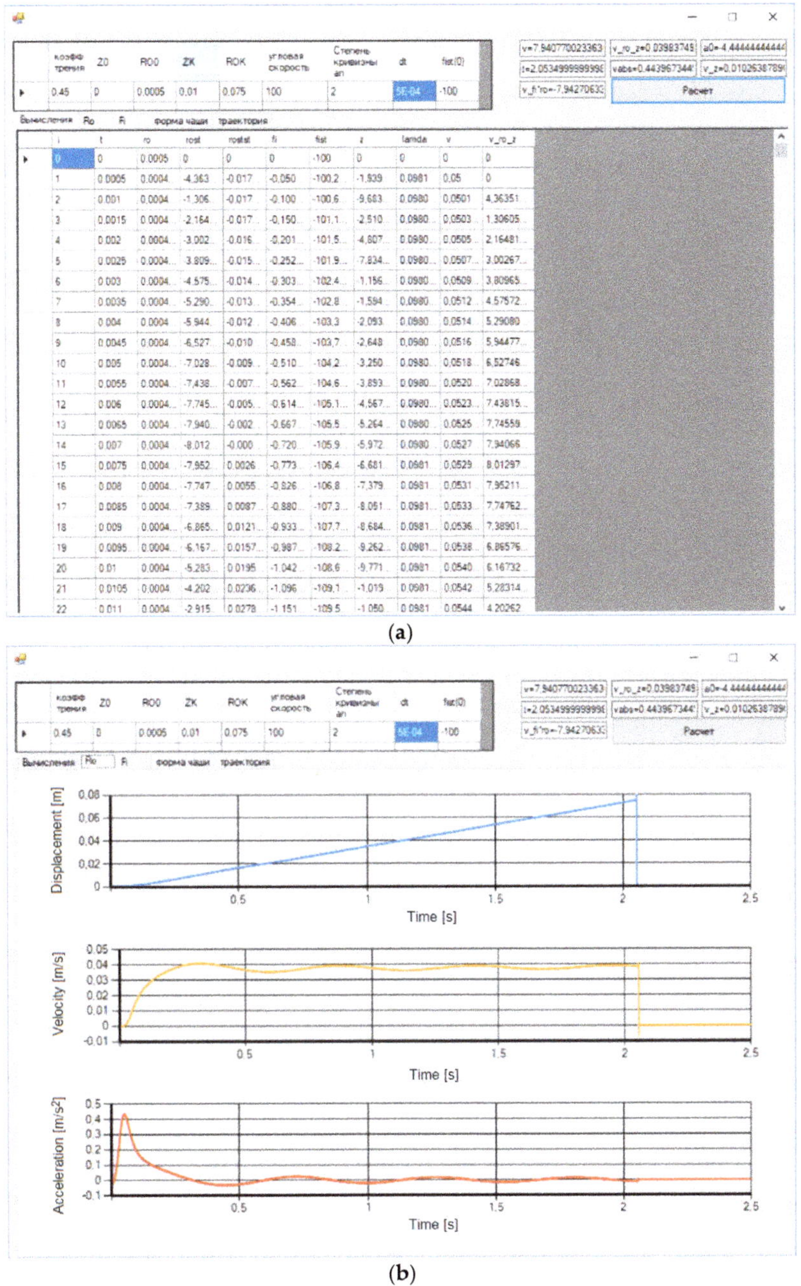

Figure 6. *Cont.*

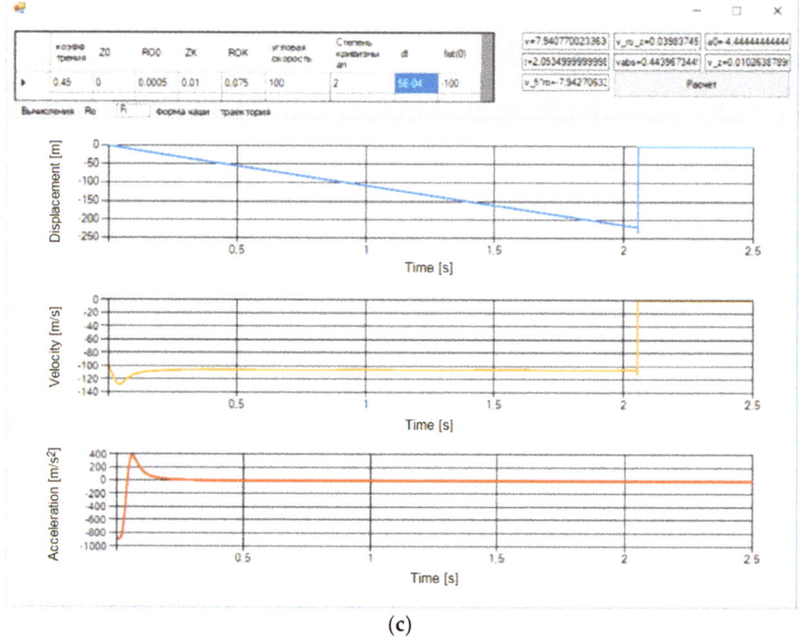

(c)

Figure 6. Example of the results of calculations of the particle motion trajectory and kinematic indicators for $n_a = 2$; (**a**) tabular data; (**b**) diagrams of particle displacement, velocity, and acceleration along the axis ρ; (**c**) diagrams of particle displacement, velocity, and acceleration along the axis z.

The ability to achieve the assumed performance parameters of the production process planned for implementation depends on the degree of reliability of the machines and technological devices included in the designed system [2]. Based on previous research [34], it can be confirmed that with the optimal values of the technological process (efficiency, time of a given operation—mixing or mining, coefficient of friction, and particle movement), it is possible to reduce energy consumption. Data on the possibility of improving the operation of production processes of various enterprises in Germany are presented by Steinhofel and others [46]. As we know, many factors influence the production process, for example, technical, economic, and social factors. Due to the above reasons, obtaining such large benefits in the whole production process, where various technological machines operate (technological operations), is difficult; nevertheless, it enables the introduction of sustainable production processes. The conducted simulation tests show possibilities of improving process efficiency and shortening the operation times, which results in economic benefits. The lower energy consumption and the increase in the efficiency of technological processes will also lead to the generation of less post-production waste (ecological benefits), which is in line with the principles of sustainable development.

4. Conclusions

In this study, a closed system of equations was obtained with a coupling equation (of constraints) that allows modeling a particle motion on a rotating surface; when numerically integrating them, it is possible to evaluate the use of a particular accelerator surface of a centrifugal rotary chopper machine to achieve the desired results in terms of downsizing, increasing productivity, or improving the quality of shredding. The use of numerical integration methods allows to consider the influence of physical and geometrical processes and kinematic parameters on the output result, for example, traveling time and velocity in cylindrical coordinates. Thus, the simulation of the particle motion along an accelerator allows choosing rational geometrical dimensions of the chopper machine and its optimal operating practice.

Supplementary Materials: The Supplementary Materials are available online at http://www.mdpi.com/2071-1050/11/18/4873/s1.

Author Contributions: Conceptualization, J.C., P.A.S. and I.I.I.; Data curation, A.V.A.; Formal analysis, A.M., A.V.A., A.Y.I. and I.I.I.; Methodology, A.V.A., P.A.S. and A.Y.I.; Resources, J.C.; Software, P.A.S.; Writing—original draft, J.C., P.A.S. and A.Y.I.; Writing—review & editing, J.C. and I.I.I.

Funding: This research received no external funding.

Acknowledgments: We are thankful to the Editor and the reviewers for their valuable comments and detailed suggestions to improve the paper.

Conflicts of Interest: The authors declare no conflict of interest.

References

1. Linke, B.S.; Corman, G.J.; Dornfeld, D.A.; Tonissen, S. Sustainability indicators for discrete manufacturing processes applied to grinding technology. *J. Manuf. Syst.* **2013**, *32*, 556–563. [CrossRef]
2. Gola, A. Reliability analysis of reconfigurable manufacturing system structures using computer simulation methods. *Eksploat. I Niezawodn. Maint. Reliab.* **2019**, *21*, 90–102. [CrossRef]
3. Wegener, K.; Bleicher, F.; Krajnik, P.; Hoffmeister, H.W.; Brecher, C. Recent developments in grinding machines. *Cirp Ann. Manuf. Technol.* **2017**, *66*, 779–802. [CrossRef]
4. Zhou, Z.; Dou, Y.; Sun, J.; Jiang, J.; Tan, Y. Sustainable Production Line Evaluation Based on Evidential Reasoning. *Sustainability* **2017**, *9*, 1811. [CrossRef]
5. Shankar, K.M.; Kumar, P.U.; Kannan, D. Analyzing the Drivers of Advanced Sustainable Manufacturing System Using AHP Approach. *Sustainability* **2016**, *8*, 824. [CrossRef]
6. Wu, G.; Duan, K.; Zuo, J.; Zhao, X.; Tang, D. Integrated Sustainability Assessment of Public Rental Housing Community Based on a Hybrid Method of AHP-Entropy Weight and Cloud Model Sustainability. *Sustainability* **2017**, *9*, 603.
7. Bardelli, M.; Cravero, C.; Marini, M.; Marsano, D.; Milingi, O. Numerical Investigation of Impeller-Vaned Diffuser Interaction in a Centrifugal Compressor. *Appl. Sci.* **2019**, *9*, 1619. [CrossRef]
8. Faria, P.M.C.; Rajamain, R.K.; Tavares, L.M. Optimization of Solids Concentration in Iron Ore Ball Milling through Modeling and Simulation. *Minerals* **2019**, *9*, 366. [CrossRef]
9. Koltai, T.; Lozano, S.; Uzonyi-Kecskés, J.; Moreno, P. Evaluation of the results of a production simulation game using a dynamic DEA approach. *Comput. Ind. Eng.* **2017**, *105*, 1–11. [CrossRef]
10. Ahmad, S.; Wong, K.Y. Development of weighted triple-bottom line sustainability indicators for the Malaysian food manufacturing industry using the Delphi method. *J. Clean. Prod.* **2019**, *229*, 1167–1182. [CrossRef]
11. Joung, C.B.; Carrell, J.; Sarkar, P.; Feng, S.C. Categorization of indicators for sustainable manufacturing. *Ecol. Indic.* **2013**, *24*, 148–157. [CrossRef]
12. Krajn, D.; Glavic, P. Indicators of sustainable production. *Clean Technol. Environ. Policy* **2003**, *5*, 279–288. [CrossRef]
13. Beekaroo, D.; Callychurn, D.S.; Hurreeram, D.K. Developing a sustainability index for Mauritian manufacturing companies. *Ecol. Indic.* **2019**, *96*, 250–257. [CrossRef]
14. Filleti, R.A.P.; Silva, D.A.L.; da Silva, E.J.; Ometto, A.R. Productive and environmental performance indicators analysis by a combined LCA hybrid model and real-time manufacturing process monitoring: A grinding unit process application. *J. Clean. Prod.* **2017**, *161*, 510–523. [CrossRef]
15. Machado, C.G.; Despeisse, M.; Winroth, M.; Ribeiro da Silva, E.H.D. Additive manufacturing from the sustainability perspective: Proposal for a self-assessment tool. In Proceedings of the 52nd CIRP Conference on Manufacturing Systems, Procedia CIRP, Ljubljana, Slovenia, 12–14 June 2019; Volume 81, pp. 482–487.
16. Zarte, M.; Pechmann, A.; Nunes, I.L. Decision support systems for sustainable manufacturing surrounding the product and production life cycle—A literature review. *J. Clean. Prod.* **2019**, *219*, 336–349. [CrossRef]
17. Piasecka, I.; Tomporowski, A.; Flizikowski, J.; Kruszelnicka, W.; Kasner, R.; Mroziński, A. Life Cycle Analysis of Ecological Impacts of an Offshore and a Land-Based Wind Power Plant. *Appl. Sci.* **2019**, *9*, 231. [CrossRef]

18. Wang, R.; Wu, Y.; Ke, W.; Zhang, S.; Zhou, B.; Hao, J. Can propulsion and fuel diversity for the bus fleet achieve the win–win strategy of energy conservation and environmental protection? *Appl. Energy* **2015**, *147*, 92–103. [CrossRef]

19. Ventura, J.A.; Kweon, S.J.; Hwang, S.W.; Tormay, M.; Li, C. Energy policy considerations in the design of an alternative-fuel refueling infrastructure to reduce GHG emissions on a transportation network. *Energy Policy* **2017**, *111*, 427–439. [CrossRef]

20. Arzoumanidis, I.; Salomone, R.; Petti, L.; Mondello, G.; Raggi, A. Is there a simplified LCA tool suitable for the agri-food industry? An assessment of selected tools. *J. Clean. Prod.* **2017**, *149*, 406–425. [CrossRef]

21. Tassielli, G.; Renzulli, P.A.; Mousavi-Avval, S.H.; Notarnicola, B. Quantifying life cycle inventories of agricultural field operations by considering different operational parameters. *Int. J. Life Cycle Assess.* **2019**, *24*, 1075–1092. [CrossRef]

22. Helleno, A.L.; Isaias de Moraes, A.J.; Simon, A.T. Integrating sustainability indicators and Lean Manufacturing to assess manufacturing processes: Application case studies in Brazilian industry. *J. Clean. Prod.* **2017**, *153*, 405–416. [CrossRef]

23. Skrucany, T.; Ponicky, J.; Kendra, M.; Gnap, J. Comparison of railway and road passenger transport in energy consumption and GHG production. In Proceedings of the Third International Conference on Traffic and Transport Engineering (ICTTE), Lucerne, Switzerland, 6–10 July 2016; pp. 744–749.

24. Skrucany, T.; Semanova, S.; Figlus, T.; Sarkan, B.; Gnap, J. Energy intensity and GHG production of chosen propulsions used in road transport. *Commun. Sci. Lett. Univ. Zilina* **2017**, *19*, 3–9.

25. Tucki, K.; Mruk, R.; Orynych, O.; Wasiak, A.; Botwinska, K.; Gola, A. Simulation of the Operation of a Spark Ignition Engine Fueled with Various Biofuels and Its Contribution to Technology Management. *Sustainability* **2019**, *11*, 2799. [CrossRef]

26. Akbar, M.; Irohara, T. Scheduling for sustainable manufacturing: A review. *J. Clean. Prod.* **2018**, *205*, 866–883. [CrossRef]

27. Singh, R.K.; Murty, H.R.; Gupta, S.K.; Dikshit, A.K. An overview of sustainability assessment methodologies. *Ecol. Indic.* **2012**, *15*, 281–299. [CrossRef]

28. Leonesio, M.; Bianchi, G.; Cau, N. Design criteria for grinding machine dynamic stability. *Procedia Cirp* **2018**, *78*, 382–387. [CrossRef]

29. Möhring, H.C.; Brecher, C.; Abele, E.; Fleischer, J.; Bleicher, F. Materials in machine tool structures. *Cirp Ann. Manuf. Technol.* **2015**, *64*, 725–748. [CrossRef]

30. Madej, O.; Kruszelnicka, W.; Tomporowski, A. Wyznaczanie procesowych charakterystyk wielokrawędziowego rozdrabniania ziaren kukurydzy. *Inżynieria I Apar. Chem.* **2016**, *4*, 144–145.

31. Stadnyk, I.; Vitenko, T.; Droździel, P.; Derkach, A. Simulation of components mixing in order to determine rational parameters of working bodies. *Adv. Sci. Technol. Res. J.* **2016**, *10*, 30–138. [CrossRef]

32. Tong, J.; Xu, S.; Chen, D.; Li, M. Design of a Bionic Blade for Vegetable Chopper. *J. Bionic Eng.* **2017**, *14*, 163–171. [CrossRef]

33. Flizikowski, J.; Topolinski, T.; Opielak, M.; Tomporowski, A.; Mrozinski, A. Research and analysis of operating characteristics of energetic biomass micronizer. *Eksploat. I Niezawodn. Maintanance Reliab.* **2015**, *17*, 19–26. [CrossRef]

34. Marczuk, A.; Misztal, W.; Savinykh, P.; Turubanov, N.; Isupov, A.; Zyryanov, D. Improving efficiency of horizontal ribbon mixer by optimizing its constructional and operational parameters. *Eksploat. I Niezawodn. –Maint. Reliab.* **2019**, *21*, 220–225. [CrossRef]

35. Rocha, A.G.; Montanhini, R.N.; Dilkin, P.; Tamiosso, C.D.; Mallmann, C.A. Comparison of different indicators for the evaluation of feed mixing efficiency. *Anim. Feed Sci. Technol.* **2015**, *209*, 249–256. [CrossRef]

36. Tropp, M.; Tomasikova, M.; Bastovansky, R.; Krzywonos, L.; Brumercik, F. Concept of deep drawing mechatronic system working in extreme conditions. *Procedia Eng.* **2017**, *192*, 893–898. [CrossRef]

37. Vitenko, T.; Droździel, P.; Rudawska, A. Industrial usage of hydrodynamic cavitation device. *Adv. Sci. Technol. Res. J.* **2018**, *12*, 158–167. [CrossRef]

38. Xiao, X.; Tan, Y.; Zhang, H.; Jiang, S.; Wang, J.; Deng, R.; Cao, G.; Wu, B. Numerical investigation on the effect of the particle feeding order on the degree of mixing using DEM. *Procedia Eng.* **2015**, *102*, 1850–1856. [CrossRef]

39. Tomporowski, A.; Flizikowski, J.; Al-Zubledy, A. An active monitoring of biomaterials grinding. *Przem. Chem.* **2018**, *97*, 250–257.

40. Camargo, I.L.; Erbereli, R.; Lovo, J.F.P.; Fortulan, C.A. Planetary Mill with Friction Wheels Transmission Aided by an Additional Degree of Freedom. *Machines* **2019**, *7*, 33. [CrossRef]
41. Druzhynin, R.A. Improving the Working Process of an Impact Centrifugal Grinding Machine. Ph.D. Thesis, Voronezh State University, Voronezh, Russia, 2014.
42. Stepanovich, S.N. Centrifugal Rotary Grain Grinders. Ph.D. Thesis, Chelyabinsk State Agroengineering University, Chelyabinsk, Russia, 2008.
43. Cho, J.; Zhu, Y.; Lewkowicz, K.; Lee, S.H.; Bergman, T.; Chaudhuri, B. Solving granular segregation problems using a biaxial rotary mixer. *Chem. Eng. Process.* **2012**, *57–58*, 42–50. [CrossRef]
44. Nikitin, N.N. *The Course of Theoretical Mechanics*; Higher School: Moscow, Russia, 1990; p. 607.
45. Timoshenko, S.P.; Young, D.H.; Weaver, W. *Vibration Problems in Engineering*; Trans. from English; D. Van Nostrand Company INC.: New York, NY, USA, 1985; p. 472.
46. Steinhöfel, E.; Galeitzke, M.; Kohl, H.; Orth, R. Sustainability Reporting in German Manufacturing SMEs. 16th Global Conference on Sustainable Manufacturing—Sustainable Manufacturing for Global Circular Economy. *Proc. Manuf.* **2019**, *33*, 610–617.

Article

Research on the Work Process of a Station for Preparing Forage

Andrzej Marczuk [1], Wojciech Misztal [1,*], Sergey Bulatov [2], Vladimir Nechayev [3] and Petr Savinykh [4]

[1] Department of Agricultural, Forestry and Transport Machines, Faculty of Production Engineering, University of Life Sciences in Lublin, 28 Gleboka Street, 20-612 Lublin, Poland; andrzej.marczuk@up.lublin.pl

[2] Department of Technical Service, SBEI HE Nizhniy Novgorod State Engineering and Economic University, 22a Oktyabrskaya Street, Knyaginino 606340, Russia; bulatov_sergey_urevich@mail.ru

[3] SBEI HE Nizhniy Novgorod State Engineering and Economic University, 22a Oktyabrskaya Street, Knyaginino 606340, Russia; nechaev-v@list.ru

[4] Federal State Budget Scientific Institution. Federal Agricultural Research Centre of the North-East named after Rudnitskiy N.V. 166a Lenin Street, Kirov 610007, Russia; peter.savinyh@mail.ru

* Correspondence: wojciech.misztal@up.lublin.pl

Received: 5 December 2019; Accepted: 31 January 2020; Published: 2 February 2020

Abstract: Forage from grain plays a special role in animal nutrition because it constitutes feed with a high content of readily available carbohydrates. Unfortunately, the equipment used to prepare forage is often manufactured without the necessary justification and confirmation of the declared sizes and indicators of the work process. This forms the basis for our theoretical and experimental studies. Research has been carried out to provide justification of the design and operating parameters of the patented station for producing forage from cereal crops. This article describes the technology for preparing forage from grain and provides a detailed description of the station used and the principle of its operation. During the experiments, we studied the influence of the angle α of setting the grid-work (plate) and the distance S from the nozzle to the grid-work on the quality of forage. Qualitative, quantitative, and energy indicators have been evaluated using up-to-date measuring instruments and equipment. The method is described, and the studied factors and evaluation criteria for the preparation of forage from grain are indicated. The forage quality results are presented, as determined by the content of whole grains in it via the residue on a sieve with a sieve size of 3 mm when preparing it with a different combination of the studied factors. The analysis of the energy consumption results of the process of preparing forage from grain under various operating conditions of the plant is shown. As a result, the optimal location parameters of the passive grinder have been found, allowing to obtain high-quality forage with minimal power consumption of the electric motor. A grid-work should be used as a grinder. Its installation angle should be 30°, and the distance between the grid-work and the nozzle should be 205 mm. With this combination of parameters, the specific energy consumption is minimal and amounts to 41.5 W·h/L.

Keywords: forage from grain; cereal grain; energy consumption

1. Introduction

Obtaining high performance indicators for farm animals is determined by a scientifically validated diet and feeding regime, which must include properly processed cereal grains [1–6]. Preparation of feed directly on the farm will give a positive result if a producer has his/her own grain raw materials. With this knowledge, agricultural producers are trying to find feed preparation equipment that is optimal in terms of quality and quantity, separately or as part of a line, depending on the volume of production [7]. These are mainly crushers or flatteners for producing dry concentrates, which are the most known and proven [8–13]. However, studies show that processing carried out using this type of equipment does not ensure full utilization of the energy value of the grain [11–17]. As previous studies [16,18–22] show, it is possible to obtain feed in the form of forage from grain with a high content of readily available carbohydrates (this kind of feed has great value, primarily for dairy cows). Unfortunately, corresponding equipment is manufactured for such needs without the necessary justification and confirmation of the declared sizes and indicators of the work process [21–24]. Recognizing the practical importance of addressing these issues, theoretical and experimental studies have been carried out to provide justification for the design and operating parameters of the plant for producing forage from cereal crops on the basis of an agreement between the leading manufacturer of animal feed equipment "Doza-Agro" LLC and the Nizhniy Novgorod State Engineering and Economic University (NNSEEU). The article presents the data reflecting the key research insights obtained in the framework of R&D.

2. Materials and Methods

To study the process of preparing forage from grain, a laboratory-scale plant has been developed consisting of a frame on wheels, a tank with a built-on passive grinder, a 1SM65-50-160/2 centrifugal pump and a NGD-1.1 disperser, material pipelines, and a control cabinet (Figure 1). The measuring station operates as follows. Water is poured into the tank based on the required volume of prepared forage at the outlet, which is 2:1 with respect to the grain. The amount of water poured was recorded by a water meter SVK 15-3-8 (Figure 2). Next, the centrifugal pump is turned on, and the heating process to 30 °C takes place. When this temperature is reached, the tank is uniformly filled with pre-processed and cleaned grain at the given water:grain ratio. Next, enzymes are added. Through special inspection ports in the built-in grinder (chopper), the uniformity and homogeneity of the working mass are visually monitored, and, upon reaching a temperature of 60 °C, the pump is switched off. The prepared mass is allowed to infuse for one hour for the more efficient operation of enzymes. The result is a high-carbohydrate forage from grain recommended for feeding farm animals as a single feed or as a part of feed mix.

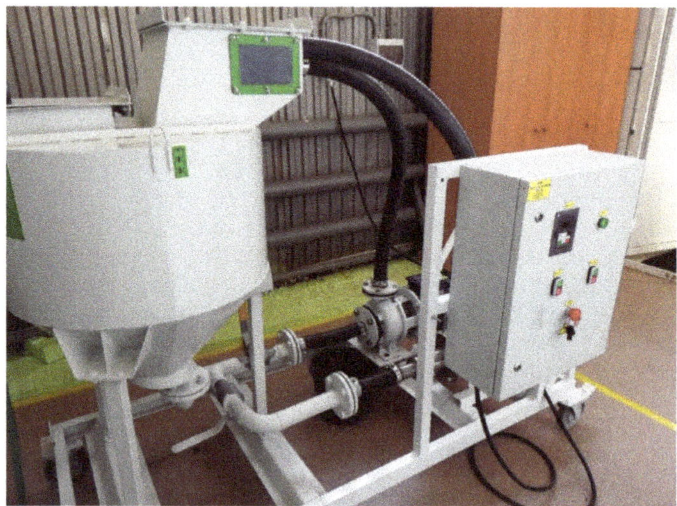

Figure 1. The measuring station for preparing forage from grain.

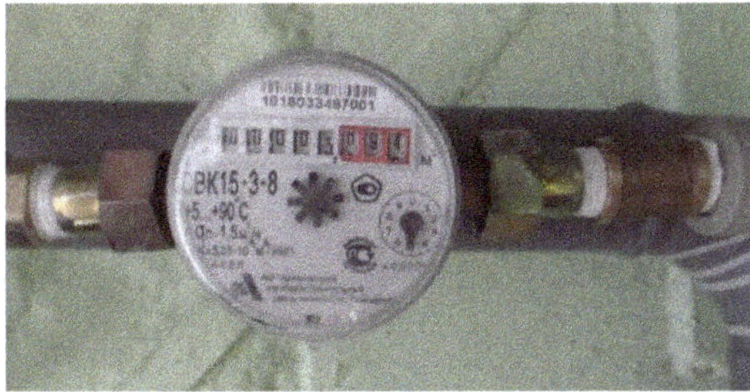

Figure 2. Water meter SVK 15-3-8 used for recording the amount of water poured into the tank of the measuring station.

While the plant pump is operating, the grain is crushed by its blades. To accelerate the process of grain grinding and to obtain a more even forage composition, it was proposed to install a passive grinder in the form of a grid-work (grill) or a plate in the upper part of the tank (Figure 3). During the experiments, the influence of the angle α of setting the grid-work (plate) and the distance S from the nozzle to the grid-work (Figure 4) on the quality of forage have been studied. It was modified by moving the grid-work (plate) between the holes of the fixture of the laboratory-scale plant (Figure 5). The distance S between the grid-work (plate) and the nozzle was adjusted by moving the grid-work (plate) along the slots in the cover (Figure 6).

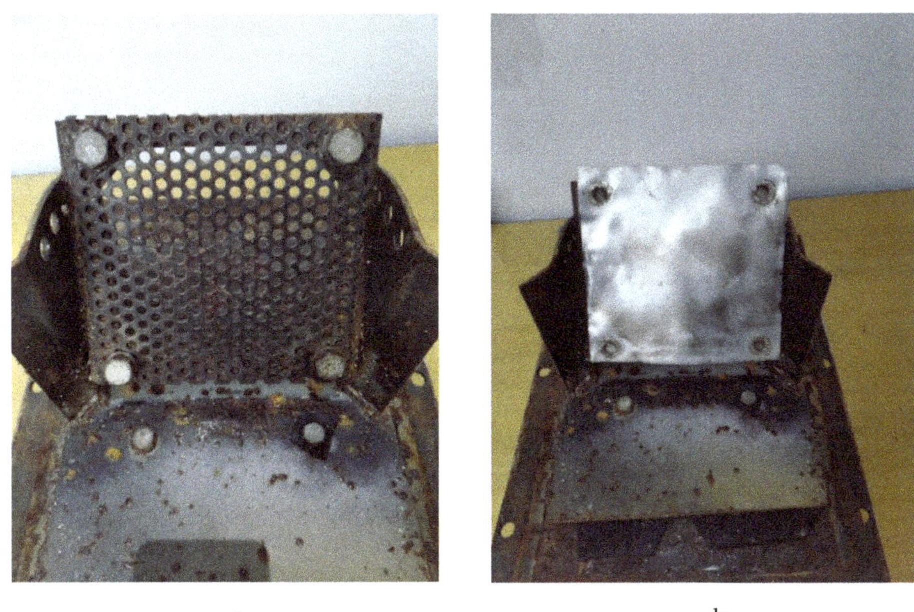

a   b

Figure 3. Passive grain grinder made in the form of (**a**) grid-work and (**b**) plate.

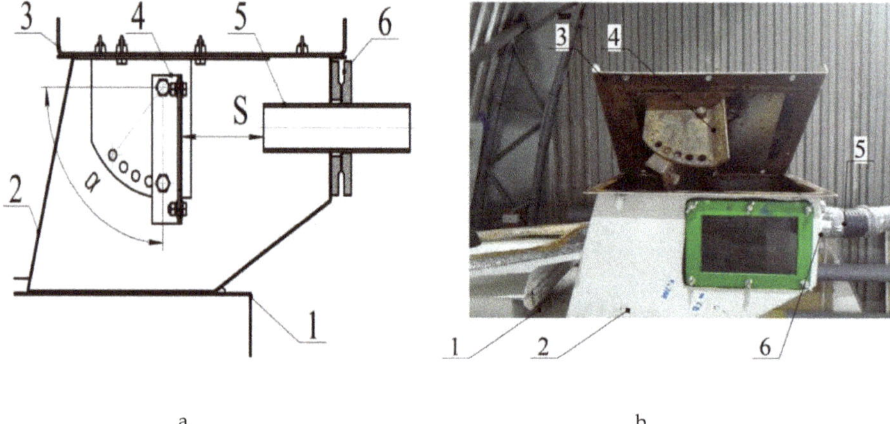

a   b

Figure 4. A device for adjusting the inclination angle of the grid-work (plate) and its distance to the nozzle. (**a**) Diagram; (**b**) General vie: (1) tank with a built-in passive grinder, (2) case of the passive grinder, (3) cover of the passive grinder, (4) grid-work (plate), (5) nozzle, (6) nozzle flange.

Figure 5. The system for adjusting the inclination angle of the grid-work (plate).

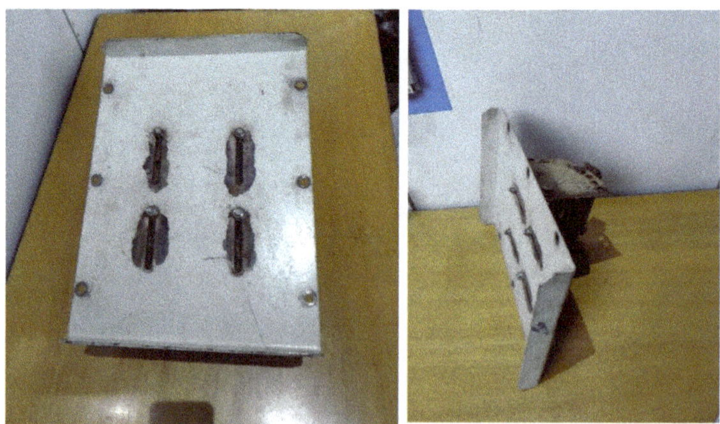

Figure 6. Cover with slots for adjusting the clearance S between the grid-work (plate) and the nozzle.

To determine the performance of the proposed plant under scientific laboratory conditions, a technique was developed to assess the influence of design and operating parameters on the work process of preparing forage from grain. Wheat was used as the source raw material. To carry out the fermentation process, the multi-enzyme composition MEC-AC-3 produced by Vostok LLC, Kirov Region (Russian Federation), was used. The volume of forage prepared was 50 L.

Routine of the experiment was as follows. First, 17 kg of grain was put into 33 L of water pre-heated to a temperature of 30 °C, and 80 g of enzyme was added, previously weighed with the accuracy of up to 0.01 g using the Biomer VK-300.01 laboratory balance (Novosibirsk) (Figure 7). From that moment, the process timing began using the HUAWEI P20 lite stopwatch (China). Every 7 min 30 s before the working mass temperature reached 60 °C, a test batch was taken from the tank: the tank cover was opened, and a 1 L volume bucket was placed under a stream and filled up. Within the same time intervals, in addition to recording the temperature values, the power consumed by the pump electric motor was recorded using the Mastech MS2203 clamp meter (Pittsburg U.S.) (Figure 8). After that, the feed sample was poured onto a sieve with a sieve size of 0.5 mm. Free water was removed by vigorous stirring with a brush. The resulting mass was poured into a plastic bag, which was assigned an information tag (Figure 9). Three to five packets with samples were obtained before the temperature reached 60 °C. After that, the pump was turned off, and one hour was timed until the forage from grain was completely prepared.

Figure 7. Weighing the multi-enzyme composition MEC-AC-3 using the VK-300.01 laboratory balance.

Figure 8. Measuring the power consumption of the electric motor using the Mastech MS2203 clamp meter.

Figure 9. Bags with samples of forage from grain.

Temperature of the forage from grain (and water) was recorded using the TST81 sensor, the data from which was transmitted to the TL-11-250 temperature controller (Figure 10). During the experiment, the dynamics of temperature changes was recorded from the digital display of the temperature controller (Figure 10).

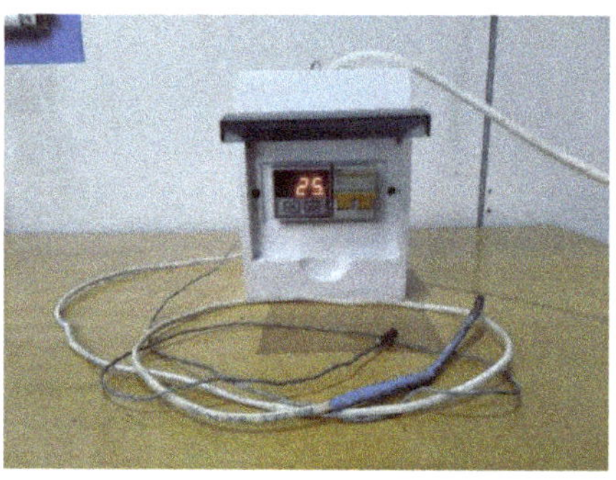

a   b

Figure 10. Temperature controller TL-11-250 with a temperature sensor TST81 (**a**) and recording the temperature of the forage from grain (**b**).

Next, a 100 g sample weight was selected from each bag and distributed on the sieve with a sieve size of 3 mm. After that, the sample was rinsed out in a bath with water. Next, the sieve residue was weighed, and the number of whole grains was counted.

An hour later, a control sample was taken from the premixed forage from grain. To determine the uniformity, screen sizing was collected using sieves of the following sizes: 2; 1.4, 1; 0.5 mm and the bottom. Then, 100 g of the sample weight was fed to the upper sieve, and then the plansifter (RL brand) was turned on for 5 min (Figure 11). First, the dry sieves were weighed, and their mass was determined. Then, they were weighed with the residue, and the difference was determined. For best data reliability, this analysis was performed in triplicate.

Figure 11. Laboratory plansifter RL brand with a set of sieves for determining the granulometric composition of the forage.

For energy estimation of the plant operation, specific energy consumption was used referred to a unit volume of the working mixture, w1:

$$w1 = \frac{W \cdot \tau}{V},$$ (1)

where

W is the average power consumption, kW;
τ is the process time, s;
V is the volume of water and working mixture, l.

3. Results

After the experiments, tables and graphs were constructed that characterize the change in the quality of forage from grain depending on the installation parameters (α and S) of the grid-work and the time t of the fermentation process (Tables 1 and 2, Figures 12–15). According to the results of experiments and calculations according to the formula (1), histogram 16 was constructed. Histogram 16a shows what full electric motor power is spent on preparing the forage. Figure 16b presents the change in the specific costs of electricity calculated using the expression (1). The constructed histograms show how the power consumption of the electric motor changes when the angle α of setting the grid-work (plate) and the distance S between the grid-work (plate) and the nozzle are modified.

Regardless of the angle α of setting the grid-work and the distance S, the number of whole grains decreases to 0 after 1350 s (22.5 min) of the plant operation (Figure 12). After 900 s (15 min) of the plant operation, the minimum number of whole grains is observed when the grid-work is installed according to options 1 and 3 (Figure 12). That is, according to the indicator m1 "the number of whole grains in the forage," we can recommend the grid-work installation scheme using options 1 and 3.

The number of whole grains reaches the value of 0 in 1350 s (in 22.5 min) when the plate is installed at an angle $\alpha = 90°$ at S = 140 mm (Figure 13). In other cases, in order to avoid the presence of whole grains in the forage, the operation time of the plant should be at least 30 min. The graph also shows that with decreasing angle α, the number of whole grains in the forage increases. This is explained by the fact that, when the angle α is decreased, the impact force of the grain on the plate goes down. From the analysis of the graphs presented in Figure 13, we can conclude that the best results of preparing forage using a plate are obtained when it is installed according to options 3, 4, and 5 (Figure 13).

The best performance in terms of the indicator "residue on the sieve with a sieve size of 3 mm" is also achieved when the plate is installed according to options 3, 4, and 5 (Figure 14). In these cases, the m2 indicator is close to zero after 1800–1850 sec of the plant operation. When the plate is installed at an angle of 30 and 45° for a time t = 1800 s, m2 equals 0.1% and 0.4%, respectively, and a zero value is reached only after 2080 and 1989 s, respectively. That is, in terms of the indicator "residue on a sieve with a diameter of 3 mm," it can be recommended to prepare forage when the plate is installed according to options 3 and 5.

The average power consumed by the engine while preparing forage did not exceed 5 kW. The lowest value was recorded when the grid-work was installed at an angle of 30° and at a distance of 205 mm from the nozzle—it was 4 kW. In general, the average engine power during the preparation of forage using a grid-work is lower than when using a plate. When using the grid-work, it did not exceed 4.5 kW, and when using the plate, it varied from 4.51 to 5 kW (Figure 16a).

Change in specific energy consumption required to prepare 1 L of forage from grain is shown in Figure 16b. In general, it can be seen that when a grid-work is used, the specific energy consumption is lower than when a plate is used. Based on the findings presented, we can recommend the use of a grid-work installed at an angle of 30° at a distance of 140 and 205 mm from the nozzle, or a plate installed at an angle of 90° at a distance of 140 mm from the nozzle. With this combination of parameters, the minimum specific energy consumption is observed, which amounts to 41.5–48.4 W·h/L. In other cases, there is an increase in this indicator to 50–52 W·h/L (Figure 15c).

Table 1. The change in the quality of forage from grain depending on the installation parameters (α and S) of the grid-work (where M is average, SD is standard deviation, and SKE is skewness factor).

Czas [s]	Number of whole grains in molasses when prepared using a grid-work (pcs.)			The residue on the sieve with a sieve size of 3 mm when prepared using a grid-work (%)		
	M	SD	SKE	M	SD	SKE
1—distance from the grid-work to the nozzle is 140 mm, grid-work inclination angle is 30 degrees						
450	916	5.18	−0.68	35.95	0.95	0.18
900	1	0.63	0	0.8	0.13	0
1350	0	0	0	0.1	0.05	0
1800	0	0	0	0	0	0
1857	0	0	0	0	0	0
2—distance from the grid-work to the nozzle is 140 mm, grid-work inclination angle is 45 degrees						
450	780	2.9	0.73	28.81	2.21	−0.01
900	2	0.63	0	1.17	0.07	−0.06
1350	0	0	0	0.43	0.1	0
1800	0	0	0	0	0	0
1984	0	0	0	0	0	0
3—distance from the grid-work to the nozzle is 205 mm, grid-work inclination angle is 30 degrees						
450	255	2.28	−0.91	7.45	0.69	−0.6
900	1	0.63	0	0.75	0.05	−1.1
1350	0	0	0	0.17	0.01	0
1800	0	0	0	0.01	0	0
1998	0	0	0	0	0	0
4—distance from the plate to the nozzle is 205 mm, grid-work inclination angle is 45 degrees						
450	653	3.52	−0.61	24.26	2.21	0.19
900	4	1.67	−1.15	1.55	0.18	−0.6
1350	0	0	0	0.07	0.01	0
1800	0	0	0	0.01	0	0
1878	0	0	0	0	0	0

Table 2. The change in the quality of forage from grain depending on the installation parameters (α and S) of the plate (where: M—average, SD—standard deviation, SKE—skewness factor).

Czas [s]	Number of whole grains in molasses when prepared using a plate [pcs.]			The residue on the sieve with a sieve size of 3 mm when prepared using a plate [%]		
	M	S	SKE	M	S	SKE
1—distance from the plate to the nozzle is 140 mm, plate inclination angle is 30 degrees						
450	615	68.77	0.56	51.21	4.86	−0.81
900	56	10.53	0.68	2.7	0.21	0.47
1350	20	3.16	0	0.59	0.04	0
1800	0	0	0	0.4	0.08	0
2080	0	0	0	0	0	0
2—distance from the plate to the nozzle is 140 mm, plate inclination angle is 45 degrees						
450	41	4	0.98	28.29	1.16	−0.44
900	9	2.28	1.21	4.9	0.3	−0.19
1350	2	0.63	0	0.67	0.06	0
1800	0	0	0	0.09	0.02	0
1989	0	0	0	0	0	0
3—distance from the plate to the nozzle is 140 mm, plate inclination angle is 90 degrees						
450	680	41.42	−2.13	44.53	3.55	−0.85
900	3	1.1	1.36	1.02	0.13	−0.1
1350	0	0	0	0.08	0.02	0
1800	0	0	0	0.04	0	0
1851	0	0	0	0.01	0	0
4—distance from the plate to the nozzle is 205 mm, plate inclination angle is 90 degrees						
450	400	4.69	−0.26	400	15.61	0.72
900	5	2.76	−0.09	5	1.67	−0.38
1350	1	5.18	0	1	0.46	0
1800	0	2.83	0	0	0	0
1852	0	6.42	0	0	0	0
5—distance from the plate to the nozzle is 305 mm, plate inclination angle is 90 degrees						
450	400	1.41	2	400	17.52	1.93
900	12	4.05	3	12	2.53	2.1
1350	1	4.69	4	1	0.38	−0.25
1800	0	3.85	5	0	0	0
1852	0	7.21	6	0	0	0

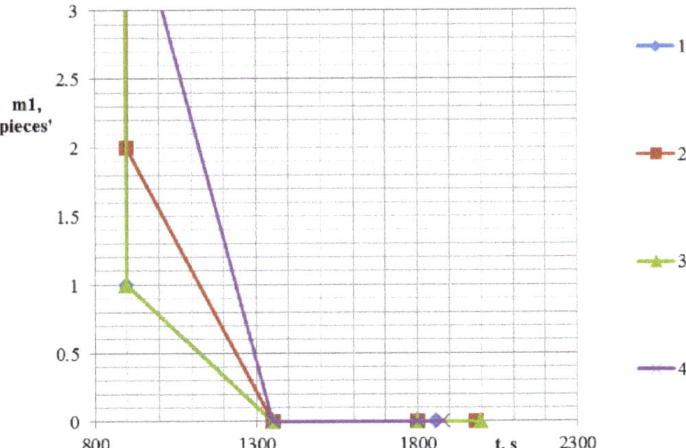

Figure 12. Number of whole grains in molasses when prepared using a grid-work: 1—the distance from the grate to the nozzle is 140 m the angle of inclination of the grate is 30 degrees; 2—the distance from the grate to the nozzle is 140 mm, the angle of inclination of the grate is 45 degrees; 3—distance between the grate and nozzle 205 mm, angle of inclination of the grate 45 degrees; 4—distance between the grate and nozzle 205 mm, angle of inclination 30 degrees.

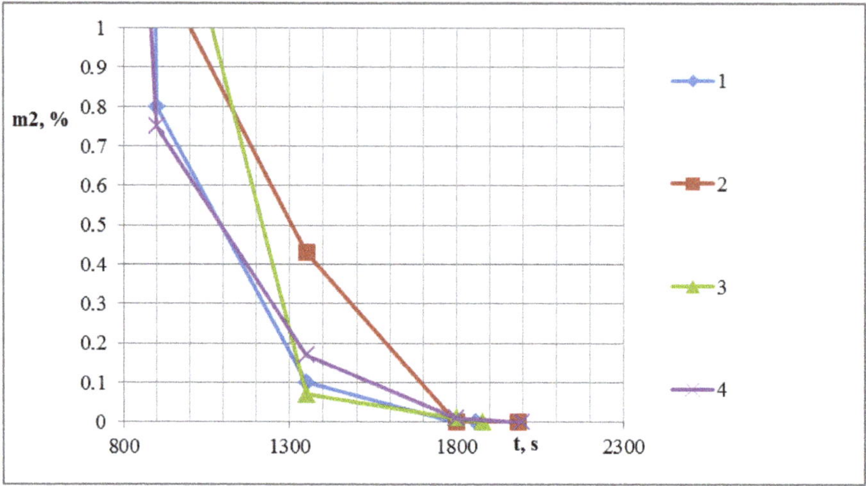

Figure 13. The residue on the sieve with a sieve size of 3 mm when preparing forage using a grid-work: 1—distance from the grid-work to the nozzle is 140 mm, grid-work inclination angle is 30 degrees; 2—distance from the grid-work to the nozzle is 140 mm, grid-work inclination angle is 45 degrees; 3—distance from the plate to the nozzle is 205 mm, grid-work inclination angle is 45 degrees; 4—distance from the grid-work to the nozzle is 205 mm, grid-work inclination angle is 30 degrees.

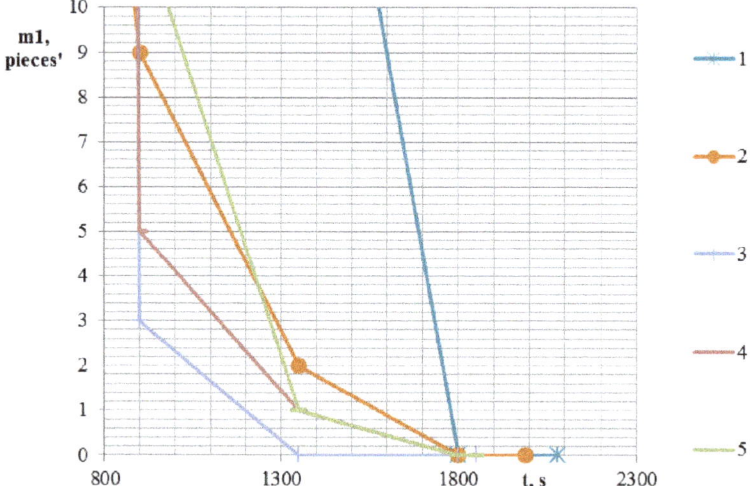

Figure 14. The number of whole grains in forage when it is prepared using a plate: 1—distance from the plate to the nozzle is 140 mm, plate inclination angle is 30 degrees; 2—distance from the plate to the nozzle is 140 mm, plate inclination angle is 45 degrees; 3—distance from the plate to the nozzle is 140 mm, plate inclination angle is 90 degrees; 4—distance from the plate to the nozzle is 205 mm, plate inclination angle is 90 degrees; 5—distance from the plate to the nozzle is 305 mm, plate inclination angle is 90 degrees.

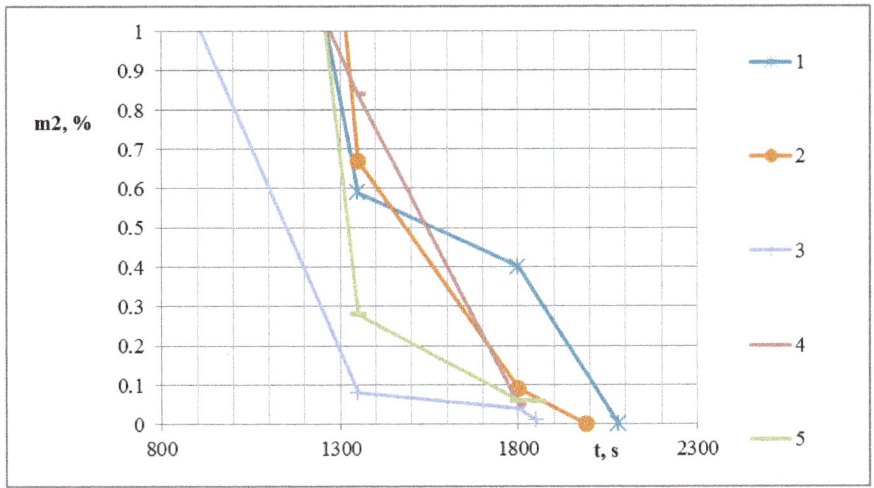

Figure 15. Residue on the sieve with a sieve size of 3 mm when preparing forage using a plate: 1—distance from the plate to the nozzle is 140 mm, plate inclination angle is 30 degrees; 2—distance from the plate to the nozzle is 140 mm, plate inclination angle is 45 degrees; 3—distance from the plate to the nozzle is 140 mm, plate inclination angle is 90 degrees; 4—distance from the plate to the nozzle is 205 mm, plate inclination angle is 90 degrees; 5—distance from the plate to the nozzle is 305 mm, plate inclination angle is 90 degrees.

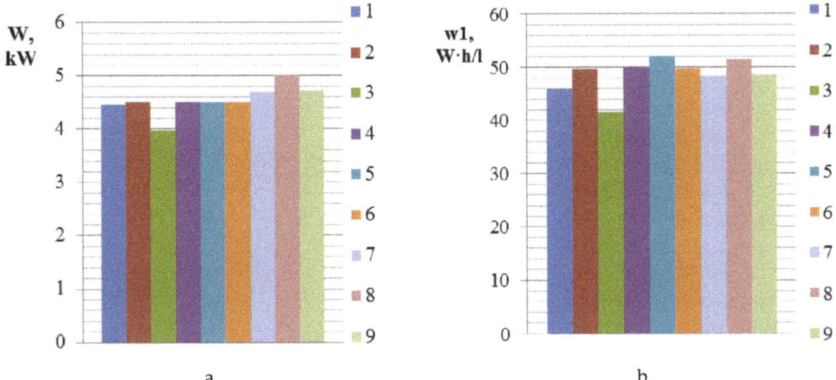

Figure 16. The influence of the studied parameters on: a—power consumption of the electric motor; b—specific energy consumption; 1—distance from the grid-work to the nozzle is 140 mm, grid-work inclination angle is 30 degrees; 2—distance from the grid-work to the nozzle is 140 mm, grid-work inclination angle is 45 degrees; 3—distance from the plate to the nozzle is 205 mm, grid-work inclination angle is 45 degrees; 4—distance from the grid-work to the nozzle is 205 mm, grid-work inclination angle is 30 degrees; 5—distance from the plate to the nozzle is 140 mm, plate inclination angle is 30 degrees; 6—distance from the plate to the nozzle is 140 mm, plate inclination angle is 45 degrees; 7—distance from the plate to the nozzle is 140 mm, plate inclination angle is 90 degrees; 8—distance from the plate to the nozzle is 205 mm, plate inclination angle is 90 degrees; 9—distance from the plate to the nozzle is 305 mm, plate inclination angle is 90 degrees.

4. Conclusions

As a result of studies of the working process of the developed patented plant for preparing forage from grain, and taking into account the aggregate estimate of the quality indicators of forage and the cost of electricity for its preparation, the optimal location parameters of the passive grinder have been found, allowing us to obtain high-quality forage with minimal power consumption of the electric motor. A grid-work should be used as a grinder. Its installation angle should be 30°, and the distance between the grid-work and the nozzle should be 205 mm. With this combination of parameters, the specific energy consumption is minimal and amounts to 41.5 W·h/L.

Author Contributions: Conceptualization, A.M. and S.B.; methodology, S.B and P.S.; software, V.N.; validation, A.M., S.B.; formal analysis, P.S.; writing—original draft preparation, S.B., P.S., V.N.; writing—review and editing, A.M., W.M. All authors have read and agreed to the published version of the manuscript.

Funding: This research received no external funding.

Acknowledgments: We are thankful to the Editor and the reviewers for their valuable comments and detailed suggestions to improve the paper.

Conflicts of Interest: The authors declare no conflict of interest.

References

1. Djuragic, O.; Levic, J.; Serdanovic, S.; Lević, L. Evaluation of homogeneity in feed by method of microtracers. *Arch. Zootech.* **2009**, *12*, 85–91.
2. Flizikowski, J.; Tomporowski, A. Movement characteristics of multi-disc cereal crusher. *Przemysł Chem.* **2013**, *92*, 498–503.
3. Flizikowski, J.; Sadkiewicz, J.; Tomporowski, A. Use characteristics of six-roll milling of grained raw materials for the chemical and food industry. *Przemysł Chem.* **2015**, *94*, 69–75.
4. Tomporowski, B.; Flizikowski, J.; Kruszelnicka, W. A new concept for a cylindrical plate mill. *Przemysł Chem.* **2017**, *96*, 1750–1755.
5. Svihus, B.; Kløvstad, K.H.; Perez, V.; Zimonja, O.; Sahlström, S.; Schüller, R.B.; Prestløkken, E. Physical and nutritional effects of pelleting of broiler chicken diets made from wheat ground to different coarsenesses by the use of roller mill and hammer mill. *Anim. Feed Sci. Technol.* **2004**, *117*, 281–293. [CrossRef]
6. Sysuyev, V.A.; Aleshkin, A.V.; Savinykh, P.A. Feed-processing machines. In *Theory, Development, Experiment*; North-Eastern Zonal Agricultural Research & Development Institute: Kirov, Russia, 2008; Volume 1.
7. Ali, E.; Morad, M.M.; Fouda, T.; Elmetwalli, A.; Derbala, A. Development of a local animal feed production line using programming control system. *MISR J. Agric. Eng.* **2016**, *33*, 181–194.
8. Bulgakov, V.; Pascuzzi, S.; Ivanovs, S.; Kaletnik, G.; Yanovich, V. Angular oscillation model to predict the performance of a vibratory ball mill for the fine grinding. *Biosyst. Eng.* **2018**, *171*, 155–164. [CrossRef]
9. Kusmierczak, S.; Majzner, T. Comprehensive approach to evaluation of degradation in chosen parts of energy equipment. *Eng. Rural Dev. Proc.* **2017**, *12*, 673–679.
10. Bulgakov, V.; Holovach, I.; Bandura, V.; Ivanovs, S. A theoretical research of the grain milling technological process for roller mills with two degrees of freedom. *INMATEH Agric. Eng.* **2017**, *52*, 99–106.
11. Savinykh, P.; Kazakov, V.; Czerniatiev, N.; Gerasimova, S.; Romanyuk, V.; Borek, K. Technology and equipment for obtaining starch syrup with ground and whole cereal grain. *Agric. Eng.* **2018**, *22*, 57–67. (In Russian)
12. Vaculik, P.; Maloun, J.; Chladek, L.; Poikryl, M. Disintegration process in disc crushers. *Res. Agric. Eng.* **2013**, *59*, 98–104. [CrossRef]
13. Yancey, N.; Wright, C.T.; Westover, T.L. Optimizing hammer mill performance through screen selection and hammer design. *Biofuels* **2013**, *4*, 85–94. [CrossRef]
14. Ghorbani, Z.; Masoumi, A.; Hemmat, A. Specific energy consumption for reducing the size of alfalfa chops using a hammer mill. *Biosyst. Eng.* **2010**, *105*, 34–40. [CrossRef]
15. Linke, B.S.; Dornfeld, D.A. Application of axiomatic design principles to identify more sustainable strategies for grinding. *J. Manuf. Syst.* **2012**, *31*, 412–419. [CrossRef]

16. Sysuev, V.; Savinyh, P.; Saitov, V.; Gałuszko, K.; Caban, J. Optimization of structural and technological parameters of a fermentor for feed heating. *Agric. Eng.* **2015**, *12*, 99. (In Russian)

17. Motovilov, K.; Motovilov, O.; Aksyonov, V. *Nanobiotechnology in the Production of Forage from Grain for Animal Husbandry: A Monograph*; Publishing House of the Novosibirsk State Agrarian University: Novosibirsk, Russia, 2015. (In Russian)

18. Donkova, N.; Donkov, S. Biotechnology for the production of sugars from grain raw materials. *Bull. Krasn. State Agrar. Univ.* **2014**, *6*, 211–213. (In Russian)

19. Volkov, V. Efficiency of up-to-date equipment for the production of forage from grain. *World Sci. Cult. Educ.* **2013**, *1*, 351–354. (In Russian)

20. Aksyonov, V. Up-dating the lines for obtaining feed forage from grain raw materials using cost-effectiveness analysis. *Bull. Krasn. State Agrar. Univ.* **2011**, *12*, 220–224. (In Russian)

21. Novosibirsk Prototype Manufacturer of Non-Standard Equipment-Selmash LLC. Available online: http://noezno.ru/equipment/korm/kip-06/kip-06 (accessed on 12 April 2019).

22. Agrotechnology. Available online: http://www.stav-agro.ru/index.php/katalog/urva-250.html (accessed on 15 May 2018).

23. Agrobase. Available online: https://www.agrobase.ru (accessed on 12 April 2019).

24. Short Feeding of Cattle. Available online: http://agrokorm.info/ru/kormoagregat-mriya-05/1/ (accessed on 12 April 2019).

 *sustainability*

Article

Ultraviolet Fluorescence in the Assessment of Quality in the Mixing of Granular Material

Dominika Barbara Matuszek

Faculty of Production Engineering, Opole University of Technology, 45-758 Opole, Poland; d.matuszek@po.edu.pl; Tel.: +48-77-4498470

Received: 30 January 2020; Accepted: 18 February 2020; Published: 19 February 2020

Abstract: The aim of the study was to determine the possibility of using ultraviolet fluorescence to evaluate the quality of the mixing process of industrial feed. A laboratory funnel-flow mixer was used for the mixing process. The studies were carried out using three different feeds for pigs. A key component in the form of ground grains of yellow maize covered with the fluorescent substance Rhodamine B was introduced into the mixer before the mixing process began. After the illumination of the sample by UV lamps, the images were taken with a digital camera. The images were analyzed in Patan® software. The information obtained on the percentage content of the key component was used for further calculations. At the same time, the tracer content was determined using the control method (weight method). The comparison of the results obtained by the two methods (statistical comparative analysis) did not indicate significant differences. Therefore, the usefulness of the proposed method to track the share of the key component by inducing it to glow in the ultraviolet light has been proven. The introduced tracer is also one of the components of the feed, which translates into the possibility of observing the material having the characteristics of a mixed material.

Keywords: mixing of granular materials; fluorescence; tracer; industrial feed; image analysis

---

1. Introduction

Mixing granular materials is a difficult and multifaceted process which, although it was already described in the 1940s, is still a current research problem. This is due, among other things, to the development and availability of new tools, thanks to which it is possible to precisely observe the process itself, as well as to describe the behavior of selected components of the mixed bed. The degree of mixing of granular components is an important quality parameter of the final product in the context of e.g., nutrition [1–3]. It seems to be particularly important to conduct research on the process of mixing components with considerable fineness, such as powders. This is due, among other things, to the scale of production of products in the form of granules or powders, which reaches up to one billion pounds a year [4–7] and the aspects of safety [8].

The correct level of mixing ingredients in the mixture is important to obtain satisfactory results (mixture of good quality) in the production of farm animals; the preparation of feed in accordance with the given recipe. Incomplete mixing may result in negative nutritional effects. The appropriate uniformity of a mixture is important also for the granulation process [1,8,9]. In industrial conditions, ensuring safe feed production requires the implementation of Quality Management Systems such as Good Hygienic Practice—GHP, Good Manufacturing Practice—GMP, and Hazard Analysis and Critical Control Points—HACCP [10]. Quality control in feed production consists of verifying the correctness of the key stages and is carried out by eliminating possible biological, chemical and physical hazards. In the HACCP system, the mixing stage is often indicated as a critical control point. The assessment of homogeneity is an important Critical-To-Quality (CTQ) factor and it must be contolled periodically. This allows us, inter alia, to determine the required mixing time to obtain a high-quality product and involves the homogeneity analysis of samples taken from the mixer or from bags containing the finished mix. Determining the exact mixing time is important due to the fact that both too long and too short mixing can cause incorrect homogeneity. Moreover, the mixing time recommended by the mixers' manufacturers may not give the declared results in industrial conditions. This is due to the multi-threading of the process of mixing granular materials, which underlies research problems. Of course, determining the effective mixing time translates into production costs. Optimizing the work of mixing devices requires quantitative means to evaluate mixing [8,11]. Moreover, even a slight change in the parameters of the mixing process, such as humidity of the mixed material, degree of filling of the mixer and others, affects the degree of mixing [5,12,13]. The quality of the feed's raw materials is another important issue. Due to its properties (biologically active material), the feed may undergo contamination with pathogenic microflora [8,14]. However, as shown by some research carried out in Poland, this type of contaminant does not exceed accepted standards, and the use of thermal treatment such as drying allows for the additional reduction of this risk [14–16].

The degree of mixing of granular materials is described by various mathematical relationships in which the main element is the content of a given ingredient in the mixed bed. The determination of the state of the mixture is usually based on its deviation from the state of segregation or the state of perfect mixture. When mixing a two-component material, the determination of the state of the mixture is quite simple and is based on the analysis of sample composition [2]. When mixing heterogeneous multicomponent systems, is often impossible to analyze the composition of the samples accurately. There are only a few items in the literature where the authors have evaluated the mixing process based on the analysis of the content of each component of the mixture. However, methods of this type are very labor- and time-consuming [17,18]. A new method for determining the mixing index based on determining the distance from the adjacent particle and coordination number concept in multiple-spouted bed was proposed by Chen et al. [19].

However, most often the condition of a multicomponent granular mixture is determined on the basis of the content of the selected key component in the taken samples. The key ingredient can be a deliberately introduced element like microtracer—colored uniformly sized iron particles or microgrits [20–24]. This solution allows the quick assessment of the homogeneity of grain mixtures in industrial conditions. However, iron filings have different characteristics from mixed components, e.g., feed components or biologically active materials. This does not allow us to use them to track the behavior of actually mixed materials.

A number of methods for assessing the homogeneity of granular mixtures (indicator methods) are based on the determination of the content of specially dyed particles or on the use of the fluorescence emission phenomenon. Many researchers have used laser-induced fluorescence (LIF method) for this purpose. Lai et al. [25,26] used it to describe the kinetics of the pharmaceutical powder mixing process. A similar methodology was also used by Karumanchi et al. [27] for determining the dead zones in the mixer during powder mixing. Meanwhile, Durao et al. [28] determined in this way the concentration of five different substances in a multicomponent vitamin preparation. The phenomenon of X-ray fluorescence (XRF) was used, among others, in the assessment of sulphur and chlorine content in feeds for ruminants. The research was conducted for the following feeds: grass, grass silage and maize silage [29]. The use of a laser does not require physical sampling, which is the main advantage of this solution. However, it is necessary to prepare the mixer properly (window to take the sample) and the possible falsification of the results with the contaminants present. In addition, the window must be kept sufficiently clean, which may not be possible if the mixed components are very fine [30].

The analysis of the distribution of the stained components is often based on computer image analysis of the mixed material. Realpe et al. [31] used image analysis to evaluate the homogeneity of colored powders mixed in a cylindrical device. Berthiaux et al. [32] used this tool to assess the homogeneity of the mixture flowing out of the mixer. The homogeneity of the mixture at the mixer outlet was tested by Muerza et al. [33] based on on-line image analysis, and Dal Grande et al. [34] proved the usefulness of the HSV color model description for the analysis of binary grain mixture images. Computer image analysis to assess the impact of vibration intensity on the segregation of a bed consisting of steel and glass balls was used by Yang [35] and then by Tai et al. [36]. Hu et al. [37] analyzed images of samples taken during mixing in a rotating conical mixer with and without a mixing blade. Liu at al. [38] used an interesting optical technique consisting of the simultaneous acquisition of images with an infrared thermal camera and an ordinary RGB camera. Red (heated to 40 and 80 °C) and white balls (at room temperature) were used in this test. Techniques of this type are a relatively cheap and good tool, especially for analyzing the surface of a mixed material or its flow, and are most often used in laboratory conditions [30,39]. The limitation of these methods results from certain difficulties in obtaining information, e.g., a lack of access to the free space of the mixed bed, covering of the tracer by loose material [30].

In the research carried out by the authors of this work, an optical method was used that combines the fluorescence of a tracer in ultraviolet light with the acquisition of images by means of a digital camera with a standard lens. The usefulness of the proposed method in the mixing of whole grains in two- and multicomponent systems was confirmed, and the results obtained were described in the works of Matuszek et al. [40–42]. Referring to the commonness of powders in industrial practice and the limitations of computer image analysis in the evaluation of their mixing [4–6,30], the developed method was tested in the mixing of multicomponent feed consisting of ground grains and additives in the form of powders—so-called micronutrients. The aim of the study is to determine the possibility of using the method based on UV-induced fluorescence to assess the mixing of multicomponent industrial ground feed. The tests were carried out under laboratory conditions; however, the mixtures were obtained from feed factories. This makes the work more functional.

2. Materials and Methods

The mixtures used for the research consisted of components used for feed production, i.e., cereal grains and feed additives aimed at improving the nutritional value or sensory attractiveness (flavors). To assess the homogeneity of this type of mixture, indicator methods based on the assessment of the content of the key component (tracer) in feed samples were used. Therefore, there was the question of choosing the component that would properly represent the state of the mixed bed. Based on the previous experience of the author of this study and the literature information, it was decided that this ingredient may be maize. The choice was dictated, among others, by the widespread use of this ingredient in feeding farm animals and the ease of grinding or coating by coloring matter. Determining the size and number of samples is another research problem. On the one hand, the sample may consist of a single grain; on the other hand, the whole mixed system. The most important thing for the sample is to be representative. In the tests, 10 samples were taken due to the construction of the mixer. The mass of a single sample was determined experimentally. Finally, one of the most frequently appearing problems, i.e., how to assess the content of a particular ingredient against the background of many components that differ in color or fineness. The authors, using a known and proven tool, namely image analysis, attempted to eliminate this disturbance–for this, the fluorescent phenomenon was used.

Feed mixes obtained from a local feed factory were used in the study. These were one of the most commonly used mixtures intended for pigs. The mixtures were multicomponent systems (10-, 13- and 14-component systems) consisting of ground products of plant origin, such as cereal grains and minerals, and chemical additives. The types and proportions of the individual components and the degree of fineness (assessment by sieve analysis) of the tested feed are shown in Table 1.

Table 1. Characteristics of the tested feed mixture.

Type of Component	Percentage of Component [%]		
	Mixture 1	Mixture 2	Mixture 3
Fodder chalk	9	1.5	7.1
Barley	-	30	-
Maize	8	9	7
Triticale	-	20	-
Wheat	-	20	-
Soya meal	65.55	12	72
Rape meal	-	5	-
Dry maize decoction	5.45	-	4.3
Sodium chloride	2.5	0.5	2
Phosphate	3.5	1	2.8
Premix	2.5	1	2
Lysine	2.5	-	1.8
Methionine	0.5	-	0.4
Threonine	0.35	-	0.3
Phytase	0.05	-	0.05
Grindazyn	0.05	-	0.05
Luctarom (aroma)	0.05	-	0.05
Neubaciol	-	-	0.15
Number of components	13	10	14
Fineness degree M [mm]	0.64	0.61	0.63

A tracer was introduced to the mixing process (before it started). The seeds of yellow maize were used as the tracer—they were subjected to grinding and then wet treated with 0.01% Rhodamine B solution ($C_{28}H_{31}ClN_2O_3$, red-violet powder, wavelength 627 nm, excitation area 553 nm). The dyeing was carried out in laboratory conditions and the obtained materials (after drying) were stored in identical conditions in tightly closed packages, protecting against light. Two types of tracer were obtained in this way—maize of average particle size $d_1 = 2.00$ mm and $d_2 = 1.25$ mm. Before starting the mixing process, the tracer was placed in the upper part of the mixer (ring 10) in the amount of 100 g (10% of the mixture). The remaining part (rings 1–9) was an industrial feed mix of 900 g (90% of the mixture). After filling the mixer, the mixing process started. Mixing was carried out in a laboratory funnel-flow mixer by means of subsequent flows from 1 to 10. The characteristics of the mixer were presented in detail in another work by the authors [43]. The mixing was completed after the 10th flow and sampling were carried out. Single samples of 10 g were taken from ten mixer locations (levels). This was possible by the special design of the mixing tanks—10 removable rings.

The samples on the Petri dishes (120 mm × 20 mm) were placed in a chamber equipped with ultraviolet fluorescent lamps (2 lamps with 15 W each, mounted inside in the upper part of chamber). The chamber was made of black material, which limited the influence of light from outside. The sample was placed on a pull-out drawer. The chamber was then closed and the lighting controlled from outside. In the upper part of the chamber was a hole for a digital camera lens. A digital camera with a standard lens, 20.1 Mpix resolution and 35 mm focal length was used. When taking pictures, the exposure correction was set to −0.5 EV. After the illumination of the sample, images with a resolution of 1600 × 1200 pixels were obtained. The images obtained were analyzed in Patan®. This program analyzes the image based on the RGB 256 scale. After the loading of the image, three areas (designation of surface fragments, assignment of pixels) responsible for individual classes were marked in the examined area (circular area): 1—tracer (fluorescent maize) and 2 and 3—background (industrial mix). Thanks to this, it was possible to capture the desired information (the tracer's share) against a multicolored sample. The results obtained referr to the percentage content of particular classes. For further research, data on the percentage share of the first class, i.e., tracer, were used. In the next stage, the manual separation of the taken samples was performed and the separated tracer was weighed on an analytical balance with accuracy of ±0.01 g. In this way, the share of the tracer was estimated using two methods: (1) computer image analysis, (2) weighing method. The second method was used as a tool to verify the results obtained by method one. The selected stages of the methodology are schematically presented in Figure 1. On the basis of the results, the arithmetic mean, the standard deviation, the difference between the results obtained by the two methods and the coefficient of variation were calculated. The coefficient of variation (CV) was treated as a parameter indicating the degree of homogeneity of the obtained mixture. The calculations were made according to the instructions of the National Veterinary Institute [44]. The coefficient of variation was in the range from 0% to 100%. According to these instructions, the mixture is considered homogeneous when the CV ≤ 15%. In this range, a good mix of the tracer in the mixture is obtained. CV > 15% means bad homogeneity of the mixture, meaning there is poor mixing of the tracer in the mixture.

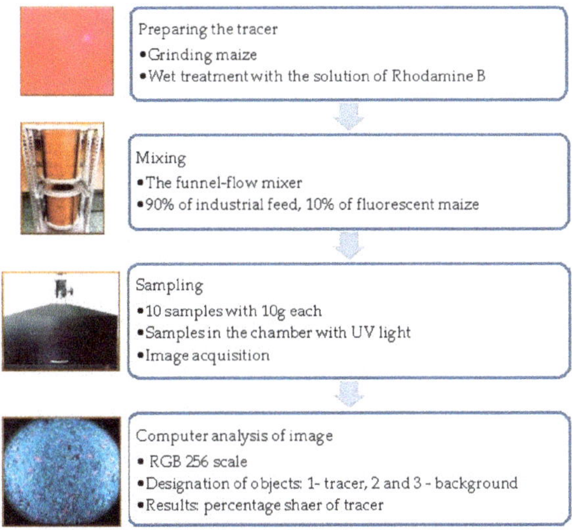

Figure 1. Schematic diagram of the methodology for the determination of fluorescent maize in the feed.

In order to verify the results obtained by the two methods, statistical comparative analysis was carried out. After checking the normality of distribution (Shapiro–Wilk test) and homogeneity of variance (Levene test), appropriate parametric (Student *t*-test) and non-parametric (U Mann-Whitney test) tests were used. The significance level $\alpha = 0.05$ was assumed for calculations. The null hypothesis regarding no differences in results obtained by the two methods was tested compared to the alternative hypothesis about the occurrence of such differences.

3. Results

3.1. Results Obtained by Means of Computer Image Analysis and the Weighing Method

The share of the tracer and the difference in results obtained by the two methods and selected statistical parameters (arithmetic mean, standard deviation, coefficient of variation) are presented in Tables 2 and 3. Graphical interpretations of the results are presented in diagrams (Figures 2–4).

Table 2. Results of the tracer's share (maize d = 2.00 mm).

Series of Tests	Method 1 [1]	Method 2 [2]	Difference [3]
Mixture 1			
1	8.00 ± 0.88	8.12 ± 0.94	0.30 ± 0.20
2	9.04 ± 0.79	9.05 ± 0.73	0.33 ± 0.18
3	8.98 ± 1.14	9.25 ± 1.06	0.41 ± 0.17
mean, %	8.67 ± 1.08	8.81 ± 1.06	0.34 ± 0.19
CV, %	10.83 ± 1.62	10.37 ± 1.66	0.55 ± 0.09
Mixture 2			
1	10.04 ± 1.20	10.03 ± 1.21	0.27 ± 0.17
2	10.38 ± 1.59	10.24 ± 1.58	0.26 ± 0.07
3	9.90 ± 1.27	9.85 ± 1.18	0.26 ± 0.09
mean, %	10.10 ± 1.41	10.04 ± 1.37	0.26 ± 0.12
CV, %	13.46 ± 1.32	13.17 ± 1.60	0.41 ± 0.16
Mixture 3			
1	9.34 ± 0.93	9.51 ± 1.04	0.41 ± 0.23
2	8.66 ± 0.92	8.84 ± 1.05	0.43 ± 0.23
3	9.45 ± 0.76	9.35 ± 0.91	0.29 ± 0.16
mean, %	9.15 ± 0.96	9.23 ± 1.06	0.38 ± 0.22
CV, %	9.55 ± 1.10	10.87 ± 0.87	0.56 ± 0.01

[1] arithmetic mean of the tracer's share obtained by computer image analysis ± standard deviation; [2] arithmetic mean of the tracer's share obtained by weighing ± standard deviation; [3] difference of results obtained by two methods ± standard deviation.

Table 3. Results of the tracer's share (maize d = 1.25 mm).

Series of Tests	Method 1 [1]	Method 2 [2]	Difference [3]
Mixture 1			
1	9.13 ± 0.87	9.37 ± 0.78	0.53 ± 0.20
2	9.49 ± 0.93	9.42 ± 0.96	052 ± 0.30
3	8.94 ± 1.09	8.89 ± 0.97	0.44 ± 0.17
mean, %	9.18 ± 1.01	9.23 ± 0.96	0.50 ± 0.23
CV, %	10.50 ± 1.19	9.83 ± 1.07	0.44 ± 0.09
Mixture 2			
1	9.39 ± 0.76	9.13 ± 0.79	0.38 ± 0.26
2	9.22 ± 0.75	9.21 ± 0.84	0.42 ± 0.26
3	9.56 ± 0.81	9.53 ± 0.89	0.33 ± 0.14
mean, %	9.39 ± 0.80	9.29 ± 0.87	0.38 ± 0.23
CV, %	8.25 ± 0.16	9.04 ± 0.28	0.58 ± 0.11
Mixture 3			
1	9.35 ± 0.75	9.61 ± 0.85	0.62 ± 0.26
2	9.09 ± 1.02	9.21 ± 0.94	0.64 ± 0.26
3	9.13 ± 0.92	9.19 ± 0.77	0.50 ± 0.25
mean, %	9.19 ± 0.92	9.33 ± 0.89	0.59 ± 0.30
CV, %	9.75 ± 1.33	9.15 ± 0.76	0.49 ± 0.06

[1] arithmetic mean of the tracer's share obtained by computer image analysis ± standard deviation; [2] arithmetic mean of the tracer's share obtained by weighing ± standard deviation; [3] difference of results obtained by two methods ± standard deviation.

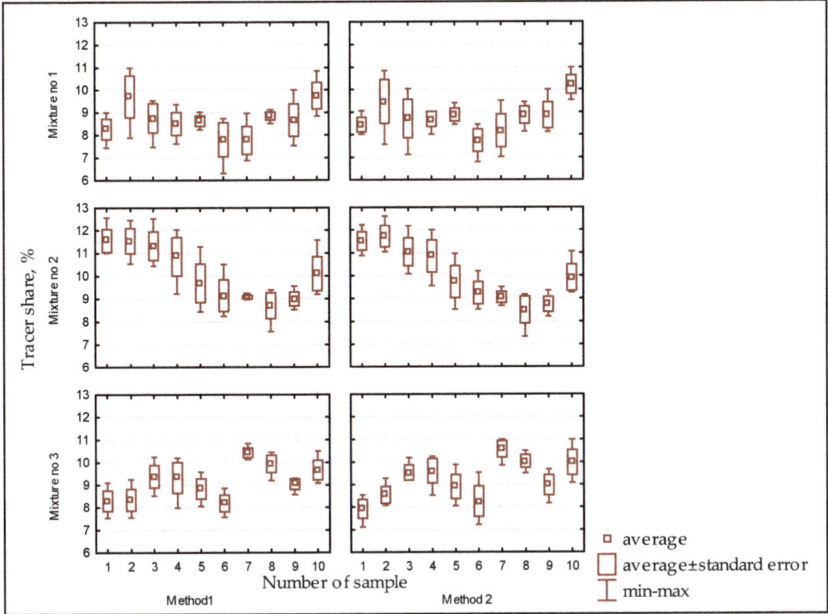

Figure 2. Results of the tracer's share (2.0 mm maize) in individual samples obtained by two methods for three compound feed mixes.

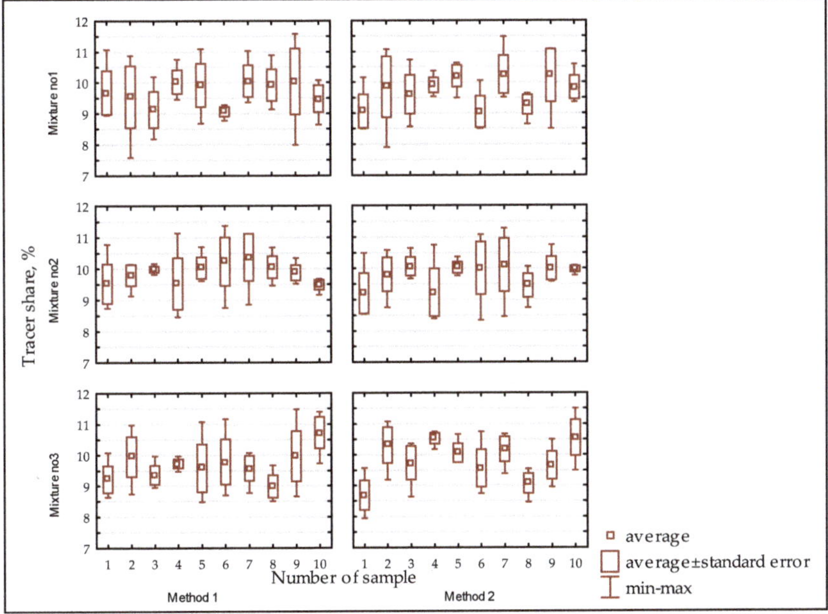

Figure 3. Results of the tracer's share (1.25 mm maize) in individual samples obtained by two methods for three compound feed mixes.

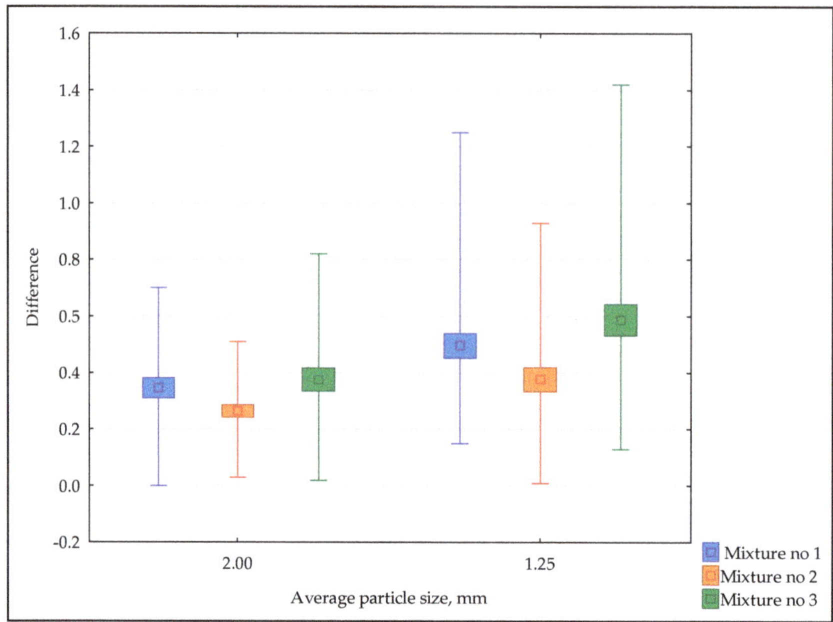

Figure 4. Box plot graph (mean, standard error, min-max) of the difference between the results obtained by the two methods for the mean tracer dimension $d_1 = 2.00$ mm and $d_2 = 1.25$ mm.

The results presented in Tables 2 and 3 show the great similarity of the results obtained by the two methods. The mean difference of the results obtained by the two methods for the tracer with mean particle size d = 2.00 ranged from 0.3 to 0.41 for mixture 1, 0.26–0.27 for mixture 2 and 0.29–0.43 for mixture 3 (Table 2). In the case of mixing with a smaller average particle size (d = 1.25 mm), slightly larger differences were obtained: 0.44–0.53 for mix 1, 0.33–0.42 for mix 2 and 0.50–0.64 for mix 3 (Table 3). In one case, a zero difference was obtained and this concerned the results of mixing feed 1 with the maize of larger average particle size. However, the highest level of the difference of 1.42 was recorded for mixing feed 3 with the d = 1.25 mm tracer. Similar values of the tracer's share obtained by method 1 and 2 in particular segments (samples) of the mixer after the mixing process was completed are also visible on the diagrams (Figures 2 and 3). By analyzing these charts, you can also track the tracer's distribution in the individual mixing zones (rings). For example, when mixing feed 2 with 2.00 mm maize, the tracer gathered in the lower part of the mixer (rings 1 to 3). Then, there is a decrease in its share (rings 4–8) and again a slight increase (rings 9 and 10). The graphical interpretation also shows that when using a tracer with an average particle size (d = 1.25 mm), larger deviations from the average value (average ± standard error) were obtained than in the case of a tracer with a smaller degree of fineness. Similarly, by analysing Figure 4, lower values of the differences obtained in mixing with the d = 2.00 mm tracer can be observed. In this case, lower mean values were obtained (0.34%, 0.26%, 0.43% respectively for mixtures 1, 2 and 3) than in tests with finer d = 1.25 mm (mean values 0.50%, 0.38%, 0.59% respectively for mixtures 1, 2 and 3). Additionally, in the same case, the values are more favorably (lower values) distributed in the min-max range (0–0.7, 0.03–0.51, 0.02–0.82 respectively for mixtures 1, 2 and 3). The values within the same range, for a series of tests with a d = 1.25 mm tracer, are 0.15–1.25, 0.01–0.93 and 0.13–1.42, respectively.

Referring to the results of the coefficient of variation, each of the mixtures achieved good homogeneity, and thus good mixing of the tracer. In each case, a coefficient of variation at CV ≤ 15% was obtained. This observation concerns the results obtained with two methods and with the use of two different tracers of different degrees of fineness.

3.2. Results of Statistical Comparative Analysis

The results of the statistical analysis are presented in Tables 4 and 5 and their graphical interpretation in Figures 5 and 6.

Table 4. Results of statistical comparative analysis (Student's *t*-test) of the results obtained by the two methods using the tracer $d_1 = 2.00$ mm.

Mixture No.	t	p
1	−0.49398	0.62318
2	0.18586	0.85320
3	−0.33460	0.73913

Table 5. Results of statistical comparative analysis (Student's *t*-test and U Mann-Whitney's test) of the results obtained by the two methods using the tracer $d_2 = 1.25$ mm.

Mixture No.	Statistical Test Result	p
1 [1]	−0.16586	0.86884
2 [2]	0.34004	0.73383
3 [2]	−0.73183	0.46427

[1] parametric Student's *t*-test. Normal distribution of variables, homogeneity of variance; [2] non-parametric U Mann-Whitney test. No normal distribution of variables, homogeneity of variance.

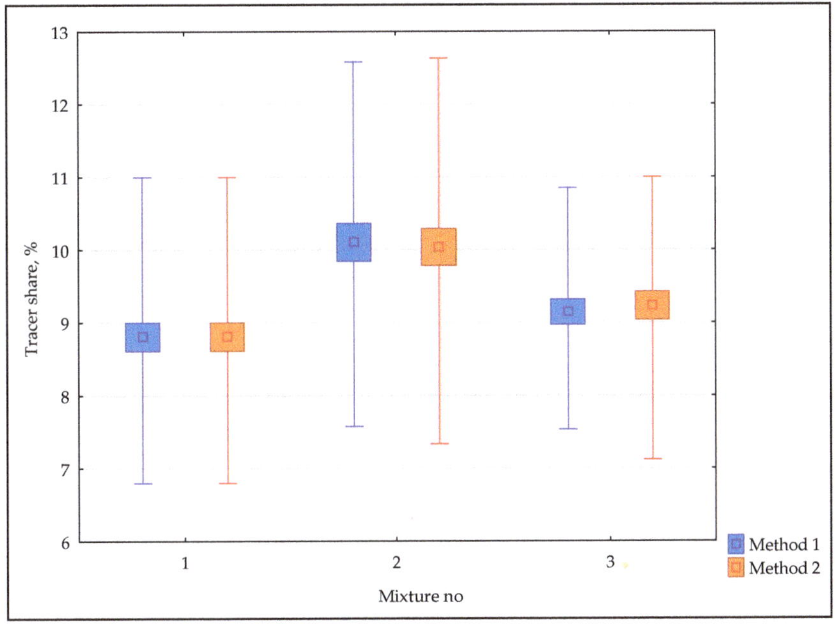

Figure 5. Box plot graph of the share of the tracer (maize d = 2.0 mm) obtained by two methods for the three analysed mixtures. Point: average, box: average ± standard error, whiskers: min-max.

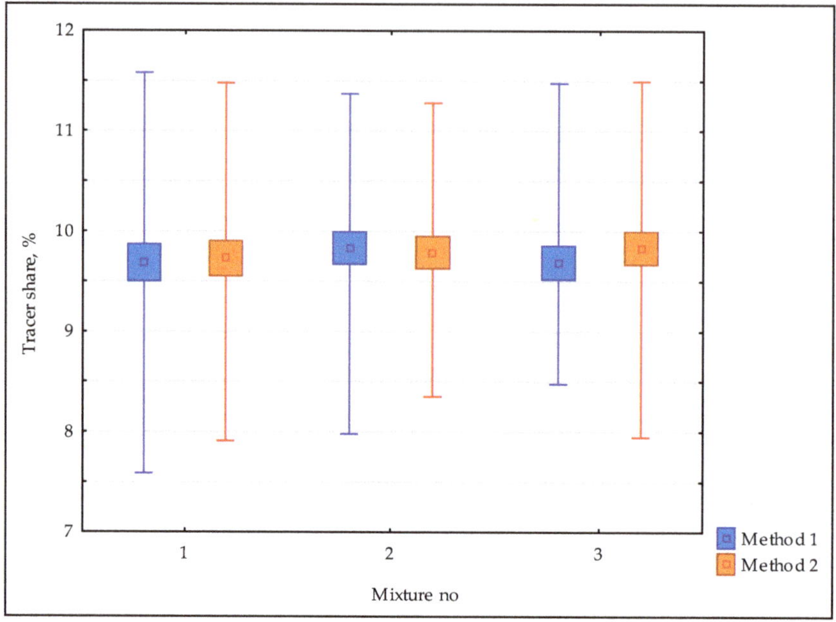

Figure 6. Box plot graph of the share of the tracer (maize d = 1.25 mm) obtained by two methods for the three analysed mixtures. Point: average, box: average ± standard error, whiskers: min-max.

The results obtained (Tables 4 and 5) do not allow us to reject the zero hypothesis about the lack of statistically significant differences in the results of the tracer's share obtained with both methods (no significant statistical differences between two methods: image analysis and weighing method). This statement applies to both tests with fluorescent maize with an average particle size of $d_1 = 2.00$ mm and $d_2 = 1.25$ mm. This confirms the preliminary observations presented in Section 3.1. The high similarity of the results obtained by the two methods is also confirmed by the graphical interpretation of statistical analysis (Figures 5 and 6).

4. Discussion

The obtained results confirm the possibility of using fluorescence induced by ultraviolet radiation to analyze the mixing process of multi-component heterogeneous feed mixtures. This method gives highly reliable results compared to the control method (weighing method), which was confirmed by appropriate statistical tests (Student's *t*-test and U Mann–Whitney test). It was observed that the differences in the results obtained by the two methods using a tracer with an average particle size of $d_1 = 2.00$ mm are smaller than in the case of a tracer with an average particle size of $d_2 = 1.25$ mm. However, in both cases, statistical comparative analysis did not indicate the significance of these differences. In addition, a mixture with an appropriate level of homogeneity was obtained in each test series.

In the proposed method, the key component (dyed maize with 0.01% Rhodamine B solution) was observed on the basis of computer image analysis (software working on the basis of RGB 256 scale) of particular mixer segments. The special design of the tank made it possible to take significant number (N = 10) of samples and thus to track the tracer distribution in the entire volume of the mixed bed. In addition, the phenomenon of fluorescence induced by ultraviolet radiation was used, which allowed us to eliminate interference during image acquisition. The main disadvantage of computer image analysis in the evaluation of mixing of granular components subjected to the milling process is the covering of the tracer by a loose material [30]. In such a situation, image analysis may give incomplete and unreliable information. The light of the tracer allowed for its "singling out" against a multicolored sample. The disadvantage presented has therefore been eliminated. Another limitation of optical methods in the mixing of granular systems is the necessity to take samples, i.e., to interfere with the mixed bed. The proposed methodology is also based on sampling. This does not explicitly exclude its use to track the tracer's behavior on line. Of course, this requires careful analysis and testing. It is worth noting, however, that non-invasive methods to characterize grain mixing require more advanced technology and can therefore be more expensive.

The authors have not yet come across a study in which a similar solution has been applied in the evaluation of the mixing of multi-component ground grain systems (degree of fineness M = 0.61–0.64 mm). Most of the studies using image analysis concerned the mixing of two-component systems differing significantly in colour, e.g., white and red balls [4,34–36,45–48]. What is more, the research carried out by the authors made use of feed mixes that are produced in a feed factory and the tracer introduced was one of the components of these mixes. Therefore, the proposed method is more likely to be used under industrial conditions. In the next stage of the research, the authors plan to verify the usefulness of the developed method by conducting tests with additional feed mixes differing in the degree of fineness and composition.

The proposed method uses a well-known computer image analysis tool. However, the use of an additional stage (illuminating the sample with ultraviolet light) allows the elimination of disturbances in the case of multi-component systems subjected to a milling process. What is more, the method allows us to assess the homogeneity of the mixture based on the ingredient used to produce feed, i.e., maize. This is important because of the properties of biologically active materials. Their parameters, such as humidity, brittleness, the ability to agglomerate and many others, can change significantly in response to the conditions of the mixing process. Materials often used by other authors (presented in introduction and discussion chapters) like steel, glass or plastic balls have definitely different properties. As presented in the introduction, slight changes in the parameters of a mixed system or mixer affect the degree of mixing or segregation. The described method is not a tool without limits. This is due to the Rhodamine B coloring agent used. This substance is a chemical reagent that has irritating properties. In laboratory conditions, when applying general precautions, it does not matter much. However, in industrial conditions, the use of this method would require appropriate handling of mixed feed (disposal). Therefore, this method is applicable only in laboratory conditions, for now.

Funding: This research received no external funding.

Conflicts of Interest: The authors declare no conflict of interest.

References

1. Neumann, K.D. Work of the Mixer is crucial for additives. *Feed Mag.* **2000**, *10*, 371–383.
2. Tasirin, S.M.; Kamarudin, S.K.; Hweage, A.M.A. Mixing behavior of binary polymer particles in bubbling fluidized bed. *J. Phys. Sci.* **2008**, *19*, 13–29.
3. Jarray, A.; Shi, H.; Scheper, B.J.; Habibi, M.; Luding, S. Cohesion-driven mixing and segregation of dry granular media. *Sci. Rep.* **2019**, *9*, 13480. [CrossRef] [PubMed]
4. Weinekötter, R.; Gericke, H. *Mixing of Solids: Particle Technology*, 1st ed.; Springer: Berlin, Germany, 2000; ISBN 978-94-015-9580-3.
5. Cleary, P.W.; Sinnott, M.D. Assessing mixing characteristics of particle-mixing and granulation devices. *Particuol. Simul. Modeling Part. Syst.* **2008**, *6*, 419–444. [CrossRef]
6. Sarkar, A.; Wassgren, C.R. Simulation of a continuous granular mixer: Effect of operating conditions on flow and mixing. *Chem. Eng. Sci.* **2009**, *64*, 2672–2682. [CrossRef]
7. Królczyk, J.B. Metrological changes in the surface morphology of cereal grains in the mixing process. *Int. Agrophys.* **2016**, *30*, 193–202. [CrossRef]
8. Matuszek, D.; Królczyk, J.B. Aspects of safety in production of feeds—A review. *Anim. Nutr. Feed Technol.* **2017**, *17*, 367–385. [CrossRef]

9. Walczyński, S.; Korol, W. Long-term monitoring of homogeneity of compound feed in the government supervision. *Krmiva Zagreb* **2008**, *50*, 311–317.

10. Zawiślak, K.; Sobczak, P.; Wełdycz, A. Mixing as CCP in the production of industrial feed. *J. Cent. Eur. Agric.* **2012**, *13*, 554–562. [CrossRef]

11. Bothe, D. Evaluating the quality of a mixture: Degree of homogeneity and scale of segregation. In *Micro and Macro Mixing. Heat and Mass Transfer*; Bockhorn, H., Mewes, D., Peukert, W., Warnecke, H.J., Eds.; Springer: Berlin, Germany, 2010; pp. 17–35. [CrossRef]

12. Liu, P.Y.; Yang, R.Y.; Yu, A.B. Self-diffusion of wet particles in rotating drums. *Phys. Fluids* **2013**, *25*, 063301. [CrossRef]

13. Li, H.; McCarthy, J.J. Controlling Cohesive Particle Mixing and Segregation. *Phys. Rev. Lett.* **2003**, *90*, 184301. [CrossRef] [PubMed]

14. Cegielska-Radziejewska, R.; Stupe, K.; Szablewsk, T. Microflora and mycotoxin contamination in poultry feed mixtures from western Poland. *Ann. Agric. Environ. Med.* **2013**, *20*, 30–35. [PubMed]

15. Sobczak, P.; Zawiślak, Z.; Żukiewicz-Sobczak, W.; Mazur, J.; Nadulski, R.; Kozak, M. The assessment of microbiological purity of selected components of animal feeds and mixtures which underwent thermal processing. *J. Cent. Eur. Agric.* **2016**, *17*, 303–314. [CrossRef]

16. Kwiatek, K.; Kukier, E. Microbiological contamination of animal feed. *Vet. Med.* **2008**, *64*, 24–26.

17. Królczyk, J.; Matuszek, D.; Tukiendorf, M. Modelling of quality changes in a multicomponent granular mixture during mixing. *Electon. J. Pol. Agric. Univ.* **2010**, *13*.

18. Królczyk, J.B. An attempt to predict quality changes in a ten-component granular system. *Teh. Vjesn.* **2014**, *21*, 255–261.

19. Chen, M.; Liu, M.; Li, T.; Tang, Y.; Liu, R.; Wen, Y.; Liu, B.; Shao, Y. A novel mixing index and its application in particle mixing behavior study in multiple-spouted bed. *Powder Technol.* **2018**, *339*, 167–181. [CrossRef]

20. Eisenberg, S.; Eisenberg, D. Closer to perfection. *Feed Manag.* **1992**, *11*, 8–20.

21. Zawiślak, K.; Grochowicz, J.; Sobczak, P. The Analysis of mixing degree of granular products with the use of Microtracers. *TEKA Kom. Mot. Energ. Roln. OL Pan* **2011**, *11*, 335–342.

22. Matuszek, D. The analysis of homogeneity of industrial fodder for cattle. *J. Res. Appl. Agric. Eng.* **2013**, *58*, 118–121.

23. Królczyk, J.B. Industrial conditions of the granular material manufacturing process. *App Mech. Mater.* **2014**, *693*, 267–272. [CrossRef]

24. Królczyk, J.B. Analysis of kinetics of multicomponent, heterogeneous granular mixtures—laminar and turbulent flow approach. *Chem. Process. Eng.* **2016**, *37*, 161–173. [CrossRef]

25. Lai, C.K.; Holt, D.; Leung, J.C.; Cooney, C.L.; Raju, G.K.; Hansen, P. Real time and noninvasive monitoring of dry powder blend homogeneity. *AIChE J.* **2001**, *47*, 2618–2622. [CrossRef]

26. Lai, C.K.; Cooney, C.L. Application of a fluorescence sensor for miniscale on-line monitoring of powder mixing kinetics. *J. Pharm. Sci.* **2004**, *93*, 60–70. [CrossRef] [PubMed]

27. Karumanchi, V.; Taylor, M.K.; Ely, K.J.; Stagner, W.C. Monitoring powder blend homogeneity using light-induced fluorescence. *AAPS Pharmscitech.* **2011**, *12*, 1031–1037. [CrossRef]

28. Durão, P.; Fauteux-Lefebvre, C.; Guay, J.M.; Batzoglou, N.; Gosselin, R. Using multiple process analytical technology probes to monitor multivitamin blends in a tableting feed frame. *Talanta* **2017**, *164*, 7–15. [CrossRef]

29. Necemer, M.; Kump, P.; Rajcevic, M.; Jacimovic, R.; Budic, B.; Ponikvar, M. Determination of sulfur and chlorine in fodder by X-ray fluorescence spectral analysis and comparison with other analytical methods. *Spectrochim. Acta* **2003**, *58*, 1367–1373. [CrossRef]

30. Nadeem, H.; Heindel, T.J. Review of noninvasive methods to characterize granular mixing. *Powder Technol.* **2018**, *332*, 331–350. [CrossRef]

31. Realpe, A.; Barrios, K.; Rozo, M. Assessment of homogenization degree of powder mixing in a cylinder rotating under cascading regime. *Int. J. Eng. Technol.* **2015**, *7*, 394–404.

32. Berthiaux, H.; Mosorov, V.; Tomczak, L.; Gatumel, C.; Demeyre, J.F. Principal component analysis for characterising homogeneity in powder mixing using image processing techniques. *Chem. Eng. Process. Process. Intensif.* **2006**, *45*, 397–403. [CrossRef]

33. Muerza, S.; Berthiaux, H.; Massol-Chaudeur, S.; Thomas, G. A dynamic study of static mixing using on-line image analysis. *Powder Technol.* **2002**, *128*, 195–204. [CrossRef]

34. Dal Grande, F.; Santomaso, A.; Canu, P. Improving local composition measurements of binary mixtures by image analysis. *Powder Technol.* **2008**, *187*, 205–213. [CrossRef]

35. Yang, S.C. Density effect on mixing and segregation processes in a vibrated binary granular mixture. *Powder Technol.* **2006**, *164*, 65–74. [CrossRef]

36. Tai, C.H.; Hsiau, S.S.; Kruelle, C.A. Density segregation in a vertically vibrated granular bed. *Powder Technol.* **2010**, *204*, 255–262. [CrossRef]

37. Hu, G.; Gong, X.; Huang, H.; Li, Y. Effects of geometric parameters and operating conditions on granular flow in a modified rotating cone. *Ind. Eng. Chem. Res.* **2007**, *46*, 9263–9268. [CrossRef]

38. Liu, X.; Gong, J.; Zhang, Z.; Wu, W. An image analysis technique for the particle mixing and heat transfer process in a pan coater. *Powder Technol.* **2016**, *295*, 161–166. [CrossRef]

39. Kingston, T.A.; Heindel, T.J. Granular mixing optimization and the influence of operating conditions in a double screw mixer. *Powder Technol.* **2014**, *266*, 144–155. [CrossRef]

40. Matuszek, D.; Wojtkiewicz, K. Application of fluorescent markers for homogeneity assessment of grain mixtures based on maize content. *Chem. Process. Eng.-Inz.* **2017**, *38*, 505–512. [CrossRef]

41. Matuszek, D.; Biłos, Ł. Use of fluorescent tracers for the assessment of the homogeneity of multicomponent granular feed mixtures. *Przem. Chem.* **2017**, *96*, 2356–2359. [CrossRef]

42. Matuszek, B.D. The use of UV-induced fluorescence for the assessment of homogeneity of granular mixtures. *Open Chem.* **2019**, *17*, 485–491. [CrossRef]

43. Matuszek, D. Modelling selected parameters of granular elements in the mixing process. *Int. Agrophys.* **2015**, *29*, 75–81. [CrossRef]

44. Kwiatek, K.; Przeniosło-Siwczyńska, M. *Instructions for Testing the Homogeneity of Medicated Feeds and Intermediate Products on the Basis of Testing the Degree of Mixing of the Active Substance*; Państwowy Instytut Weterynaryjny: Puławy, Poland, 2007.

45. Weinekötter, R.; Reh, L. Characterization of particulate mixtures by in-line measurments. *Part. Part. Syst. Char.* **1994**, *11*, 284–290. [CrossRef]

46. Boss, J.; Krótkiewicz, M.; Tukiendorf, M. The application of picture analysis as a method to evaluate the quality of granular mixture during funnel-flow mixing. *Agric. Eng.* **2002**, *4*, 27–32.

47. Arratia, P.E.; Duong, N.H.; Muzzio, F.J.; Godpole, P.; Lange, A.; Reynolds, S. Characterizing mixing and lubrication in the Bohle bin blender. *Powder Technol.* **2006**, *161*, 202–208. [CrossRef]

48. Brone, D.; Muzzio, F.J. Enhanced mixing in double-cone blenders. *Powder Technol.* **2010**, *110*, 179–189. [CrossRef]

Article

Risk Assessment for Social Practices in Small Vegetable farms in Poland as a Tool for the Optimization of Quality Management Systems

Marcin Niemiec [1,*], Monika Komorowska [2], Anna Szeląg-Sikora [3], Jakub Sikora [3], Maciej Kuboń [3], Zofia Gródek-Szostak [4] and Joanna Kapusta-Duch [5]

[1] Department of Agricultural and Environmental Chemistry, Faculty of Agriculture and Economics, University of Agriculture in Krakow, Mickiewicz Ave. 21, 31-120 Krakow, Poland

[2] Department of Vegetable and Medicinal Plants, Faculty of Biotechnology and Horticulture, University of Agriculture in Krakow, 54 29 Listopada Ave., 31-426 Krakow, Poland

[3] Institute of Agricultural Engineering and Informatics, University of Agriculture in Krakow, 116b Balicka St., 30-149 Krakow, Poland

[4] Department of Economics an Organization of Enterprises, Cracow University of Economics, 27 Rakowicka St., 31-510 Krakow, Poland

[5] Faculty of Food Technology, Department of Human Nutrition, University of Agriculture in Krakow, 122 Balicka St., 30-149 Krakow, Poland

[*] Correspondence: marcin1niemiec@gmail.com

Received: 3 June 2019; Accepted: 15 July 2019; Published: 18 July 2019

Abstract: Globalization of the food market is associated with the possibility of selling products into newer markets. However, it is also associated with the necessity to ensure proper quality products. Quality defined by the ISO 9001:2015 standard consists of factors that are part of customers' expectations concerning the safety of products and the technology of their manufacture. Currently, consumers are looking for products with defined and reproducible sensory properties, in which the content of harmful substances is below the critical values specified by legislation. This is observable particularly in developed countries. The second quality factor is the use of a production technology where negative environmental impacts are reduced. Recently, issues associated with protecting workers' rights and social needs have also become very important. In successive versions of quality management systems, such as GLOBAL G.A.P. or SAI Platform, social issues are becoming more and more important. The aim of this study was to assess the role of risk analysis for social practices in small farms in building a quality management system. Surveys were conducted in 2018. The surveys covered 62 vegetables or fruit farms with a cultivated area of up to 20 ha. Their lack of staff was due to the character of production. Where mechanic production is possible in small farms, family members can secure workforce demand. To achieve the research objective, a risk analysis was carried out for the implementation of social practices according to the guidelines of the ISO 31000:2018 standard. The criteria and inventory of identified risks were carried out, based on the guidelines of GLOBAL G.A.P. Risk Assessments on Social Practice (GRASP). Based on the identified risks, the areas relating to social practices, which require improvement in order to satisfy compliance with the GLOBAL G.A.P. standard, were indicated. The results of the conducted research pointed to a high risk of good social practices not being carried out and not meeting compliance with the requirements of the GLOBAL G.A.P. standard. The most important identified problems are associated with the deficiency of competent workers as well as the lack of facilities where workers can rest, eat and drink. A considerable problem is the conformity of employment contracts with local legislation and ensuring that work time and rest time are consistent with the law. In conditions of small farms in Poland, the problem with ensuring compliance with the standard in question is often the small number of workers. Creating an organized quality management system in the area of social practices is difficult in these cases, and sometimes even impossible.

Keywords: GLOBAL G.A.P.; GRASP; quality management systems; certification; primary production; social practice

---

1. Introduction

The conditions of the modern food market of products are associated with the necessity to increase the effectiveness of the use of the means of production and to reduce the negative impact on the natural environment. The need for rationalization of land use, work, and depletable environmental resources result, on one hand, from the need to lower the costs of food production, and on the other hand, from the needs of the consumer who seeks food with specific quality. Quality is one of the most important factors that influence consumers' choices. It is very often more important than product price, particularly in developed countries. It is one of the most important factors of achieving a competitive advantage. According to the used definition in management systems, quality relates to the degree of customer satisfaction. In the case of agricultural products, the idea of quality has been changing. This has influenced societies and the economic potential of consumers [1].

At the beginning of the first half of the 20th century, quality of agri-food products was identical with their price. In that period, the most important problem in the world was to provide developing societies with an adequate amount of food. Increasing the production efficiency was based mainly on the intensification of fertilization and on increasing the amount of production-boosting chemicals. The consequence of such a strategy of production (both plant and animal) was the emission of a considerable amount of pollutants into the environment, which led to soil and water degradation, as well as air pollution. The second, and very important, effect of agricultural intensification was the deterioration in the quality of products in terms of their chemical composition. The biggest problems included high content of pesticides, nitrates, and heavy metals in plant products, and in the case of animal products, a high content of hormones and pharmaceuticals. Due to the development of agricultural sciences and changes in consumer awareness, the quality of products is more and more associated with the technology of their production. This quality factor has been expressed in the idea of sustainable development of agriculture. Quality is also associated with the way a product is packed and presented [2].

Traditional agriculture has a negative impact on the natural environment—both water, soil, air, and on consumer health [3]. In the 1990s, formalized quality systems in primary production began to develop in developed countries. Those systems took into account the production principles leading to an improvement in food safety at the stage of producing raw materials intended for food or fodder purposes. In that period, the concepts of sustainable production systems for biomass crops began to appear [4]. The fundamental assessment criterion of a given technology in this case is the energy fixation potential of plants, taking into account the expenditure incurred on production. Another important aspect is the issue of the possibility of producing food in areas intended for cultivation of energy crops. According to the adopted ideology, the introduction of specific principles of a system should lead to an improvement in the quality of products, in terms of their safety and chemical composition [5]. Moreover, the principles of modern systems of quality management in agriculture have involved issues associated with rational utilization of soil resources, water resources, and issues connected with decreasing energy consumption in the entire production process [6,7]. The efficient implementation of quality management systems in food production is also associated with creating no-waste or low-waste technologies, as pointed out by Sikora et al. [8]. More and more attention has been paid to the optimization of logistic processes, not only at the production stage, but in the entire supply chain. Optimization of logistic processes is an important part of improving economic efficiency, which was highlighted by Xiao et al. [8], as well as Cupiał et al. [9]. As regards to food safety, many countries have safety assurance systems, created and administered at a national level. They can vary both in scope, as well as in the level of requirements, which result from economic, cultural, climatic, or political conditions [10]. The differences concern mainly environmental and societal issues in production. Therefore, there is a threat that products generated in compliance with local law will not fulfill qualitative criteria requested by consumers in target countries [11]. This applies particularly to the issues connected with environmental and social aspects in developing countries. Food production, compliant with local law of these countries, is very often insufficient to satisfy a conscious consumer. Therefore, one of the main factors of developing agricultural production and the possibility of exporting products, is the implementation of quality management systems in primary production.

One of the most popular quality management standards in primary production in agriculture is the GLOBAL G.A.P. (Good Agriculture Practice) system. It is an independent, optionally-implemented system of assuring product quality and safety in primary agricultural production. The standard was introduced into the market in 1997 under the name EUREPGAP. It was elaborated by members of EUREP (Euro-Retailer Produce Working Group) organization. The purpose of the standard was to develop principles that would be common for the entire primary production, aimed at ensuring compliance with Good Agricultural Practice (GAP) and ensuring food safety. The GLOBAL G.A.P. standard was introduced in exchanged for EUREPGAP on 7 September 2007.

One of the primary objectives of the standard was to minimize the use of fertilizers and plant protection products, so as to limit the negative impact of agriculture on the environment, and also to make sure that good soil culture on areas intended for agricultural production is maintained. One of the most frequently described problems associated with implementation of quality management systems in primary production is a failure to adapt them to small farms [12,13]. Inadequate support from state institutions and non-governmental organizations is a factor that limits the development of quality management systems [14]. To accommodate these problems, the certification system and the manner of implementing the principles of the GLOBAL G.A.P. standard at a farm level were adapted to small farms. Certification is possible within option 1 and 2. In the case of option 1, an individual producer intends to certify their own products in compliance with the requirements of GLOBAL G.A.P., in the scope associated with specific activities. Option 2 relates to the certification of a group of producers (e.g., cooperatives or organizations of producers). In this case, certification concerns agricultural products, in accordance with the requirements of the used range of GLOBAL G.A.P., complemented with requirements pertaining to managing through the implementation of the Quality Management System (QMS). In this option, a certificate is issued for the leading organization, which guarantees that the requirements of the standard are observed by producers that are members of the group. Due to the higher effectiveness of means of production, producer groups have a bigger and bigger share in primary production in Poland. A properly organized quality management system at the level of the producer group allows for the efficient management of quality for all members [5,15]. Within the certification of a producer group, an audit is conducted for the management system in the context of system tools, and the efficiency of communication with particular group members, as well as the effectiveness of identifying non-conformities and monitoring corrective actions. Such a solution makes it possible to minimize the risks associated with the fragmentation of farms and the scattering of production sites [16]. Due to these factors, small producers can be included into the global supply chain, which facilitates the development of small family farms. Including small farms as a part of supply chain, not only on local markets, but also on international scale, is an important part of a sustainable development of rural areas. This is of great importance in developing countries, where plants are cultivated on small areas [17]. However, such an approach associated with certification poses a threat to product quality in the case of an ineffective system of controlling the members of the producer group by head office, which implements the quality management system. GLOBAL G.A.P. is a standard that is based on ethical principles of the producers participating in the system [18]. The aim of this study was to assess the role of risk analysis for social practices in small farms, in building a quality management system.

2. Research Methods

To meet the set objective, surveys were conducted in 2018 in small family farms in Poland, in the following provinces: Świętokrzyskie (11 farms), Mazovia (18 farms), Łódź (19 farms), Lublin (6 farms), and Wielkopolska (8 farms). The research was conducted using the direct interview method. The study involved farms, in which the buyers of products reported the need to implement the GRASP standard. The surveys covered 62 vegetables or farms that grow berry plants, with a cultivated area of up to 20 ha. Among the examined farms, 37 of them grew vegetables, while 25 were growing berry plants. Nearly 30% of the surveyed farms declared that they did not employ workers. These farms were excluded from further analysis, due to the lack of possibility of attaining the GLOBAL G.A.P. GRASP standard. Almost half of the farms employed local workers and family members. Approximately 10% of the farms hired only foreign workers, and 15% hired foreign and local workers Table 1. Other farms employ seasonal workers, mainly citizens of the Ukrainian Republic.

Table 1. Employment structure in the surveyed farms.

Research Group No.	Specification	Number of Farms	Average Farm Area in the Group
1	Lack of workers	17	5.19
2	Local workers, close or extended family	28	9.32
3	Foreign workers	7	14.89
4	Foreign and local workers	10	15.26

Workers were employed on employment contracts only in 10 farms. For separated research groups, a risk analysis was carried out for the implementation of social practices according to the guidelines of the ISO 31000:2018 standard. Inventory of identified risks was carried out based on the guidelines of GLOBAL G.A.P.

In the conducted research, the strategic goal was to adjust the management policy of social practices on the farm to GLOBAL G.A.P. GRASP standards. Based on the conditions in the surveyed farms, (the size and assortment of production, infrastructure equipment, cultural factors, and the mentality of farmers). The matrix used for risk analysis is included in Table 2. According to this table, the level of risk in a three-level scale is defined as small, medium, and large. The identified risks are presented in Table 3. Two values have been identified for each identified risk factor; the probability of risk occurrence and the threat to the strategic goal (compliance with the GLOBAL G.A.P. GRASP standard) in the case of risk.

Table 2. Matrix used for risk assessment.

The Probability of Occurrence of a Risk Factor	The Consequences of Risk		
	Low	Medium	High
Low	L	L	M
Medium	L	M	H
High	M	H	H

H—high risk, M—medium risk, L—low risk.

3. Results and Discussion

The implementation of quality management systems, both at state and private level, is widely considered to be a factor that increases the profitability of farms, as well as a method for reducing the environmental effects associated with agricultural production. Bibliographic data presented by other researchers point to a relationship between the implementation of a quality management system and the amount of income [18,19]. Positive effects are perceived also in relation to the effectiveness of using natural resources in agricultural production. One of the weak points of quality management systems in primary production is associating them with the amount of earnings and workers' lifestyle. Oya et al. [18] draw attention to this problem and point to the need of putting more emphasis on social issues in food production at farm level. Despite the fact that quality systems in primary production, such as Organic Farming, GLOBAL G.A.P., Integrated Production, or sales network quality systems take social aspects into account in their principles. Nevertheless, they can rarely guarantee that farm owners will satisfy workers' social needs. That is the reason why, in many cases, it is essential to certify producers for compliance with the principles of social systems [20]. In recent years, social aspects have permanently entered the range of parameters that are part of the notion of food product quality. The basic components of the quality of agri-food products include:

1. Product safety associated with microbiological, chemical, and physical threats
2. Confirmation that the production principles applied in the used technology are compliant with the principles of sustainable development.
3. Confirmation that, during production, basic principles associated with hygiene and safety of workers have been met.
4. Confirmation that, during production, social practices consistent with principles of good social practices and with principles of international labor conventions have been applied.

All these factors shape the quality of agricultural products and their implementation is connected with incurring costs for technical infrastructure, consultancy, and for administering the system. The surveyed farms are associated in producer groups, and the implementation of the GRASP standard was necessary in order to win a new market in Great Britain. Prior to making a decision about implementing the GRASP standard, a risk analysis was carried out. GRASP is a voluntary, additional module, which not part of the accredited GLOBALG.A.P. certification. It completes the requirements of the standard with respect to good social practices. GRASP certification is only possible when producers are certified for compliance with the GLOBALG.A.P. system or certified in compliance with an equivalent system subjected to benchmarking. The interpretation of GRASP control points depends on the country in which activities are conducted. According to the principles of the standard, the requirements of national legislation replace the GRASP requirements when appropriate legal regulations are more demanding of the GRASP requirements. If there are no legal regulations or if the law is not as demanding, GRASP principles provide minimum compliance criteria.

The foundation of the GRASP standard is a structural organization of a farm that will allow it to eliminate the influence of the management on the workers' representatives. According to the standard, an enterprise must have people acting as the workers' representatives. The workers' representatives are obliged to represent their interests. These representatives should be independent and not work in a position associated with managing the company. The introduced system should guarantee regular meetings between the workers' representatives and the Board, where workers' issues are addressed. Transcripts of these meetings should be made available in company documentation. Workers' representatives should be chosen through voting. They should be chosen by workers and recognized by the Board. A representative can be nominated only in exceptional circumstances. The voting or nomination must take place in the current year or season. Workers' representatives should be aware of their role and rights, and they should be able to have discussions with the Board regarding complaints and suggestions. They should also have knowledge about current legislation and principles of international labor conventions. A quality management system in a farm should be armed with an effective procedure of filing complaints and motions. The manner of filing complaints should be clear and should not generate sanctions for the person who files a complaint or for the workers' representatives who speak for the said person.

Table 3. Risk assessment for social practice in the researched farms.

Specification	Group of Farms			
	1	2	3	4
Selection of the workers' representative	n.a.	H	H	H
Establishing the date of selecting workers	n.a.	H	H	H
Adapting the issue of employee discrimination	n.a.	L	L	M
Adapting the issues relating to contracts with workers	n.a.	H	L	M
Adapting the issue of minimum wages and equal salary	n.a.	H	L	M
Adapting the issues connected with documenting the work time, as well as the level of remuneration connected therewith, overtime hours	n.a.	H	M	M
Adapting the issues relating to employing minors	n.a.	H	L	M
Adapting the issues connected with ensuring proper social conditions for workers living on the farm	n.a.	L	M	H
Total risk	n.a.	H	M	M

H—high risk, M—medium risk, L—low risk, n.a.—not applicable.

Assuring the above identified principles of the standard in the surveyed farms is problematic for objective reasons. In the farms which employ workers, nominating a potential worker's representative eventuated as the biggest problem. The organizational structure of most farms are based on the management and production workers. Among the seasonal workers, it is very difficult to find a person who can assume the responsibility of representing workers' interest and who is knowledgeable of the principles of labor law that are in force in Poland and who knows about international labor conventions. In the majority of the surveyed farms, the only workers who possess the necessary knowledge are those connected with the owner. Establishing the date of selecting the workers' representatives might also be problematic. Therefore, the risk in this respect was assessed as high Table 3. The harvest season is short in the majority of the surveyed farms, and before the season, a limited number of workers are hired. If the election is carried out at the beginning of the vegetation season, then the workers hired in the harvest season will not have the ability to choose their representative. If the election is carried out after hiring pickers, in the period from the beginning of the vegetation season to harvest, the workers' representatives will not operate. In both cases there is a risk that the principles of the standard will not be observed and the certificate will not be granted.

Another principle of the GRASP standard is signing and implementing a declaration to ensure that good social practices and the observance of human rights for all employees. Such a declaration should contain the obligation to comply with the ILO (International Labor Organization) convention with respect to discrimination, legal working age, and the eliminating forced labor. The declaration should have an obligation regarding the acceptance of the rights to freedom of association and to equal remuneration.

In all the studied farms, the risk associated with signing and implementing a declaration of good social practices was assessed as low. Medium risk was determined in farms that employ both foreigners and local population Table 3. An identified risk associated with this field of activities were issues of equal remuneration and minimum wages. To this extent, the functioning systems generally needed improvement, although, to a smaller extent in farms employing foreign workers Table 3.

One of the most significant sources of risk to implementing the GRASP standard was the issue of contracts with workers. According to the principles of the standard, each worker should be employed on a contract consistent with the legislation of the country where activities are carried out. The contract must be signed, should include a full name, nationality, description of work position, date of birth, date of commencing work, working time, remuneration and employment period. For foreign workers—their legal status and work permit. In the case of farms that employ only foreign workers, the risk for the area in question was estimated as low, due to Polish law regarding hiring foreigners. In the case of seasonal local workers, issues associated with employment contracts were assessed at a high level Table 3. During harvest and other activities associated with production, neighbors or both nuclear or extended family members are often hired. Due to the seasonal nature of production and problems with acquiring workers, the turnover of people employed in the farm is very high. In such a situation, the reorganization of the human resources management system would be very difficult and entail incurring additional costs. In addition, adapting employment to legal regulations in many cases may be connected with difficulty in finding employees.

Employing minors on farms is part of a tradition of small farms in many countries [21]. Among the surveyed farms, in group no. 2 that uses local labor force, minors are very frequently hired as help. This particularly applies to producers growing strawberries and raspberries. Because of the harvest period which takes place during school summer holidays, pupils are often hired for the harvest. In these farms, resigning from this source of labor is associated with an increase in costs incurred on production and with the risk of a shortage of workers.

Assurances of adequate social conditions for workers living on the premises of the farm is key to good social practices in agriculture. The parameters of assessment of housing conditions for seasonal workers should be selected by taking into account the workers' nationality and cultural identity that lead to their specific needs [22]. Regardless of the workers background, providing basic social needs is essential to ensuring proper quality of life and conditions for rest. Ensuring adequate housing conditions in farming, particularly in the case of seasonal workers, is a crucial factor associated with social practices on farms, and regardless of the place of agricultural production, which was highlighted by Vallejos et al. [23], Rima et al. [24]. On the other hand, the possibility of living on a farm is a factor which makes it easier to obtain seasonal workers. As a result of the conducted risk analysis, in compliance with the guidelines of the ISO 31000:2018 standard, the risks associated with the scope of these activities was assessed as low in farms using local labor force and medium in the group of farms that offer apartments to their employees Table 3. All the surveyed farms, that offer apartments to their employees, carry out actions for the improvement of housing conditions as this is what employees expect. In the case of farms with low income, adjusting housing conditions to the requirements of the standard can be a considerable strain.

The implementation of quality systems in primary production is associated with the necessity of incurring high costs, consisting of charges incurred on the certification, consultancy, infrastructural changes, as well as changes in production technology, which are frequently related with increasing cost intensity. These factors restrict, and sometimes prevent, the implementation of quality management systems, particularly in countries where small farms with a small production scale are dominant. A consequence of the development of quality management systems in primary production, and of increasing market demand for certified products can be the restructuring of agriculture towards creating large commercial farms, where the implementation of a quality management system is easier. Such changes might be disadvantageous in social and societal terms, particularly in countries where the efficient functioning of small farms is part of local tradition [25,26]. Despite the image benefits and facilitation of selling products from certified farms, economic factors play a key role when deciding to implement a quality management system [27,28]. One of the main problems of the certification of quality management systems is their mal-adjustment to the producer market based on small farms [12,29]. The results of the conducted surveys point to a high risk associated with implementation of the GRASP standard in the group of the smallest farms which use local labor force as well as family. In these cases, the benefits of owning a certificate may not cover the costs arising from implementation of the standard. In bigger farms, which employ foreign workers, the risk associated with the implementation was assessed as medium Table 3. When analyzing the current situation in the market, at a high level of risk, implementation of the GLOBAL G.A.P. GRASP standard is unreasonable. In the case of the medium risk, the decision about implementing the standard should be made basing on a risk analysis carried out for an individual farm. Due to the changing requirements of retail chains, one should expect an increase in farmers interest in the GLOBAL G.A.P. GRASP. Based on the obtained results, it was found that they need to continue in the conditions of small commercial farms.

4. Summary

The results of the conducted surveys point to a substantial level of risk associated with the implementation of the GRASP standard in small vegetable and horticultural farms in Poland. The effective implementation of the standard can be problematic, due to there being no one who fulfils the criteria for workers' representatives. Another problem is establishing a proper date for choosing the workers' representative because of considerable staff rotations during the vegetation season. The obligation to conclude employment contracts or civil law agreements, and to create a system for recording workers' work time is a problem for a lot of farms. Ensuring proper housing conditions is a factor preventing the implementation of the GRASP standard for some farms. In approximately 35% of the surveyed farms, the workers are people from local communities who do not want to work based on civil law agreements. It is the traditional model of purchasing labor in small farms in Poland. Adapting the system to the requirements of the standard would involve the necessity to build, from scratch, a system for managing human resources in farms. This may be problematic, not only in terms of costs, but also because it would require changing the mentality. The results of the conducted surveys indicate a high risk associated with reorganizing the surveyed farms. Based on the conducted risk analysis, it was established that the implementation of the standard in the surveyed farms, which use the local labor force, is currently unreasonable. In the case of bigger farms, which use foreign workers, making a decision about certification should be preceded by an individual risk analysis carried out for a specific farm.

Author Contributions: Conceptualization, M.N., Z.G.-S. and D.K.; methodology, M.K., A.S.-S. and J.K.-D.; resources, M.N., M.K., A.S.-S. and J.S.; formal analysis, M.N., D.K. and M.K.; investigation, D.K., Z.G.-S., M.N. and J.S.; resources, M.N., M.K., M.K. and J.K.-D.; data curation, A.S.-S. and D.K.; writing—M.N., M.K., Anna S.-S. and M.K.; visualization, D.K., J.S. and M.N.; funding acquisition, M.N.

Funding: This research was funded by the Ministry of Science and Higher Education of the Republic of Poland.

Conflicts of Interest: The authors declare no conflict of interest. The funders had no role in the design of the study; in the collection, analyses, or interpretation of data; in the writing of the manuscript, or in the decision to publish the results.

References

1. Wencong, L.; Godwin Seyram Agbemavor Kwasi, H. Transition of small farms in Ghana: Perspectives of farm heritage, employment and networks. *Land Use Policy* **2019**, *81*, 434–452. [CrossRef]
2. Kocira, S.; Kuboń, M.; Sporysz, M. Impact of information on organic product packagings on the consumers decision concerning their purchase. In Proceedings of the 17th International Multidisciplinary Scientific GeoConference SGEM 2017, Sofia, Bulgaria, 27 June–6 July 2017; Volume 17, pp. 499–506. [CrossRef]
3. Chowaniak, M.; Klima, K.; Niemiec, M. Impact of slope gradient, tillage system, and plant cover on soil losses of calcium and magnesium. *J. Elementol.* **2016**, *21*, 361–372. [CrossRef]
4. Ivanyshyn, V.; Nedilska, U.; Khomina, V.; Klymyshena, R.; Hryhoriev, V.; Ovcharuk, O.; Hutsol, T.; Mudryk, K.; Jewiarz, M.; Wróbel, M.; et al. Prospects of growing miscanthus as alternative source of biofuel. *Renew. Energy Sources Eng. Technol. Innov.* **2018**, 801–812. [CrossRef]
5. Szeląg-Sikora, A.; Niemiec, M.; Sikora, J.; Chowaniak, M. Possibilities of Designating swards of grasses and small-seed legumes from selected organic farms in Poland for feed. In Proceedings of the IX International Scientific Symposium "Farm Machinery and Processes Management in Sustainable Agriculture", Lublin, Poland, 22–24 November 2017; pp. 365–370.

6. Niemiec, M.; Komorowska, M.; Szeląg-Sikora, A.; Kuzminova, N. Content of Ba, B, Sr and as in water and fish larvae of the genus Atherinidae L. sampled in three bays in the Sevastopol coastal area. *J. Elem.* **2018**, *23*, 1009–1020. [CrossRef]

7. Niemiec, M.; Sikora, J.; Szeląg-Sikora, A.; Kuboń, M.; Olech, E.; Marczuk, A. Applicability of food industry organic waste for methane fermentation. *Przem. Chem.* **2017**, *69*, 685–688. [CrossRef]

8. Sikora, J.; Niemiec, M.; Szelag-Sikora, A.; Kuboń, M.; Olech, E.; Marczuk, A. Biogasification of wastes from industrial processing of carps. *Przem. Chem.* **2017**, *96*, 2275–2278.

9. Cupiał, M.; Szeląg-Sikora, A.; Niemiec, M. Optimization of the machinery park with the use of OTR-7 software in context of sustainable agriculture. *Agric. Agric. Sci. Proc.* **2015**, *7*, 64–69. [CrossRef]

10. Shukl, S.; Prakash Singh, S.; Shankar, R. Evaluating elements of national food control system: Indian context. *Food Control* **2018**, *90*, 121–130. [CrossRef]

11. Mzoughi, N. Farmers adoption of integrated crop protection and organic farming: Do moral and social concerns matter? *Agric. Econ.* **2011**, *70*, 1536–1545. [CrossRef]

12. Azhar, B.; Prideaux, M.; Razi, N. Sustainability certification of food. *Ref. Modul. Food Sci. Encycl. Food Secur. Sustain.* **2019**, *2*, 538–544. [CrossRef]

13. Tran, D.; Daisaku, D. Impacts of sustainability certification on farm income: Evidence from small-scale specialty green tea farmers in Vietnam. *Food Policy* **2019**, *83*, 70–82. [CrossRef]

14. Gródek-Szostak, Z.; Szeląg-Sikora, A.; Sikora, J.; Korenko, M. Prerequisites for the cooperation between enterprises and business supportinstitutions for technological development. *Bus. Non-Profit Organ. Facing Increased Compet. Grow. Cust. Demand* **2017**, *16*, 427–439.

15. Szeląg-Sikora, A.; Niemiec, M.; Sikora, J. Assessment of the content of magnesium, potassium, phosphorus and calcium in water and algae from the black sea in selected bays near Sevastopol. *J. Elem.* **2016**, *21*, 915–926. [CrossRef]

16. Tey, Y.S.; Rajendran, N.; Brindal, M.; Ahmad Sidique, S.F.; Nasir, M.; Shamsudin, N.M.; Alias, S.; Radam, A.; Hadi, A.H.I. A review of an international sustainability standard (GlobalGAP) and its local replica (MyGAP). *Outlook Agric.* **2016**, *45*, 67–72. [CrossRef]

17. Holzapfel, S.; Wollni, M. Global GAP Certification of small-scale farmers sustainable? Evidence from Thailand. *J. Dev. Stud.* **2014**, *50*, 731–746. [CrossRef]

18. Oya, C.; Schaefera, F.; Skalidou, D. The effectiveness of agricultural certification in developing countries: A systematic review. *World Dev.* **2018**, *112*, 282–312. [CrossRef]

19. Byerlee, D.; Rueda, X. From public to private standards for tropical commodities: A century of global discourse on land governance on the forest frontier. *Forests* **2015**, *6*, 1301–1324. [CrossRef]

20. Raynolds, L.T. Fair-trade labour certification: The contested incorporation of plantations and workers. *Third World Q.* **2017**, *38*, 1473–1492. [CrossRef]

21. Hamenoo, E.S.; Dwomoh, E.A.; Dako-Gyeke, M. Child labor in Ghana: Implications for children'seducation and health. *Child Youth Serv. Rev.* **2018**, *93*, 248–254. [CrossRef]

22. Xiao, Y.; Benoît, C.; Norris, B.N.; Lenzen, M.; Norris, G.; Murray, J. How social footprints of nations can assist in achieving the sustainable development goals. *Ecol. Econ.* **2017**, *135*, 55–65. [CrossRef]

23. Vallejos, Q.M.; Quandt, S.A.; Grzywacz, J.G.; Isom, S.; Chen, H.; Galván, L.; Whalley, L.; Chatterjee, A.B.; Arcury, T.A. Migrant farmworkers' housing conditions across an agricultural season in North Carolina. *Am. J. Ind. Med.* **2011**, *54*, 533–544. [CrossRef] [PubMed]

24. Habib, R.R.; Mikati, D.; Hojeij, S.; El Asmar, K.; Chaaya, M.; Zurayk, R. Associations between poor living conditions and multi-morbidity among Syrian migrant agricultural workers in Lebanon. *Eur. J. Public Health* **2016**, *26*, 1039–1044. [CrossRef] [PubMed]

25. Kibet, N.; Obare, G.A.; Lagat, K.J. Risk attitude effects on Global-GAP certification decisions by smallholder French bean farmers in Kenya. *J. Behav. Exp. Financ.* **2018**, *18*, 18–29. [CrossRef]

26. Partzsch, L.; Kemper, L. Cotton certification in Ethiopia: Can an increasing demand for certified textiles create a "fashion revolution"? *Geoforum* **2019**, *99*, 111–119. [CrossRef]

27. Ibanez, M.; Blackman, A. Is Eco-Certification a win-win for developing country agriculture? Organic coffee certification in Colombia. *World Dev.* **2016**, *82*, 14–27. [CrossRef]

28. Staricco, J.I.; Ponte, S. Quality regimes in agro-food industries: A regulation theory reading of Fair Trade wine in Argentina. *J. Rural. Stud.* **2015**, *38*, 65–76. [CrossRef]

29. Brown, S.; Getz, C. Privatizing farm worker justice: Regulating labor through voluntary certification and labeling. *Geoforum* **2008**, *39*, 1184–1196. [CrossRef]

Article

Assessment of the Properties of Rapeseed Oil Enriched with Oils Characterized by High Content of α-linolenic Acid

Agnieszka Sagan [1], Agata Blicharz-Kania [1,*], Marek Szmigielski [1], Dariusz Andrejko [1], Paweł Sobczak [2], Kazimierz Zawiślak [2] and Agnieszka Starek [1]

[1]   Department of Biological Bases of Food and Feed Technologies, University of Life Sciences in Lublin, 20-612 Lublin, Poland; agnieszka.sagan@up.lublin.pl (A.S.); marek.szmigielski@up.lublin.pl (M.S.); dariusz.andrejko@up.lublin.pl (D.A.); agnieszka.starek@up.lublin.pl (A.S.)

[2]   Department of Food Engineering and Machines, University of Life Sciences in Lublin, 20-612 Lublin, Poland; pawel.sobczak@up.lublin.pl (P.S.); kazimierz.zawislak@up.lublin.pl (K.Z.)

*   Correspondence: agata.kania@up.lublin.pl; Tel.: 81-531-96-46

Received: 25 September 2019; Accepted: 11 October 2019; Published: 13 October 2019

Abstract: Functional foods include cold-pressed oils, which are a rich source of antioxidants and bioactive n-3 and n-6 polyunsaturated fatty acids. The aim of this study was to assess the quality of rapeseed oils supplemented with Spanish sage and cress oils. Seven oil mixtures consisting of 70% of rapeseed oil and 30% of sage and/or cress oil were prepared for the analyses. The oil mixtures were analyzed to determine their acid value, peroxide value, oxidative stability, and fatty acid composition. In terms of the acid value and the peroxide value, all mixtures met the requirements for cold-pressed vegetable oils. The enrichment of the rapeseed oil with α-linolenic acid-rich fats resulted in a substantially lower ratio of n-6 to n-3 acids in the mixtures than in the rapeseed oil. The mixture of the rapeseed oil with the sage and cress oils in a ratio of 70:10:20 exhibited higher oxidative stability than the raw materials used for enrichment and a nearly 20% α-linolenic acid content. The oils proposed in this study can improve the ratio of n-6:n-3 acids in modern diets. Additionally, mixing the cress seed oils with rapeseed oil and chia oil resulted in a reduction in the content of erucic acid in the finished product. This finding indicates that cress seeds, despite their high content of erucic acid, can be used as food components. The production of products with a positive effect on human health is one of the most important factors in the sustainable development of agriculture.

Keywords: cold-pressed oils; functional food; oxidative stability; rapeseed oil; Spanish sage seed oil; cress seed oil

---

1. Introduction

The results available from health and nutrition research increase public awareness of the relationship between health and diet. Therefore, functional food is becoming increasingly popular among consumers. The idea of functional food originates from Japan, where a national program of studies on the relationship between food science and medicine was introduced in 1984. Functional food is defined as follows: "a food can be regarded as 'functional' if it is satisfactorily demonstrated to affect beneficially one or more target functions in the body, beyond adequate nutritional effects, in a way that is relevant to either an improved state of health and well-being and/or reduction of risk of disease. Functional foods must remain foods and they must demonstrate their effects in amounts that can normally be expected to be consumed in the diet: these are not pills or capsules, but part of a normal food pattern" [1–3]. The production of such food may be one of the most important aspects of strategies for sustainable development. The essence of the sustainable development of food production is the production of products with a positive effect on human health [4].

Functional foods include cold-pressed oils, as they are a rich source of antioxidants such as tocopherols and polyphenols, and hence exhibit high antioxidant activity. These also contain polyunsaturated fatty acids from the n-3 and n-6 groups as well as sterols, which exert a bioactive effect [5,6]. Seeds or fruits of plants that are traditionally regarded as oil-bearing species, e.g., rape, soybean, sunflower, olives, etc., are used for pressing oils. Also, oils pressed from unusual plant materials such as nuts and sage or garden cress seeds are gaining popularity.

Rapeseed oil is one of the most frequently consumed vegetable oils, and it is one of the most valuable edible fats, mainly due to the high content (approx. 90%) of 18-carbon unsaturated acids. It is produced from rapeseed varieties with a low level of erucic acid and is characterized by the presence of fatty acids that are desirable in human and animal diets and are present in ratios that are similar to those in the most valuable oils, e.g., olive oil. Furthermore, it is rich in many bioactive compounds, whose presence in food is the current focus of attention. These are primarily antioxidant compounds. The oil is also a source of essential unsaturated fatty acids from the n-6 and n-3 groups. The content of linoleic and α-linolenic acids in rapeseed oil is usually approx. 20% and 10%, respectively [7,8]. It also exhibits good oxidative stability, which is better than that of soybean and sunflower oils [9]. The oxidative stability of rapeseed oil can be improved by supplementation with natural antioxidants contained in spices (dried rosemary, oregano) and in resin from trees of the *Burseraceae* family [10,11]. Research was also conducted on the stability of mixtures of rapeseed oil with linseed oil. Mixtures of rapeseed oil with 25% and 50% linseed oil were found to have the best properties [12].

It is important to maintain an appropriate n-6:n-3 ratio of polyunsaturated fatty acids. Most reports in the literature indicate that the n-6:n-3 ratio should range from 1:1 to 4:1. However, modern diets usually contain up to 20-fold higher levels of n-6 than n-3 fatty acids [13]. Therefore, in terms of sustainable development strategies, it is vital to enrich food with n-3 acids, e.g., α-linolenic acid, which originates from vegetable oils such as Spanish sage or garden cress oils [14,15]. The production of products that have a positive effect on human health is one of the most important element in the sustainable development of agriculture.

Seeds of Spanish sage (*Salvia hispanica* L.) contain large amounts of fat (30%–40%), which is rich in essential fatty acids, in particular α-linolenic acid. This may account for 68% of all fatty acids [14,16]. Sage seeds are also characterized by a high level of antioxidants, which may be helpful in preventing diseases related to oxidative DNA damage [17,18].

Garden cress (*Lepidium sativum* L.) is a good source of n-3 fatty acids. It is an annual plant from the family Brassicaceae (*Cruciferae*). Its seeds contain large amounts of fat (28%). Its main fatty acids are oleic acid (30%) and α-linolenic acid (32%) representing the n-3 group. Cress seeds also contain a high level of polyphenols and are a source of bioactive glycosides [15,19]. It has been found that the isoflavone glycoside isolated from *L. sativum* seeds can improve liver function and the serum lipid profile. It can also reduce the generation of free radicals through induction of an antioxidant defense mechanism and acts as a potential antioxidant against paracetamol poisoning [20].

Polyunsaturated fatty acids from the n-3 group enhance the nutritional value of food products but also increase the susceptibility to fat oxidation. Spanish sage seeds are rich in essential fatty acids; yet, the low oxidative stability of the oil pressed from this raw material is a shortcoming. Additionally, the nutritional value of cress seed oil may be reduced by the elevated content of erucic acid.

The aim of this study was to verify the quality of a composition of rapeseed oil supplemented with oils from Spanish sage and cress seeds in various proportions.

2. Materials and Methods

2.1. Research Material

The research material included seven different oil mixtures based on rapeseed oil. Seeds of winter oilseed rape (*Brassica napus* L.), Spanish sage (*Salvia hispanica* L.), and garden cress (*Lepidium sativum* L.) were used to obtain the experimental oils. The seeds were analyzed for moisture [21] and fat content [22]. The moisture contents were approximately 7% and the fat contents were typical for the seeds of the plants (Table 1).

Table 1. Humidity and fat content in the seeds used for oil pressing and pressing yields.

Seeds	Humidity (%)	Fat Content (%)	Pressing Yield (%)
Rape	6.97 [a] ± 0.01	38.40 [b] ± 0.06	79.0
Spanish sage	7.11 [b] ± 0.01	36.59 [a] ± 0.31	60.7
Garden cress	7.61 [c] ± 0.03	20.53 [c] ± 0.01	41.9

[abc]—Statistically significant differences in columns ($p \leq 0.05$). The results are expressed as mean ± SD (n = 3).

Individual batches of seeds were cold pressed in a Farmet DUO screw press (Czech Republic) with a capacity of 18–25 kg·h^{-1}, engine power 2.2 kW, and screw speed 1500 rpm equipped with a 10-mm diameter nozzle. Before starting, the press was heated to 50 °C. After pressing, the oil was left for natural sedimentation for 5 days in refrigerated conditions (10 ± 1 °C) and then decanted. The pressing efficiency was calculated based on the weight of the seeds used for the process, the fat content of the seeds, and the weight of the pressed oil. The following formula was used to calculate the pressing efficiency (W):

$$W = \frac{m_o \cdot 10^4}{f \cdot m_s} \ (\%) \tag{1}$$

where:

m_o—weight of oil (kg),
m_s—weight of seeds (kg),
f—fat content in seeds (%)

The highest yield of 79% was obtained from the rapeseeds, whereas the lowest value, i.e., approximately 42%, was noted for the cress seeds (Table 1).

The acid value (AV), peroxide value (PV), oxidative stability, and fatty acid composition in the pressed oil were also determined.

2.2. Research Methods

The oils were mixed in 500-ml amber glass bottles at appropriate weight ratios. Seven mixtures of oil marked from M1 to M7 were prepared for the analyses with 70% of rapeseed oil (R) and 30% of Spanish sage and/or cress oils. The proportions of the individual oils and the symbols denoting the individual mixtures are presented in Table 2. The oil mixtures were analyzed for their acid value (AV), peroxide value (PV), oxidative stability, and fatty acid composition.

Table 2. Experimental setup.

Oil	Proportions of the Individual Oils in the Mixtures (%)						
	M1	M2	M3	M4	M5	M6	M7
Rapeseed oil	70	70	70	70	70	70	70
Spanish sage seed oil	-	5	10	15	20	25	30
Cress seed oil	30	25	20	15	10	5	-

The acid value was determined with the titration method using a cold solvent and expressed in mg KOH per 1 g oil [23]. The peroxide value was estimated by titration with iodometric determination of the end point and expressed in mmol O_2 per 1 kg of oil [24]. Oxidative stability was assessed in the Rancimat accelerated oxidation test [25]. The test was carried out using a Metrohm 893 Professional Biodiesel Rancimat device. Oil samples (3.00 ± 0.01 g) were weighed, placed in a measuring vessel, and exposed to air at a flow rate of 20 l/h at 120 °C. The results were expressed as the induction time determined from the curve inflection point using the StabNet1.0 software provided by the manufacturer. The fatty acid composition was determined using gas chromatography (a Bruker 436GC chromatograph with an FID detector) following the relevant standards [26,27]. Fatty acid methyl esters were separated on a BPX 70 capillary column (60 m × 0.25 mm, 25 μm) with nitrogen as the carrier gas.

2.3. Statistical Analysis

All determinations were made in triplicate. The arithmetic mean of the replicates was assumed as the result. The results were analyzed statistically using the StatSoft Polska STATISTICA 10.0 program. The significance of differences between the mean values of the determined parameters was verified using Tukey's test. The calculations were made at the significance level $\alpha = 0.05$.

3. Results and Discussion

The determined acid and peroxide values, which determine the quality of pressed oils, are presented in Table 3.

Table 3. Chemical properties of the analyzed oils.

Oil	Acid Number (mg KOH·g^{-1})	Peroxide Number (mmol O$_2$·kg^{-1})	Induction Time (h)
Rapeseed oil	0.66 [a] ± 0.02	1.85 [a] ± 0.25	4.40 [a] ± 0.09
Spanish sage seed oil	4.77 [b] ± 0.02	1.00 [b] ± 0.01	0.65 [c] ± 0.07
Cress seed oil	0.64 [a] ± 0.03	1.24 [ab] ± 0.07	2.67 [b] ± 0.04

[abc]—Statistically significant differences in columns ($p \leq 0.05$). The results are expressed as mean ± SD (n = 3).

The highest acid value (4.77 mg KOH·g^{-1} fat) was determined for the chia seed oil. This value was only slightly higher than the requirements specified for this parameter (LK $\leq$ 4 mg·g^{-1}) in the Codex Alimentarius [28]. As demonstrated by Krygier et al. [29], the acid value of oil depends on the quality of seeds used for pressing and increases with the increasing amounts of damaged seeds, which is explained by the authors by the greater activity of lipolytic enzymes present in such material.

The peroxide value, which reflects the amount of primary oxidation products, ranged from 1.00 to 1.85 mmol O$_2$·kg^{-1} in the analyzed oils and did not exceed the maximum value of 10 mmol O$_2$·kg^{-1} specified in the Codex Alimentarius [28].

Sample gas chromatograms of each oil are shown in Figure 1. The analyzed rapeseed oil had a typical fatty acid composition [30]. The content of unsaturated fatty acids exceeded 91% (Table 4).

Oleic acid was the main fatty acid (55.22%). The rapeseed oil contained the highest amount of linoleic acid (24.24%) and the lowest level of α-linolenic acid (10.34%). The ratio of n-6 to n-3 acids of 2:3 was equally low.

The Spanish sage oil had a high level of unsaturated fatty acids (89.43%), which is consistent with results reported by other authors [31,32]. The main acids were represented by α-linolenic acid (63.40%) followed by linoleic acid (20.80%). Such a high amount of α-linolenic acid resulted in a very low ratio of n-6 to n-3 acids in chia seed oil, i.e., 0.3.

The cress seed oil exhibited the greatest variety of fatty acids. Palmitic acid (9.02%), stearic acid (3.06%), and arachidonic acid (3.64%) represented saturated fatty acids. Mono-unsaturated fatty acids were represented by oleic acid (20.68%) as well as eicosenoic acid (13.8%) and erucic acid (5.98%). The content of essential unsaturated fatty acids (linoleic and linolenic) accounted for 9.08% and 30.03% of the sum of fatty acids, respectively. The lower content of linoleic acid than that determined in the other oils resulted in a low ratio of n-6 to n-3 acids, as in the Spanish sage oil (0:3). Additionally, there were small amounts of eicosadienoic, eicosatrienoic, lignoceric, and nervonic acids, which were not present in the other oils. The results obtained in the present study are similar to those reported by other authors investigating cress seeds [33,34], although the analyzed oil had almost two-fold higher content of erucic acid than the value determined by Moser et al. [34]. Due to its properties, erucic acid is classified as an anti-nutritional compound. The presence of erucic acid in cress oil is a drawback and limits the application of cress seeds as a raw material for oil pressing.

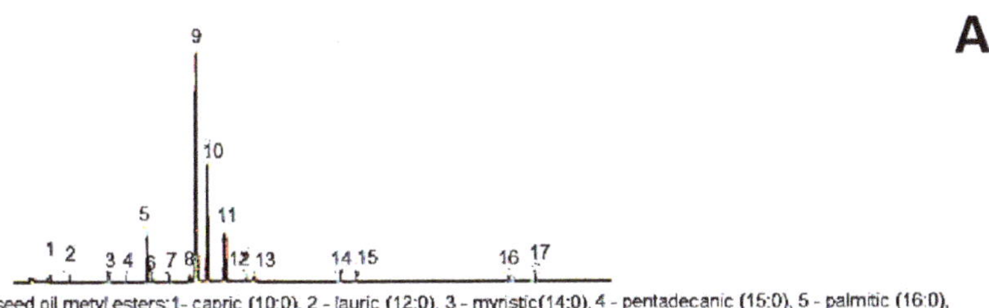

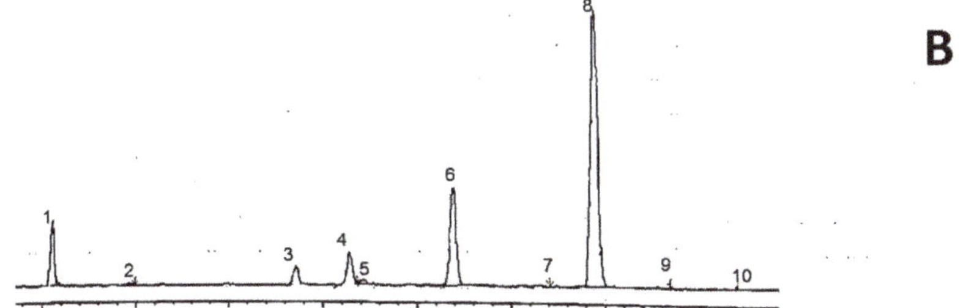

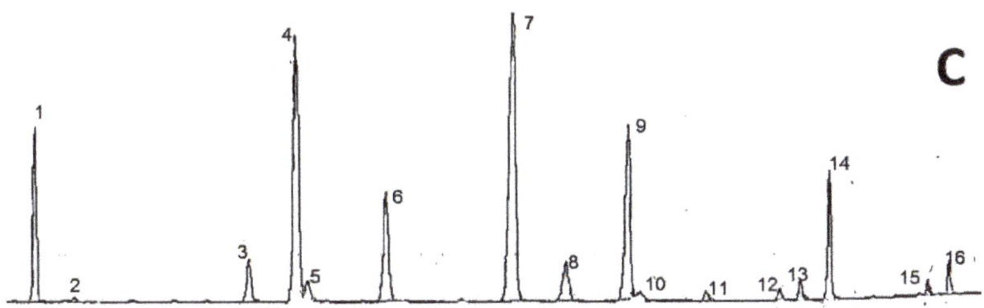

Figure 1. Typical gas chromatogram of: A—rapeseed oil, B—Spanish sage seed oil, C—cress seed oil.

Table 4. Fatty acid composition of the oils (% of FA sum).

Fatty Acids	Rapeseed Oil	Spanish Sage Seed Oil	Cress Seed Oil
Myristic acid 14:0	0.25 ± 0.01	-	-
Palmitic acid 16:0	6.06 ± 0.18	7.46 ± 0.02	9.02 ± 0.17
Palmitoleic acid16:1	0.28 ± 0.01	0.15 ± 0.01	0.18 ± 0.01
Stearic acid 18:0	2.08 ± 0.09	2.79 ± 0.01	3.06 ± 0.05
Oleic acid 18:1	55.22 ± 0.85	4.71 ± 0.02	20.68 ± 0.02
Linoleic acid18:2	24.24 ± 1.13	20.80 ± 0.03	9.08 ± 0.01
α- linolenic acid 18:3 (n-3)	10.34 ± 0.91	63.40 ± 0.02	30.03 ± 0.25
γ- linolenic acid18-3 (n-6)	-	0.19 ± 0.01	-
Arachidic acid 20:0	0.27 ± 0.01	0.32 ± 0.01	3.64 ± 0.01
Eicosaenoic acid 20:1	1.00 ± 0.03	0.18 ± 0.01	13.83± 0.11
Eicosadienoic acid 20:2	-	-	0.54 ± 0.01
Eicosatrienoic acid 20:3	-	-	0.75 ± 0.01
Behenic acid 22:0	0.23 ± 0.01	-	1.11 ± 0.03
Erucic acid 22:1	0.03 ± 0.01	-	5.98 ± 0.12
Lignoceric acid 24:0	-	-	0.58 ± 0.01
Nervonic acid 24:1	-	-	1.52 ± 0.08
ΣSFA [1]	8.89	10.57	17.41
ΣMUFA [2]	56.53	5.04	42.19
ΣPUFA [3]	34.58	84.39	40.40
n-6/n-3	2.3	0.3	0.3

[1] SFA—saturated fatty acids; [2] MUFA—monounsaturated fatty acids; [3] PUFA—polyunsaturated fatty acids.

The oils used in the analyses had different values of oxidative stability (Table 3). The longest induction time was recorded for the rapeseed oil (4.40 h). It was similar to values reported by other authors [30,35]. The oxidative stability of vegetable oils is determined by the fatty acid composition: the greater the number of unsaturated bonds, the greater the susceptibility to oxidation. The chia seed oil exhibited the highest content of polyunsaturated fatty acids (over 84%) and hence the lowest oxidative stability (0.65 h).

The chemical properties of the oil mixtures are shown in Figures 2–4 and Table 5. The oils differed in their acid value and oxidative stability. The acid value ranged from 0.79 to 2.00 mg·g^{-1} and increased with higher amounts of the Spanish sage oil.

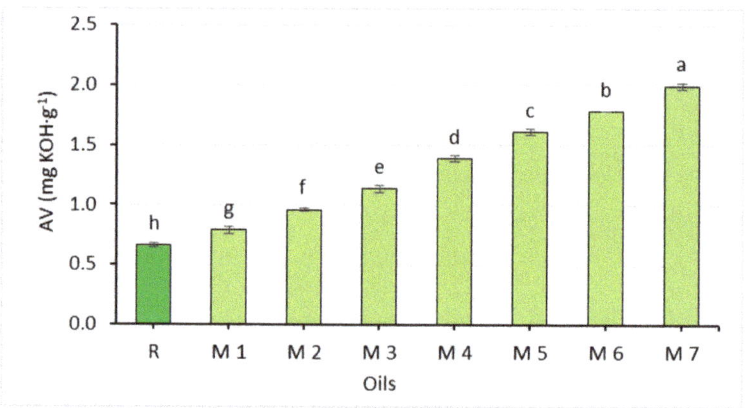

Figure 2. Comparison of the acid value of rapeseed oil (R) and oil mixtures (abc—statistically significant differences, $p \leq 0.05$).

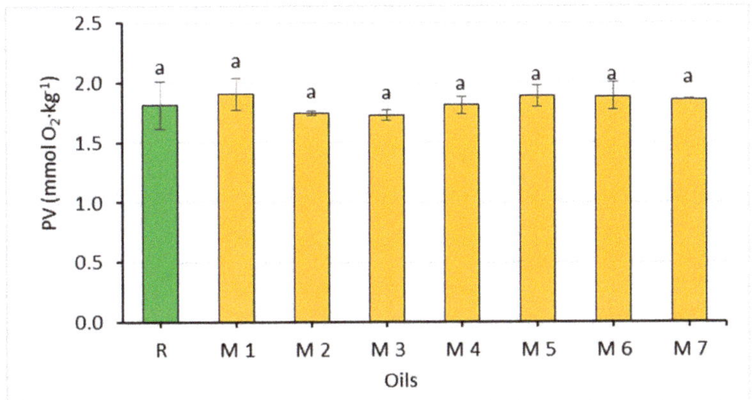

Figure 3. Comparison of the peroxide value of rapeseed oil (R) and oil mixtures (abc—statistically significant differences, $p \leq 0.05$).

Table 5. Fatty acid composition of the oil mixtures (% of FA sum).

Fatty Acids	Oil Mixtures						
	M1	**M2**	**M3**	**M 4**	**M5**	**M6**	**M7**
Myristic acid 14:0	0.13	0.18	0.18	0.18	0.18	0.16	0.18
Palmitic acid 16:0	7.01	6.73	6.62	6.85	6.68	6.60	6.45
Palmitoleic acid16:1	0.28	0.28	0.27	0.26	0.26	0.26	0.27
Stearic acid 18:0	2.27	2.18	2.23	2.35	2.34	2.12	2.33
Oleic acid 18:1	44.94	44.05	43.37	42.36	41.59	40.93	40.15
Linoleic acid 18:2	19.83	20.31	20.74	21.40	22.07	22.65	23.02
α- linolenic acid 18:3 (n-3)	16.18	17.82	19.72	21.26	22.92	24.61	26.30
γ- linolenic acid 18-3 (n-6)	-	0.01	0.02	0.03	0.04	0.05	0.06
Arachidic acid 20:0	1.29	1.11	1.02	0.75	0.61	0.45	0.29
Eicosaenoic acid 20:1	4.62	4.30	3.52	2.80	2.09	1.44	0.77
Eicosadienoic acid 20:2	0.16	0.14	0.11	0.08	0.05	0.03	-
Eicosatrienoic acid 20:3	0.23	0.19	0.15	0.12	0.09	0.04	-
Behenic acid 22:0	0.52	0.46	0.36	0.32	0.25	0.22	0.16
Erucic acid 22:1	1.81	1.69	1.25	0.92	0.62	0.33	0.02
Lignoceric acid 24:0	0.18	0.15	0.12	0.09	0.07	0.03	-
Nervonic acid 24:1	0.55	0.40	0.32	0.23	0.14	0.08	-
ΣSFA [1]	11.40	10.81	10.53	10.54	10.13	9.58	9.41
ΣMUFA [2]	52.20	50.72	48.73	46.57	44.70	43.04	41.21
ΣPUFA [3]	36.40	38.47	40.74	42.89	45.17	47.38	49.38
n-6/n-3	1.2	1.1	1.1	1.0	1.0	0.9	0.9

[1] SFA—saturated fatty acids; [2] MUFA—mono-unsaturated fatty acids; [3] PUFA—polyunsaturated fatty acids.

Since the oils used to prepare the mixtures did not differ significantly in their peroxide value, the values of this parameter were similar in the oil mixtures as well (1.73–1.91 mmol $O_2 \cdot kg^{-1}$). All mixtures met the requirements of acid and peroxide values for cold-pressed vegetable oils [28].

The fatty acid profile of the prepared oil mixtures reflected their raw material composition (Table 5). The highest content of saturated and mono-unsaturated fatty acids (11.40% and 52.20%, respectively) was determined for the M1 mixture consisting of 70% of rapeseed oil and 30% of cress seed oil, which was reflected in it showing the highest oxidative stability of all the oil mixtures. The induction time was only 0.73 h shorter than that of the rapeseed oil. The content of α-linolenic acid in the oil mixtures ranged from 16.18% to 23.02%.

The highest value for α-linolenic acid was determined for the M7 mixture containing 30% Spanish sage oil. Enrichment with this acid resulted in a substantially smaller ratio of n-6 to n-3 acids, i.e., in the range of 1.2 to 0.9, in the oil mixtures than in the rapeseed oil. As suggested by nutritional recommendations, there is a need to increase the intake of long-chain fatty acids from the n-3 group in the diet [36,37]. The oil mixtures proposed in this study can be used to enrich food with this component. According to the EU Commission Regulation of 2014 [38], the maximum allowable level of erucic acid in vegetable oils and fats is 50 g·kg^{-1}, i.e., 5%. By mixing the cress seed oil (containing 5.98% of this acid) with the rapeseed and Spanish sage oils, the erucic acid content was reduced below 2% in all of the analyzed mixtures.

The oxidative stability of the oils, which was expressed as the induction time ranged from 3.67 to 1.92 h and decreased with the increasing content of the Spanish sage oil. However, there were no statistically significant differences in the values for this parameter between mixtures with 5 and 10%, 15 and 20%, as well as 20 and 25% of the chia seed oil (Figure 4).

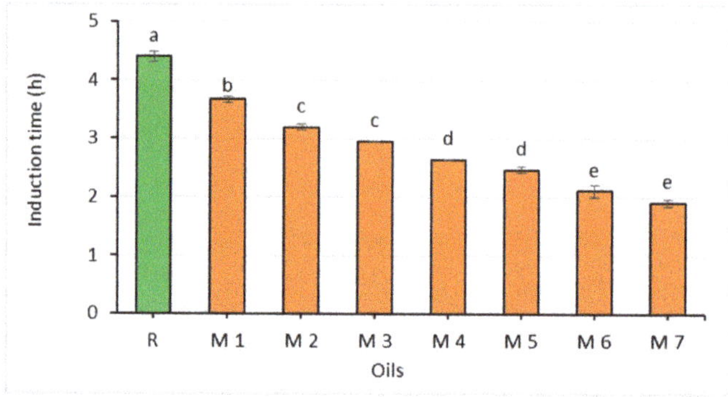

Figure 4. Comparison of the oxidative stability of rapeseed oil (R) and oil mixtures (abc—statistically significant differences, $p \leq 0.05$).

Fat oxidation susceptibility increases proportionally to the number of unsaturated bonds in individual fatty acids [30]. The dependence of oxidative stability of the experimental oil mixtures on the content of polyunsaturated fatty acids is presented graphically (Figure 5) as a linear function.

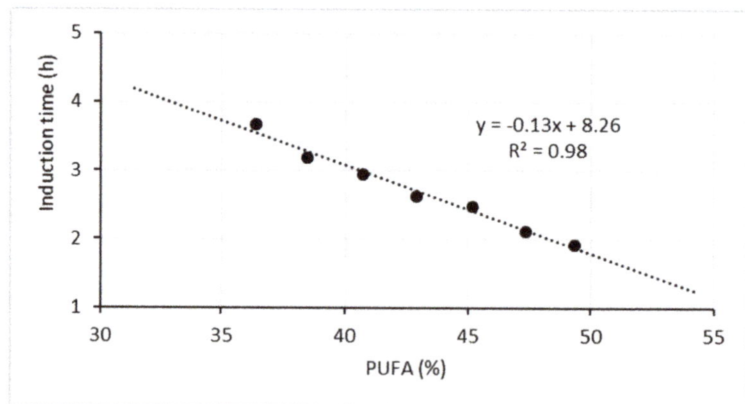

Figure 5. Correlation between the oxidative stability of the oil mixtures and the content of polyunsaturated fatty acids.

As shown by the equation, a 5% increase in the PUFA content in the oil mixtures reduces oxidative stability by shortening the induction time by 0.65 h.

4. Conclusions

The oil mixtures proposed in this study are characterized by higher contents of n-3 fatty acids compared to rapeseed oil, and can therefore improve the n-6 to n-3 ratio in the modern diet. This is beneficial in terms of consumers' health and complies with the principles of sustainable development of food production. The regression equation, developed using the study results, facilitates the precise determination of the properties of rapeseed, cress, and chia oil mixtures. The results indicated that a mixture of rapeseed, chia, and cress oils at a ratio of 70:10:20 exhibits higher oxidative stability than the individual enrichment raw materials, and a nearly 20% level of α-linolenic acid.

Furthermore, the mixture of cress seed oil with rapeseed and Spanish sage oils ensured reduced erucic acid content in the finished product. Hence, cress seeds, which are characterized by a high content of erucic acid can be used in the food industry, which is important for the sustainable development of agriculture and extensive use of its resources. This issue is also related to food security, which is one of the specific objectives of the strategy for sustainable development of rural areas, agriculture and fisheries for the years 2012–2020 in Poland. Within this objective, one of the priorities is the production of high-quality, agri-food products, which are safe for consumers [39].

Author Contributions: Conceptualization, A.S. (Agnieszka Sagan), D.A. and K.Z.; methodology, A.S. (Agnieszka Sagan), M.S. and D.A.; investigation, M.S., P.S. and A.S. (Agnieszka Starek); data curation, A.S. (Agnieszka Sagan), A.B.-K. and M.S.; writing—original draft preparation, A.S. (Agnieszka Sagan), A.B.-K. and M.S.; writing—review and editing, D.A., P.S. and K.Z.; visualization, A.S. (Agnieszka Sagan), A.B.-K. and A.S. (Agnieszka Starek); supervision, D.A.

Funding: This research received no external funding.

Conflicts of Interest: The authors declare no conflicts of interest.

References

1. Cencic, A.; Chingwaru, W. The role of functional foods, nutraceuticals, and food supplements in intestinal health. *Nutrients* **2010**, *2*, 611–625. [CrossRef] [PubMed]
2. Bigliardi, B.; Galati, F. Innovation trends in the food industry: The case of functional foods. *Trends Food Sci. Tech.* **2013**, *31*, 118–129. [CrossRef]
3. Ozen, A.E.; Pons, A.; Tur, J.A. Worldwide consumption of functional foods: A systematic review. *Nutr. Rev.* **2012**, *70*, 472–481. [CrossRef] [PubMed]
4. Tiwari, B.K.; Norton, T.; Holden, N.M. Introduction. In *Sustainable Food Processing*; Tiwari, B.K., Norton, T., Holden, N.M., Eds.; John Wiley & Sons: Hoboken, NJ, USA, 2014; pp. 1–7.
5. Obiedzińska, A.; Waszkiewicz-Robak, B. Oleje tłoczone na zimno jako żywność funkcjonalna. *Zywn Nauk. Technol. Ja.* **2012**, *1*, 27–44.
6. Orsavova, J.; Misurcova, L.; Ambrozova, J.V.; Vicha, R.; Mlcek, J. Fatty acids composition of vegetable oils and its contribution to dietary energy intake and dependence of cardiovascular mortality on dietary intake of fatty acids. *Int. J. Mol. Sci.* **2015**, *16*, 12871–12890. [CrossRef]
7. Gugała, M.; Zarzecka, K.; Sikorska, A. Prozdrowotne właściwości oleju rzepakowego. *Postępy Fitoter.* **2014**, *2*, 100–103.
8. Wang, Y.; Meng, G.; Chen, S.; Chen, Y.; Jiang, J.; Wang, Y.-P. Correlation analysis of phenolic contents and antioxidation in yellow- and black-seeded *Brassica Napus*. *Molecules* **2018**, *23*, 1815. [CrossRef]
9. Wroniak, M.; Łukasik, D.; Maszewska, M. Porównanie stabilności oksydatywnej wybranych olejów tłoczonych na zimno z olejami rafinowanymi. *Zywn Nauk. Technol. Ja.* **2006**, *1*, 214–221.
10. Krajewska, M.; Ślaska-Grzywna, B.; Szmigielski, M. The effect of the oregano addition on the chemical properties of cold-pressed rapeseed oil. *Przem. Chem.* **2018**, *97*, 1953–1956.
11. Starek, A.; Sagan, A.; Kiczorowska, B.; Szmigielski, M.; Ślaska-Grzywna, B.; Andrejko, D.; Kozłowicz, K.; Blicharz-Kania, B.; Krajewska, M. Effects of oleoresins on the chemical properties of cold-pressed rapeseed oil. *Przem. Chem.* **2018**, *97*, 771–773.

12. Marciniak-Łukasiak, K.; Zbikowska, A.; Krygier, K. Wpływ stosowania azotu na stabilność oksydacyjna mieszanin oleju rzepakowego z olejem lnianym. *Żywn Nauk Technol. Ja.* **2006**, *13*, 206–215.

13. Asif, M. Health effects of omega-3,6,9 fatty acids: *Perilla frutescens* is a good example of plant oils. *Orient Pharm. Exp. Med.* **2011**, *11*, 51–59. [CrossRef] [PubMed]

14. Kulczyński, B.; Kobus-Cisowska, J.; Taczanowski, M.; Kmiecik, D.; Gramza-Michałowska, A. The chemical composition and nutritional value of chia seeds—Current state of knowledge. *Nutrients* **2019**, *11*, 1242. [CrossRef] [PubMed]

15. Zia-Ul-Haq, M.; Ahmad, S.; Calani, L.; Mazzeo, T.; del Rio, D.; Pellegrini, N.; de Feo, V. Compositional study and antioxidant potential of *ipomoea hederacea* jacq. and *lepidium sativum* l. seeds. *Molecules* **2012**, *17*, 10306–10321. [CrossRef] [PubMed]

16. Grimes, S.J.; Phillips, T.D.; Hahn, V.; Capezzone, F.; Graeff-Hönninger, S. Growth, yield performance and quality parameters of three early flowering chia (*Salvia hispanica* l.) genotypes cultivated in southwestern Germany. *Agriculture* **2018**, *8*, 154. [CrossRef]

17. Guevara-Cruz, M.; Tovar, A.R.; Aguilar-Salinas, C.A.; Medina-Vera, I.; Gil-Zenteno, L.; Hernandez-Viveros, I.; Lopez-Romero, P.; Ordaz-Nava, G.; Canizales-Quinteros, S.; Guillen Pineda, L.E.; et al. A dietary pattern including nopal, chia seed, soy protein, and oat reduces serum triglycerides and glucose intolerance in patients with metabolilic syndrome. *J. Nutr.* **2012**, *142*, 64–69. [CrossRef]

18. Nowak, K.; Majsterek, S.Z.; Ciesielska, N.; Sokołowski, R.; Klimkiewicz, K.; Zukow, W. The role of chia seeds in nutrition in geriatric patients. *J. Educ. Health Sport* **2016**, *6*, 35–40.

19. Fan, Q.L.; Zhu, Y.D.; Huang, W.H.; Qi, Y.; Guo, B.L. Two new acylated flavonol glycosides from the seeds of *Lepidium sativum. Molecules* **2014**, *19*, 11341–11349. [CrossRef]

20. Sakran, M.; Selim, Y.; Zidan, N. A new isoflavonoid from seeds of *Lepidium sativum* L. and its protective effect on hepatotoxicity induced by paracetamol in male rats. *Molecules* **2014**, *19*, 15440–15451. [CrossRef]

21. PN-EN ISO 665:2004. *Nasiona Oleiste. Oznaczanie Wilgotności i Zawartości Substancji Lotnych*; Polski Komitet Normalizacyjny: Warsaw, Poland, 2004.

22. PN-EN ISO 659:2010. *Nasiona Oleiste. Oznaczanie Zawartości Oleju (Metoda Odwoławcza)*; Polski Komitet Normalizacyjny: Warsaw, Poland, 2010.

23. PN-EN ISO 660:2010. *Oleje i Tłuszcze Roślinne oraz Zwierzęce—Oznaczanie Liczby Kwasowej i Kwasowości*; Polski Komitet Normalizacyjny: Warsaw, Poland, 2010.

24. PN-EN ISO 3960:2012. *Oleje i Tłuszcze Roślinne oraz Zwierzęce—Oznaczanie Liczby Nadtlenkowej—Jodometryczne (Wizualne) Oznaczanie Punktu Końcowego*; Polski Komitet Normalizacyjny: Warsaw, Poland, 2012.

25. Mathäus, B. Determination of the oxidative stability of vegetable oils by rancimat and conductivity and chemiluminescence measurements. *J. Am. Oil Chem. Soc.* **1996**, *73*, 1039–1043.

26. PN-EN ISO 5508:1996. *Oleje i Tłuszcze Roślinne oraz Zwierzęce—Analiza Estrów Metylowych Kwasów Tłuszczowych Metodą Chromatografii Gazowej*; Polski Komitet Normalizacyjny: Warsaw, Poland, 1996.

27. PN-EN ISO 5509:2001. *Oleje i Tłuszcze Roślinne oraz Zwierzęce—Przygotowanie Estrów Metylowych Kwasów tłuszczowych*; Polski Komitet Normalizacyjny: Warsaw, Poland, 2001.

28. FAO/WHO. Codex Standard for Named Vegetable Oils. In *Codex Alimentarius*; ALINORM: Santa Croce, Italy, 2009.

29. Krygier, K.; Wroniak, M.; Grześkiewicz, S.; Obiedziński, M. Badanie wpływu zawartości nasion uszkodzonych na jakość oleju rzepakowego tłoczonego na zimno. *Oilseed Crops* **2000**, *41*, 587–596.

30. Cichosz, G.; Czeczot, H. Stabilność oksydacyjna tłuszczów jadalnych—Konsekwencje zdrowotne. *Bromatol. Chem. Toksyk.* **2011**, *44*, 50–60.

31. Krajewska, M.; Zdybel, B.; Andrejko, D.; Ślaska-Grzywna, B.; Tańska, M. Właściwości chemiczne wybranych olejów tłoczonych na zimno. *Acta Agroph.* **2017**, *24*, 579–590.

32. Segura-Campos, M.R.; Ciau-Solis, N.; Rosado-Rubio, G.; Chel-Guerrero, L.; Betancur-Ancona, D. Physicochemical characterization of chia (*Salvia hispanica*) seed oil from Yucatán, México. *Agric. Sci.* **2014**, *5*, 220–226.

33. Gokavi, S.S.; Malleshi, N.G.; Guo, M. Chemical composition of garden cress (*Lepidium sativum*) seeds and its fractions and use of bran as a functional ingredient. *Plant Food Hum. Nutr.* **2004**, *59*, 105–111. [CrossRef]

34. Moser, B.R.; Shah, S.N.; Winkler-Moser, J.K.; Vaughn, S.F.; Evangelista, R.L. Composition and physical properties of cress (*Lepidium sativum* L.) and field pennycress (*Thlaspi arvense* L.) oils. *Ind. Crop. Prod.* **2009**, *30*, 199–205. [CrossRef]

35. Kruszewski, B.; Fąfara, P.; Ratusz, K.; Obiedziński, M. Ocena pojemności przeciwutleniającej i stabilności oksydacyjnej wybranych olejów roślinnych. *Zesz. Probl. Post. Nauk Roln.* **2013**, *572*, 43–52.
36. Marciniak-Łukasik, K. Rola i znaczenie kwasów tłuszczowych omega-3. *Zywn Nauk. Technol. Ja.* **2011**, *6*, 24–35.
37. Sawada, N.; Inoue, M.; Iwasaki, M.; Sasazuki, S.; Shimazu, T.; Yamaji, T.; Takachi, R.; Tanaka, Y.; Mizokami, M.; Tsugane, S. Consumption of n-3 fatty acids and fish reduces risk of hepatocellular carcinoma. *Gastroenterology* **2012**, *142*, 1468–1475. [CrossRef]
38. Commission Regulation (EU) No 696/2014 of 24 June 2014 amending Regulation (EC) No 1881/2006 as regards maximum levels of erucic acid in vegetable oils and fats and foods containing vegetable oils and fats. *Off. J. EU* **2014**, *L184*, 1–2.
39. Żmija, D. Zrównoważony rozwój rolnictwa i obszarów wiejskich w Polsce. *Studia Ekonomiczne* **2014**, *166*, 149–158.

 *sustainability*

Article

Evaluation of Dust Concentration During Grinding Grain in Sustainable Agriculture

Paweł Sobczak [1], Jacek Mazur [1,*], Kazimierz Zawiślak [1], Marian Panasiewicz [1], Wioletta Żukiewicz-Sobczak [2], Jolanta Królczyk [3] and Jerzy Lechowski [4]

[1] Department of Food Engineering and Machines, University of Life Sciences in Lublin, 20-612 Lublin, Poland
[2] Department of Public Health, Pope John Paul II State School of Higher Education in Biala Podlaska, 21-500 Biala Podlaska, Poland
[3] Faculty of Mechanical Engineering, Opole University of Technology, 45-271 Opole, Poland
[4] Department of Biochemistry and Toxicology, University of Life Sciences, 20-950 Lublin, Poland
*   Correspondence: jacek.mazur@up.lublin.pl; Tel.: +48-818-397-13

Received: 22 July 2019; Accepted: 20 August 2019; Published: 22 August 2019

Abstract: This work analyses the organic dust concentration during a wheat grinding process which was carried out using two types of grinders: A hammer mill and a roller mill. DustTrak II aerosol monitor was used to measure the concentration of the dust PM10 (particles with the size smaller than 10 µm), PM4.0, and PM1.0. An increase of the grain moisture to 14% resulted in the reduction in PM10 when grinding grain using the hammer mill. An inverse relationship was obtained when grain was ground using the roller mill. A smaller amount of the fraction below 0.1 mm was observed for larger diameter of the holes in the screen and smaller size of the working gap in the roller mill. For both mills, the obtained concentration of the PM10 fraction dust exceeded the acceptable level. To protect farmers health, it is necessary to use dust protection equipment or to modify the grinding technology by changing the grain moisture content and/or the grinding parameters.

Keywords: grinding; organic dust; sustainable agriculture

1. Introduction

Management and processing of agricultural goods in a sustainable agriculture farm is of particular importance in terms of further use of these products. In agricultural holdings, straight after harvest, cereals usually have a higher moisture content than the one necessary for successful storage. This is the result of varied harvest periods. Installation of a specialized drying facility is one of the ways to lower the moisture content, however, this generates additional costs and entails higher energy consumption [1,2]. There are approaches to reduce costs of drying by using the heat produced by combustion of pellets derived from by-products of agricultural processing, e.g., Straw, rapeseed oil cakes, soybean hulls [3,4]. Moreover, technologies are also developed for processing agricultural goods, including cereals, directly after harvesting without drying [5]. In most cases such raw materials are used directly as livestock fodder. Grinding cereals and other seeds is the key production step in animal fodder production. Several factors, including moisture content and the type of raw material, its hardness and brittleness, as well as chemical composition influence the course, efficiency and effectiveness of grinding. In addition to physical properties, the milling process is determined by both structural and operating characteristics of a grinder. One of the consequences of using this group of machines is their wear, which also affects the degree of fragmentation of raw material. Among the elements of grinding machines that are affected by use wear; one should mention hammers, screens, and rollers. Irrespective of the milling method, organic dust is produced when cereal raw materials are ground. For health reasons, of particular importance is the continuous monitoring of aerosol concentration in the air and the use of equipment protecting against dust. Dust is one of the main harmful factors occurring in the work environment of farmers. Harmful influence of dust on the human body may cause many diseases, including pneumoconiosis and cancer. According to the data from the literature of the subject [6]; a human inhales daily approximately 12.3 m^3 of air, which is a mixture of gas and particles of liquids as well as solids. Dust suspended in the air is a mixture with varied chemical composition and physical characteristics. Organic dust present in the air with a particle diameter greater than 10 μm quickly settles on surface and is called deposited dust. At the same time, smaller fractions are suspended in the air. PM10 fraction refers to particles with the size smaller than 10 μm, while PM1.0 to the particles with the diameter smaller than 1 μm. Dust with dimensions smaller than 10 μm (PM10) enters the respiratory system and those with particle size smaller than 1 μm, may penetrate alveoli and thus enter bloodstream and all other systems [7–9]. As evidenced by studies, dangerous mycotoxins enter the human body together with inhaled organic dust [10,11]. The presence of dust during the grinding process is very common. Primarily, particles of a greater size are present (PM10), but there are also those with smaller particle size (PM2.5). As the result of their further spread, and frequently mutual collision, their additional fragmentation takes place, which increases the amount of fine fraction PM2.5 and very fine fraction PM1.0. Therefore, a PM4.0 fraction concentration might be a good determinant of dust contamination in its initial phase.

The European Commission takes into account mainly dust particles of PM10 and PM2.5 fraction [12]. According to the data from 2018, it is estimated that in Poland's agriculture and waste disposal there is approx. 16% with PM10 particulate size and approx. 15.2% with PM2.5 particulate size in the overall amount of generated dusts [13]. According to the data of the World Health Organization [14]; for a PM10 acceptable level of 24 h continuous concentration of dust is 50 μg·m^{-3}, while in the case of PM2.5 the acceptable level is 25 μg·m^{-3}. According to epidemiological research [12], it is difficult to identify the threshold values of dust concentration below which no adverse effects on human organism are observed. Many research works were performed on the harmfulness of suspended particulate matter on the human body [15–18]. In the research by Zwoździak et al. [19] concentration of PM10, PM2.5, and PM1 at school was examined by comparing it with the results of the children examined using a spirometer. The average concentration of PM10 was equal to 115.2 μg·m^{-3}, while that of PM2.5 was 46.4 μg·m^{-3}. A short-term cause and effect relationship was documented between the concentration of PM2.5 fraction inside the school and the change of values of the parameters measured in spirometry.

Health statistics show that most of occupational diseases reported in Polish farmers is caused by pathogens present in organic dusts. In Poland, lung diseases are more common in farmers than in the rest of the population, just as in other countries. Therefore, the problem is serious, and there is a need to take appropriate preventive measures [10,20–22].

The studies carried out were intended to assess concentration of organic dust produced during grinding grain in a hammer mill and roller mill, i.e., the most popular grinding devices in agricultural holdings. The test presented in the paper were carried out for the grain with three different levels of moisture content in order to assess the impact of moisture content on the level of dust contamination.

2. Materials and Methods

The test material was wheat of Zyta variety. The wheat grain is the typical kind of grain used in the farms. The initial moisture content of the wheat, the so-called storage humidity, was 9% [23]. The studies were carried out also on wheat with the moisture content of 14% and 18%. For this purpose, the test material was moistened to the desired moisture content by adding the appropriate amount of water and storing in cold storage conditions for 24 h in order to balance the moisture content. The quantity of water needed for moistening was calculated using the following formula:

$$M_W = \frac{x_2 - x_1}{100 - x_2} \cdot M_N$$

where:

M_w—amount of water necessary for moistening,

M_N—moistened grain mass,
X_1—initial moisture content,
X_2—desired moisture content.

The first moisture content (9%) it is mainly moisture of grain after drying or storage, moisture 14% this is a maximum safe moisture content for storage and 18% and more this is a moisture content of grain after harvesting. Some farmers processed the grain immediately after harvesting. The wheat grain is the typical kind of grain used in the farms.

Grinding was performed using two different types of mills, i.e., A hammer mill with interchangeable screens and a roller mill with the diameter of rollers equal to 200 mm. In the hammer mill three different screens with mesh with the diameter of openings respectively: $\varphi = 3$, $\varphi = 5$ and $\varphi = 8$ mm were used, while in the case of the roller mill three different working gaps were set, i.e., 0.4, 0.7, and 1.0 mm. The operating parameters of machines are different according to the destination of the ground grain. The disintegration at holes at screen size of 3 mm is typical feed for pigs, 5 mm is typical for poultry and 8 mm is designed for hens in the hammer mill. The size of gap was set according to the same destiny of feed in roller mill.

Concentration of organic dust, as measured using a DustTrak II monitor, with which three interchangeable size-selective inlets were used, allowing the measuring of the concentration of dust with particulate matter size: PM10, PM4.0 and PM1.0. The duration of the measurement was set to 1 min. During this time, 30 measurements were made, i.e., one every 2 s. The device used to measure the dust concentration was installed at a distance of 1 m from the grinding mill, at an employee head height, i.e., 1.5 m above the ground. Each time the device was set in the same place. The measurements were taken at the same distance from the milling device, thus simulating the typical position of personnel handling the grinder (Figure 1). Additionally, a control test was also made by measuring dust concentration in the room when grinding process was not taking place. In the discussed study, the PM4.0 was used instead of the typical value of PM2.5, taking into account a phenomenon typically occurring during the process of dynamic milling that is mutual collision of particles and resulting from it gradual formation of a certain quantity of finer particles [24–26].

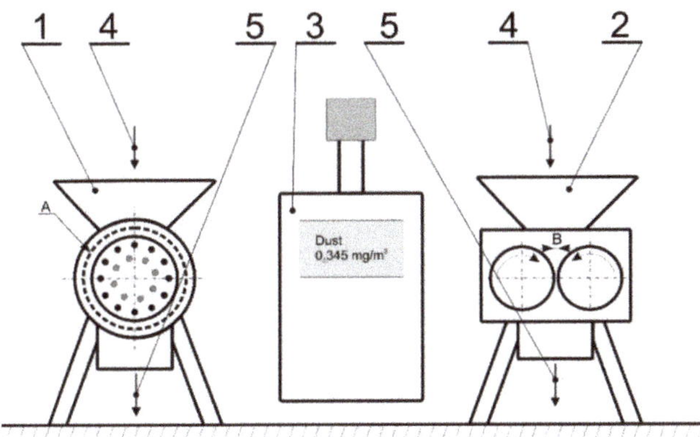

Figure 1. The layout of the dust measuring station: 1—hammer mill, 2—roller mill, 3—DustTrak II dust monitor, 4—inlet of the raw material prior to grinding, 5—outlet of the raw material after grinding, A—screen with interchangeable mesh with circular openings, B—adjustable working gap between the rollers.

The obtained wheat middlings were analyzed on sieves in order to determine the finest fraction, i.e., with the size smaller than 0.1 mm, as well as the average particle size. The study was performed on a Retsch sieve separator in accordance with PN-R-64798: 2009 standard [27].

Designations presented in Table 1 were used for descriptive purposes in this work.

Table 1. Description of the designations used in this work.

Designation	Description	Designation	Description
W9-h3	Grain moisture content of 9%, the diameter of openings in the screen of the mill equal to 3 mm	W9-h0.4	Grain moisture content of 9%, the working gap between rollers of the mill equal to 0.4 mm
W9-h5	Grain moisture content of 9%, the diameter of openings in the screen of the mill equal to 5 mm	W9-h0.7	Grain moisture content of 9%, the working gap between rollers of the mill equal to 0.7 mm
W9-h8	Grain moisture content of 9%, the diameter of openings in the screen of the mill equal to 8 mm	W9-h1	Grain moisture content of 9%, the working gap between rollers of the mill equal to 1 mm
W14-h3	Grain moisture content of 14%, the diameter of openings in the screen of the mill equal to 3 mm	W14-h0.4	Grain moisture content of 14%, the working gap between rollers of the mill equal to 0.4 mm
W14-h5	Grain moisture content of 14%, the diameter of openings in the screen of the mill equal to 5 mm	W14-h0.7	Grain moisture content of 14%, the working gap between rollers of the mill equal to 0.7 mm
W14-h8	Grain moisture content of 14%, the diameter of openings in the screen of the mill equal to 8 mm	W14-h1	Grain moisture content of 14%, the working gap between rollers of the mill equal to 1 mm
W18-h3	Grain moisture content of 18%, the diameter of openings in the screen of the mill equal to 3 mm	W18-h0.4	Grain moisture content of 18%, the working gap between rollers of the mill equal to 0.4 mm
W18-h5	Grain moisture content of 18%, the diameter of openings in the screen of the mill equal to 5 mm	W18-h0.7	Grain moisture content of 18%, the working gap between rollers of the mill equal to 0.7 mm
W18-h8	Grain moisture content of 18%, the diameter of openings in the screen of the mill equal to 8 mm	W18-h1	Grain moisture content of 18%, the working gap between rollers of the mill equal to 1 mm

3. Results and Discussion

Prior to the commencement of the research on the grinding process the average concentration of organic dust in the room was determined. The concentrations were as follows: For PM1.0—26 $\mu g \cdot m^{-3}$, for PM4.0—35 $\mu g \cdot m^{-3}$, and for PM10—53 $\mu g \cdot m^{-3}$. The same analysis was also performed 5 min after completion of the grinding process and turning off the machinery. The concentration of dust was as follows: For PM1.0—42 $\mu g \cdot m^{-3}$, for PM4.0—328 $\mu g \cdot m^{-3}$, and for PM10—101 $\mu g \cdot m^{-3}$.

The average values of organic dust concentration in the course of grinding with the hammer mill are shown in Figure 2, while Figure 3 presents the corresponding values in the case of the roller mill. In addition, in Tables 2–4 respectively, the maximum and minimum concentration of dust particulates during the grnding process is shown.

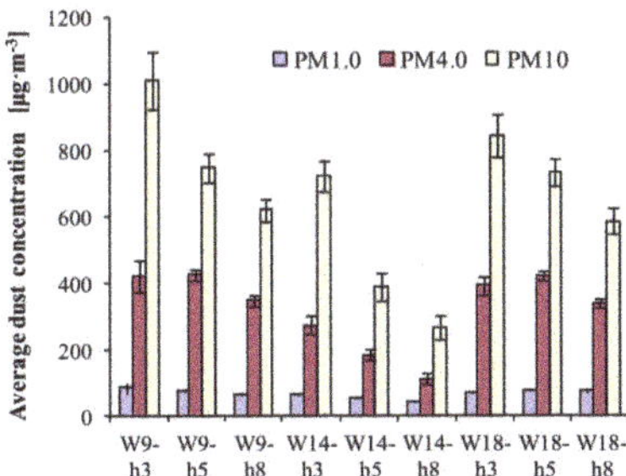

Figure 2. The distribution of organic dust concentration during grinding using the hammer mill in relation to the moisture content and the size of the openings in the screen.

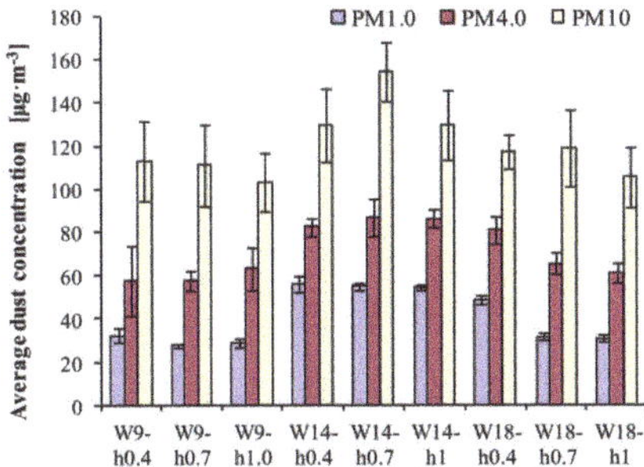

Figure 3. Distribution of organic dust concentration during grinding using the roller mill depending on the moisture content and the size of the working gap between the rollers.

In the case of the hammer mill, at a constant moisture content, a significant decrease of concentration of the organic particulate matter of the PM10 fraction was observed for increased diameter of screen openings. In the case of the roller mill, a similar trend can be observed for increased working gap, however, only for the raw material with the moisture content of 9%. It should be noted that in this case the trend is noticeable, though not confirmed statistically. Additionally, in the case of the hammer mill, for the moisture content of 9% and 14%, a tendency can be observed of lowering the concentration of PM1.0 fraction with the increase of the diameter of screen openings, although statistical analysis does not confirm the importance of these differences at 0.05 significance level. In the case of the roller mill, a similar relationship was noted for the raw material with the moisture content of 18% and increased size of the working gap between the rollers. For a particulate matter of the PM4.0 fraction produced in the course of grinding using the hammer mill, a decrease of dust concentration was observed for larger mesh openings in the case of moisture content of 14%. A similar tendency is noticeable for the roller mill for the increased size of the working gap in the case of the raw material with the moisture content of 18%.

Table 2. Statistical analysis of dust concentration PM1.0 during grinding.

Mill	Moisture Content of the Raw Material [%]	Screen [mm]	Max. [$\mu g \cdot m^{-3}$]	Min. [$\mu g \cdot m^{-3}$]	Dusts Average [$\mu g \cdot m^{-3}$]	Dusts—S.D.	Homogeneous Group
	-	Control	50	33	38.7	4.4	
	9	3	121	69	84	14.3	
		5	82	69	77.5	3.7	d
		8	77	59	65.3	4	c
Hammer mill	14	3	71	57	63.6	4.4	c
		5	66	51	55.9	2.9	b
		8	49	39	42.2	2	b
	18	3	77	63	69.4	3.6	
		5	83	72	77.7	2.6	d
		8	92	66	74	5.5	d
	-	Control	39	25	27.1	3	a
	9	0.4	39	28	32.3	3.2	
		0.7	29	26	27.7	0.9	a
		1	36	26	28.6	2	a
Roller mill	14	0.4	68	51	56.2	3.8	
		0.7	58	52	54.9	1.6	b
		1	56	52	54.2	1	b
	18	0.4	52	43	48.6	1.9	
		0.7	38	29	31.6	1.9	
		1	34	28	30.7	1.6	a

Table 3. Statistical analysis of dust concentration PM4.0 during grinding.

Mill	Moisture Content of the Raw Material [%]	Screen [mm]	Max. [$\mu g \cdot m^{-3}$]	Min. [$\mu g \cdot m^{-3}$]	Dusts Average [$\mu g \cdot m^{-3}$]	Dusts—S.D.	Homogeneous Group
	-	Control	140	106	120.4	7.5	b
	9	3	530	344	421.6	46.9	d
		5	463	404	423.6	15.5	d
		8	375	307	345.8	17.1	c
Hammer mill	14	3	354	234	272.2	26.7	
		5	213	156	183.1	15.6	
		8	131	81	110.7	15	b
	18	3	445	342	391.5	28.9	
		5	451	391	421	15	d
		8	359	296	335	13.6	c
	-	Control	55	37	41.9	3.7	
	9	0.4	105	42	57.7	15.8	
		0.7	65	49	57.6	4.5	
		1	84	48	63.1	9.5	a
Roller mill	14	0.4	90	75	82.3	4.1	
		0.7	104	74	86.8	8.7	
		1	94	78	86.2	3.9	
	18	0.4	94	68	81	6.1	
		0.7	75	56	65.4	5	
		1	69	53	61.3	4.6	a

Table 4. Statistical analysis of dust concentration PM10 during grinding.

Mill	Moisture Content of the Raw Material [%]	Screen [mm]	Max. [mg·m^{-3}]	Min. [mg·m^{-3}]	Dusts Average [mg·m^{-3}]	Dusts—S.D.	Homogeneous Group
Hammer mill	-	Control	294	158	200.3	30.4	
	9	3	1200	832	1011.9	87.7	
		5	837	670	746.6	43.8	c
		8	698	552	620.3	32.9	
	14	3	795	647	720.9	48.1	c
		5	460	297	387.2	40	
		8	314	184	262.8	35.7	
	18	3	1020	702	842.3	63.5	
		5	811	637	730.9	41	c
		8	667	525	583.2	39.4	
Roller mill	-	Control	86	54	68.8	8.8	a
	9	0.4	160	78	113.1	18.7	
		0.7	157	80	111	18.7	
		1	150	90	102.8	13.6	a,b
	14	0.4	165	106	129.3	16.9	b
		0.7	178	134	154.5	13.5	
		1	161	100	129.6	16.1	
	18	0.4	129	104	117.4	8	
		0.7	144	86	119	18	
		1	128	72	105.3	14	b

When grinding in the hammer mill, for each of the tested sizes of mesh openings in the screen, the highest value of organic dust concentration of PM10 fraction was measured for the grain with initial moisture content of 9%, while the lowest concentration characterized the grain with an initial moisture content of 14%. The highest value of concentration of the PM10 fraction was recorded for the moisture content of 9% and the screen openings size equal to 3 mm (1102 μg·m^{-3}), while the lowest concentration was measured for the moisture content of 14% and the screen openings size of 8 mm (263 μg·m^{-3}). In the case of the PM4.0 fraction, the highest concentrations of organic dust was observed for the raw material with the initial moisture content of 9% and 18%. It reached a similar value both for the openings with the diameter of 3 as well as 5 mm (approx. 420 μg·m^{-3}). For all tested initial moisture contents, the use of screens with the openings size of 8 mm resulted in the decrease of concentration of the organic dust by approx. 24% (to the level of approx. 340 μg·m^{-3}). In the case of the very fine particulate matter PM1.0, the highest value of concentration was recorded for the input material with moisture content of 9% and the screen openings size of 3 mm (84 μg·m^{-3}). For the entire range of tested moisture content of the input material, the lowest value of the PM1.0 fraction concentration was observed for grain with the moisture content of 14% (42 μg·m^{-3}). Using the roller mill reduced the amount of produced PM10, PM4.0, and PM1.0 dust as compared to the amount obtained during grinding of the raw material in the hammer mill. In this case, for all examined mill gaps between the rollers, the highest value of the PM10 fraction concentration was measured for grain with the input moisture content of 14%. The highest concentration was equal to 155 μg·m^{-3} for the gap of 0.7 mm, followed by the gaps of 0.4 and 1 mm (approximately 129 μg·m^{-3}). The lowest value of dust concentration of the PM10 fraction was recorded during grinding grain with the initial moisture content of 9% for the mill gap of 1 mm; this value was 103 μg·m^{-3}. When grinding grain using the roller mill the concentration of dust of the PM4.0 fraction was also the highest for the input material with the moisture content of 14% (87 μg·m^{-3} for the mill gap of 0.7 mm). In contrast, the lowest concentration of organic dust of the PM4.0 fraction was recorded during grinding grain with the moisture content of 9% (the lowest concentration of 57 μg·m^{-3} was measured for the mill gap of 0.7 mm). For the roller mill the highest concentration of the finest particulate matter of the PM1.0 fraction was recorded for the initial humidity of 14%, and mill gap of 0.4 mm (52 μg·m^{-3}). As in the above described variants, considering the analyzed mill gaps, also in the case of the finest particulate matter the lowest values of organic dust concentration were observed for grain with the initial moisture content of 9% (the lowest concentration equal to 28 $\mu g \cdot m^{-3}$ was measured for the gap of 0.7 mm). In the research by Dacarro [28], the average concentration of organic dust of 790 $\mu g \cdot m^{-3}$ was obtained during the grinding stage of grain processing in mills, while the highest concentration of inhaled dust was obtained at the cleaning stage, when it was equal to 2763 $\mu g \cdot m^{-3}$, with the concentration of dust of the smallest particulate size (respirable dust) was, respectively: 321 and 1400 $\mu g \cdot m^{-3}$.

The use of the roller mill, instead of the hammer mill in the process of grinding grain with moisture content of 9%, resulted in an average reduction of the PM10 fraction by approx. 724%, the PM4.0 fraction by approx. 671%, and the PM1.0 fraction by 256%, and these were the largest decreases registered among the analyzed moisture content levels. For grain with the moisture content of 14% which is characterized by the lowest dynamics of the decrease these values were, respectively: PM10—337%, PM4.0—223%, and PM1.0—98%. Moreover, in the case of the last one even an approx. 22% increase in the amount of PM1.0 fraction was recorded for the roller mill with the mill gap of 1 mm in comparison to the amount of dust produced by the hammer mill with the screen openings with the diameter of 8 mm.

Concentration of organic dust originating from the production of industrial fodder was determined in the research by Sobczak [29], when concentration of this type of dust was measured at various stages of production. The research showed that the highest average concentration was recorded during grinding grain, and for the PM10 fracture it was equal to 1250 $\mu g \cdot m^{-3}$, however, neither the machinery nor the grinding parameters were provided. In the study presented here, the maximum instantaneous concentration of the organic dust for the PM10 fraction was equal to 1200 $\mu g \cdot m^{-3}$, and it was recorded while grinding grain using the hammer mill with the screen openings with the diameter of 3 mm and input material moisture content of 9%. In addition, the research done by Karpaciński [30] confirms the high concentration of organic dust in grain processing plants. This study was performed in Canadian mills, where the determined concentration of dust was higher than 10,000 $\mu g \cdot m^{-3}$, which is significantly above the acceptable level. Concentration of organic dust in grain milling industry and herbal plants during the packing stage is much lower, and amounts approximately to 100 $\mu g \cdot m^{-3}$ for dust of the PM1.0 fraction [31,32].

Table 5 shows the average size of particles after the grinding process, and the share of the finest fraction (below 0.1 mm). The average size of particles depended on grain moisture content and the choice of grinding method. The highest amount of the fine fraction was obtained during grinding grain using the hammer mill with the screen of 3 mm mesh size and the material moisture content of 9%. In this case, the share of the finest fraction was equal to 3.98%. Analogously, when using the roller mill with the smallest tested mill gap, i.e., 0.4 mm, the amount of the finest dust fraction was equal to 2.78%. When grinding grain with higher moisture content the amount of the finest fraction decreased. The average size of particles obtained during grinding grew with the increase of grain moisture content. The largest average size of particles was obtained after grinding grain in the roller mill with the mill gap between the rollers of 1 mm and grain moisture content of 18%. For the hammer mill, the largest average size of particles was obtained when grinding grain with the moisture content of 18% and using the screen with the mesh having the diameter of the openings equal to 8 mm. A research was conducted analyzing the average particle size and production of fine fraction below 0.2 mm when using a roller mill for grinding hard and soft grain into flour. Sprouted grain with balanced moisture content of 12% was used as the soft grain [21]. The study showed that grinding sprouted grain produced greater amount of fine fraction, which might result from the bonding strength of starch molecules present in germinated material. In the study presented here, the change of moisture content, and therefore the change of grain hardness, had an opposite effect, which was caused by lack of enzymatic reactions that take place during germination.

Table 5. The average size of particles and the share of the finest fraction in ground middlings from hammer and roller mill.

Hammer mill	W9-h3	W9-h5	W9-h8	W14-h3	W14-h5	W14-h8	W18-h3	W18-h5	W18-h8
The share particles smaller than 0.1 mm [%]	3.98	1.08	0.6	1.99	0.51	0.31	1.78	0.34	0.06
The average size of particles [mm]	0.563	0.975	1.094	0.651	1.067	1.309	0.609	1.106	1.693
Roller mill	W9-h0.4	W9-h0.7	W9-h1	W14-h0.4	W14-h0.7	W14-h1	W18-h0.4	W18-h0.7	W18-h1
The share particles smaller than 0.1 mm [%]	2.78	1.57	1.7	1.26	0.46	0.24	0.38	0.22	0.14
The average size of particles [mm]	1.003	1.334	1.942	1.928	2.521	2.877	2.775	2.955	2.997

4. Conclusions

The average concentration of organic dust during wheat grinding depends on the type of the milling machinery and the moisture content of the input material. The highest concentration of organic dust was recorded when grinding the raw material with the lowest tested moisture content, i.e., 9%, which can be related to the hardness and at the same time brittleness of grain that during the grinding process breaks more easily into small fractions. An increase of grain moisture to 14% resulted in the reduction in particulate matter concentration (PM10) inhaled by employees when grinding grain using the hammer mill. An inverse relationship was obtained when grain was ground using the roller mill, which may be the result of different construction of working parts. For both the mills, the obtained concentration of the PM10 fraction dust was high, and in every case, it exceeded the acceptable level, which for most European countries is 100 $\mu g \cdot m^{-3}$ [33]. However, when comparing both types of the mills, statistically smaller concentration of organic dust was produced when grinding using the roller mill. The sieve analysis revealed that the amount of the finest fraction, i.e., below 0.1 mm, was smaller when grinding using the roller mill as compared with the hammer mill. The amount of the finest fraction depends on the size of the openings in the screen mesh, i.e., the larger the diameter of the holes in the screen, the smaller the amount of the fraction below 0.1 mm. Similar is the relationship with the width of the mill gap in the roller mill, i.e., the smaller the gap, the higher the share of the finest fraction. In some cases, instantaneous maximum concentration of organic dust during the process of grinding grain exceeded even ten times the accepted limit values. Having in mind the health safety of farmers, it is necessary to use dust protection equipment or to modify the grinding technology by changing the grain moisture content or choosing appropriate grinding parameters.

Author Contributions: All authors contributed equally to this paper.

Funding: This research received no external funding.

Conflicts of Interest: The authors declare no conflict of interest.

References

1. Amantea, R.P.; Fortes, M.; Ferreira, W.R.; Santos, G.T. Energy and exergy efficiencies as design criteria for grain dryers. *Dry. Technol.* **2018**, *36*, 491–507. [CrossRef]
2. Lutfy, O.F.; Noor, S.B.M.; Abbas, K.A.; Marhaban, M.H. Some control strategies in agricultural grain driers: A review. *J. Food Agric. Environ.* **2008**, *6*, 74–85.
3. Niedziółka, I.; Szpryngiel, M.; Kachel-Jakubowska, M.; Kraszkiewicz, A.; Zawiślak, K.; Sobczak, P.; Nadulski, R. Assessment of the energetic and mechanical properties of pellets produced from agricultural biomass. *Renew. Energy* **2015**, *76*, 312–317. [CrossRef]

4. Kraszkiewicz, A.; Kachel-Jakubowska, M.; Niedziółka, I.; Zaklika, B.; Zawiślak, K.; Nadulski, R.; Sobczak, P.; Wojdalski, J.; Mruk, R. Impact of various kinds of straw and other raw materials on physical characteristics of pellets. *Rocz. Ochr. Środowiska* **2017**, *19*, 270–287.

5. Zawiślak, K. Przetwarzanie Ziarna Kukurydzy na cele Paszowe. In *Rozprawy Naukowe*; Akademia Rolnicza w Lublinie: Lublin, Poland, 2006; pp. 1–95. Available online: http://yadda.icm.edu.pl/yadda/element/bwmeta1. element.agro-article-2e48e26a-bef0-41d8-b73f-ecd603e23ba6.

6. Mahajan, S.P. *Air Pollution Control Commonwealth of Learning*; Ramachandra, T.V., Ed.; Capital Publishing Company: New Delhi, India, 2006.

7. Kuskowska, K.; Dmochowski, D. Analiza rozkładu stężeń pyłu zawieszonego frakcji PM_{10}, $PM_{2.5}$ i $PM_{1.0}$ na różnych wysokościach Mostu Gdańskiego. *Zesz. Nauk. SGSP* **2016**, *53*, 101–119.

8. Frąk, M.; Majewski, G.; Zawistowska, K. Analysis of the quantity of microorganisms adsorbed on particulate matter PM_{10}. *Sci. Rev.–Eng. Environ. Sci.* **2014**, *64*, 140–149.

9. Warren, K.J.; Wyatt, T.A.; Romberger, D.J.; Ailts, I.; West, W.W.; Nelson, A.J.; Nordgren, T.M.; Staab, E.; Heires, A.J.; Poole, J.A. Post-Injury and Resolution Response to Repetitive Inhalation Exposure to Agricultural Organic Dust in Mice. *Safety* **2017**, *3*, 10. [CrossRef] [PubMed]

10. Żukiewicz-Sobczak, W.; Cholewa, G.; Krasowska, E.; Chmielewska-Badora, J.; Zwoliński, J.; Sobczak, P. Grain dust originating from organic and conventional farming as a potential source of biological agents causing respiratory diseases in farmers. *Postepy Dermatol. I Alergol.* **2013**, *6*, 358–364. [CrossRef]

11. Niculita-Hirzel, H.; Hantier, G.; Storti, F.; Plateel, G.; Roger, T. Frequent Occupational Exposure to Fusarium Mycotoxins of Workers in the Swiss Grain Industry. *Toxins* **2016**, *8*, 370. [CrossRef]

12. European Environment Agency. *EEA 2017. Air Quality of Europe*; EEA: København, Denmark, 2017; ISSN 1977-8449. Available online: https://www.eea.europa.eu/publications/air-quality-in-europe-2017. (accessed on 26 February 2019).

13. Klimat dla Polski – Polska dla klimatu: 1988-2018-2050. Available online: www.kobize.pl/pl/fileCategory/id/ 1/opracowania (accessed on 26 February 2019).

14. WHO. *WHO Air Quality Guidelines for Particulate Matter, Ozone, Nitrogen Dioxide and Sulfur Dioxide*; Global Update 2005; WHO: Geneva, Switzerland, 2006; Available online: http://whqlibdoc.who.int/hq/2006/WHO_ SDE_PHE_OEH_06.02_eng.pdf (accessed on 26 February 2019).

15. Ashmore, M.R.; Dimitroulopoulou, C. Personal exposure of children to air pollution. *Atmos. Environ.* **2009**, *43*, 128–141. [CrossRef]

16. Chen, C.; Zhao, B. Review of relationship between indoor and outdoor particles: I/O ratio, infiltration factor and penetration factor. *Atmos. Environ.* **2011**, *45*, 275–288. [CrossRef]

17. Grahame, T.J.; Schlesinger, R.B. Evaluating the health risk from secondary sulfates in Eastern North American regional ambient air particulate matter. *Toxicology* **2005**, *17*, 15–27. [CrossRef] [PubMed]

18. Pawłowski, L. How heavy metals affect sustainable development. *Rocz. Ochr. Środowiska (Annu. Set Environ. Prot.)* **2011**, *13*, 51–64.

19. Zwoździak, A.; Sówka, I.; Fortuna, M.; Balińska-Miśkiewicz, W.; Willak-Janc, E.; Zwoździak, J. Wpływ stężeń pyłów (PM_1, $PM_{2.5}$, PM_{10}) w środowisku wewnątrz szkoły na wartości wskaźników spirometrycznych u dzieci. *Rocz. Ochr. Środowiska (Annu. Set Environ. Prot.)* **2013**, *15*, 2022–2038.

20. Żukiewicz-Sobczak, W.; Cholewa, G.; Krasowska, E.; Chmielewska-Badora, J.; Zwoliński, J.; Sobczak, P. Rye grains and the soil derived from under the organic and conventional rye crops as a potential source of biological agents causing respiratory diseases in farmers. *Postępy Dermatol. I Alergol.* **2013**, *6*, 373–380. [CrossRef] [PubMed]

21. Dziki, D.; Laskowski, J. Study to analyze the influence of sprouting of the wheat grain on the grinding process. *J. Food Eng.* **2010**, *96*, 562–567. [CrossRef]

22. Hameed Hassoon, W.; Dziki, D. The Study of Multistage Grinding of Rye. In Proceedings of the IX International Scientific Symposium; Farm Machinery and Processes Management in Sustainable Agriculture, Lublin, Poland, 22–24 November 2017. [CrossRef]

23. Polski Komitet Normalizacyjny. *PN-EN ISO 712:2009. Ziarno Zbóż i Przetwory Zbożowe. Oznaczanie Wilgotności*; Polski Komitet Normalizacyjny: Warszawa, Poland, 2009.

24. Shieh, J.Y.; Ku, C.H.; Christiani, D.C. Respiratory effects of the respirable dust ($PM_{4.0}$). *Epidemiology* **2004**, *15*, 4–166. [CrossRef]

25. De Lima Gondim, F.; Lima, Y.C.; Melo, P.O.; dos Santos, G.R.; Serra, D.S.; Araújo, R.S.; de Oliveira, M.L.M.; Lima, C.C.; Cavalcante, F.S.A. Exposure to $PM_{4.0}$ from the combustion of cashew nuts shell in the respiratory system of mice previously exposed to cigarette smoke. *Int. J. Recent Sci. Res.* **2017**, *8*, 16762–16769. [CrossRef]

26. Josino, J.B.; Serra, D.S.; Gomes, M.D.M.; Araújo, R.S.; de Oliveira, M.L.M.; Cavalcante, F.S.Á. Changes of respiratory system in mice exposed to $PM_{4.0}$ or TSP from exhaust gases of combustion of cashew nut shell. *Environ. Toxicol. Pharmacol.* **2017**, *56*, 1–9. [CrossRef]

27. Polski Komitet Normalizacyjny. *PN-R-64798:2009. Pasze–Oznaczanie Rozdrobnienia*; Polski Komitet Normalizacyjny: Warszawa, Poland, 2009.

28. Dacarro, C.; Grisoli, P.; Del Frate, G.; Villani, S.; Grignani, E.; Cottica, D. Micro-organisms and dust exposure in an Italian grain mill. *J. Appl. Microbiol.* **2004**, *98*, 163–171. [CrossRef]

29. Sobczak, P.; Zawiślak, K.; Żukiewicz-Sobczak, W.; Wróblewska, P.; Adamczuk, P.; Mazur, J.; Kozak, M. Organic dust in feed industry. *Pol. J. Environ. Stud.* **2015**, *24*, 5–2177. [CrossRef]

30. Karpaciński, E.A. Exposure to inhalable flour dust in Canadian flour mills. *Appl. Occup. Environ. Hyg.* **2003**, *18*, 1022–1030. [CrossRef] [PubMed]

31. Buczaj, A. Poziom zapylenia w wybranych zakładach przemysłu zbożowego w województwie lubelskim. *Inżynieria Rol.* **2011**, *1*, 7–13.

32. Zawiślak, K.; Sobczak, P.; Kozak, M.; Mazur, J.; Panasiewicz, M.; Żukiewicz-Sobczak, W.; Wojdalski, J.; Mieszkalski, L. Microbiological analysis and concentration of organic dust in an herb processing plant. *Pol. J. Environ. Study* **2019**, *28*, 1–7. [CrossRef]

33. Directive 2008/50/EC of the European Parliament and of the Council of 21 May 2008 on ambient air quality and cleaner air for Europe OJ L. *Off. J. Eur. Union* **2008**, *152*, 1–44.

Article

Compaction Process as a Concept of Press-Cake Production from Organic Waste

Paweł Sobczak [1], Kazimierz Zawiślak [1], Agnieszka Starek [2,*], Wioletta Żukiewicz-Sobczak [3], Agnieszka Sagan [2], Beata Zdybel [2] and Dariusz Andrejko [2]

[1] Department of Food Engineering and Machines, University of Life Sciences in Lublin, Akademicka 13, 20-612 Lublin, Poland; pawel.sobczak@up.lublin.pl (P.S.); kazimierz.zawislak@up.lublin.pl (K.Z.)
[2] Department of Biological Bases of Food and Feed Technologies, University of Life Sciences in Lublin, Akademicka 13, 20-612 Lublin, Poland; agnieszka.sagan@up.lublin.pl (A.S.); beata.zdybel@up.lublin.pl (B.Z.); dariusz.andrejko@up.lublin.pl (D.A.)
[3] Pope John Paul II State School of Higher Education in Biala Podlaska, Sidorska 95/97, 21-500 Biala Podlaska, Poland; wiola.zukiewiczsobczak@gmail.com
* Correspondence: agnieszka.starek@up.lublin.pl; Tel.: +48-81-531-9647

Received: 31 January 2020; Accepted: 18 February 2020; Published: 19 February 2020

Abstract: As a result of agri-food production large amounts of organic waste are created in the form of press cakes. Until now, they were mainly used as animal fodder and also utilized for biofuels production. No other way usage has been found yet. A large quantity of these by-products is usually discarded in open areas, which leads to potentially serious environmental problems. The rich chemical composition of these waste products makes it possible to use them for producing other food products valuable for consumers. Based on the test results obtained, it can be stated that moisture content of press cakes is varied and depends on the input material. However, appropriately composed mixtures of various waste products and a properly conducted compaction process allows for obtaining a new product with functional properties. In addition, application of honey powder and starch tablet coating creates a product of resistant to compression and cutting. Results seem to have commercial importance, as they demonstrate that properly processed by-products can be used in food preparations as dietary supplements.

Keywords: press cakes; compaction; disposal; sustainable development; modern products

1. Introduction

In the food industry, during processing of raw materials, waste products are generated. Their substantial amounts accumulated in a short time causes serious problems for processing plants, primarily due to the instability of these by-products, including the microbiological one. Therefore, they should be used as an intermediate for further processing. In general, around the world waste management is carried out to convert into useful components as much waste products as possible without endangering the natural environment [1–5].

A particularly large amount of waste is produced during cold-press edible oil extraction. A common method of using press cakes from rapeseed, flaxseed, and sunflower extraction, is their utilisation for animal nutrition as a high protein vegetable fodder or as a component for enriching the protein content in the production of maize silage, which is intended for high-yielding cows. The use of press cakes reduces the cost of animal nutrition products [6–8]. They can also be successfully used as organic fertilizer rich in minerals (Ca, Mg, P, Fe, and Mn) and organic compounds, as well as for energy purposes [9–12] after a compaction process that turns them into pellets [13,14]. Press cakes can be used for manufacturing formulated products such as functional foods. Compaction could be a particularly important method for processing press cakes in the areas where there is a surplus of them, there is no possibility for their fast processing into fodder, and transport over long distances is not economically viable [15].

Production of coconut oil also entails generation of waste. The research done by Ramachandran et al. [16,17] confirms that these press cakes are characterised not only by a clearly sweet taste, but also high content of amino acids (mainly arginine). They are also used as a potential raw material in bioprocesses, as they are an excellent substrate for growing microorganisms (substrates for the production of phytase based on the use of Rhizopus spp. strain).

Avocado oil, pressed from fruit pulp, is nowadays used as a dietary supplement reinforcing body resistance and improving skin appearance. Moreover, it is used for production of cosmetic goods [18]. At the same time avocado pit, until lately considered inedible, has recently been the subject of scientific interest. As reported by Soong et al. [19] and Wang et al. [20] this hard part of the fruit contains many more polyphenols, exhibiting intense antioxidant activity, than the pulp itself. According to the ORAC scale for determining the free radical absorbance capacity of antioxidants, the pit of the fruit has high antioxidant potential, equal to 428.8 µmolTE/g for the Hass variety. For comparison, the same quantity of pulp only has a potential of 11.6 µmolTE/g. Avocado extracts, thanks to their polyphenol content, expressed as gallic acid equivalents, are already used in cosmetic care [21], as well as a means to delay the oxidation of sunflower oil [22]. Moreover, many studies confirm [23–26] that avocado pit has cytotoxic properties, inter alia, against breast, lung, colon, and prostate cancer cells.

Individual types of waste are increasingly used in various industries [27,28]. In the scientific literature, however, there are no reports on their comprehensive use to obtain a valuable products intended for human consumption. In addition, there are difficulties in the proper preparation of this organic waste (ensuring adequate moisture content, disintegration and chemical composition) and their appropriate concentration [14,15]. Therefore, the aim of the study is to determine the potential use of organic waste and combine it in such a way as to obtain a new, stable product with pro-health properties.

2. Materials and Methods

2.1. Research Material

Materials used for the study consisted of flax, sunflower, pumpkin, and coconut press cakes, and powder avocado pit created during cold-press oil production. In order to obtain a uniform product all raw materials were mixed and subjected to compaction in a chamber equipped with a piston (Figure 1), where they were formed into a tablet. Before blending the compound press cakes were ground to a uniform size. The percentage share of particular components constituting the blend is shown in Table 1. The press cakes were sampled immediately after cold-press oil extraction using a screw press (model DUO Farmet, Česká Skalice, Czech Republic), while the avocado pit was cut into smaller pieces, dried in an SLN 15 STD type laboratory drying oven, manufactured by the POL-EKO Company (Wodzislaw Śląski, Poland), at a temperature of 40 °C for 1.5 hours, and then ground in a mill so that powder was obtained.

Table 1. Percentage share of individual raw materials in the compound.

Type of Raw Material	Compound 1	Compound 2
Flax press cake	30	29
Sunflower press cake	20	20
Pumpkin press cake	30	29
Avocado pit	5	5
Coconut press cake	15	15
Honey powder	0	2

Shares of particular components were selected having in mind the possibility of obtaining value products in form of a stable tablet. To Compound no. 2 a 2% of dried honey was added. The purpose of this admixture was to obtain a stable tablet form by combining particles of honey with the remaining raw materials.

Then the tablets were divided into two groups and one of them was coated with a thin layer of starch solution. For this purpose a 40% solution of potato starch was prepared which was heated to the temperature of 40 °C in order to obtain lower viscosity. Surface of the tablets was coated using a spray nozzle. Afterwards, the tablets were left to dry at room conditions for 24 hours.

2.2. Measurement of Chemical and Physical Properties of Post-Production Waste

Fat content was determined in the obtained press cakes as well as in the fragmented and dried avocado pit using an automatic Soxhlet apparatus (Tecator Soxtec System HT 1043 extraction unit, Gemini, Apeldoorn, Sweden). Total protein content measurements were carried out according to the Kjeldahl method using a Kjeltec 8400 automatic distiller (Foss, Foss Anatytical AB, Höganäs, Sweden). The total protein content was calculated using a 6.25 conversion factor. Determination of the ash content in the press cakes was done in a muffle furnace according to PN-EN ISO 18122:2016-01 [29] standard. Analysis of mineral composition included the determination of the amount of: calcium, copper, iron, potassium, magnesium, and zinc by using ICP OES spectrometer (SpectroBlue, SPECTRO Analytical Instruments GmbH, Kleve, Germany). Analytical curves were prepared by dilutions of VHG SM68-1-500 Element Multi Standard 1 in 5% HNO_3. Moisture content of the raw materials was determined using a laboratory drying oven with forced air circulation. Samples of raw materials were placed in the oven chamber and then dried at the temperature of 105 °C until constant weight was achieved, in accordance with PN-EN-ISO 18134-3:2015-11 [30] standard.

2.3. Methodology of the Tablet Forming Process

The tablet forming process was carried out with a ZO20/TN2S machine (Zwick.Roell AG, BT1-FR0.5TN.D14, Ulm, Germany) using the designed attachment shown in Figure 1.

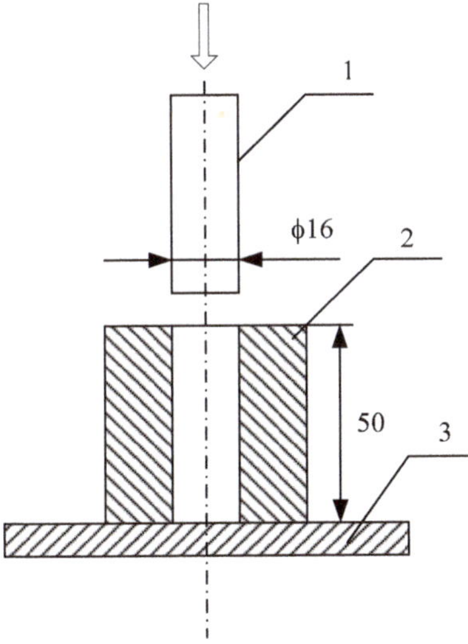

Figure 1. Workstation for forming tablets out of the compound: 1—piston, 2—feed hopper, 3—base.

The pressing force exerted by the piston was constant, equal to 20 kN. The head pressing the piston, allowed for the formation of a single agglomerate with the weight of 2 g. Compaction work and specific density of the obtained tablets was designated by means of testXpert software, which is used to operate the Zwick devices.

2.4. Methodology for Assessing Strength of the Tablets

The tablets (Figure 2) were analysed by examining the cutting force and hardness. The stress tests were carried out using a Micro Stable Pro TA.XT Plus system (Stable Micro Systems Ltd, Surrey, United Kingdom).

Figure 2. Tablets obtained out of the compound.

The method of placing a single tablet for the cutting force test and hardness test is depicted in Figure 3. In the cutting test a knife with the cutting edge angle of 45° was used. The test was run until the tablet was cut into half. The hardness test was made using a flat head and recording the compression force until reaching half the diameter of the tablet. The tests were performed in 10 repetitions.

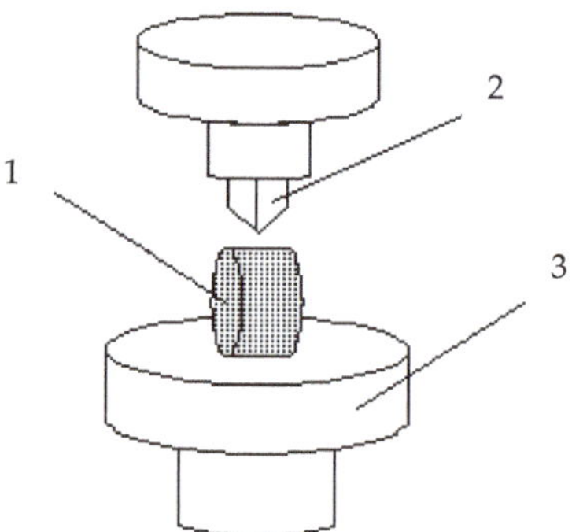

Figure 3. Method of placing the sample for the cutting force test: 1—sample, 2—cutting knife, 3—measuring table.

2.5. Statistical Analysis

Results were statistically analysed with Statistica 10.0 software (Statistica 10, StatSoft Inc., Tulsa, OK., U.S.A.), using one-way analysis of variance. The significance of differences between the mean values was tested using Tukey's procedure.

3. Results and Discussion

3.1. Characterization of Chemical Properties of Post-Production Waste

The results of carried out experiments are shown in Tables 2 and 3.

Table 2. Chemical composition of the press cakes and obtained compounds.

	By-products					Compound	
Discriminant	Flax Press Cake	Sunflower Press Cake	Pumpkin Press Cake	Coconut Press Cake	Avocado Pit	1	2
Protein [%]	32.47 ± 0.11	37.65 ± 0.36	53.98 ± 0.07	17.40 ± 0.41	4.78±0.06	36.2± 0.26	36.1± 0.23
Fat [%]	11.52 ± 0.07	31.42 ± 0.25	14.23 ± 0.05	15.02 ± 0.04	1.30±0.21	18.1± 0.04	17.1± 0.06
Ash [%]	5.56 ± 0.02	4.69 ± 0.01	8.09 ± 0.01	4.13 ± 0.01	2.30±0.01	5.75± 0.01	5.70± 0.01

Means of three determinations ± SD (standard deviation).

Table 3. Content of selected elements in the press cake.

	Content of Selected Elements [ppm]					
By-Products	**Ca**	**Cu**	**Fe**	**K**	**Mg**	**Zn**
Flax press cake	3489.355	11.808	135.916	8067.798	4035.281	69.089
Sunflower press cake	1421.465	23.737	66.131	<0.005	3708.333	111.364
Pumpkin press cake	602.551	6.378	134.694	<0.005	6707.143	155.102
Coconut press cake	394.628	20.661	75.000	10017.355	2021.694	32.645
Avocado pit	913.545	7.977	31.909	7143.500	759.023	8.568

By-products resulting from processing of pumpkin seeds had the highest protein content of 53.98%. In the flax and sunflower press cakes the value of this parameter ranged between 32.39 and 37.91%, while in the case of coconut press cakes its quantity was about half the amount. The avocado pit was characterised by the lowest protein content, equal to 4.78%.

Sunflower press cakes had the highest fat content of 31.415%. The coconut, pumpkin, and flax press cakes were characterised by a significantly lower fat content, i.e. between 15.02 and 11.52%. Ground avocado pit contained a negligible amount of fat equal to 1.30%.

The ash content in the analysed samples varied between 2.30% (for the avocado pit) and 8.09% (for the pumpkin seed press cakes).

As it can be seen in Table 3, the calcium content varied very significantly. Flax press cake contained the highest amount of this element, i.e. 3489 ppm. In the by-products of coconut oil extraction process, calcium content was about 74% lower. Copper content in the analysed production by-products was less varied and fell within the range from 6.378 ppm (for the pumpkin press cake) to 23.737 ppm (for the sunflower press cake). Iron content in the waste originating from flax and pumpkin processing was at a similar levels. The analysis showed that avocado pit was characterised by the lowest amount of this element. Potassium content was also determined in the tested by-products. Coconut press cake had the highest amount of this element, equal to 10017.355 ppm. Concentration of potassium in the waste resulting from flax seed oil extraction and in the avocado pit was lower respectively by approx. 20% and 29%. In the remaining by-products subject to the study amount of potassium was below limit of quantification. Magnesium content in the analysed samples varied within the range between 759.023 and 6707.143 ppm. The lowest content of this element was measured for the avocado pit. Pumpkin press cake was characterized by a several times higher magnesium content. For the entire studied material the highest zinc content was measured for the pumpkin press cake and sunflower press cake. In turn, the lowest amount of this element was recorded in the case of avocado pit.

The chemical composition of the press cakes can be highly variable, depending on the quality of seeds, method of oil extraction, storage parameters, and so on. As confirmed by the studies done by other authors these oil pressing by-products are characterized by high nutritional value. For example, the by-product obtained in the course of oil extraction from dried pumpkin seeds has a high amount of valuable protein and hydrolysates produced from it have antioxidant and functional properties [31]. Research conducted by Salgado et al. [32] shows that sunflower press cakes are also a source of protein with high solubility in water, good physicochemical properties, and high antioxidant activity. The data presented by Ramachandran et al. [17] demonstrate that coconut press cake contains a high level of residual oil consisting of saturated short-chain fatty acids. Moreover, it is characterised by a high content of protein and crude fibre. Literature on chemical composition of sesame press cake [33,34], rapeseed press cake [35], Camelina sativa, and flax press cake [36] also provides a lot of information about their high nutritional value, and confirms that in future by-products will be an important resource for use as a food ingredient for direct human consumption.

3.2. Analysis of the Compaction Process

Before the compaction process moisture content of individual components as well as that of the compound was measured (Figure 4).

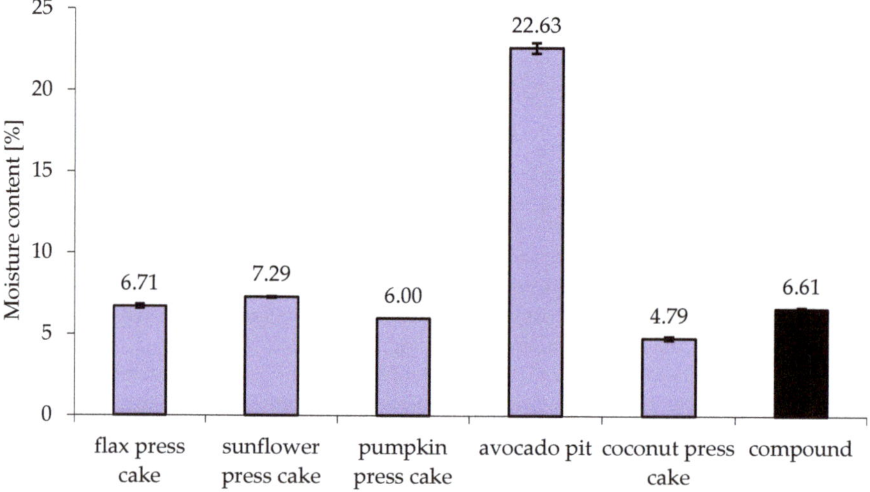

Figure 4. Moisture content of the individual press cakes and the compound.

Avocado pit press cake had the highest moisture content (22.63%). For the remaining raw materials moisture content varied within the range between 4.79% (for the coconut press cake) and 7.29% (for the sunflower press cake). The obtained compound had the moister content of 6.61%.

Moisture content of press cakes is varied and depends on the input material. As reported by research moisture content for oil press cakes ranges from 5 to 9% [13]. Wojdalski et al. [15] conducted research on the compaction process of apple pomace resulting from juice production, for which, at the temperature of 20 °C and relative humidity of the air equal to 31.7%, the total compaction energy in the case of unfragmented apple pomace ranged from 66.60 to 150.00 J/g, while for pellets made from fragmented pomace this parameter was between 34.79 J/g and 149.95 J/g. Specific density of the obtained tablets was respectively within the range of 1114.0-1166.3 and 1114.0-1168.1 g/dm^3. In the case of examined here compounds a significantly smaller compaction work, i.e. of approx. 9.7 J/g, was necessary to obtain tablets. This is due to high fat content, which for compound 1 was equal to 18.1%, while for compound 2 to 17.1% (Figure 5).

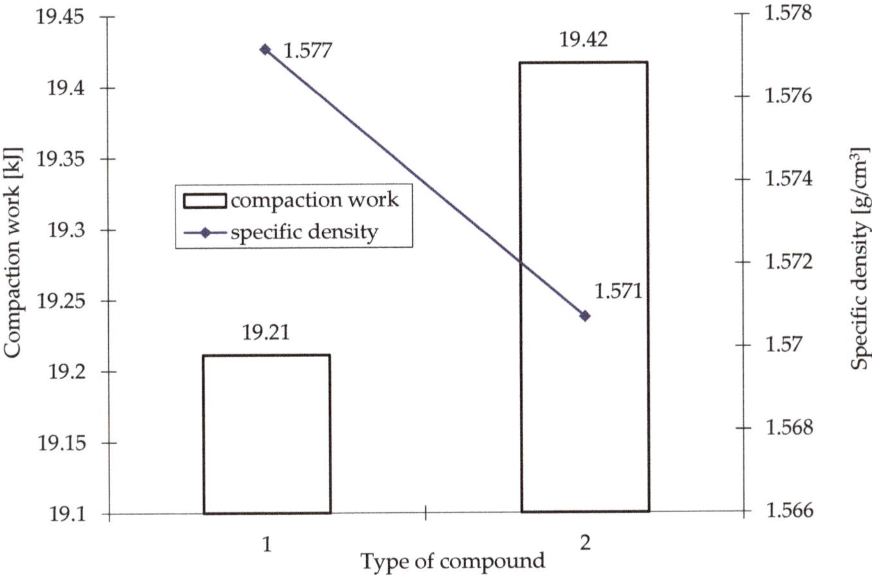

Figure 5. Compaction work and specific density obtained after forming tablets out of the compound.

3.3. Evaluation of the Obtained Tablets

In accordance with the accepted methodology the obtained tablets were subject to stress tests in order to determine the impact of adding honey and coating technology. Figures 6 and 7 and Table 4 show results of respectively the cutting test and hardness test.

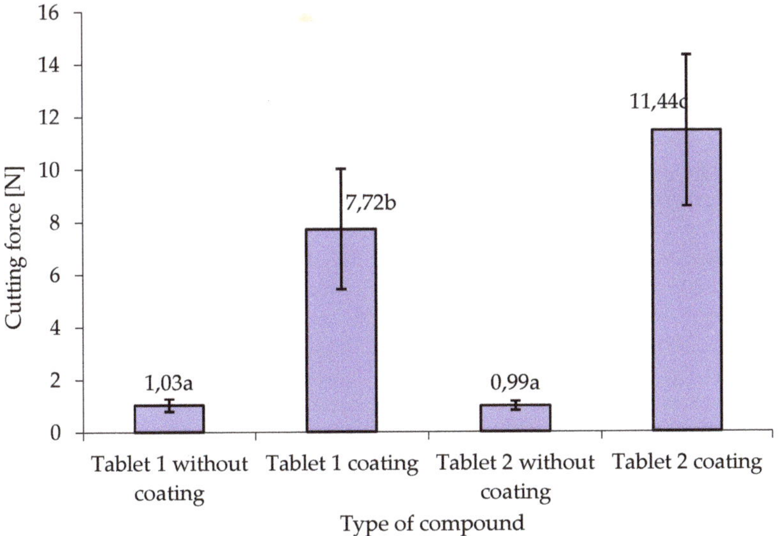

Figure 6. Maximum cutting force for individual tablets.

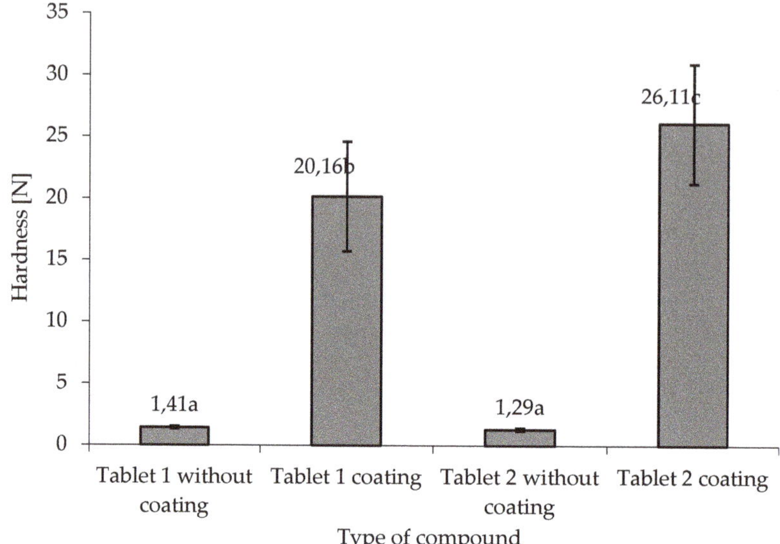

Figure 7. Hardness of the obtained tablets.

Table 4. Analysis of variance (ANOVA).

Source Variables	SS	df	MS	F Value	Probability
Compaction work [J/g]	0.1652	1	0.1652	3.578845	0.063046
Specific density [g·dm^{-3}]	659.469	1	659.469	1.038803	0.311938
Cutting force [N]	570,115	3	190.038	61.1161	0
Hardness [N]	3598.742	3	1199.581	117.406	0

As confirmed by the study tablets with starch coating are characterized by greater resistance to cutting. The maximum cutting force was recorded for tablet 2, i.e. the one with honey powder admixture. Temperature increase caused by friction force resulting from compression transforms powdered honey into a viscous liquid which reinforces produced tablet. In order to obtain high quality agglomerate it is necessary to combine this raw material with other raw materials that provide appropriate strength properties [37,38].

A similar relationship was obtained during the tablet hardness tests. The tablets with addition of honey powder and coated with starch were characterised by highest hardness (26.11N). Statistical analysis of the results confirmed the significance of the differences between the tablets with honey powder admixture and those without it. This greater hardness of the tablets with honey results from its bonding with the other components constituting the compound, as well as with the applied starch coating. Tablets without coating were characterised by low hardness of 1.3–1.4 N, regardless of the admixture of honey powder. Studies on the impact of coating on the properties of final product were done for many sectors of the food industry. As demonstrated by the research done by Fijałkowska et al. [39] after applying a coating of thermally hardened starch on dried apple a slight increase in the hardness of dried apple was recorded as compared with the control sample [40]. Another advantage is extending the freshness of a variety of fruits by applying a coating and thus controlling the water content in the fruits, as well as their colour and firmness [41].

The presented diagram (Figure 8) illustrates the technologies of producing a new product that can be used in food preparations as dietary supplements.

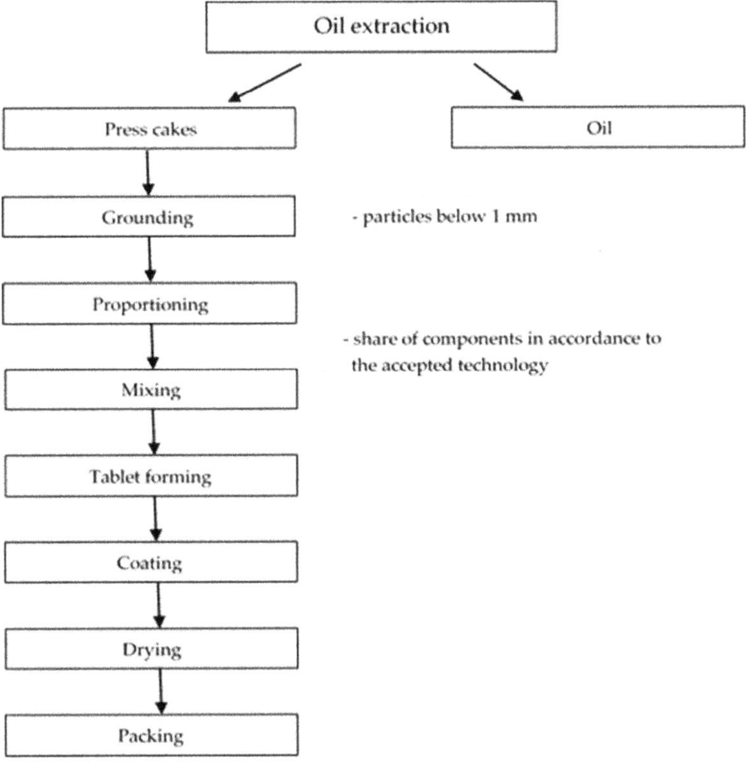

Figure 8. Prospects for using by-products of oil production.

4. Conclusions

The potential of some organic wastes is not always fully appreciated. They can be useful basic raw materials which can be sources of valuable, previously unknown or overlooked ingredients. The studies carried out suggest the possibility for using post-production press cakes in the manufacture of food products immediately after oil pressing (without the need of transport of post-production press cakes). The products obtained have a high protein content and significant amounts of minerals. An appropriate compaction process of the ingredients and application of a coating also allows for obtaining tablet of appropriate quality. Products with starch (40% solution of potato starch) coating and enriched with 2% honey powder are characterized by greater resistance to cutting and better hardness.

The proposal for organic waste compaction process resulting from cold-press oil extraction presented here can significantly increase profitability of processing raw materials. Moreover, manufacture of new products having the characteristics of pro-health food will also enhance the food market and make it more attractive. Through the admixture of various flavor coatings (honey, chocolate, caramel) they could become a product for direct consumption.

Author Contributions: Conceptualization, P.S., K.Z., and D.A.; methodology, P.S., and D.A.; formal analysis, A.S. (Agnieszka Sagan), P.S.; investigation, P.S., A.S. (Agnieszka Starek), B.Z. and W.Ż.-S.; data curation, P.S., A.S. (Agnieszka Starek); writing—original draft preparation, A.S. (Agnieszka Starek), P.S.; writing—review and editing, A.S. (Agnieszka Starek), A.S. (Agnieszka Sagan); visualization, P.S., A.S. (Agnieszka Starek); supervision, P.S. All authors have read and agreed to the published version of the manuscript.

Funding: This research received no external funding.

Acknowledgments: The samples of chemical analysis have been examined at the Regional Research Center for Environment, Agricultural and Innovative Technologies, Pope John II State School of Higher Education in Biała Podlaska.

Conflicts of Interest: The authors declare no conflict of interest.

References

1. Darlington, R.; Staikos, T.; Rahimifard, S. Analytical methods for waste minimization in the convenience food industry. *Waste Manag.* **2009**, *4*, 1274–1281. [CrossRef]
2. Garcia-Garcia, G.; Woolley, E.; Rahimifard, S. Identification and analysis of attributes for industrial food waste management modelling. *Sustainability* **2019**, *11*, 2445. [CrossRef]
3. Martin, M.; Danielsson, L. Environmental implications of dynamic policies on food consumption and waste handling in the European union. *Sustainability* **2016**, *8*, 282. [CrossRef]
4. Shin, S.G.; Han, G.; Lim, J.; Lee, C.; Hwang, S.A. Comprehensive microbial insight into two-stage anaerobic digestion of food waste-recycling wastewater. *Water Res.* **2010**, *44*, 4838–4849. [CrossRef] [PubMed]
5. Strotmann, C.; Göbel, C.; Friedrich, S.; Kreyenschmidt, J.; Ritter, G.; Teitscheid, P.A. Participatory approach to minimizing food waste in the food industry—A manual for managers. *Sustainability* **2017**, *9*, 66. [CrossRef]
6. Ahern, N.A.; Nuttelman, B.L.; Klopfenstein, T.J.; MacDonald, J.C.; Erickson, G.E.; Watson, A.K. Comparison of wet and dry distillers grains plus solubles to corn as an energy source in forage-based diets. *Prof. Anim. Sci.* **2016**, *32*, 758–767. [CrossRef]
7. Kaczmarek, P.; Korniewicz, D.; Lipiński, K.; Mazur, M. Chemical composition of rapeseed products and their use in pig nutrition. *Pol. J. Nat. Sci.* **2016**, *4*, 545–562.
8. Świerczewska, E.; Mroczek, J.; Niemiec, J.; Słowiński, M.; Jurczak, M.; Siennicka, A.; Kawka, P. Broiler chick performance and meat quality depending on the type of fat in feed mixtures. *J. Anim. Feed Sci.* **1997**, *6*, 379–389. [CrossRef]
9. Appels, L.; Lauwers, J.; Degrève, J.; Helsen, L.; Lievens, B.; Willems, K.; Dewil, R. Anaerobic digestion in global bio-energy production: Potential and research challenges. *Renew. Sustain. Energy Rev.* **2011**, *15*, 4295–4301. [CrossRef]
10. Maćkowiak, C.; Igras, J. Chemical composition of sewage sludge and food industry wastes of fertilisation value. *Inż. Ekol.* **2005**, *10*, 70–77.
11. Singh, R.P.; Ibrahim, M.H.; Esa, N.; Iliyana, M.S. Composting of waste from palm oil mill: A sustainable waste management practice. *Rev. Environ. Sci. Biotechnol.* **2010**, *9*, 331–344. [CrossRef]
12. Yaakob, Z.; Mohammad, M.; Alherbawi, M.; Alam, Z.; Sopian, K. Overview of the production of biodiesel from waste cooking oil. *Renew. Sustain. Energy Rev.* **2013**, *18*, 184–193. [CrossRef]
13. Kraszkiewicz, A.; Kachel-Jakubowska, M.; Niedziółka, I.; Zaklika, B.; Zawiślak, K.; Nadulski, R.; Mruk, R. Impact of various kinds of straw and other raw materials on physical characteristics of pellets. *Annu. Set Environ. Prot.* **2017**, *19*, 270–287.
14. Zawiślak, K.; Sobczak, P.; Kraszkiewicz, A.; Niedziółka, I.; Parafiniuk, S.; Kuna-Broniowska, I.; Tanaś, W.; Żukiewicz-Sobczak, W.; Obidziński, S. The use of lignocellulosic waste in the production of pellets for energy purposes. *Renew. Energy* **2020**, *145*, 997–1003. [CrossRef]
15. Wojdalski, J.; Grochowicz, J.; Ekielski, A.; Radecka, K.; Stępniak, S.; Orłowski, A.; Kosmala, G. Production and Properties of Apple Pomace Pellets and their Suitability for Energy Generation Purposes. *Annu. Set Environ. Prot.* **2016**, *18*, 89–111.
16. Ramachandran, S.; Roopesh, K.; Nampoothiri, K.M.; Szakacs, G.; Pandey, A. Mixed substrate fermentation for the production of phytase by *Rhizopus* spp. using oilcakes as substrates. *Process Biochem.* **2005**, *40*, 1749–1754. [CrossRef]
17. Ramachandran, S.; Singh, S.K.; Larroche, C.; Soccol, C.R.; Pandey, A. Oil cakes and their biotechnological applications—A review. *Bioresour. Technol.* **2007**, *98*, 2000–2009. [CrossRef]
18. Flores, M.; Saravia, C.; Vergara, C.E.; Avila, F.; Valdés, H.; Ortiz-Viedma, J. Avocado Oil: Characteristics, Properties, and Applications. *Molecules* **2019**, *24*, 2172. [CrossRef]
19. Soong, Y.Y.; Barlow, P.J. Antioxidant activity and phenolic content of selected fruit seeds. *Food Chem.* **2004**, *88*, 411–417. [CrossRef]

20. Wang, W.; Bostic, T.R.; Gu, L. Antioxidant capacities, procyanidins and pigments in avocados of different strains and cultivars. *Food Chem.* **2010**, *122*, 1193–1198. [CrossRef]

21. Alex, S.; Nogent, L.; Caroline, B.; Sophie, L.B.; Philippe, M. Avocado Flesh and/or Skin Extract Rich in Polyphenols and Cosmetic, Dermatological and Nutraceutical Compositions Comprising Same. United States Patent Application No. 16/117, 511, 30 August 2018.

22. Segovia, F.; Hidalgo, G.; Villasante, J.; Ramis, X.; Almajano, M. Avocado seed: A comparative study of antioxidant content and capacity in protecting oil models from oxidation. *Molecules* **2018**, *23*, 2421. [CrossRef]

23. Alkhalf, M.I.; Alansari, W.S.; Ibrahim, E.A.; ELhalwagy, M.E. Anti-oxidant, anti-inflammatory and anti-cancer activities of avocado (*Persea americana*) fruit and seed extract. *J. King Saud Univ. Sci.* **2018**, *31*, 1358–1362. [CrossRef]

24. Dabas, D.; Elias, R.J.; Ziegler, G.R.; Lambert, J.D. In Vitro Antioxidant and Cancer Inhibitory Activity of a Colored Avocado Seed Extract. *Int. J. Food Sci.* **2019**. [CrossRef]

25. Lara-Marquez, M.; Spagnuolo, P.A.; Salgado-Garciglia, R.; Ochoa-Zarzosa, A.; Lopez-Meza, J.E. Cytotoxic Mechanism of Long-chain Lipids Extracted from Mexican Native Avocado Seed (*Persea americana var. drymifolia*) on Colon Cancer Cells. *FASEB J.* **2018**, *32*, 804–832.

26. Vo, T.S.; Le, P.U. Free radical scavenging and anti-proliferative activities of avocado (*Persea americana* Mill.) seed extract. *Asian Pac. J. Trop. Med.* **2019**, *9*, 91.

27. Berbel, J.; Posadillo, A. Review and analysis of alternatives for the valorisation of agro-industrial olive oil by-products. *Sustainability* **2018**, *10*, 237. [CrossRef]

28. Zafeiriou, E.; Arabatzis, G.; Karanikola, P.; Tampakis, S.; Tsiantikoudis, S. Agricultural commodities and crude oil prices: An empirical investigation of their relationship. *Sustainability* **2018**, *10*, 1199. [CrossRef]

29. *Biopaliwa Stałe—Oznaczanie Zawartości Popiołu*; Polish Committee for Standardization: Warsaw, Poland, 2016; PN-EN ISO 18122:2016-01.

30. *Biopaliwa Stałe—Oznaczanie Zawartości Wilgoci—Metoda Suszarkowa—Część 3: Wilgoć w Próbce do Analizy Ogólnej*; Polish Committee for Standardization: Warsaw, Poland, 2015; PN-EN-ISO 18134-3:2015-11.

31. Popović, L.; Peričin, D.; Vaštag, Ž.; Popović, S.; Krimer, V.; Torbica, A. Antioxidative and functional properties of pumpkin oil cake globulin hydrolysates. *J. Am. Oil Chem. Soc.* **2013**, *90*, 1157–1165. [CrossRef]

32. Salgado, P.R.; Molina Ortiz, S.E.; Petruccelli, S.; Mauri, A.N. Sunflower protein concentrates and isolates prepared from oil cakes have high water solubility and antioxidant capacity. *J. Am. Oil Chem. Soc.* **2011**, *88*, 351–360. [CrossRef]

33. Mohdaly, A.A.; Smetanska, I.; Ramadan, M.F.; Sarhan, M.A.; Mahmoud, A. Antioxidant potential of sesame (*Sesamum indicum*) cake extract in stabilization of sunflower and soybean oils. *Ind. Crop. Prod.* **2011**, *34*, 952–959. [CrossRef]

34. Yasothai, R. Chemical composition of sesame oil cake—Review. *Int. J. Sci. Environ. Technol.* **2014**, *3*, 827–835.

35. Amarowicz, R.; Fornal, J.; Karamac, M. Effect of seed moisture on phenolic acids in rapeseed oil cake. *Grasas Aceites* **1995**, *46*, 354–356. [CrossRef]

36. Terpinc, P.; Čeh, B.; Ulrih, N.P.; Abramovič, H. Studies of the correlation between antioxidant properties and the total phenolic content of different oil cake extracts. *Ind. Crop. Prod.* **2012**, *39*, 210–217. [CrossRef]

37. Miranda, T.; Arranz, J.I.; Montero, I.; Román, S.; Rojas, C.V.; Nogales, S. Characterization and combustion of olive pomace and forest residue pellets. *Fuel Process. Technol.* **2012**, *103*, 91–96. [CrossRef]

38. Obidziński, S. Analysis of usability of potato pulp as solid fuel. *Fuel Process. Technol.* **2012**, *1*, 67–74. [CrossRef]

39. Fijałkowska, A.; Witrowa-Rajchert, D.; Weroński, A. The influence of raw material coating on drying process and reconstitution properties of dried product. *Zesz. Probl. Postęp. Nauk Roln.* **2012**, *571*, 39–47.

40. Krochta, J.M.; De Mulder-Johnston, C. Edible and biodegradable polymer films: Challenges and opportunities. *Food Technol.* **1997**, *51*, 61–74.

41. García, M.A.; Martino, M.N.; Zaritzky, N.E. Plasticized starch-based coatings to improve strawberry (Fragaria×ananassa) quality and stability. *J. Agric. Food Chem.* **1998**, *46*, 3758–3767. [CrossRef]

sustainability

Article

Waste Management in Dairy Cattle Farms in Aydın Region. Potential of Energy Application

Gürel Soyer [1] and Ersel Yilmaz [2],*

1   11th Regional Directorate, General Directorate of State Hydraulic Works, Ministry of Agriculture and Forestry, 22100 Edirne, Turkey; gurel.soyer@dsi.gov.tr
2   Department of Biosystems Engineering, Aydın Adnan Menderes University, 09020 Aydın, Turkey
*   Correspondence: eyilmaz@adu.edu.tr

Received: 12 December 2019; Accepted: 18 February 2020; Published: 21 February 2020

Abstract: In this paper, the dairy cattle waste management systems on farms in Aydın region in Turkey were investigated. Number of farms and livestock herd size, type of barn, type of machinery and farm labour force were studied. The collection, management and storage systems of manure produced in dairy cattle farms were taken into consideration. Additionally, biogas amount, which is produced from animal waste, was calculated for all districts of Aydın by using the number of livestock animals and various criteria such as the rate of dry matter. Results show that the typical and representative farm in the Aydın region is facility with a total head over 100 heads. 89.6% of the farms have heads in the range of 100 to 200. The amount of biogas that can be produced from all manure collected in Aydın region in the biogas plants is approximately 160,438 m^3/day (based on 0.5 m^3/day biogas per cattle), which would produce around 100 GWh/year that can be used for own needs of farms owners.

Keywords: animal waste; biogas; dairy cattle farms; energy potential; waste management

---

1. Introduction

Nowadays, expansion and intensification of large-scale animal feeding operations has resulted in an increase in the size of farms and in the amount of waste produced from farms causing serious problems such as a negative impact on environment and public health in rural areas.

By the end of 2014, according to FAOSTAT [1], 24.99 billion animals were produced on farms all over the World. The livestock sector is one of the fastest growing parts of the agricultural economy. In recent years there has been an increasing demand for cattle production. The large cattle producers are Brazil about 218 million, India 186 million and China 83 million heads [1].

In Turkey, the greatest livestock production belongs to cattle farms, with about 17 million heads of cattle being bred in 2018, resulting in an increase of 33% compared to 2010. Dairy cattle produced about 22 million tonnes of milk and 1 million tonnes of meat in 2018 [2]. Table 1 presents the total amount of animal production from species across years in Turkey.

The breeding and agricultural activities, especially livestock production on an industrial scale, are seen as one of the main sources of natural environment pollution [3,4]. Depending on the farming system, animal farms generate solid (dung) and liquid (liquid manure) animal excrement. In this day and age, no-mulch systems are becoming more and more popular, particularly for livestock production on a large scale. The excrement in this system is so-called liquid manure, i.e., liquid, or a semiliquid mixture of faeces, urine, water and feed leftovers.

Table 1. Total amount of animal production in Turkey by species [1].

Animal	Unit	2008	2010	2012	2014
Buffaloes	Head	84,705	87,207	107,435	121,826
Camels	Head	1057	1041	1315	1442
Cattle	Head	11,036,753	10,723,958	13,914,912	14,223,109
Sheep	Head	25,462,292	21,794,508	27,425,233	31,140,244
Chickens	Head ×1000	269,368	229,969	253,712	293,728
Goats	Head ×1000	6,286,358	5,128,285	8,357,286	10,344,936
Turkeys	Head ×1000	2675	2755	2761	2990

It is estimated that the cattle residues produced in Turkey reached the value of 1.3×10^6 tonnes/year in 2012 [5]. The amount of wet manure from animals could be a major problem for farms. If the wet manure cannot be utilized properly, it can create pollution risk with a potentially disastrous impact on the environment.

Manure management depends on many factors such as the size of the herd and type of manure, as well as available labour, soil type, climate and region [6,7]. Additionally, intensive animal production can be significantly problematic with respect to manure storage and removal [3].

Effluents of unproperly stored manure can flow directly or indirectly into surface waters in open lagoons. As a result, gaseous emissions and odours can also be released upon decomposition of manure, with negative consequences for farmers' fields and livestock farms [8,9]. Fangueiro [10] reported that greenhouse gas (NH_3, N_2O, CH_4) emissions during storing depend on type of manure, i.e., emission from separated solids, are typically higher than from liquid or unseparated manure. Animal manure contains a wide range of micro-organisms which could be a source of hazards to humans and animals. These micro-organisms can cause food contamination and epidemics and are dangerous to public health [3,11]. Therefore, sustainable manure management systems on farms must minimize risks for the environment associated with storage, handling and utilization of manure.

Animal manure contains essential nutrients such as nitrogen, phosphorus, potassium and can be applied to land as a natural fertilizer [7,8], which is the most common method of manure application. Organic matter improves the physical and biological properties of soil, as well as aeration and soil water infiltration [12].

However, in recent years, we have observed a large problem of environmental pollution caused by nitrates connected with irrational use of natural fertilizers in agriculture [13,14]. The manure contains large amount of N in organic form and converted to inorganic form through mineralization process which is ultimately a serious risk to the environment. Manure is applied to the soil at one time (usually by spreading out on the field), so more leaching occurs as compared to chemical fertilizer and the N content may reach the ground and surface waters [15].

Animal manure can also be used as substrate for biogas production in the process anaerobic digestion [16–18]. Biogas is a product of methane fermentation of organic fraction of many types of biomass.

The methane fermentation process consists of four phases (hydrolysis, acidogenesis, acetogenesis and methanogenesis) [19]. The main stages of anaerobic digestion are presented in Figure 1.

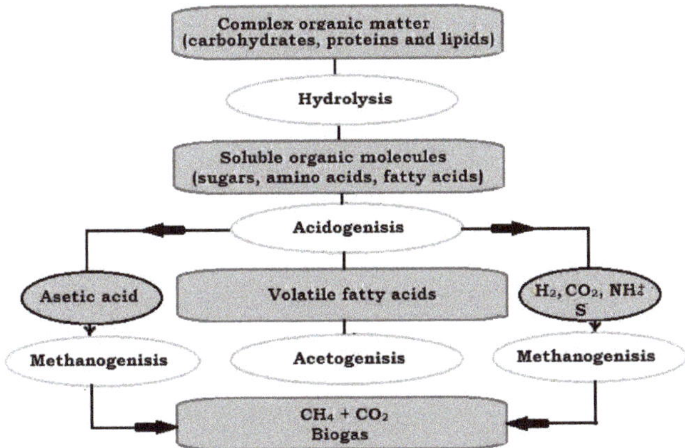

Figure 1. Stages of anaerobic digestion (methane fermentation process) [20].

The composition of biogas is different and depends on the applied substrates; however, typically it consists mostly of CH_4 (40%–70%) and CO_2 (15%–60%), as well as other compounds in small amounts: H_2O (2%–7%), N_2 (2%–5%), O_2 (0%–2%) < 1% H_2, NH_3 (0%–1%) and H_2S (0.005%–2%) [8,21].

In the process of biochemical transformations in the absence of oxygen instead of biogas is also produces the nutrient rich organic fertilizer which is easy assimilated by the plants, with a reduction in the odours and the disease-causing agents [19].

The biogas energy potential of Turkey was found to be 2.18 billion m^3 based on the animal numbers in the last agricultural census. The total biogas potential originates from 68% cattle, 5% small ruminants, and 27% poultry. The biogas energy equivalent of Turkey is approximately 49 PJ [5]. After comparing the biogas potential for animal manure of Turkey with that for different countries (Germany 20.6 billion m^3, Poland 6.4 billion m^3, Italy 1.9 billion m^3 and Sweden 7.04 billion m^3) [22], Turkey has a high biogas potential, which is associated with the increasing production in the livestock sector.

As of now, only 7% of this potential is used. There are 19 biogas power plants that produce electricity from animal manure in Turkey. The total installed power capacity of the biogas power plants is 43.41 MW_e. The range of the installed power capacity is from 0.33 to 6.40 MW_e [23].

The collection, storage and utilization of animal manure are the major problems for local livestock farmers. Problems and strategies with respect to manure management should be taken care of on a local scale and adapted to the existing conditions in a given area. There are several studies focused on cattle in Turkey [24–27]. However, region-based studies are few and limited [28,29].

The aim of this study was to investigate the collection and management of manure in the cattle farms in Aydın region. Number of farms, livestock herd size, type of barns, type of machinery for collecting manure, farm labour force and manure management were also studied to evaluate the possibility of using manure as a feedstock for biogas production for energy generation.

2. Materials and Methods

2.1. Study Area

Aydın province is located in Aegean Region of western Turkey (Figure 2). The Aegean Region has a typical Mediterranean climate with hot-dry summer and warm-rainy winter. The average annual temperature is 17.6 °C, 26.77 °C in summer and 9.33 °C in winter.

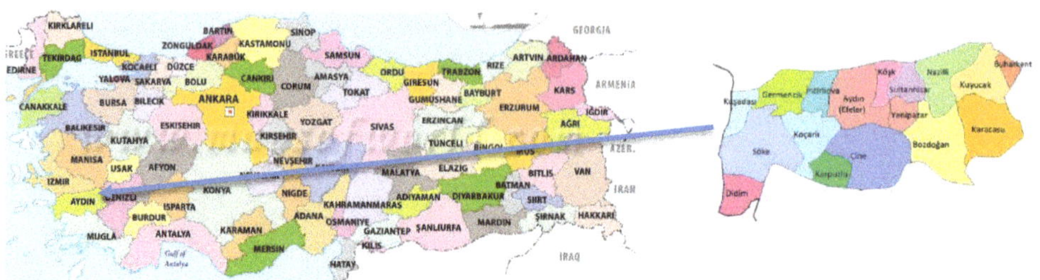

Figure 2. Turkey and Aydın province [30].

The relative humidity of the air is between 48% and 55% and the average rainfall is on the level of 647 mm [31].

The area of Aydın province with 17 districts is 8007 km^2. The population was 1,097,746 with density of 140 people/km^2 in 2018. The cultivated area is about 395,494 ha corresponding to 49.3% of soil sources of Aydın and 75,000 ha of cultivated area is used fo cereal production. The main agricultural products in Aydın province are fig, olive, chestnut, cotton and fruits [32].

2.2. Methods

The aim of this study was to investigate manure collection and management in cattle farms in the Aydın region of Turkey and determine the energy potential of the waste generated on farms.

In this study, a survey was conducted by interviewing the owners of 87 farms located in 17 districts of Aydın province and each farm was photographically documented.

The survey included general topics presented below:

- education level of farmers and the possibility of using new technologies,
- livestock herd size,
- type of barns,
- type of machinery for collecting manure,
- manure storage systems,
- methods of manure application.

In addition, the energy parameters of manure waste as a potential substrate for the biogas production were also examined. The tests were carried out in accordance with the following standards:

- moisture—EN ISO 18134:2015,
- ash—EN ISO 18122:2015,
- organic matter—EN 12176:2004,
- high heating value (HHV)—EN 14918:2009, ISO 18125:2015 using IKA C 5000 Calorimeter,
- elementary analysis (C, H, N, S, O)—EN ISO 16948:2015 using Elementary Vario Macro Cube analyser.

Based on the data obtained from 87 farms livestock size in the study region, potential of biogas and electricity production were calculated using equation 1 and 2 below:

Biogas production (BP) [33]:

$$BP = N_c \times C, \tag{1}$$

where: N_c—the number of cattle, C—production of manure per day/cattle (on the basis of an assumption of 0.4 m^3/day/cattle [34].

Electricity production (EP) [21]:

$$EP = BP \times LHVx\%CH_4, \tag{2}$$

where: $\%CH_4$—methane content in the biogas (on the basis of an assumption it is 62%), LHV_{CH4}—low heating value of methane (21 MJ/m^3)—1.7 kWh in cogeneration process: 1.7 kWh electricity and 2 kWh heat) [34].

3. Results and Discussion

In Aydın province, many of farms are located in the districts: Efeler (18 farms), Cine and Kuyucak (12 farms) and Söke (10 farms). Table 2 presents number of cattle in districtes of Aydın.

Table 2. Total amount of cattle in Aydın's districts [35].

Aydın's Districts	Number of Cattle, Head
Efeler	34,300
Bozdoğan	26,244
Buharkent	2025
Çine	62,376
Didim	3047
Germencik	19,144
İncirliova	9048
Karacasu	10,219
Karpuzlu	27,027
Koçarlı	23,953
Köşk	8757
Kuşadası	1283
Kuyucak	21,713
Nazilli	26,000
Söke	24,145
Sultanhisar	4595
Yenipazar	17,000
Total	**320,876**

Farms in Aydın province usually have more than 100 cattle and the number of animals in the 89.6% of the farms range between 100 and 200 heads.

The study shows that 69.55% of the farmers are under age 50. The ages of the youngest and oldest farmers are 28 and 74, respectively. The percentage of owners who have a university degree was 14.9%, whereas most of the owners have an elementary school degree (43.7%). Only farmers with higher education showed an interest in application of new technologies.

Manure storage type is of importance in terms of impacts on gaseous emissions and the flexibility it offers for land application and hence the potential for nutrient losses to ground and surface waters.

Generally, owners of farms have noticed the problem of disposal of manure, the facility must minimize the impact on water quality, especially on groundwater and surface water. It is indicated that the manure storage facility should be located at least 100 m away from the water resources [36].

In the Aydın region, the distance between open-air manure storage and water resources, as well as source of drinking water supplies, is 96 m on average. Çayır and Atılgan [37] examined about 74 farms in Burdur province and determined the distances to be 1–10 m in 39 farms, 11–20 m in 20 farms and 21–30 m in 10 farms, and 31 m or more in 5 of 74 farms. According to Mutlu [38], Jacopson et al. [39] and Nizam et al. [40], this distance should be much longer.

In the study area, manure storage facilities are located in the open area. The most common type of manure storage is midden (60%), and 30% and 10.3% of farms store the manure on flat ground and on leak-proof pits, respectively. Figure 3 presents the types of manure storage used in Aydın region. Manure stored on flat ground is shown in Figure 4.

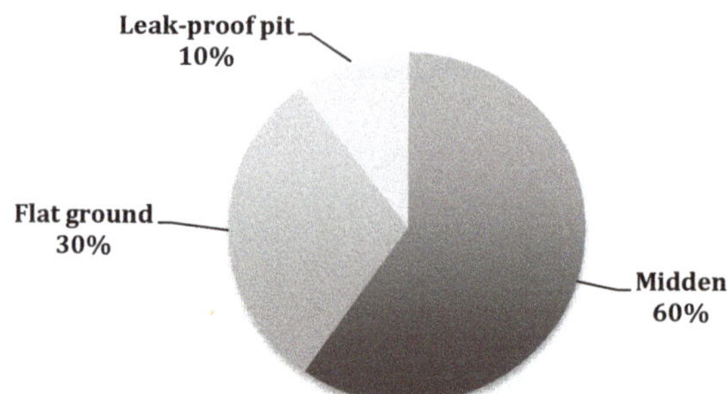

Figure 3. Types of storage used in Aydın region.

Figure 4. Manure storage on flat ground.

Study shows that 66.6% of farms do not have manure storage pits on protected ground (concrete floor). Therefore, there is a danger of contamination of ground water with nitrogen compounds.

For example, according to the survey by Smith et al. [6], manure is stored in concrete floor compounds (40%) and temporary field heaps (60%) in England. Loyon [41] stated that 23% of facilities for storing manure are covered in France.

Manure is usually stored for many months, and during its decomposition, manure emits unpleasant gases such as ammonia and hydrogen sulphide and impacts the health and comfort of surrounding people. Another problem is to minimize odour from manure storage locations as well as from open barns, which depends on the size of the intensive livestock operation, the type of livestock or manure management system and storage time.

The results of this study show that 48.2% of farms have closed-wall barns, 41.4% semi-open barns and rest of them have open sheltered barns. Semi-open barns are shown in Figure 5.

The conventional method of handling manure has been to use sufficient bedding to keep the manure relatively dry and then to move it out of the confinement area and deposit it into a manure pile [42].

In large production units, manure is handled both mechanically and hydraulically. Mechanical removal of the wastes is normally done with tractors, manure spreaders or scrapers with permanently installed equipment, such as shuttle conveyors, floor augers or pumps.

The information collected from the dairy farms assessed in this study showed that 67.8% of the farms used tractor shovels for the collection of manure produced in barns. The percentage of manual collection was 14.9%, and there were only 9 farms (10.4% of the farms evaluated) in which the manure was collected with scrapers equipped with chain. 89.7% of the farms do not have any impermeable manure pits.

Figure 5. Semi-open barns.

The most common waste management strategy on farms is to apply the manure to the land. Atılgan et al. [43] divided it according to content of solids, i.e., above 25% solid fertilizer; 10–20% semi-solid and 0–10% content of soils is called liquid manure. All produced manure in the studied farms is used in agriculture as fertilizer, mostly for their own purposes, and only 12.6% of farms sell it. The studied farms utilize only the solid manure, which provides minimum benefit because of the loss of organic nitrogen content during long storage, while it can also cause serious environmental pollution.

It is stated that the main source of nitrate contamination in groundwater is agricultural fertilizers.

In the Aydın region the high level of nitrite (0–124 mg/L) [44] in groundwater used for the irrigation was noted. The average level of nitrite in the surface water is in the range of 0.01–0.7 mg/L, nitrate 1.20–3.70 mg/L, ammonia 004–5.20 mg/L [45]. According to the World Health Organization (WHO) [46], the standard nitrate level in drinking water is 50 mg/L. Groundwater is used as irrigation (from 8–32 m deep) and drinking (>32 m deep). Elevated nitrate concentrations in groundwater can cause public health problems. Now Turkey has updated regulations aiming to combat agricultural nitrate pollution in rivers and soil. The revised rules include procedures and principles for determining, reducing and eliminating nitrate pollution [47].

Due to the increase in the Turkish population, and therefore the increase in demand from the animal sector, contamination from pollutants may also cumulatively increase in the next years. Regulations are required in order to control manure management, especially the localization of manure storage and type of floor construction for its temporary storage, as well as the limits for the use of manure as fertilizer.

Baytekin [33] claims that, under normal conditions, a healthy cow produces 40–45 kg of manure per day. According to this value, the total manure amount obtained from the research area is as presented in Table 3.

Table 3. Total amount of produced manure in farms.

	Daily Tonnes	Weekly Tonnes	Monthly Tonnes	Annual Tonnes
Production of Manure *	13,637	95,460	409,116	4,977,588

* calculated from obtained data.

The production of biogas from manure, in particular, is one of the alternative utilization methods of organic wastes that can be implemented in this region. This study also attempts to identify the biogas potential of the Aydın region basing on obtained data of animal manure production.

As a first step for this application, the energy parameters of manure, as a potential feedstock for the biogas production were tested. Obtained energy parameters of manure is presented in Table 4.

Table 4. Energy properties of manure.

Parameter	Moisture % wt	Dry mass % wt	Ash %wt	Organic Matter %wt	C % dm	H % dm	N % dm	S % dm	O * % dm	HHV MJ/kg
Value	79.9	20.1	11.50	74.62	48.09	7.13	2.14	0.27	42.66	16.48

wt—weight percentage; dm—dry mass; HHV—High Heating Value; *—calculated on the basis of the obtained difference.

One of the key parameters in terms of the efficiency of biogas production is associated with the high content of organic matter in wastes, and this determines the course of the fermentation process and the volume of the biogas [21]. The tested samples have a high content of organic matter (74.62%), which is comparable with data obtained by Zue et al. [48], it may range from 68% to 76%. For methane fermentation, especially for the growth of microorganisms, one of the important factors is the ratio of C:N (optimum 20–30:1) [49]. The ratio of C:N in the waste (24:1) is adequate in this respect.

For the Aydın province, based on the amount of produced waste, it is possible to obtain about 160,438 m^3 biogas/day, assuming 0.5 m^3/day biogas per cattle. Table 5 contains total amount of produced biogas in Aydın and LPG equivalent.

Table 5. Total amount of produced waste and biogas in Aydın and its LPG equivalent.

Production of Biogas *	Daily m^3	Weekly m^3	Monthly m^3	Annual m^3	Equivalent of LPG m^3/year
	160,438	1,123,066	4,813,140	58,559,870	93,895,792

* calculated from obtained data.

It gives production of electricity on level 99,552 MWh annually. In Aydın in 2012 electricity consumption was 1,860,667 MWh. In the case of the use of biogas, which can be substituted for conventional fuels, 5.4% of electricity can be covered by biogas.

There is only one biogas power plant—Efeler BGEPP—in the Aydin region, with a max installed power of 4.8 MW$_e$. Because of the distributed allocation of small-capacity livestock farms in the region, and also due to the low interest of farmers in installing their own small biogas installations, it is proposed to build centralized facilities.

Considering the use of only half of the manure generated in the region, it is possible to install 7 biogas power plants with a capacity 4.8 MW$_e$ in different locations. Because of the topography and the distances between farm locations, the Aydın region can be divided into four districts, in which 3-4 biogas power plants can be installed with capacities similar to the Efeler BGEPP facility.

There are some funds that support investments in renewable energy sources in Turkey such as: *Renewable Energy Resources Support Mechanism (YEKDEM)*, coordinated by the Energy and Natural Resources Ministry, the regional scale *Agriculture and Rural Development Support Institution (TKDK)*, supported by the Agriculture and Forestry Ministry, and also local development agencies subordinate to teh Ministry of Development, for example, the South Aegean Development Agency (GEKA) serving, i.a., the Aydın region.

According to the Turkish National Energy and Mining Strategy, it is a top priority for Turkey to generate 30% of its electricity from local and renewable resources by 2023. The costs of achieving this target by 2023 are estimated to require investment in renewable energy generation of around 21 billion USD (1,5 billion USD/year) [50]. In the case of Turkey, which is a net energy importer, 73% of its energy needs come from foreign suppliers, and investments in a local and secure energy supply is main pillar of the energy sector.

Biogas, biomass, and geothermal energy resources are expected to comprise a considerable part of RES with the rapid growth in utilization of these resources in the market [51].

Biogas can be used for heating and electricity production, providing local autonomy for the region in the face of the increasing cost of fossil fuels.

Manure storage facilities on farms should be considered to be a temporary solution, and farmers should have knowledge about the negative influence on the environment caused by improper treatment of manure. Education and financial support in changing the approach to animal waste management can be a key factor.

The conversion of animal waste to biogas through anaerobic digestion processes can provide added value to manure as an energy resource and reduce the environmental problems associated with animal waste. It is worth mentioning that dairy cattle manure is endowed with considerable biogas production, offering numerous benefits with respect to environmental, agricultural and socio-economic standards.

Author Contributions: Conceptualization, E.Y.; methodology, E.Y. formal analysis, E.Y.; investigation, G.S.; resources, G.S.; writing—original draft preparation, G.S.; writing—review and editing, E.Y.; supervision, E.Y. All authors have read and agreed to the published version of the manuscript.

Funding: This study was funded by Scientific Research Fund of Adnan Menderes University (Project No: ZRF12049).

References

1. FAOSTAT Website for Statistics. Available online: http://faostat3.fao.org/home (accessed on 20 January 2020).
2. TUIK (Turkish Statistical Institute). *Livestock Statistics*; TUIK (Turkish Statistical Institute): Ankara, Turkey, 2019.
3. Malomo, G.A.; Madugu, A.S.; Bolu, S.A. *Sustainable Animal Manure Management Strategies and Practices*; Intech Open: London, UK, 2018.
4. Delgad, J.A.; Nearing, M.A.; Rice, C.W. Chapter Two—Conservation practices for climate change adaptation. *Adv. Agron.* **2013**, *121*, 47–115.
5. Acaroglu, M.; Aydogan, H. Biofuels energy sources and future of biofuels energy in Turkey. *Biomass Bioenergy* **2012**, *3*, 69–76. [CrossRef]
6. Smith, K.A.; Williams, A.G. Production and management of cattle manure in the UK and implications for land application practice. *Soil Use Manag.* **2016**, *32*, 73–82. [CrossRef]
7. Mac-Safley, L.M.; Boyd, W.H.; Schmidt, A.R. Agricultural waste management systems. In *Agricultural Waste Management Field Handbook*; Hickman, D., Owens, L., Pierce, W., Self, S., Eds.; USDA: Washington, WA, USA, 2011; pp. 9–31.
8. Font-Palma, C. Methods for the treatment of cattle manure—A review. *J. Carbon Res.* **2019**, *5*, 27. [CrossRef]
9. Chandra, M.; Manna, M.; Naidux, R.; Sahu, A.; Bhattacharjya, S.; Wanjari, R.H.; Patra, A.K.; Chaudharijj, S.K.; Majumdar, K.; Khanna, S.S. Bio-waste management in subtropical soils of India: Future challenges and opportunities in agriculture. *Adv. Agron.* **2018**, *152*. [CrossRef]
10. Fangueiro, D.; Dave, J.C.; Chadwick, R.; Trindade, H. Effect of cattle slurry separation on greenhouse gas and ammonia emissions during storage. *J. Environ. Qual.* **2008**, *37*, 2322–2331. [CrossRef]
11. Manyi-Loh, C.E.; Mamphweli, S.N.; Meyer, E.L.; Makaka, G.; Simon, M.I.; Okoh, A.I. An overview of the control of bacterial pathogens in cattle manure. *Int. J. Environ. Res. Public Health* **2016**, *13*, 843. [CrossRef]
12. Francis, J.; Larney, D.; Angers, A. The role of organic amendments in soil reclamation: A review. *Can. J. Soil Sci.* **2012**, *92*, 913–938.
13. European Council. European Council Directive of 12 December 1991 concerning the protection of waters against pollution caused by nitrates from agricultural sources (91/676/EEC). *Off. J. Eur. Commun. L.* **1991**, *375*, 1–8.
14. Hokeem, K.R.; Akhtor, J.; Sobir, M. *Soil Science: Agricultural and Environmental Prospective*; Springer: Berlin/Heidelberg, Germany, 2016.
15. Webb, J.; Pain, B.F.; Bittman, S.; Morgan, J. The impacts of manure application methods on emissions of ammonia, nitrousoxide and on crop response—A review. *Agric. Ecosyst. Environ.* **2010**, *137*, 39–46. [CrossRef]
16. Abdeshahian, P.; Lim, J.S.; Hoa, W.S.; Hashim, H. Potential of biogas production from farm animal waste in Malaysia. *Renew. Sustain. Energy Rev.* **2016**, *60*, 714–723. [CrossRef]
17. Sun, Q.; Li, H.; Yan, J.; Liu, L.; Yu, Z.; Yu, X. Selection of appropriate biogas upgradingtechnology-a review of biogas cleaning, upgrading and utilisation. *Renew. Sustain. Energy Rev.* **2015**, *51*, 521–532. [CrossRef]

18. Cucui, G.; Ionescu, C.A.; Goldbach, I.R.; Coman, M.D.; Marin, E.L.M. Quantifying the economic effects of biogas installations for organic waste from agro-industrial sector. *Sustainability* **2018**, *10*, 2582. [CrossRef]

19. Gould, M.C. Bioenergy and anaerobic digestion. In *Bioenergy*; Elsevier Inc.: Amsterdam, The Netherlands, 2015; Volume 18, pp. 297–317.

20. Rameshprabu, R.; Natthawud, D. Biological purification processes for biogas using algae cultures: A review. *Int. J. Sust. Green Energy* **2015**, *4*, 20–32.

21. Pawlita-Posmyk, M.; Wzorek, M. Assessment of application of selected waste for production of biogas. In Proceedings of the E3S Web of Conferences, International Conference Energy, Environment and Material Systems (EEMS 2017), Polanica-Zdrój, Poland, 13–15 September 2017; Volume 19, p. 02017. [CrossRef]

22. Karaca, C. Determination of biogas production potential from animal manure and GHG emission abatement in Turkey. *Int. J. Agric. Biol. Eng.* **2019**, *11*, 205–210. [CrossRef]

23. Available online: https://www.enerjiatlasi.com/elektrik-uretimi (accessed on 20 January 2020).

24. Avcioğlu, A.O.; Türker, U. Status and potential of biogas energy from animal wastes in Turkey. *Renew. Sustain. Energy Rev.* **2012**, *16*, 1557–1561. [CrossRef]

25. Acaroglu, M. The potential of biomass and animal waste of Turkey and the possibilities of these as fuel in thermal generating stations. *Energy Source* **1999**, *21*, 339–346.

26. Kaygusuz, K.; Türker, M.F. Biomass energy potential in Turkey. *Biomass Bioenergy* **2002**, *26*, 661–678. [CrossRef]

27. Balat, M. Use of biomass sources for energy in Turkey and a view to biomass potential. *Biomass Bioenergy* **2005**, *29*, 32–41. [CrossRef]

28. Kizilaslan, H.; Onurlubas, H.E. Potential of production of biogas from animal origin waste in Turkey (Tokat Provincial Example). *J. Anim. Vet. Adv.* **2004**, *9*, 1083–1087. [CrossRef]

29. Eryilmaz, T.; Yesilyurt, M.K.; Gokdogan, O.; Yumak, B. Determination of biogas potential from animal waste in Turkey: A case study for Yozgat Province. *Eur. J. Sci. Technol.* **2015**, *2*, 106–111.

30. World Map. Available online: https://www.mapsofworld.com (accessed on 20 January 2020).

31. Meteoroloji Genel Müdürlüğü. Available online: https://meteor.gov.tr (accessed on 26 January 2020).

32. TÜİK. *Selected Indicators in Aydın. No 4038:189*; TÜİK: Ankara, Turkey, 2016.

33. Baytekin, H. *Bitkisel Üretimde Çiftlik Gübresi ve Biyogaz Kompostu Kullanımının Yaygınlaştırılması. Türk—Alman Biyogaz Projesi*; Çevre ve Şehircilik Bakanlığı: Ankara, Turkey, 2012.

34. Demirer, G.N.; Chen, S. Anaerobic digestion of dairy manure in a hybrid reactor with biogas recirculation. *World J. Microb. Biot.* **2015**, *21*, 1509–1514. [CrossRef]

35. Soyer, G. *Aydin ili süt sığırcılığı işletmelerinde elde edilen gübrenin değerlendirilmesi üzerine bir çalışmà, Yüksek Lisans Tezi*; Fen Bilimleri Enstitüsü, Adnan Menderes Üniveritesi: Aydın, Turkey, 2014.

36. Camberato, J.; Lıppert, B.; Chastaın, J.; Plank, O. Land Application of Animal Manure. In *Agricultural uses of by-Products and wastes American Chemical Society*; Washington, DC, USA, 1996; Available online: http://hubcap.clemson.edu (accessed on 26 January 2020).

37. Çayır, A.; Atılgan, A. Büyükbaş Hayvan Barınaklarındaki Gübrelikler ve Su Kaynaklarına Olan Durumlarının İncelenmesi. *Süleyman Demirel Üniversitesi Ziraat Fakültesi Dergisi* **2012**, *7*, 1–9.

38. Mutlu, A. *Adana İli ve Çevresindeki Hayvancılık Tesislerinde ortaya Çıkan Atıkların Yarattığı Çevre KirliliğiÜzerinde Bir Çalışma. Yüksek Lisans Tezi, Ç.Ü*; Fen Bilimleri Enstitüsü: Adana, Turkey, 1999.

39. Jacobson, L.D.; Moon, R.; Bicudo, J.; Yanni, K.; Noll, S. *Generic Environmental Impact Statement on Animal Agriculture. A Summary of the Literature Related to Air Quality and Odor of Animal Science*; University of Minnesota: Twin cities, MN, USA, 1999.

40. Nizam, S.; Armağan, G. Aydın İlinde Pazara Yönelik Süt Sığırcılığı İşletmelerinin Verimliliklerinin Belirlenmesi. *ADÜ Ziraat Fakültesi Dergisi* **2006**, *3*, 53–60.

41. Loyon, L. Overview of manure treatment in France. *Waste Manag.* **2017**, *61*, 516–520. [CrossRef]

42. *Managing Manure Nutrients at Concentrated Animal Feeding Operations*; U.S. Environmental Protection Agency: Austin, TX, USA, 2004.

43. Atılgan, A.; Erkan, M.; Saltuk, B. Akdeniz Bölgesindeki Hayvancılık İşletmelerinde Gübrenin Yarattığı Çevre Kirliliği. *Ekoloji* **2006**, *15*, 1–7.

44. Öztürk, S. *Determination of Ground Water Pollution Degrees in Some Pilot Areas Where Intensive Irrigated Agriculture is Practiced in Söke Plain of Aydın*; Enstitute of Applied Science, Adnan Menderes Üniversitesi: Aydın, Turkey, 2009.

45. Yildiz, B. *A Research on Determination of Polluter Parameters according to Receiver Environmental Conditions*; Yüksek Lisans Tezi, Enstitute of Applied Science, Adnan Menderes Üniversitesi: Aydın, Turkey, 2020.

46. WHO: World Health Organization. *Guidelines for Drinking-Water Quality*, 4th ed.; WHO: Geneva, Switzerland, 2011.

47. Evci, B. *Implementation of the Nitrate Directive in Turkey*; Ministry of Agriculture and Rural Affairs: Ankara, Turkey, 2007. Available online: https://www.slideshare.net/iwlpcu/implementation-of-the-nitrate-directive-in-turkey-evci (accessed on 15 February 2020).

48. Zhu, L.; Yan, C.; Li, Z. Microalgal cultivation with biogas slurry for biofuel production. *Bioresour. Technol.* **2016**, *220*, 629–636. [CrossRef]

49. Matheri, A.N.; Ndiweni, M.; Belaid, N.S.; Muzenda, E.; Hubert, R. Optimising biogas production from anaerobic co-digestion of chicken manure and organic fraction of municipal solid waste. *Renew. Sustain. Energy Rev.* **2017**, *80*, 756–764. [CrossRef]

50. Uğurlu, A. An overview of Turkey's renewable energy trend. *J. Energy Syst.* **2017**, *1*, 148–157. [CrossRef]

51. Guide to Investing in Turkish Renewable Energy Sector, Presidency of the Republic of Turkey Investment Office. Available online: https://www.invest.gov.tr (accessed on 15 February 2020).

 *sustainability*

Article

Impact of Integrated and Conventional Plant Production on Selected Soil Parameters in Carrot Production

Anna Szeląg-Sikora [1,*], Jakub Sikora [1], Marcin Niemiec [2], Zofia Gródek-Szostak [3], Joanna Kapusta-Duch [4], Maciej Kuboń [1], Monika Komorowska [5] and Joanna Karcz [1]

[1] Faculty of Production and Power Engineering, University of Agriculture in Krakow, ul. Balicka 116B, 30-149 Kraków, Poland; Jakub.Sikora@ur.krakow.pl (J.S.); Maciej.Kubon@ur.krakow.pl (M.K.); jwieczorek2@poczta.onet.pl (J.K.)

[2] Faculty of Faculty of Agriculture and Economics, University of Agriculture in Krakow, al. Mickiewicza 21, 31-121 Kraków, Poland; Marcin.Niemiec@ur.krakow.pl

[3] Department of Economics and Enterprise Organization, Cracow University of Economics, ul. Rakowicka 27, 31-510 Krakow, Poland; grodekz@uek.krakow.pl

[4] Faculty of Food Technology, University of Agriculture in Krakow, ul. Balicka 122, 30-149 Kraków, Poland; Joanna.Kapusta-Duch@ur.krakow.pl

[5] Faculty of Biotechnology and Horticulture, University of Agriculture in Krakow, al. Mickiewicza 21, 31-121 Kraków, Poland; m.komorowska@ogr.ur.krakow.pl

* Correspondence: Anna.Szelag-Sikora@urk.edu.pl; Tel.: +48-12-662-46-81

Received: 20 August 2019; Accepted: 8 October 2019; Published: 12 October 2019

Abstract: Currently, the level of efficiency of an effective agricultural production process is determined by how it reduces natural environmental hazards caused by various types of technologies and means of agricultural production. Compared to conventional production, the aim of integrated agricultural cultivation on commercial farms is to maximize yields while minimizing costs resulting from the limited use of chemical and mineral means of production. As a result, the factor determining the level of obtained yield is the soil's richness in nutrients. The purpose of this study was to conduct a comparative analysis of soil richness, depending on the production system appropriate for a given farm. The analysis was conducted for two comparative groups of farms with an integrated and conventional production system. The farms included in the research belonged to two groups of agricultural producers and specialized in carrot production.

Keywords: soil fertility; integrated agricultural production; conventional agricultural production; management

---

1. Introduction

In an agronomic sense, the agricultural system is defined as a way to manage the space for the production of plant and animal products. The agricultural system also includes the processing of primary products [1,2]. In modern agriculture, there are three basic management systems [3,4]: conventional, ecological, and integrated. The basis for this distinction is the extent of dependence of agriculture on the industrial means of production, mainly mineral fertilizers and pesticides, and the degree of their impact on the natural environment [5]. Conventional production is a management method aimed at maximizing profits. It is based on increasing the use of means of production and minimizing the number of agrotechnical operations in order to maximize profits [6,7]. Integrated agriculture is a form of alternative farming, which is based on harmonizing the premises of conventional agriculture with elements of biological plant protection in order to increase the safety of food products. This form of agriculture treats the farm as an agricultural ecosystem (agro-ecosystem). Its main goal is to maintain a high level of agricultural production, while protecting the population of beneficial organisms that inhabit the ecosystem and preventing soil degradation. The integrated production (IP) of plants is a more restrictive system, in terms of both environmental protection and product safety. To ensure compliance with the principles of IP, quantitative and qualitative restrictions on the use of pesticides, as well as quantitative and technological restrictions related to the use of fertilizers, are introduced. Restrictions related to the use of agrochemicals require more precise application that takes into account a wide range of agrotechnical, climatic, and habitat factors. This requires greater knowledge and experience of producers [8–10]. Due to the smaller number of restrictions related to fertilization and protection, conventional agriculture is a greater burden on the natural environment and is therefore much less effective in achieving ecological goals. However, it should be noted that in both integrated and conventional agriculture, production capacities are not yet fully utilized. Similarly, in both systems, there are opportunities to better achieve environmental goals [11,12].

Created in 2001, the European Initiative for the Sustainable Development of Agriculture (EISA) was developed to promote and defend consistent principles of integrated production in the European Union. One of the first tasks of this organization was to create the Common Codex of Integrated Farming. The document, which presented EISA's policy in terms of integrated agriculture, was published in 2006 and reviewed in 2012. FAO (Food and Agriculture Organization) used the latter version to determine sustainable practices in agriculture [13]. The state of research on implementations of integrated agriculture systems in many Western European countries is advanced [14]. Research by Dutch scientists shows that an integrated farm can achieve income at 94% of the income of a conventional farm. In Germany, where the average area of an integrated farm in Germany is approx. 17 ha [15], the implementation of an integrated system is carried out by, e.g., the Institute for Plant Protection in Stuttgart. In Poland, the integrated production system is currently regulated by the provisions of art. 5 of the Act on Plant Protection of 18 December 2003 (Journal of Laws [Dz. U.] of 2008 No. 133, item 849, as amended) and the Regulation of the Minister of Agriculture and Rural Development of 16 December 2010 on integrated production (Journal of Laws [Dz. U.] of 2010 No. 256, item 1722, as amended). Since 14 June 2007, due to the decision of the Minister of Agriculture and Rural Development, integrated production, as understood by art. 5, par. 1 of the Act on Plant Protection, has been recognized as a national food quality system. Detailed guidelines for the production technology of each plant species are included in methodologies developed by the Main Inspectorate of Plant Health and Seed Inspection. A producer wishing to join the integrated production system is obliged to continue agricultural production based on the methodologies approved by the Chief Inspector of Plant Health and Seed Inspection. Each methodology contains practical information about the planting, care, and harvesting of the crop. At the request of the plant producer, a certificate of integrated production is issued by the regional inspector consistent with the place of cultivation, along with an integrated production trademark signed with the producer's number.

The aim of this research was to assess the soil properties on farms using integrated crop production and on conventional farms.

2. Materials and Methods

2.1. Material

The research objects were two producer groups, varying in terms of the available land resources, direction of production, and the number of members. The main grouping factor was the type of production, i.e., integrated production (22 farms) and conventional production (8 farms). On conventional farms, fertilization was carried out without reference to actual nutritional needs under given agrotechnical and environmental conditions. Therefore, the amounts of biogen introduced into the soil were much higher than the nutritional needs of plants. Plant nutrient balance was not maintained on conventional farms.

The soil sampling scheme included taking 20 primary samples per pooled sample, with one pooled sample per max. 4 ha. The weight of a single sample was approx. 0.5 kg.

2.2. Analytical Methods

Within the assumed objective, in 2016, tests were carried out on 22 farms producing in accordance with the IP standard (Integrated Plant Production) and on eight conventional farms carrying out intensive production with no certified quality management system. All integrated farms were subject to the control of a certification body and were certified based on inspections. The examined farms were located in the Małopolskie (22) and Śląskie (8) provinces and their area ranged from 30 to 90 ha. A soil sample from the 0–20 cm layer was collected from each farm, in accordance with the PN-R-04031:1997 standard at the end of the growing season. The collected soil samples were dried and sieved with a 1 mm mesh sieve. Next, their parameters determining the greatest agricultural usefulness, including the pH, organic carbon content, assimilable forms of phosphorus, potassium, calcium, and magnesium, were designated. The content of assimilable forms of phosphorus and potassium was determined by the Egner Riehm method. The content of the remaining macro- and micronutrients was determined by atomic emission spectrometry (ICP-OES), following prior extraction with acetic acid at a concentration of 0.03 mol·dm^{-3}. Soil pH was determined using the potentiometric method, in a KCl suspension at 1 mol·dm^{-3}.

3. Results and Discussion

Carrot (*Daucus carota* L.) is a two-year plant belonging to the celery family (*Apiaceae*), formerly umbellifers (*Umbelliferae*). Carrots are characterized by a high capacity for an excessive accumulation of heavy metals, especially cadmium and lead [16,17]. For this reason, the soil with the lowest content of these elements should be selected for the cultivation of this plant. Growing carrots in the first year after fertilizing with manure is not recommended as it fosters an increased accumulation of nitrates, resulting in distorted and forking roots, which significantly worsen the quality of the produce. In rational pest management, the plant should not be cultivated in monoculture, as well as following other umbelliferous plants [18].

Phosphorus is the basic fertilizer macronutrient that must be delivered to agroecosystems. Intensification of plant production has led to a high demand for this element. As a consequence, the plants' ability to nourish with this element through the soil's ecosystem has become impaired. Phosphorus is taken in the form of phosphoric acid (V) ions. In the plant cell, the element is a component of nucleotides, phospholipids, and adenosine triphosphate (ATP), the latter of which plays a fundamental role as an energy carrier in the plant cell. It participates in the activation of enzymes by their phosphorylation or dephosphorylation. The availability of phosphorus in the initial stage of plant development affects the proper development of roots and thus results in drought and nutrient deficiency resistance. The deficiency of this macronutrient negatively affects the growth and development of plants, which leads to reduced crops and the deterioration of their quality, both sensory and technological. Very often, growth inhibition of lateral shoots is observed. Purple spots appear on the leaves, which, over time, become deformed and dry. The plant blooms; however, it does not bear fruit. Fertilization with phosphorus is carried out based on the soil's content of this element, or when its deficiency in the plant is observed. Phosphorus is a macronutrient, which is very often deficient in agriculturally used soils. The reason for the low content of phosphorus is due, on the one hand, to the insufficient level of fertilization with this element, and on the other hand, to processes related to the chemisorption of this element [11]. Therefore, in addition to application of the phosphate fertilizer itself, the management of the fertilization process includes the control and maintenance of soil properties at an optimum level. The most important parameters affecting the availability of phosphorus for plants are the soil's pH and its content of organic matter [19–21]. Phosphorus is best absorbed by plants at a soil pH of 6–7. In a strongly acidic environment, phosphorus is rebound by combining with aluminum, iron, and manganese cations. On the other hand, at a very high pH, calcium phosphate precipitates. In line with the principles of the development of sustainable agriculture, primary production management should aim not only at the intensification of production, but also at the quality of yield, as well as at reducing the negative impact of agriculture on individual elements of the environment [22,23]. The proper management of phosphorus in agrocenoses is associated with maintaining an adequate amount of assimilable forms of this element in the soil. Both a too high and too low content of the element in the soil is unfavorable for plants and the environment. Carrot is not a plant with a high demand for phosphorus. The optimum content of phosphorus in soils intended for carrot cultivation varies between 40 and 60 mg $P \cdot dcm^{-3}$ [18]. According to the methodology of integrated carrot production, the level of phosphate fertilization at average soil fertility should amount to about 60–80 kg of $P_2O_5 \cdot ha^{-1}$.

The average phosphorus content in the studied soils from integrated carrot farms was 118.0 $mg \cdot kg^{-1}$, ranging from 47.87 to 275.2 $mg \cdot kg^{-1}$. In the soils of farms producing carrots using the conventional method, an average of 29.5 $mg \cdot kg^{-1}$ more of this element was found, i.e., 147.5 $mg \cdot kg^{-1}$. No low or very low phosphorus content was found in the soil samples collected from carrot producing farms. In the group of farms with integrated carrot production implemented, an average phosphorus content was observed in 18% of soil samples, whereas in 23% of cases, the content of this element was found at a high level. Approx. 60% of the studied soils contained bioavailable phosphorus compounds at a very high level. In the group of conventional vegetable farms, a high content of phosphorus was found in only one case. On the other hand, the remaining soils contained a very high amount of this element. From the point of view of the rationalization of phosphorus management, it is not beneficial to maintain very high concentrations of this element in the soil. The phosphorus not used by plants undergoes immobilization processes, and as a result of erosion, it enriches aquatic ecosystems, leading to the intensification of eutrophication processes. In addition, very high levels of phosphorus in the soil can lead to a reduced absorption of certain micronutrients, for example, zinc. According to the methodology of integrated carrot production, for carrot cultivation, the optimal content of assimilable phosphorus in the soil should be between 40 and 60 $mg \cdot dcm^{-3}$ [18], which gives approximately 30–75 $mg \cdot kg^{-1}$. At higher contents of this element, fertilization with phosphorus should be limited. The results of our own research indicate that on approx. 60% of farms carrying out integrated carrot production, and on all conventional farms studied, the content of available phosphorus forms was higher than recommended in the integrated carrot production methodology. On the other hand, in conventional farms, these quantities were much higher (Table 1).

Table 1. Content of available forms of phosphorus in the soils of the studied farms ($mg \cdot kg^{-1}$).

	Min.	Average	Max.	Median	Standard Deviation
Integrated vegetable farms	47.87	118.0	275.2	116.2	69
Conventional vegetable farms	134.6	147.5	174.5	151.4	13.4

Potassium belongs to the group of macronutrients. One of its most important functions in plants is the regulation of water management and maintenance of cellular turgor. As an activator of many enzymes, it is responsible for regulatory functions and is involved in the synthesis of both simple and complex proteins and sugars. This element has an active part in the transport of nitrate (NO_3^-) and phosphate (PO_4^{3-}) ions, as well as assimilates. It increases plant resistance to frost, diseases, and pests. Potassium is responsible for the proper growth of apple fruits, as well as their color and firmness. A good supply of this element strengthens plants' resistance to drought. Over 50% of agricultural land is characterized by a potassium deficit [24–26], which is why rational fertilization with this element plays a strategic role in the development of modern agriculture. According to the principles of integrated carrot production, the optimal content of potassium in the soil should be 120–150 $mg \cdot dm^3$. The number of doses of fertilization with this element should be determined on the basis of a chemical analysis of the soil. However, when the analysis shows that the potassium content is equal to or higher than the optimal content, then it is reasonable to reduce the level of fertilization with this element, or in the case of a very high content, to discontinue fertilization altogether in a given year [24]. With a potassium content below optimal, the element should be supplemented with doses higher than that required by the plants. For a fruiting orchard, the level of fertilization should amount to 50–80 kg $K_2O\cdot ha^{-1}$ [18]. The greatest demand for this nutrient occurs in the period of setting of the fruit and its intensive growth. Later, potassium takes part in retarding the growth of tree shoots and entering winter dormancy.

According to the methodology of integrated carrot production, the optimal content of assimilable phosphorus in the soil for carrot cultivation should be between 120 and 150 $mg\cdot dcm^{-3}$ [18], which gives approx. 80–100 $mg\cdot kg^{-1}$. The results of our own research show that on 18% of farms carrying out integrated carrot production, and on all conventional farms, the content of assimilable forms of potassium in soil was much higher than recommended by the methodology (Table 2).

Table 2. Content of assimilable forms of potassium in the soils of the studied farms ($mg\cdot kg^{-1}$).

	Min.	Average	Max.	Median	Standard Deviation
Integrated vegetable farms	48.35	57.9	148.6	52.10	31.32
Conventional vegetable farms	86.2	100.9	248.3	109.6	74.80

Boron is a common element worldwide; however, in agroecosystems, its deficit is observed more and more often [27]. It is most common in the form of boric acid and belongs to the group of micronutrients essential for living organisms [28]. Plants collect boron from the soil through the root system, in the form of anion $H_2BO_3^-$ or in the form of undissociated boric acid molecules (H_2BO_3). Only a part of this element is fully available to plants; usually no more than 5%–6% of the total boron content in the soil, and sometimes even less than 1%. Carrot is very sensitive to the deficiency of some micronutrients, especially boron. The deficiency of this element is most often observed in alkaline soils. The effect of the boron deficit is the stunting of the plant growth cone and the appearance of black spots on carrot roots after washing.

The average content of available boron forms in soil on integrated carrot farms was 2.960 $mg\cdot kg^{-1}$, whereas on conventional farms, it was 1.70 $mg\cdot kg^{-1}$. Soil analysis carried out on integrated fruit farms demonstrated that the average content of available boron was 0.70 $mg\cdot kg^{-1}$, while for conventional production farms, this value fluctuated at 0.66 $mg\ B\cdot kg^{-1}$. The boron contents indicated the possibility of a deficit of this element in the agroecosystem [29].

Calcium plays a very important role in the production of vegetables and fruits. As a nutrient that is not very mobile in the plant, it is absorbed into fruits and vegetables in small quantities, causing a need for its urgent replenishment [30]. For this reason, even a high content of calcium in the soil may not provide a sufficient level of plant nutrition. Therefore, foliar feeding of the apple tree is a necessary element of integrated production, and is an integral part of the full fertilization program. Symptoms of calcium deficiency on vegetative parts of plants are rare, appearing in the form of brightening apical leaves with yellow spots. With a deficit of calcium, apples are small and tend to crack and cork. They are sensitive to sunburn. Regardless of the use of calcium in apple cultivation, its optimal content in soil is a prerequisite for cultivating fruit plants. It improves the soil structure and prevents it from crusting, regulates its pH, and supports soil microbes by accelerating the distribution of organic matter and facilitating the development of the root system. According to the methodology of integrated carrot production, the optimum level of this component in the soil is 1000–2000 $mg\cdot dm^{-3}$ [18], which amounts to approx. 666–1333 Ca $mg\cdot kg^{-1}$ of soil. The available amount of this element in the soil on farms carrying out integrated carrot production was 1095 $mg\cdot kg^{-1}$ on average, whereas for farms from the conventional group, the average amount of available calcium forms in the soil was 255.9 $mg\cdot kg^{-1}$ (Table 3). An appropriate content of calcium in carrot increases the strength of cell walls, making the roots less susceptible to cracking.

Table 3. Calcium content in the tested soil samples (mg·kg^{-1}).

	Min.	Average	Max.	Median	Standard Deviation
Integrated vegetable farms	416.4	1095	2119	768.4	536
Conventional vegetable farms	174.6	255.9	474.1	204.4	300.8

Copper plays an important role in plant organisms, impacting the regulation of cellular metabolism [29]. This element controls the transport of electrons in the process of photosynthesis, takes an active part in nitrogen transformations and in the synthesis of proteins and vitamin C, and binds and neutralizes free radicals. Plants require small amounts of copper for proper development. In the majority of soils in Poland, there are shortages of copper. On average, the largest amount of assimilable forms of copper was recorded in the soil from vegetable farms with an integrated production profile (0.54 mg·kg^{-1}). Soils collected from vegetable and conventional farms contained 0.39 mg Cu·kg^{-1} on average (Table 4).

Table 4. Copper content in the tested soil samples (mg·kg^{-1}).

	Min.	Average	Max.	Median	Standard Deviation
Integrated vegetable farms	0.25	0.54	1.60	0.50	0.32
Conventional vegetable farms	0.23	0.39	0.73	0.35	0.20

The iron content in the soils of Poland is very diverse, ranging from 0.8% to 1.8% [31]. As a rule, heavy soils contain much more of this element than sandy soils. The physiological functions of this metal in plants are related to the activation of oxidation-reduction reactions associated with many metabolic processes, such as respiration, photosynthesis, or the transformation of nitrogen compounds. The symptom of iron deficiency in plants is iron chlorosis of leaves, which first appears on the youngest leaves. The most important causes of iron deficiency in agroecosystems include intensive cultivation, large temperature fluctuations in the growing season, and intensive exposure. The iron content varies considerably in the individual organs. Its concentration in plant tissues ranges from 50 to 300 ppm [32]. The iron is taken up by plants in the ionic form of Fe^{2+} and the form of chelates. In the case of this micronutrient, its deficiencies are most often associated with soil properties. At a pH level above 6, and with the presence of large quantities of other macro- and micronutrients, the ability of its assimilation by plants may be limited. The average content of available forms of iron in the soil (Table 5) on the farms producing carrots using the integrated method was 0.750 mg·kg^{-1}, while on conventional farms, the average was 1.330 mg·kg^{-1}.

Table 5. Iron content in the tested soil samples (mg·kg^{-1}).

	Min.	Average	Max.	Median	Standard Deviation
Integrated vegetable farms	0.28	0.75	1.65	0.67	0.42
Conventional vegetable farms	0.73	1.33	2.43	1.01	0.49

Most Polish soils are characterized by a low magnesium content [33,34]. The reason for the deteriorating deficit of this element in the soils is their acidification and low content of organic matter [35,36]. Magnesium is an element that is easily washed into deeper soil layers. It is estimated that its annual leaching oscillates between 10 and 40 kg MgO·ha^{-1} [37]. Magnesium is taken up by plants in accordance with its osmotic concentration, i.e., passive movement from the soil water. A high content of dissolved magnesium in the soil water allows it to be better absorbed by the roots. In order to prevent soil degradation and to ensure the supply of this element to plants, its content in soil should be maintained at the level of 60 to 80 mg·dm^3, i.e., approx. 40–53 mg·kg^{-1} [18]. Magnesium is the basic ingredient of chlorophyll. It determines the course of photosynthesis and energy transformations taking place in the plant, as well as the synthesis of proteins, carbohydrates, and fats. Magnesium is an activator of many enzymes. It plays an important role in the construction of cell walls, and thus increases the resistance of plants to diseases. Magnesium deficiency is most often observed in young trees, as chlorosis and necrosis between the main nerves of the leaves, or yellow-purple spots on the lamina. Magnesium deficiency accelerates the generative development of plants and early maturing of fruits. In addition, plants are less resistant to low temperatures. The average magnesium content in the soils in which the carrots were grown in the integrated system was 33.63 mg·kg^{-1} (Table 6). On conventional farms, however, a slightly smaller amount of this element was found, i.e., 31.90 mg·kg^{-1}.

Table 6. Magnesium content in the tested soil samples (mg·kg^{-1}).

	Min.	Average	Max.	Median	Standard Deviation
Integrated vegetable farms	20.85	33.63	36.95	32.07	3.30
Conventional vegetable farms	29.56	31.90	36.15	3.02	2.05

Manganese is a micronutrient that is essential for the life of plants as it contributes to the processes of nitrogen absorption and protein synthesis, vitamin C synthesis, respiration, and photosynthesis. Manganese deficiency leads to excessive iron uptake by plants. The range of manganese content in the soil varies from 20 to 5000 mg·kg^{-1} and it occurs in several forms of mineral and organic compounds [37]. The absorption of manganese for plants is strongly correlated with the pH of the soil. Acidic soil promotes solubility of this element. In most cases, acidic soils demonstrate no need for manganese fertilization. Symptoms of manganese deficiency are similar to those of iron deficiency; however, yellowing of leaves starts from the margin of the lamina and develops in a V-shaped direction towards the midrib. In the case of an iron deficit, all veins remain green, while with manganese deficiency, the final vein segments discolor. Most often, manganese deficiency is observed in older leaves. In the case of apple trees, the symptoms of manganese chlorosis occur on long and short shoots. The fruits are small and not very juicy and they quickly lose the green color of the skin. In the case of carrot cultivation, manganese is not as important an element as in the cultivation of apple trees. However, the deficiency of manganese in carrot causes retarded growth, and thus a reduction in yield [38].

The average content of assimilable manganese in the soil recorded for samples from conventional vegetable farms was 41.02 mg·kg^{-1}, while in the integrated production group, it was 33.66 mg·kg^{-1} (Table 7).

Table 7. Manganese content in the tested soil samples (mg·kg^{-1}).

	Min.	Average	Max.	Median	Standard Deviation
Integrated vegetable farms	5.20	33.56	87.60	29.90	80.0
Conventional vegetable farms	10.95	41.02	97.83	37.54	66.0

Zinc and its compounds are characterized by a good solubility. Its best solubility occurs in acidic and slightly acidic soils [39]. Organic matter and soil minerals bind zinc ions present in the soil. On Polish farms, zinc is an element often overlooked in the process of fertilization, because farmers believe this element has little impact on the yield increase [40,41].

The results of the conducted research indicate that the average content of available zinc forms in the studied soil samples on conventional vegetable farms was 0.710 mg·kg^{-1} (Table 8), while in soils from integrated farms, this value fluctuated at 0.580 mg·kg^{-1}.

Table 8. Zinc content in the tested soil samples (mg·kg^{-1}).

	Min.	Average	Max.	Median	Standard Deviation
Integrated vegetable farms	0.10	0.58	1.85	0.52	0.41
Conventional vegetable farms	0.20	0.71	1.78	0.56	0.65

Zinc deficiency in agroecosystem plants is often observed at a very high level of phosphorus fertilization, which has been pointed out by many researchers who studying this problem [42]. On the studied farms, the results of our own research indicate a too high level of phosphorus fertilization, inconsistent with the plants' demand. This may lead to an insufficient supply of zinc, especially under intense production conditions.

pH is one of the most important soil parameters, determining its fertility. It is influenced by external and internal factors determined by the applied production techniques. External factors include the type of parent rock, acid rain, and the leaching of alkaline cations, whilst internal factors are fertilization and liming treatments [19]. The pH measured in the aqueous soil suspension indicates the active acidity created by the hydrogen ions found in the soil solution, while the pH measured in the potassium chloride suspension (KCl) also takes into account the acid ions associated with the sorption complex [43].

This parameter determines the conditions of plant growth and development, as well as the direction and speed of biological and physicochemical processes in the soil [43–45]. It is the basic factor affecting the uptake of nutrients by plants, as well as the transformation of nitrogen and phosphorus compounds in the soil. The optimal pH of mineral soil for carrot cultivation is within the range of pH 6–7 [18]. The results of our own research (Table 9) indicate that the soil pH on 27% of farms producing carrot with the integrated methodology was below the optimal values. As a result, on some farms, the necessity of soil liming was identified, while on others, liming was only recommended. Almost 90% of conventional farms also had soil pH below optimal values, thus the need for liming was identified. Only one farm in this group was characterized by a soil with a pH level optimal for carrot production.

The basic element that significantly impacts the formation of soil properties is organic matter consisting of carbohydrates, proteins, fats, and humus. Hummus is part of the organic matter that impacts soil fertility, and is characteristic for each soil [46,47]. The content of organic matter in the soil depends on, e.g., the climate, terrain, parent rock, and water conditions prevailing in the area. In addition, the amount and type of organic matter in the soil are impacted by anthropogenic factors such as indirect or direct human influence on the environment, as well as its flora and fauna. The classification of soils according to the content of humus in soil is presented in Table 10.

The results of tests for humus content in the soils of vegetable and fruit farms indicate large differences between them (Table 11). In the group of integrated vegetable farms, 13.6% of the samples were classified as humus-deficient, 63.6% were low-humus soils, and 22.7% were medium-humus soils. In the above group of farms, there was not a single farm with humus soil. On conventional farms, no humus-deficient soil samples were identified: 62.5% of samples were low-humus soils, 25% were medium-humus soils, and 12.5% were humus soils.

Table 9. Liming demand of individual farms.

	Type of Soil	Reaction	
		pH	**Liming**
Farms producing carrots in the integrated system			
1	III	4.24	necessary
2	III	6.66	unnecessary
3	III	6.3	unnecessary
4	IV	4.89	necessary
5	III	6.85	unnecessary
6	III	5.21	necessary
7	III	6.65	unnecessary
8	III	6.94	unnecessary
9	III	7.63	unnecessary
10	III	6.4	unnecessary
11	IV	7.05	unnecessary
12	III	5.14	necessary
13	III	6.54	unnecessary
14	IV	6.94	unnecessary
15	IV	6.33	unnecessary
16	III	6.75	unnecessary
17	IV	7.06	unnecessary
18	IV	6.92	unnecessary
19	IV	5.28	necessary
20	III	6.68	unnecessary
21	IV	4.69	necessary
22	III	6.94	unnecessary
Farms producing carrots in the conventional system			
1	III	5.65	necessary
2	III	4.89	necessary
3	IV	4.92	necessary
4	III	6.93	unnecessary
5	III	5.16	necessary
6	III	4.85	necessary
7	IV	4.99	necessary
8	III	4.8	necessary

Table 10. Soil classification according to the content of humus in the soil [48].

Humus-Deficient Soils	less than 1%
Soils with Low Humus Level	1.0%–2.0%
Soils with Medium Humus Level	2.1%–3.0%
Humus Soils	above 3.0%

Table 11. The content of humus in the soils of vegetable farms, based on the obtained test results.

	Farms Producing Carrots in the Integrated System	Farms Producing Carrots in the Conventional System
Humus-deficient soils **<1%**	3 farms (13.6%)	-
Soils with low humus level **1.0%–2.0%**	14 farms (63.6%)	5 farms (62.5%)
Soils with medium humus level **2.1%–3.0%**	5 farms (22.7%)	2 farms (25%)
Humus soils **(3.0%)**	-	1 farm (12.5%)

Research carried out in recent years has shown a decrease in humus content in Polish soils [46]. This is related to the disturbance of hydrographic conditions and intensive soil use. Low levels of humus in Polish soils, as well as risks associated with mineralization, can cause large emissions of carbon dioxide from soils [1,46]. The mechanism that allows humus depletion to be counteracted is the development of agri-environmental programs, under which farmers receive subsidies for the cultivation of after and intercrops improving the balance of organic matter in the soils of their farms.

4. Conclusions

The following conclusions can be drawn from this study:

1. The rational management of plant nutrients and maintenance of appropriate soil parameters are strategic elements of the quality management system in plant production;
2. The results of our own research indicate that on approx. 60% of farms carrying out integrated carrot production, and on all conventional farms studied, the content of available phosphorus forms was higher than recommended in the integrated carrot production methodology. On the other hand, on conventional farms, these quantities were much higher (Table 1);
3. One of the goals of integrated production is to improve soil properties. In the majority of both integrated and conventional farms, balanced fertilization was not implemented due to irrational fertilization with potassium. The result of such a management strategy may negatively impact both the soil and economic efficiency of production [11,14,31,49]. The calcium content in the tested soil samples varied significantly within the compared production systems. Unfavorable values, i.e., below 1000 (mg·kg^{-1}), were observed on farms with the conventional production system;
4. The results of our own research indicate a very small variability in the amount of available forms of iron in individual samples within the research groups. The value of the coefficient of variation in the group of vegetable farms carrying out production in accordance with the principles of integrated and conventional agriculture was 56% and 36%, respectively;
5. The results of the conducted research indicate that on each of the studied farms, both the integrated and conventional group, the soil had a magnesium deficit. A too low magnesium content in the soil can cause plant metabolism disorders;
6. Comparative analysis indicates an insufficient effectiveness of the integrated production system in terms of soil resource management. However, compared to conventional farms, soil resource management on integrated farms follows the concept of sustainable agriculture more closely. The implementation of obligatory consulting on practical aspects of fertilization should impact optimization of the production process.

Author Contributions: Conceptualization, A.S.-S., M.N., and J.S.; methodology, A.S.-S., M.N., and J.K.-D.; resources, J.K., A.S.-S., M.N., and J.S.; formal analysis, A.S.-S., M.N., J.S., and M.K.; investigation, A.S.-S., Z.G.-S., and M.K.; data curation, A.S.-S. and M.N.; writing, A.S.-S. and M.N.; visualization, J.K.-D., J.S., and M.K.

Funding: This research received no external funding.

Conflicts of Interest: The authors declare no conflicts of interest.

References

1. Szeląg-Sikora, A.; Niemiec, M.; Sikora, J.; Chowaniak, M. Possibilities of Designating swards of grasses and small-seed legumes from selected organic farms in Poland for feed. In Proceedings of the IX International Scientific Symposium "Farm Machinery and Processes Management in Sustainable Agriculture", Lublin, Poland, 22–24 November 2017; pp. 365–370.

2. Holzapfel, S.; Wollni, M. Global GAP Certification of small-scale farmers sustainable? Evidence from Thailand. *J. Dev. Stud.* **2014**, *50*, 731–746. [CrossRef]

3. Mzoughi, N. Farmers adoption of integrated crop protection and organic farming: Do moral and social concerns matter? *Agric. Econ.* **2011**, *70*, 1536–1545. [CrossRef]

4. Niemiec, M.; Komorowska, M.; Szeląg-Sikora, A.; Sikora, J.; Kuboń, M.; Gródek-Szostak, Z.; Kapusta-Duch, J. Risk Assessment for Social Practices in Small Vegetable farms in Poland as a Tool for the Optimization of Quality Management Systems. *Sustainability* **2019**, *11*, 3913. [CrossRef]

5. Szeląg-Sikora, A.; Niemiec, M.; Sikora, J. Assessment of the content of magnesium, potassium, phosphorus and calcium in water and algae from the black sea in selected bays near Sevastopol. *J. Elem.* **2016**, *21*, 915–926. [CrossRef]

6. Rivas, J.; Manuel Perea, J.; De-Pablos-Heredero, C.; Angon, E.; Barba, C.; García, A. Canonical correlation of technological innovation and performance in sheep's dairy farms: Selection of a set of indicators. *Agric. Syst.* **2019**, *176*, 102665. [CrossRef]

7. Erbaugha, J.; Bierbaum, R.; Castillejac, G.; da Fonsecad, G.A.B.; Cole, S.; Hansend, B. Toward sustainable agriculture in the tropics. *World Dev.* **2019**, *121*, 158–162. [CrossRef]

8. Niemiec, M.; Komorowaka, M.; Szeląg-Sikora, A.; Sikora, J.; Kuzminova, N. Content of Ba, B, Sr and as in water and fish larvae of the genus Atherinidae, L. sampled in three bays in the Sevastopol coastal area. *J. Elem.* **2018**, *23*, 1009–1020. [CrossRef]

9. Niemiec, M.; Sikora, J.; Szeląg-Sikora, A.; Kuboń, M.; Olech, E.; Marczuk, A. Applicability of food industry organic waste for methane fermentation. *Przem. Chem.* **2017**, *69*, 685–688. [CrossRef]

10. Kuboń, M.; Krasnodębski, A. Logistic cost in competitive strategies of enterprises. *Agric. Econ.* **2010**, *56*, 397–402.

11. Niemiec, M. Efficiency of slow-acting fertilizer in the integrated cultivation of Chinese cabbage. *Ecol. Chem. Eng.* **2014**, *21*, 333–346.

12. Cassman, K.; Dobermann, A.; Walters, D. Agroecosystems, nitrogen-use efficiency, and nitrogen. *J. Hum. Environ.* **2002**, *31*, 132–140. [CrossRef] [PubMed]

13. Meziere, D.; Lucas, P.; Granger, S.; Colbach, N. Does Integrated Weed Management affect the risk of crop diseases? A simulation case study with blackgrass weed and take all disease. *Eur. J. Agron.* **2013**, *47*, 33–43. [CrossRef]

14. Helander, A.; Delin, K. Evaluation of farming systems according to valuation indices developed within a European network on integrated and ecological arable farming systems. *Eur. J. Agron.* **2004**, *21*, 53–67. [CrossRef]

15. Musshoff, O.; Hirschauer, N. Adoption of organic farming in Germany and Austria: An integrative dynamiv investment perspective. *Agric. Econ.* **2008**, *39*, 135–145. [CrossRef]

16. Bizkarguenaga, E.; Zabaleta, I.; Mijangos, L.; Iparraguirre, A.; Fernández, L.A.; Prieto, A.; Zuloaga, O. Uptake of perfluorooctanoic acid, perfluorooctane sulfonate and perfluorooctane sulfonamide by carrot and lettuce from compost amended soil. *Sci. Total Environ.* **2016**, *571*, 444–451. [CrossRef]

17. Kapusta-Duch, J.; Szeląg-Sikora, A.; Sikora, J.; Niemiec, M.; Gródek-Szostak, Z.; Kuboń, M.; Leszczyńska, T.; Borczak, B. Health-Promoting Properties of Fresh and Processed Purple Cauliflower. *Sustainability* **2019**, *11*, 4008. [CrossRef]

18. *Methodology of Integrated Carrot Production*; Polish Institute of Plant Protection and Fertilization (PIORIN): Warszawa, Poland, 2016. Available online: www.piorin.gov.pl (accessed on 15 June 2019).

19. Qian, X.; Gu, J.; Sun, W.; Li, D.; Fu, X.; Wang, J.; Gao, H. Changes in the soil nutrient levels, enzyme activities, microbial community function, and structure during apple orchard maturation. *Appl. Soil Ecol.* **2014**, *77*, 18–25. [CrossRef]

20. Higgs, B.; Johnston, A.E.; Salter, J.L.; Dawson, C.J. Some Aspects of Achieving Sustainable Phosphorus Use in Agriculture. *J. Environ. Qual.* **2000**, *17*, 80–87. [CrossRef]

21. Ayaga, G.; Todd, A.; Brookes, P.C. Enhanced biological cycling of phosphorus increases its availability to crops in low-input sub-Saharan farming systems. *Soil Biol. Biochem.* **2006**, *38*, 81–90. [CrossRef]

22. Skafa, L.; Buonocorea, E.; Dumonteta, S.; Capone, R.; Franzesea, P.P. Food security and sustainable agriculture in Lebanon: An environmental accounting framework. *J. Clean. Prod.* **2019**, *209*, 1025–1032. [CrossRef]

23. Yu, J.; Wu, J. The Sustainability of Agricultural Development in China: The Agriculture–Environment Nexus. *Sustainability* **2018**, *10*, 1776. [CrossRef]

24. Malik, M.A.; Marschner, P.; Khan, K.S. Addition of organic and inorganic P sources to soil e Effects on P pools and microorganisms. *Soil Biol. Biochem.* **2012**, *49*, 106–113. [CrossRef]

25. Gródek-Szostak, Z.; Malik, G.; Kajrunajtys, D.; Szelag-Sikora, A.; Sikora, J.; Kuboń, M.; Niemiec, M.; Kapusta-Duch, J. Modeling the Dependency between Extreme Prices of Selected Agricultural Products on the Derivatives Market Using the Linkage Function. *Sustainability* **2019**, *11*, 4144. [CrossRef]

26. Kocira, S.; Kuboń, M.; Sporysz, M. Impact of information on organic product packagings on the consumers decision concerning their purchase. *Int. Multidiscip. Sci. GeoConf. SGEM* **2017**, *17*, 499–506. [CrossRef]

27. Brown, P.; Bellaloui, N.; Wimmer, M.A.; Bassil, E.S.; Ruiz, J.; Hu, H.; Pfeffer, H.; Dannel, F. Boron in plant biology. *Plant Biol.* **2002**, *4*, 205–223. [CrossRef]

28. Itaktura, T.; Sasai, R.; Itoh, H. Precipitation recovery of boron from wastewater by hydrothermal mineralization. *Water Res.* **2005**, *39*, 2543–2548. [CrossRef]

29. Dirceu, M.; Hippler, F.; Boaretto, R.; Stuchi, E.; Quaggio, J. Soil boron fertilization: The role of nutrient sources and rootstocks in citrus production. *J. Integr. Agric.* **2017**, *16*, 1609–1616.

30. Danis, T.G.; Karagiozoglou, D.T.; Tsakiris, I.N.; Alegakis, A.K.; Tsatsakis, A.M. Evaluation of pesticides residues in Greek peaches during 2002–2007 after the implementation of integrated crop management. *Food Chem.* **2011**, *126*, 97–103. [CrossRef]

31. Courtney, R.G.; Mullen, G.J. Soil quality and barley growth as influenced by the land application of two compost types. *Bioresour. Technol.* **2008**, *5*, 2913–2918. [CrossRef]

32. Lemberkovics, E.; Czinner, E.; Szentmihalyi, K.; Bals, A.; Szoke, E. Comparitive evaluation of Helichrysi flos herbat extracts as dietary sources of plant polyphenols, and macro- and microelements. *Food Chem.* **2002**, *78*, 119–127. [CrossRef]

33. Wang, H.Q.; Zhao, Q.; Zeng, D.H.; Hu, Y.L.; Yu, Z.Y. Remediation of a Magnesium-Contaminated Soil by Chemical Amendments and Leaching. *Land Degrad. Dev.* **2015**, *15*, 613–619. [CrossRef]

34. Qadir, M.; Schubert, S.; Oster, J.D.; Sposito, G.; Minhas, P.S.; Cheraghi, S.M.A.; Murtaza, G.; Mirzabaev, A.; Saqib, M. High-magnesium waters and soils: Emerging environmental and food security constraints. *Sci. Total Environ.* **2018**, *642*, 1108–1117. [CrossRef] [PubMed]

35. Kuboń, M.; Sporysz, M.; Kocira, S. Use of artificial of clients of organic farms. In Proceedings of the 17th International Multidisciplinary Scientific GeoConference SGEM 2017, Bulgaria, Balkans, 27 June–6 July 2017; Volume 17, pp. 1099–1106.

36. Gródek-Szostak, Z.; Szelag-Sikora, A.; Sikora, J.; Korenko, M. Prerequisites for the cooperation between enterprises and business supportinstitutions for technological development. In *Business and Non-Profit Organizations Facing Increased Competition and Growing Customers' Demands*; Wyższa Szkoła Biznesu—National-Louis University: Nowy Sącz, Poland, 2017; Volume 16, pp. 427–439.

37. Wanli, G.; Hussain, N.; Zongsuo, L.; Dongfeng, Y. Magnesium deficiency in plants: An urgent problem. *Crop J.* **2016**, *4*, 83–91.

38. Serpil, S. Investigation of Effect of Chemical Fertilizers on Environment. *Apcbee Procedia* **2012**, *1*, 287–292.

39. Chen, C.; Zhang, H.; Gray, E.; Boyd, S.; Yang, H.; Zhang, D. Roles of biochar in improving phosphorus availability in soils: A phosphate adsorbent and a source of available phosphorus. *Geoderma* **2016**, *276*, 1–6.

40. Domańska, J. Soluble forms of zinc in profiles of selected types of arable soils. *J. Elem.* **2009**, *14*, 55–62. [CrossRef]

41. Van Oort, F.; Jongmans, A.G.; Citeau, L.; Lamy, I.; Chevallier, P. Microscale Zn and Pb distribution patterns in subsurface soil horizons: An indication for metal transport dynamics. *Europ. Soil Sci.* **2006**, *57*, 154–166. [CrossRef]

42. Zhu, G.; Smith, E.; Smith, A. Zinc (Zn)—Phosphorus (P) interaction in two cultivars of spring wheat (*Triticum aestivum* L.) differing in P uptake efficiency. *Ann. Bot.* **2001**, *88*, 941–945. [CrossRef]

43. Han, T.; Cai, A.; Liu, K.; Huang, J.; Wang, B.; Li, D. The links between potassium availability and soil exchangeable calcium, magnesium, and aluminum are mediated by lime in acidic soil. *J. Soils Sediments* **2019**, *19*, 1382–1392. [CrossRef]

44. Ji, C.-J.; Yang, Y.-H.; Han, W.-X.; He, Y.-F.; Smith, J.; Smith, P. Climatic and Edaphic Controls on Soil pH in Alpine Grasslands on the Tibetan Plateau, China: A Quantitative Analysis. *Pedosphere* **2014**, *24*, 39–44. [CrossRef]

45. Sikora, J.; Niemiec, M.; Szeląg-Sikora, A.; Kuboń, M.; Olech, E.; Marczuk, A. Biogasification of wastes from industrial processing of carps. *Przem. Chem.* **2017**, *96*, 2275–2278.

46. Klimkowicz-Pawlas, A.; Smreczak, B.; Ukalska-Jaruga, A. The impact of selected soil organic matter fractions on the PAH accumulation in the agricultural soils from areas of different anthropopressure. *Environ. Sci. Pollut. Res.* **2017**, *24*, 10955–10965. [CrossRef] [PubMed]

47. Li, L.-J.; Zhu-Barker, X.; Ye, R.; Doane, T.A.; Horwath, W.R. Soil microbial biomass size and soil carbon influence the priming effect from carbon inputs depending on nitrogen availability. *Soil Biol. Biochem.* **2018**, *119*, 41–49. [CrossRef]

48. Mocek, A.; Drzymała, S.; Maszner, P. *Geneza, Analiza i Klasyfikacja Gleb*; Wydawnictwo Akademia Rolnicza w Poznaniu: Poznan, Finland, 1997; p. 416.

49. Domagała-Świątkiewicz, I.; Gąstoł, M. Soil chemical properties under organic and conventional crop management systems in south Poland. *Biol. Agric. Hortic.* **2013**, *29*, 12–28. [CrossRef]

Article

Survivability of Probiotic Bacteria in Model Systems of Non-Fermented and Fermented Coconut and Hemp Milks

Agnieszka Szparaga [1], Sylwester Tabor [2], Sławomir Kocira [3,*], Ewa Czerwińska [4], Maciej Kuboń [2], Bartosz Płóciennik [1] and Pavol Findura [5]

[1] Department of Agrobiotechnology, Koszalin University of Technology, Racławicka 15–17, 75-620 Koszalin, Poland; agnieszka.szparaga@tu.koszalin.pl (A.S.); b.plociennik2@gmail.com (B.P.)

[2] Department of Production Engineering, Logistics and Applied Computer Science, Agricultural University Krakow, Balicka 116B, 30-149 Krakow, Poland; sylwester.tabor@urk.edu.pl (S.T.); maciej.kubon@ur.krakow.pl (M.K.)

[3] Department of Machinery Exploitation and Management of Production Processes, University of Life Sciences in Lublin, Akademicka 13, 20-950 Lublin, Poland

[4] Department of Biomedical Engineering, Koszalin University of Technology, Śniadeckich 2, 75-453 Koszalin, Poland; ewa.czerwinska@tu.koszalin.pl

[5] Department of Machines and Production Biosystems, Slovak University of Agriculture in Nitra, Tr. A. Hlinku 2, 949 76 Nitra, Slovakia; pavol.findura@uniag.sk

* Correspondence: slawomir.kocira@up.lublin.pl; Tel.: +48-81-531-97-35

Received: 14 August 2019; Accepted: 27 October 2019; Published: 1 November 2019

Abstract: This study aimed at determining the survivability of probiotic bacteria cultures in model non-dairy beverages subjected or not to the fermentation and storage processes, representing milk substitutes. The experimental material included milks produced from desiccated coconut and non-dehulled seeds of hemp (*Cannabis sativa* L.). The plant milks were subjected to chemical and microbiological evaluation immediately after preparation as well as on day 7, 14, and 21 of their cold storage. Study results proved that the produced and modified plant non-dairy beverages could be the matrix for probiotic bacteria. The fermentation process contributed to increased survivability of *Lactobacillus casei* subsp. *rhamnosus* in both coconut and hemp milk. During 21-day storage of inoculated milk substitutes, the best survivability of *Lactobacillus casei* was determined in the fermented coconut milk. On day 21 of cold storage, the number of viable *Lactobacillus casei* cells in the fermented coconut and hemp milks ensured meeting the therapeutic criterion. Due to their nutritional composition and cell count of bacteria having a beneficial effect on the human body, the analyzed groceries—offering an alternative to milk—represent a category of novel food products and their manufacture will contribute to the sustainable development of food production and to food security assurance.

Keywords: probiotic; non-dairy beverages; survivability; fermentation; bacteria; coconut; hemp; sustainable food production

---

1. Introduction

Sustainable food production should be considered through the perspective of a better understanding of food security. In recent years, many researchers and policy makers have focused only on the physical availability of food, owing to the sufficient agricultural production [1,2]. This has partly been driven by widespread claims that we need to boost the global food production to feed the world in 2050 [3]. However according to the FAO (Food and Agriculture Organization of the United Nations) definition [4]: "food security exists when all people, at all times, have physical and economic access to sufficient, safe and nutritious food that meets their dietary needs and food preferences for an active and healthy life". Hence, it needs to be emphasized that the sustainable food production is inevitably related to the food security in its three aspects: food security, food safety, and food quality, without which the development of the food industry sector would not be possible [5,6]. Due to the current dynamic development of sciences related to food and human nutrition, a correlation has been confirmed between the health status and nutritional patterns. A well-balanced diet is the key factor in diseases prevention and treatment. The growing nutritional interests and awareness of consumers have prompted many producers to manufacture functional food [7–9]. A functional food definition covers certain strains of microorganisms being constituents of food of plant and animal origin that contain physiologically active compounds. These compounds are beneficial for human health and help minimizing the risk of chronic diseases development [10]. One of the multiple examples of functional food products are these containing microorganisms endogenous to the human gastrointestinal tract and exhibiting a positive effect on human health [11]. So far, the greatest part of probiotic products has been offered by fermented beverages made of animal milk. Currently, research is underway into other products that may be matrix for probiotic bacteria [12,13]. Consumers avoiding milk because of allergies or lactose intolerance, and consumers following a vegan diet can replace milk with other plant-based substitutes. Beverages derived from soybeans have for many years been the predominant equivalents of milk. Today, coconut, almonds, hemp, and various cereals (e.g., oats, buckwheat, and rice) are also used to produce plant-based beverages [14,15]. A drawback of these products is however their specific taste that does not suit to everyone. A solution to this problem is offered by lactic acid fermentation, which imparts a characteristic, pleasant after-taste to these products and contributes to the improvement of the digestibility [13,16]. Numerous attempts have recently been undertaken to ferment vegan beverages serving as milk substitutes using various strains of probiotic bacteria, which was expected to additionally increase their health value [16,17]. However, most of the study results reported in literature concern the feasibility of producing fermented soybean milk [13–16]. This is related to the fact, that manufacture of high quality plant-based beverages containing probiotic bacteria poses a serious challenge [18,19]. According to Yuliana et al. [20], the production of coconut-based beverages is difficult because of the suppressed growth and survivability of these probiotic microorganisms, compared to dairy beverages. Difficulties in the manufacture and fortification of hemp milk were encountered by Batkiene et al. [21]. The first ones were related to the stability of produced emulsions, whereas the latter ones to the survivability of probiotic bacteria during storage. Worthy of notice is that the production of hemp-based products has increased in recent years due to the confirmed nutritional value and low allergenicity of seeds of this plant [22]. This has been feasible owing to new varieties characterized by a low concentration of a psychoactive compound delta-9-tetrahydrocannabinol (THC) [23] and to cultivations with the use of elite category sowing material [24].

The current definition of a probiotic means those microbial strains that positively affect consumer health when taken in the right amount [25]. Accordingly to FAO/WHO guidelines, the count of probiotic bacteria cannot be less than the value corresponding to 10^6 cfu per 1 mL of a product through the entire period of its storage till the end of its shelf life. This value has been deemed the therapeutic minimum [26–28].

The main problems associated with the fermentation of plant beverages are related to the sensory quality of the final product and to the resistance of probiotic microorganisms. Producers encounter difficulties with the physical stability caused by milk coagulation (it occurs at the beginning or in the course of storage). The appearance of these products resembles that of low-fat yoghurt [29–32]. Additional problems concern the survivability of probiotic bacteria, which is dependent on multiple factors, including e.g., presence of other microorganisms in the product, time and conditions of strains culture and product storage, product processing technology or pH value [13,26,33,34].

Considering the above, the major objective of this study was to determine the survivability of probiotic bacterial cultures in model fermented and non-fermented stored plant beverages being milk substitutes, because today the plant-based alternative milks provide a huge perspective for the sustainable development of the healthy food market and should therefore be widely scrutinized. Evaluation of the effect of production and processing techniques, and also of fortification techniques, of plant-based beverages may serve to develop a nutritionally complete beverage with a high overall acceptability and health values.

2. Materials and Methods

2.1. Plant Material and Beverages Production

The experimental material included non-dairy beverages produced from the following plant raw materials: desiccated coconut (Bakalland, Warszawa, Poland) and non-dehulled seeds of hemp (*Cannabis sativa* L.; Sante, Warszawa, Poland; Figure 1). Seeds and desiccated coconut were ground in a WZ-1 laboratory mill (Sadkiewicz Instruments, Bydgoszcz, Poland). Chemical composition of analyzed material is presented in Table 1.

Figure 1. Raw materials used for non-dairy beverages production: (**a**) hemp seeds and (**b**) desiccated coconut.

Table 1. Contents of protein, lipids, and sugars of desiccated coconut and non-dehulled hemp seeds (nutritional information available on respective product labels).

Product	Protein	Lipids	Sugars
		(%)	
Desiccated coconut	5.6	63.2	5.9
Non-dehulled seeds of hemp	25.2	36.1	5.4

Milk substitutes to be analyzed were produced according to the schemes presented in block diagrams in Figures 2 and 3.

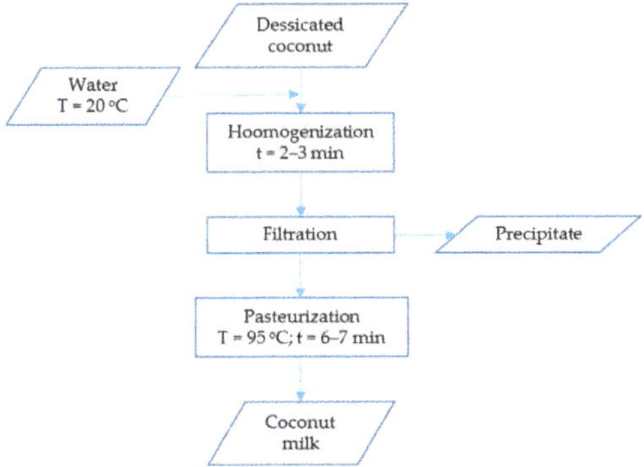

Figure 2. Technological scheme of coconut milk production (own study based on Blasco [35]).

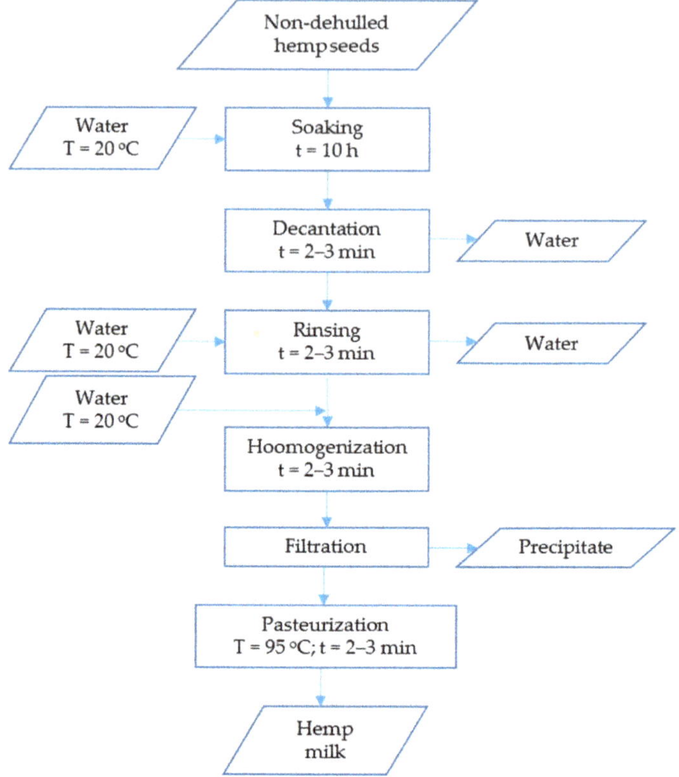

Figure 3. Technological scheme of hemp milk production (own study based on Szakuła [36]).

Immediately after production, the plant-based milks were subjected to the physicochemical analysis to determine content of their protein, lipids, and carbohydrates, and their acidity. Their microbiological status was assessed as well.

Models of non-dairy beverages intended for the determination of counts of viable bacterial cells during storage were divided into four groups, each containing three samples—two groups included beverages with probiotic addition and two groups—beverages without the probiotic. Both, the samples supplemented and not supplemented with a probiotic monoculture were fermented in a laboratory incubator (Binder BD 260) at a temperature of 37 °C for 6 h, and then stored at a temperature of 4 °C. The remaining samples were cold stored. Duration of the fermentation process was chosen based on results of a study conducted by Zielińska et al. [37], who demonstrated that intensive proliferation of *Lactobacillus casei* subsp. *rhamnosus* cells proceeded till the 6th hour of the process. Optimal parameters of the fermentation process, established by Zielińska et al. [38], allow producing plant-based fermented beverages with sensory quality acceptable by consumers [39]. Based these findings, investigations have not assumed own sensory evaluation.

2.2. Probiotic Microorganisms

The study was conducted with probiotic bacterial strain *Lactobacillus casei* subsp. *rhamnosus* LCR 3013 in the form of a lyophilizate (Serowar, Szczecin, Poland). Before analyses, the strain was stored at a temperature of −18 °C to preserve its properties. The bacterial culture was activated by transferring 0.1 g of the lyophilized strain to 5 mL of an MRS broth (Merck, Warszawa, Poland). Next, the suspension was incubated at a temperature of 37 °C for 24 h. The resultant culture was centrifuged at 10,000× *g* for 5 min. Centrifugation was repeated after rinsing the resultant precipitate with a physiological saline solution. Cell biomass suspended in a physiological saline solution (5 mL) and having the optical density of 1° McF (Densimat, bioMérieux, Grassina FI, Italy) was added to 100 mL of plant milk (5% *v/v*), to achieve bacterial cell count of approximately 10 log (cfu/mL) [21].

2.3. Chemical Evaluation of Non-Dairy Beverages

The produced non-dairy beverages were analyzed for content of: protein with the Kjeldahl's method [40] (Büchi Distilation Unit, K314) and lipids with the Gerber's method [41]. Their active acidity (pH) was measured as well using a VOLTCRAFT KBM-110 m (Conrad Electronic SE, Hirschau, Germany) with a pH electrode [42].

Additional analyses were conducted to determine the content of reducing sugars. They were made with the colorimetric method using 3,5-dinitrosalicilic acid (DNS) [43]. DNS acid (1950 µL) were added to 50 µL of the analyzed non-dairy beverages, and the mixture was incubated in a water bath at a temperature of 99 °C for 10 min. After cooling the mixture, 900 µL of DNS acid were added to 100 µL of mixture sample. Absorbance was measured at $\lambda = 540$ nm (UV-vis spectrophotometer, VWR UV-6300PC, USA) and results of these measurements were converted based on the standard curve into equivalents of glucose (g/L) contained in non-dairy beverages.

2.4. Microbiological Status of Plant Beverages

The produced plant beverages were pasteurized and afterwards subjected to a microbiological analysis based on the pour-plate Koch's method [44] and the sterile serial dilutions method (from 10^{-1} to 10^{-8}). For this purpose, 1 mL of each of the two subsequent dilutions (10^{-6} and 10^{-8}) and 9 mL of each of the appropriately selected medium were transferred onto Petri dishes and left to solidify [16]. Nutrient agar (BTL, Łódź, Poland) was used to isolate mesophilic and psychrophilic bacteria, whereas Sabouraud agar enriched with chloramphenicol (BTL, Łódź, Poland) was used for fungi and yeast isolation from the beverages. Samples were incubated at a temperature of 30 °C for 48 h (mesophiles) and at 20 °C for 72 h (psychrophiles), and at 20 °C for 5 days (fungi and yeast) [44]. Total bacterial count (TBC) was determined as well. Bacterial cultures were inoculated with the pour-plate method in three replications for each sample. Plates with inoculates were incubated at a temperature of 37 °C for 48 h under anaerobic conditions using anaerostats with anaerocult A inserts (Merck, Darmstadt, Germany). Nutrient broth agar (BTL, Łódź, Poland) was used for inoculations. After completed incubation, results were converted into the number of colony forming units per 1 mL of product (cfu/mL). Dilutions of 10^{-6} and 10^{-8} were used for analyses and for TBC determination in each sample. The above analyses were carried out for plant beverages without probiotic strain addition.

2.5. Microbiological Analyses of Counts of Viable Bacterial Cells During Storage of Fermented and Non-Fermented Non-Dairy Beverages

Analyses of the survivability of the probiotic strain *Lactobacillus casei* subsp. *rhamnosus* in fermented and non-fermented models of coconut and hemp beverages (with added starter monoculture of probiotic bacteria) were conducted immediately after their preparation, after their fermentation as well as on day 7, 14, and 21 of their storage at a temperature of 4 °C. The maximal cold storage time assumed in the study was selected based on results of a research conducted by Gustaw et al. [45] into the survivability of *Lactobacillus casei* strain in fermented beverages with the addition of selected protein preparations.

Having been diluted in sterile water, the analyzed samples were transferred onto sterile Petri dishes (1 mL of sample from each dilution of 10^{-6} and 10^{-8}), to which 9 mL of the selective MRS Agar medium (by de Man, Rogosa and Sharpe) [46] (BTL, Poland) were added afterwards. Next, the samples were incubated at a temperature of 30 °C for 72 h. The total count of viable lactic acid bacteria per 1 mL of the sample was computed according to the following formula:

$$N = n_c \cdot d_r, \tag{1}$$

where: N—number of viable bacterial cells (cfu/mL), n_c—number of bacterial colonies, and d_r—dilution rate.

2.6. Statistical Analysis

All experiments were carried out in three replications. Results were expressed as arithmetic means. The Shapiro–Wilk test was used to evaluate the normal distribution of data. Results were analyzed using a one-way analysis of variance (ANOVA). The significance of differences between mean values was estimated based on Tukey confidence intervals, at a $p < 0.05$. Values followed by different small letters are significantly different at $p < 0.05$ (effect of storage). Values followed by different big letters are significantly different at $p < 0.05$ (effect of treatment). The standard deviation (±SD) value was determined for all reported mean values. The statistical analysis was performed using Statistica 13.3 software (StatSoft, Kraków, Poland).

3. Results and Discussion

3.1. Evaluation of the Produced Non-Dairy Beverages

One of the key technological aspects in probiotic food production is to maintain optimal conditions that would ensure the proper growth and viability of potentially probiotic bacteria during fermentation and storage [47]. This may be accomplished through, i.e., the appropriate choice of a carrier and product supplementation with nutrients [48]. Hence, raw materials of plant origin need to be analyzed for the content of nutrients indispensable for probiotic bacteria metabolism, and for the effect of environment on their survivability [27,49].

In order to identify factors that affect probiotics survivability, a study with non-dairy beverages produced from desiccated coconut and hemp seeds under laboratory conditions was conducted. It needs to be emphasized that the nutritional value of non-dairy beverages is largely determined by their protein content [28,50]. In the study, protein content was determined at 3.23 g/100g in coconut beverage and at 6.96 g/100 g in hemp beverage (Table 2).

Table 2. Contents of protein, lipids, and glucose, and active acidity of coconut and hemp milks.

Non-Dairy Beverage	Protein	Lipids	Reducing Sugars	Active Acidity
	(% ± SD)		(g glucose/L ± SD)	(pH ± SD)
Coconut milk	3.23 ± 0.28	21.08 ± 0.41	34.53 ± 0.39	6.15 ± 0.15
Hemp milk	6.96 ± 0.19	18.02 ± 0.54	30.21 ± 0.33	6.81 ± 0.11

Differences in protein content were determined depending on the plant raw material used for beverages production. Discrepancies were also noted when comparing protein content determined in the study and these declared by selected producers of plant beverages intended for the European market [13], i.e., obtained results were higher than protein content declared by producers of coconut and hemp beverages. These differences might be due to the various quality of raw materials used for beverages production and to treatments applied in the production process (e.g., heat treatment) that contribute to a decrease in total protein concentration.

Hoffman and Kostyra [15] evaluated plant-based milk substitutes in terms of their nutritional value and demonstrated that only the beverage made of soybean seeds equaled milk in this respect. The other plant materials had significantly lower content of protein, i.e., two-fold lower—quinoa, and three-fold lower—a mixed beverage made of soybean, rice, and oats. Beverages made of coconuts and almonds were characterized by trace amounts of protein.

This indicates that although coconut-based milk substitutes provide very low amounts of protein, they may offer an alternative to consumers who seek for gluten-free food.

Results of study confirm findings reported by Sethi et al. [13], who demonstrated that plant beverages are inexpensive substitutes of milk, especially for consumers allergic to milk, but are not comparable nor equal with it in terms of their nutritional value, as hemp milk (living harvest) provides barely 2 g and coconut milk less than 1 g of protein in 240 mL of the product.

Coconut beverages are food products characterized by a high content of lipids, including significant content of the following saturated fatty acids: lauric (50%) and myristic (6%–7%). The unsaturated fatty acids of coconut include oleic acid (monounsaturated) and linolic acid (polyunsaturated) [51]. In turn, the hemp beverage contains approximately 80% of essential unsaturated fatty acids (EFAs), including linolic acid (56%) and α-linolenic acid (19%). According to dieticians, the optimal ratio of these acids should reach 3:1, as is the case with the hemp beverage [52]. Figure 4 presents content of lipids in the analyzed raw beverages made of coconut and hemp seeds. As it results from our study, the coconut beverage contained 21% and the hemp beverage contained 18% of lipids.

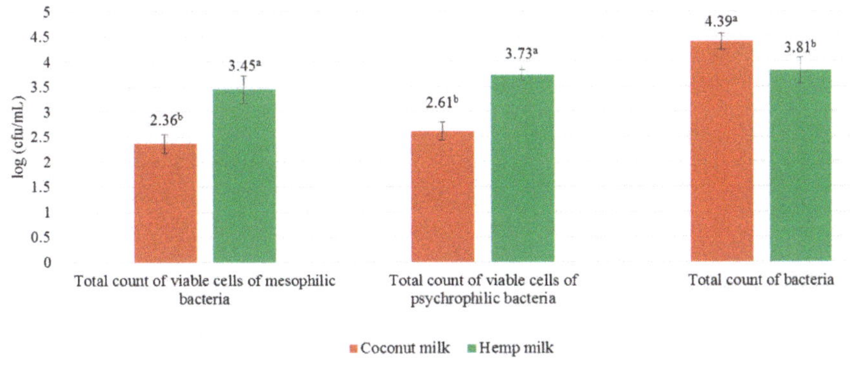

Figure 4. Total count of microorganisms isolated from pasteurized non-diary beverages (a, b—statistically significant differences, $p \leq 0.05$).

Due to the composition of the plant milks, their fatty acid profile differed significantly from that of milk. Milk contains approximately 1.2% of saturated fatty acids, whereas their content in its substitutes does not exceed 0.7%. Similar conclusions were formulated by Belewu and Belewu [53], who determined lipids content in the produced coconut milk at 24.10%. In turn, Sethi et al. [13] demonstrated that the analyzed plant-derived milks were characterized by a similar concentration of lipids reaching 6 g in hemp milk (living harvest) and 5 g in coconut milk per 240 mL of the product.

Content and types of carbohydrates in fermented beverages have a significant impact on the development and activity of health-promoting bacteria. Reducing sugars present in the medium may be good sources of carbon necessary for probiotics metabolism [54]. This was confirmed by Jurkowski and Błaszczyk [55], who demonstrated monosaccharides to be indispensable substrates during the fermentation process by lactic acid bacteria. Hence, control of their content seems necessary in the production of probiotic foods. In the study, the content of reducing sugars was expressed as glucose concentration in the produced non-dairy beverages (Figure 4), and reached 34.52 g glucose/L in coconut milk and 30.21 g glucose/L in hemp milk. Data obtained demonstrate that the coconut milk is a better raw material for the production of probiotic non-dairy beverages because it is richer in compounds necessary for fermentation (e.g., glucose). This is in line with results reported by Quasem et al. [17] who analyzed a sesame beverage as a bacteria matrix. Differences in the content of reducing sugars are one of the factors, which determine the possibility of using plant-based milk as a natural medium for lactic acid bacteria development. According to Sethi et al. [13], however, the choice of a matrix for these beneficial microorganisms may also be driven by the total content of carbohydrates. In hemp milk, their concentration reaches barely 1 g, whereas in coconut milk it is high and reaches 7 g/240 mL of product.

The content of acids is a factor that determines food freshness, but at the same time it largely affects its color and taste. Acids in beverages and other food products, contribute to control microflora growth [56]. In our study, the active acidity of coconut beverage (before fermentation) was determined at pH 6.15 (Figure 4). This value is similar to results reported by other authors. For instance, when investigating the effect of strawberry beverage supplementation with a soybean protein isolate, Dłużewska et al. [57] observed changes in the real acidity, values of which ranged from pH 5.82 to pH 6.61. In turn, acidity of a coconut beverage analyzed by Belewu and Belewu [53] reached pH 6.23. A negligibly higher active acidity (pH 6.81) was determined in our study for the hemp seed beverage. Many scientists have reported on the decreased acidity of the medium in the case of fermented products, which appeared to result from the accumulation of lactic acid caused by the activity of microorganisms.

3.2. Microbiological Purity of Non-Dairy Beverages

From the perspective of health safety, pasteurization is an indispensable process during the manufacture of plant-based beverages. It contributes to eradication of pathogenic microorganisms and to inactivation of certain spore-forms. Therefore, in the present study, we evaluated its effectiveness (Figure 4), which is consistent with provisions of Commission Regulations (EC) 2073/2005 and 229/2019 [58], according to which the safety of food is mainly ensured by a preventive approach including, e.g., control of heat treatment effectiveness.

It was demonstrated that the type of raw material influenced on the presence of bacteria, both the mesophilic and psychrophilic ones, in the produced non-dairy beverages. The determined count of mesophilic bacteria ranged from 2.36 log (cfu/mL; coconut milk) to 3.45 log (cfu/mL; hemp milk). A more numerous group of the isolated microorganisms turned out to be the psychrophilic bacteria, with counts ranging from 2.6 to 3.7 log (cfu/mL). In contrast, no fungi or yeast were detected. The total bacterial count in the produced non-dairy beverages reached 3.81 log (cfu/mL) in hemp milk and 4.39 cfu/mL in coconut milk. It was, therefore, concluded that the short-term temperature increase to 95 °C did not contribute to the complete neutralization of the microflora of the non-dairy beverages. Thus, it seems necessary to develop some other method that would be more effective in ensuring the appropriate microbiological purity of plant-based beverages. Lee et al. [59] evaluated the effect of increased hydrostatic pressure coupled with high temperature on counts of viable, spore-forming cells of pathogenic microorganisms. They reported a significant decrease in the number of active resting spores to a negligible level (below 1 cfu/mL) upon the coupled use of pressure of 207 MPa and temperature of 90 °C.

3.3. Evaluation of the Effect of Fermentation Process on Chemical and Microbiological Properties of Non-Dairy Beverages

Immediately after the addition of the inoculum from *Lactobacillus casei* subsp. *rhamnosus* lyophilizate, the total count of viable lactic acid bacteria (LAB) cells reached 11.72 log (cfu/mL) in coconut beverage and 8.41 log (cfu/mL) in hemp beverage. The fermentation process (37 °C/6 h) caused an increase in bacteria count in the samples to 13.26 and 10.92 log (cfu/mL), respectively (Figure 5). Initially, the difference in the count of viable LAB cells could be due to the viability of probiotics themselves in the food matrix, which may be affected by pH (initial pH values were at 6.12 for coconut milk and 6.79 for hemp milk), oxygen level, and presence of competing microorganisms (bacterial cells and their resting spores undamaged during pasteurization: 4.39 log (cfu/mL) in coconut milk and 3.81 log (cfu/mL) in hemp milk) [60–62]. Therefore, LABs resistance to inconvenient conditions appears to be an important technological trait, which enables selecting strains for untypical food matrix like, e.g., non-dairy beverages [63]. Initial differences in the count of probiotic bacterial cells in both analyzed types of plant milks could be due to the fact that viable, metabolically active cells may rapidly lose their capability for growth, and this dormancy state of a part of the population may occur especially when the cells are exposed to unbeneficial factors. For explicit confirmation of these assumptions, fluorescent techniques should be employed that allow monitoring subtle changes in the dynamics of proliferation and decay of microorganisms that may be in the viable but non-culturable (VBNC) or active but non-culturable (ABNC) state, i.e., in the state of bacteria transition to the dormancy state under unfavorable colonization conditions [63–66].

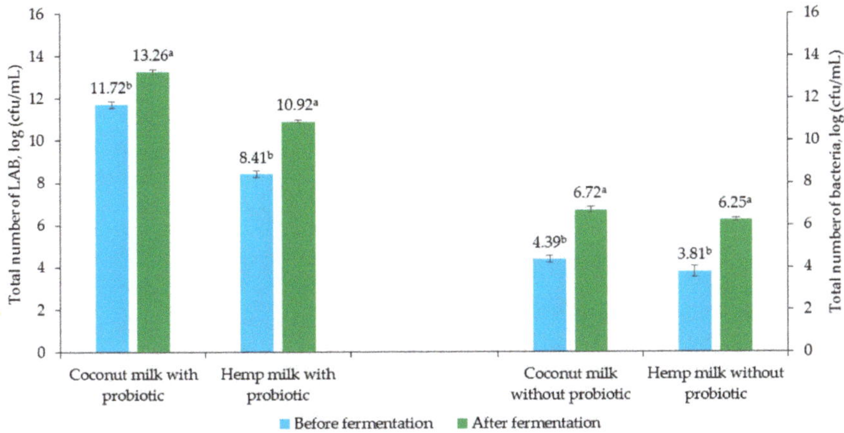

Figure 5. Total count of viable bacterial cells in the analyzed non-dairy beverages before and after fermentation (a, b—statistically significant differences, $p \leq 0.05$).

A similar dependency was demonstrated in the study conducted by Zaręba [67], who noted that 4-h fermentation process of soybean milk contributed to the proliferation of *Lactobacillus* species bacteria by approximately 0.5 log (cfu/mL). A study conducted by Bartkiene et al. [21] also showed that *L. casei* cell count in fermented hemp milk reached 8.78 log (cfu/mL) and was higher compared to the count determined before the fermentation process (8 log (cfu/mL)).

The analyzed models of non-dairy beverages differed in terms of the total bacteria count. TBC was also determined in the control samples (without the probiotic). Before fermentation it reached 4.39 log (cfu/mL) in coconut milk and 3.81 log (cfu/mL) in hemp milk; whereas after fermentation for the respective values were at 6.72 and 6.25 log (cfu/mL; Figure 5). Presumably, these were resting spores that had survived pasteurization and whose proliferation was promoted by fermentation temperature (37 °C) being optimal for their growth. These speculations may be confirmed by results reported earlier by Czaczyk et al. [68], who noticed the greatest growth of *Bacillus* ssp. bacilli under these conditions. Similar observations were made by Huy et al. [69]. The activity of microorganisms during incubation is also affected by the type and amount of nutrients available in the medium. However, considering the "Microbiological Limits for Assessment of Microbiological Quality of Ready-to-eat Foods" [70], the criterion related to the microbiological quality did not exceed the maximum value of 7 log (cfu/mL), set by the International Commission for Microbiological Specification of Food.

Monosaccharides, including mainly glucose, present in plant beverages represent a good source of carbon to bacteria. The physicochemical analysis conducted in the study allowed concluding that coconut milk (34.53 g glucose/L) was a better source of these compounds compared to hemp beverage (30.21 g glucose/L). The content of reducing sugars decreased significantly after the fermentation process (Figure 6).

The non-dairy beverages with probiotic addition were characterized by a greater reduction in glucose content after fermentation, i.e., to 23.05 g glucose/L in coconut milk and to 19.79 g glucose/L in hemp milk. In turn, milks without the probiotic bacteria were characterized by noticeably higher glucose content, which decreased due to the activity of undesirable microorganisms.

Figure 7 presents results of measurements of active acidity of the non-dairy beverages before and after fermentation.

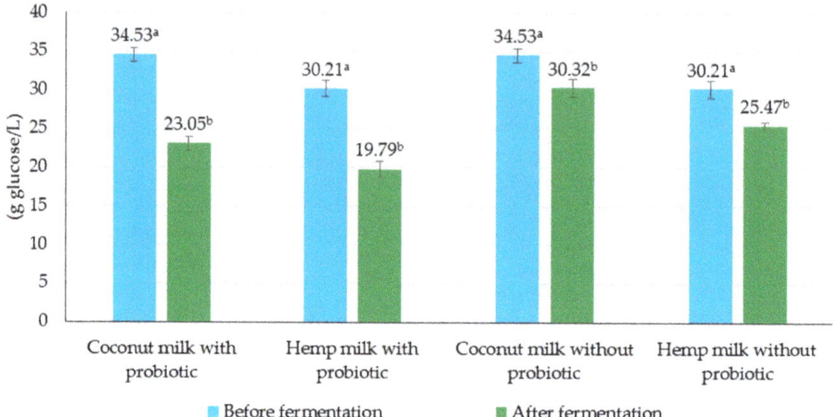

Figure 6. Reducing sugars concentration in the analyzed non-dairy beverages before and after fermentation (a, b—statistically significant differences, $p \leq 0.05$).

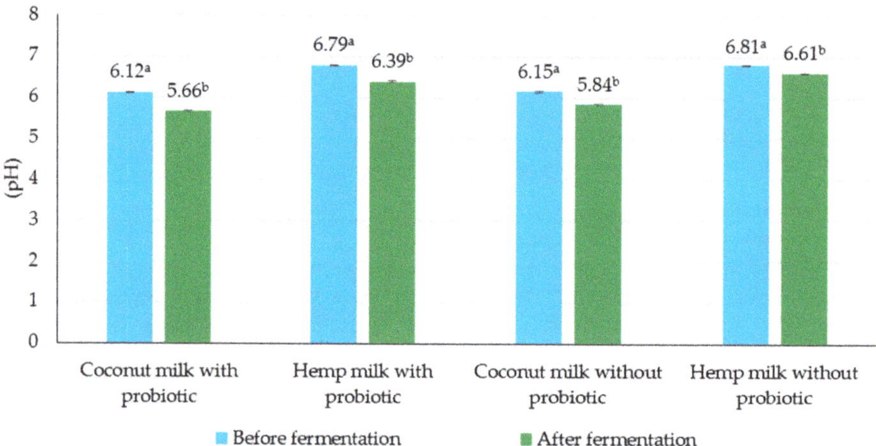

Figure 7. Active acidity of the analyzed non-dairy beverages before and after fermentation (a, b—statistically significant differences, $p \leq 0.05$).

The greatest differences in the active acidity, before and after fermentation, were demonstrated in the fermented probiotic beverage made of coconut, i.e., pH 6.12 and pH 5.66, respectively. In turn, the smallest decrease in the pH value (by 0.21) was noted in the fermented hemp seed milk without the probiotic. It may, therefore, be concluded that the acidity level is determined by the accumulation of organic acids caused by monosaccharides metabolism. So negligible pH changes may be due to the fact that *Lactobacillus casei* subsp. *rhamnosus* strain used in the study is a representative of facultatively heterofermentative bacteria. Apart from lactic acid, these bacteria are capable of producing CO_2, acetic acid (aerobic conditions), acetic aldehyde, and/or ethanol (anaerobic conditions) [71–73]. During the heterofermentation process, glucose degradation proceeds accordingly to the pentose phosphate pathway, and the capability of lactic acid bacteria for this fermentation results from a lack of certain enzymes like, e.g., triphosphate isomerase and aldolase [74].

3.4. Assessment of the Quality of the Produced Non-Dairy Beverages During Cold Storage

Changes of the active acidity of the non-dairy beverages during 21-day cold storage were presented in Table 3. As expected, storage time had a significant effect on the acidity level of milk substitutes, causing vast differences in pH values of the fermented samples of coconut and hemp milks between day 1 and 21 of storage.

Table 3. Changes in the active acidity of fermented and non-fermented non-dairy beverages during 21-day storage.

Treatment	Storage Time			
	Day 1	**Day 7**	**Day 14**	**Day 21**
	Fermented			
CP	5.61 ± 0.02 [aF]	5.11 ± 0.02 [bG]	4.93 ± 0.02 [cB]	4.81 ± 0.02 [dB]
HP	6.47 ± 0.01 [aC]	6.08 ± 0.02 [bA]	5.91 ± 0.02 [cA]	5.78 ± 0.02 [dA]
C	5.84 ± 0.01 [aE]	4.85 ± 0.01 [bH]	4.08 ± 0.01 [cG]	3.58 ± 0.02 [dG]
H	6.62 ± 0.02 [aB]	5.49 ± 0.02 [bD]	4.39 ± 0.02 [cF]	3.41 ± 0.01 [dH]
	Non-Fermented			
CP	6.15 ± 0.02 [aD]	5.31 ± 0.02 [bE]	4.88 ± 0.02 [cC]	4.21 ± 0.01 [dC]
HP	6.81 ± 0.02 [aA]	5.87 ± 0.02 [bB]	4.65 ± 0.01 [cD]	3.95 ± 0.02 [dE]
C	6.15 ± 0.01 [aD]	5.25 ± 0.02 [bF]	4.53 ± 0.02 [cE]	4.01 ± 0.02 [dD]
H	6.81 ± 0.02 [aA]	5.71 ± 0.01 [bC]	4.35 ± 0.02 [cF]	3.70 ± 0.02 [dF]

CP—Coconut beverage with probiotic; HP—Hemp beverage with a probiotic; C—Coconut beverage without probiotic; H—Hemp beverage without probiotic. Means in the rows, followed by different small letters (a–d) are significantly different at $p < 0.05$ (effect of storage). Means in the columns, followed by different big letters (A–F) are significantly different at $p < 0.05$ (effect of treatment).

Usually, the active acidity (pH) of fermented plant-based beverages should not be lower than 4.0 throughout the storage period [75]. Results presented in Table 3 indicate that the pH value remained above 4.0 within fourteen days of cold storage of the analyzed milk substitutes. However, a pH decline was recorded at the end of the storage period in the case of the samples not inoculated with the probiotic monoculture. This is in agreement with results reported by Paseephol and Sherkat [76] and by Colakoglu and Gursoy [77]. Guo et al. [78] reported that the pH value of fermented buffalo milk containing *Lactobacillus casei* decreased from 5.02 (day 1) to 4.00 (day 30 of storage) [78]. In turn, Akalin et al. [79] demonstrated that the probiotic bacteria decreased the pH value of various yoghurts from 4.51 to 4.40 after 28 days of their cold storage [79].

Data obtained in the study indicate that pH values of all beverage samples decreased during cold storage. This dependency may be explained by the persistent metabolic activity of the probiotic monoculture, which was also noticed by Bonczar et al. [80] during cold storage of fermented beverages. When comparing pH values over the storage period, these researchers observed decrease in all samples. Similar conclusions were drawn by Bartkiene et al. [21], who analyzed hemp milk in a 15-day storage model. The pH value of the fermented hemp milk with the addition of a probiotic culture of *L. casei* decreased slightly from 5.15 in the first day to 4.77 in the last day of cold storage.

The study demonstrated also a decrease in protein content of the produced non-dairy beverages along with storage time (Table 4).

The analysis of the nutritional value of non-fermented milk substitutes demonstrated that, after 21 days of storage, the total protein content was higher in the non-dairy beverages supplemented with probiotic bacteria.

A similar tendency was noted during storage of fermented non-dairy beverages (i.e., decreased protein concentration). However, the conducted analyses showed a higher total protein content in the fermented than in the non-fermented beverages, and lower in the non-dairy beverages fortified with the probiotic monoculture.

Table 4. Protein content of fermented and non-fermented non-dairy beverages during 21-day storage.

Treatment	Storage Time			
	Day 1	**Day 7**	**Day 14**	**Day 21**
Fermented				
CP	3.02 ± 0.21 [aB]	2.95 ± 0.23 [aB]	2.84 ± 0.15 [aB]	2.67 ± 0.26 [aB]
HP	6.68 ± 0.26 [aA]	6.41 ± 0.28 [aA]	6.33 ± 0.21 [aA]	6.19 ± 0.24 [aA]
C	3.02 ± 0.19 [aB]	2.96 ± 0.17 [aB]	2.87 ± 0.18 [aB]	2.71 ± 0.18 [aB]
H	6.68 ± 0.29 [aA]	6.45 ± 0.25 [aA]	6.34 ± 0.23 [aA]	6.22 ± 0.23 [aA]
Non-Fermented				
CP	3.23 ± 0.22 [aB]	3.17 ± 0.19 [aB]	3.10 ± 0.21 [aB]	3.01 ± 0.24 [aB]
HP	6.96 ± 0.27 [aA]	6.91 ± 0.25 [aA]	6.82 ± 0.23 [aA]	6.68 ± 0.26 [aA]
C	3.23 ± 0.18 [aB]	3.15 ± 0.20 [aB]	3.09 ± 0.20 [aB]	2.98 ± 0.18 [aB]
H	6.96 ± 0.27 [aA]	6.89 ± 0.22 [aA]	6.78 ± 0.21 [aA]	6.63 ± 0.25 [aA]

CP—Coconut beverage with probiotic; HP—Hemp beverage with a probiotic; C—Coconut beverage without probiotic; H—Hemp beverage without probiotic. Means in the rows, followed by different small letters (a) are significantly different at $p < 0.05$ (effect of storage). Means in the columns, followed by different big letters (A,B) are significantly different at $p < 0.05$ (effect of treatment).

According to Bernat et al. [81], the significant decrease in protein content is due to the fact that during the fermentation and storage of beverages, the bacterial starters could hydrolyze proteins to synthesize amino acids necessary for their nutrition. Investigations conducted by the aforementioned authors demonstrated also that fermented plant-based substitutes of milk had by approximately 17% lower content of β-glucan than their non-fermented counterparts, while this compound is capable of proteins crosslinking in food [82].

Changes observed in lipids content of the analyzed non-dairy beverages resembled these of protein (Table 5). In the entire period of cold storage, lipids concentration decreased negligibly in all milks. It may, thus, be concluded that the study demonstrated stability of this component in the produced and modified plant-based beverages.

Table 5. Lipids content in fermented and non-fermented non-dairy beverages during 21-day storage.

Treatment	Storage Time			
	Day 1	**Day 7**	**Day 14**	**Day 21**
Fermented				
CP	21.06 ± 0.06 [aA]	20.98 ± 0.06 [abA]	20.87 ± 0.06 [bcA]	20.75 ± 0.06 [cA]
HP	18.01 ± 0.06 [aB]	17.95 ± 0.06 [abB]	17.81 ± 0.06 [bcB]	17.66 ± 0.06 [cB]
C	21.06 ± 0.07 [aA]	20.96 ± 0.07 [aA]	20.89 ± 0.07 [abA]	20.73 ± 0.06 [bA]
H	18.01 ± 0.06 [aB]	17.89 ± 0.05 [abB]	17.78 ± 0.06 [bB]	17.62 ± 0.06 [cB]
Non-Fermented				
CP	21.08 ± 0.02 [aA]	21.01 ± 0.06 [aA]	20.94 ± 0.06 [aA]	20.84 ± 0.06 [bA]
HP	18.02 ± 0.02 [aB]	17.97 ± 0.06 [aB]	17.92 ± 0.06 [aB]	17.69 ± 0.06 [bB]
C	21.08 ± 0.02 [aA]	21.02 ± 0.06 [aA]	20.93 ± 0.06 [abA]	20.79 ± 0.06 [bA]
H	18.02 ± 0.02 [aB]	17.95 ± 0.06 [aB]	17.89 ± 0.06 [aB]	17.66 ± 0.06 [bB]

CP—Coconut beverage with probiotic; HP—Hemp beverage with a probiotic; C—Coconut beverage without probiotic; H—Hemp beverage without probiotic. Means in the rows, followed by different small letters (a–c) are significantly different at $p < 0.05$ (effect of storage). Means in the columns, followed by different big letters (A,B) are significantly different at $p < 0.05$ (effect of treatment).

As reported by Bernat et al. [82], who analyzed the microstructure of oat milk during storage, the similar concentration of lipids may be due to their embedding in a polysaccharide network. Stability of such system is additionally associated with the cross-linking properties of β-glucans [75]. Furthermore, almost all lipid droplets are retained in the polysaccharide-protein matrix, which is responsible for the physical stability of plant milk. It has been demonstrated that certain proteins may be attached to lipid globules, thereby ensuring protection of emulsions against destabilization processes [82].

Changes in concentrations of individual nutrients in plant-based beverages during cold storage may also be due to the fermentation process, which contributes to decreased content of carbohydrates and also of some non-digestible poly- and oligo-saccharides, to the improvement of protein quality, to the facilitated synthesis of selected amino acids, and to the improved availability of vitamins. In addition, it ensures optimal pH conditions for the enzymatic degradation of many compounds being important growth factors for the potentially probiotic bacteria [83].

The *Lactobacillus* strains used in the study are complex microorganisms that need carbohydrates, amino acids, B-group vitamins, nucleic acids, and minerals for their proper growth [84]. For this reason, it can be concluded that the fermentation of plant-based beverages may offer an inexpensive method for the synthesis of substrates in the product that would promote the growth of beneficial microorganisms [85].

3.5. Quantitative Analysis of Viable Bacterial Cells During Storage of Fermented and Non-Fermented Non-Dairy Beverages

An important aspect determining the quality of health-promoting fermented beverages is the analysis of changes in the number of viable lactic acid bacteria in 1 mL of a product over the entire period of its shelf life [86]. Al-Otaibi [87] emphasized that 6 log (cfu/mL) of viable probiotic cells should be consumed every day to ensure health benefits to consumers. Both types of the fermented non-dairy beverages met this requirement regarding viability of *L. casei* till the end of the storage period. On the first day of storage (after fermentation), the number of active bacterial LAB cells ranged from 10.92 log (cfu/mL) in hemp milk to 13.26 log (cfu/mL) in coconut milk (Table 6).

Table 6. Changes in the number of *Lactobacillus casei* subsp. *rhamnosus* during 21-day storage (4 °C) of fermented and non-fermented non-dairy beverages.

Treatment	Storage Time			
	Day 1	**Day 7**	**Day 14**	**Day 21**
	Fermented			
CP	13.26 ± 0.14 [aA]	11.46 ± 0.18 [bA]	11.26 ± 0.17 [bA]	9.41 ± 0.24 [cA]
HP	10.92 ± 0.10 [aC]	10.31 ± 0.25 [aB]	8.28 ± 0.33 [bB]	7.35 ± 0.26 [cB]
	Non-Fermented			
CP	11.72 ± 0.04 [aB]	6.81 ± 0.13 [bC]	5.42 ± 0.15 [cC]	3.12 ± 0.13 [dC]
HP	8.41 ± 0.18 [aD]	6.35 ± 0.16 [bC]	4.53 ± 0.08 [cD]	3.54 ± 0.20 [dC]

CP—Coconut beverage with probiotic; HP—Hemp beverage with a probiotic. Means in the rows, followed by different small letters (a–d) are significantly different at $p < 0.05$ (effect of storage). Means in the columns, followed by different big letters (A–D) are significantly different at $p < 0.05$ (effect of treatment).

Opposite observations were made in the case of non-fermented models of plant-based beverages. The number of LAB cells isolated from these samples on the first day of storage reached 11.72 log (cfu/mL) in coconut milk and 8.41 log (cfu/mL) in hemp milk. These values decreased to 3.12 log (cfu/mL) and 3.54 log (cfu/mL), respectively, after 21 days of cold storage (Table 6).

Bakirci and Kavaz [88] reported that the total counts of *Lactobacillus acidophilus*, *Bifidobacterium* ssp., and *Streptococcus thermophilus* decreased during cold storage of banana yoghurts, but remained at the required level (above 6 log (cfu/mL)) until day 14. In addition, a few other authors demonstrated that lactic acid bacteria (*L. debrueckii* spp. *bulgaricus* and *S. thermophilus*) survived well in yoghurt throughout its shelf life [79,89]. Olson and Aryana [90] showed that the number of *Lactobacillus acidophilus* strain cells decreased in natural yoghurt from 6.84 to 4.43 log (cfu/mL) over an 8-week storage period. Results of own study pointed to a higher survivability of potentially probiotic bacteria, compared to that reported by Mousavi et al. [91] for a probiotic juice from pomegranate. These authors observed reductions in counts of *Lactobacillus delbrueckii* and *Lactobacillus plantarum* by three logarithmic cycles after 14 days of cold storage. High survival rates of *Lactobacillus casei* under cold storage conditions were also demonstrated by Pereira et al. [92], who investigated fermentation and survivability of this probiotic in a juice from cashew apple.

The reduced count of *L. casei* subsp. *rhamnosus* during cold storage may be due to the production by these microorganisms of agents exhibiting anti-microbial activity like e.g., organic acid, bacteriocins, and hydrogen peroxide [93]. The concentration of hydrogen peroxide has an important impact, because *L. casei* subsp. *rhamnosus* do not produce the catalase enzyme. Accumulation of metabolism products during storage of non-dairy beverages may lead to the transition of *L. casei* subsp. *rhamnosus* at VBNC (viable but non-culturable). At this stage, the cells have an "unsatisfactory" physiological state, which means that they are alive but do not divide, and as a result do not have the ability to grow and reproduction [63]. Al-Otaibi [87] noticed that the number of bifidobacteria in eight commercial fermented dairy products decreased significantly since the day of manufacture till the end of cold storage (5 °C). This author reported also that the count of viable bacteria maintained at 10^6 cfu/mL till the end of the storage period in only two of the analyzed products. Reduction in the number of viable health-promoting bacteria can also be caused by decreased acidity, presence of post-production acid [94] sensitivity to oxygen [95] and metabolites, i.e., hydrogen peroxide and ethanol, and also to bacteriocins produced by lactic acid bacteria [96].

The determination of the number of active *L. casei* cells in the analyzed non-dairy beverages is difficult due to the presence of other microorganisms [97]. For this reason, simultaneous analyses were carried out for control samples (without probiotic). In their case, the total bacterial count was observed to decrease. In contrast, interesting seem to be changes in the viability of microorganisms (other than LABs) in the non-fermented control beverages, in which acidity approximating the neutral pH contributed to the development of undesirable microorganisms. However, the total bacterial count in the beverages not fortified with the probiotic monoculture, fitted within the range from 3.82 log (cfu/mL) to 3.93 log (cfu/mL) in fermented non-dairy beverages and from 5.92 to 6.67 log (cfu/mL) the non-fermented ones (Table 7). Thus, the criterion related to the microbiological quality did not exceed the maximum value of 7 log (cfu/mL).

Table 7. Changes in the total number of bacteria during 21-day storage (4 °C) of fermented and non-fermented non-dairy beverages.

Treatment	Storage Time			
	Day 1	**Day 7**	**Day 14**	**Day 21**
	Fermented			
C	6.72 ± 0.17 [aA]	6.23 ± 0.30 [aA]	5.54 ± 0.24 [bA]	3.82 ± 0.18 [cB]
H	6.25 ± 0.13 [aB]	5.69 ± 0.29 [aA]	4.37 ± 0.20 [bB]	3.93 ± 0.27 [bB]
	Non-Fermented			
C	4.39 ± 0.17 [bC]	4.41 ± 0.15 [bB]	4.79 ± 0.19 [bB]	5.67 ± 0.18 [aA]
H	3.81 ± 0.14 [bD]	3.86 ± 0.08 [bB]	5.76 ± 0.18 [aA]	5.92 ± 0.16 [aA]

C—Coconut beverage without probiotic; H—Hemp beverage without probiotic. Means in the rows, followed by different small letters (a–c) are significantly different at $p < 0.05$ (effect of storage). Means in the columns, followed by different big letters (A–D) are significantly different at $p < 0.05$ (effect of treatment).

Results of the present study indicate that cell viability was maintained at a satisfactory level throughout the storage period of fermented non-dairy beverages fortified with the probiotic monoculture. According to Pereira et al. [98], sugars, proteins, and lipids are only some of the factors that may affect the growth of probiotic bacteria and their survival rates in food products. Hence, the relative stability observed in content of nutrients in the plant-based beverages contributed indirectly to ensuring the therapeutic minimum of the analyzed products throughout their storage period.

4. Conclusions

The conducted study proved that the produced and modified plant-based beverages could serve as a food matrix for probiotic bacteria. The growth and survivability of probiotic bacteria in food products was determined by many factors including e.g., storage conditions, medium acidity, and sensitivity of oxygen and metabolites. The fermentation process contributed to the increased survival rates of *Lactobacillus casei* subsp. *rhamnosus* in both coconut and hemp milk. During 21-day storage of inoculated milk substitutes, the highest survivability of *Lactobacillus casei* subsp. *rhamnosus* was demonstrated in the fermented coconut milk (9.41 log (cfu/mL)). On day 21 of cold storage, the number of viable *Lactobacillus casei* subsp. *rhamnosus* cells in fermented coconut and hemp milk ensured meeting the therapeutic minimum (>6 log (cfu/mL)). Due to their nutrients composition and number of bacterial cells exhibiting a positive effect on a human body, the analyzed non-dairy beverages, offering an alternative to milk, represent a category of novel food products, and their manufacture will contribute to the sustainable development of food production and to the assurance of food safety.

Author Contributions: A.S. and S.K. conceived and designed the research. A.S., S.K., B.P. and E.C. performed the experiments. A.S., S.T., S.K., E.C., M.K., and B.P. prepared the materials. A.S., S.T., S.K., M.K. and P.F. analyzed the data. A.S., E.C. and S.K. wrote the paper. A.S., E.C., S.T., S.K., M.K. and P.F. revised the manuscript. All authors read and approved the final manuscript.

Funding: This research received no external funding.

Conflicts of Interest: The authors declare no conflict of interest.

References

1. Dam, L.R.; Boafo, Y.A.; Degefa, S.; Gasparatos, A.; Saito, O. Assessing the food security outcomes of industrial crop expansion in smallholder settings: Insights from cotton production in Northern Ghana and sugarcane production in Central Ethiopia. *Sustain. Sci.* **2017**, 677–693. [CrossRef]

2. Szparaga, A.; Kocira, S. Generalized logistic functions in modelling emergence of *Brassica napus* L. *PLoS ONE* **2018**, *13*, e0201980. [CrossRef] [PubMed]

3. Alexandratos, N.; Bruinsma, J. World Agriculture: Towards 2015/2030: The 2012 Revision. Available online: http://www.fao.org/3/a-ap106e.pdf (accessed on 24 July 2019).

4. FAO. Food security information for action programme. In *Food Security Concepts and Frameworks*; FAO: Rome, Italy, 2008.

5. Blay-Palmer, A.; Sonnino, R.; Custot, J. A food politics of the possible? Growing sustainable food systems through networks of knowledge. *Agric. Hum. Values* **2016**, *33*, 27–43. [CrossRef]

6. Szparaga, A.; Stachnik, M.; Czerwińska, E.; Kocira, S.; Dymkowska-Malesa, M.; Jakubowski, M. Multi-objective optimization of osmotic dehydration of plums in sucrose solution and its storage using utopian solution methodology. *J. Food Eng.* **2019**, *245*, 104–111. [CrossRef]

7. Saluk-Juszczak, J.; Kołodziejczyk, J.; Babicz, K. Functional food—A role of nutraceuticals in cardiovascular disease prevention. *Kosmos* **2010**, *59*, 527–538.

8. Martirosyan, D.; Singh, J. A new definition of functional food by FFC: What makes a new definition unique? *Funct. Food Health Dis. J.* **2015**, *5*, 209–223. [CrossRef]

9. Dymkowska-Malesa, M.; Szparaga, A.; Czerwińska, E. Evaluation of polychlorinated biphenyls content in chosen vegetablesfrom Warmia and Mazury region. *Rocz. Ochr. Srodowiska* **2014**, *16*, 290–299.

10. Mullin, G.; Delzenne, N.M. Functional foods and dietary supplements in 2017: Food for thought. *Curr. Opin. Clin. Nutr. Metab. Care* **2017**, *20*, 453–455. [CrossRef]

11. Douglas, L.C.; Sanders, M.E. Probiotics and prebiotics in dietetics practice. *J. Am. Diet. Assoc.* **2008**, *108*, 510–521. [CrossRef]

12. Kapka-Skrzypczak, L.; Niedźwiecka, J.; Wojtyła, A.; Kruszewski, M. Probiotics and prebiotics as a bioactive component of functional food. *Pediatr. Endocrinol. Diabetes Metab.* **2012**, *18*, 79–83.

13. Sethi, S.; Tyagi, S.K.; Anurag, R.K. Plant-based milk alternatives an emerging segment of functional beverages: A review. *J. Food Sci. Technol.* **2016**, *53*, 3408–3423. [CrossRef] [PubMed]

14. Kandylis, P.; Pissaridi, K.; Bekatorou, A.; Kanellaki, M.; Koutinas, A. Dairy and non-dairy probiotic beverages. *Curr. Opin. Food Sci.* **2016**, *7*, 58–63. [CrossRef]

15. Hoffman, M.; Kostyra, E. Sensory quality and nutritional value of vegan substitutes of milk. *PTPS* **2015**, *1*, 52–57.

16. Zielińska, D. *Lactobacillus* Strain Survival Study in Fermeted Soy Beverage. *Zywnosc-Nauka Technol. Jakosc* **2006**, *49*, 121–128.

17. Quasem, J.M.; Mazahreh, A.S.; Abu-Alruz, K. Development of Vegetable Based Milk from Decorticated Sesame (Sesamum Indicum). *Am. J. Appl. Sci.* **2009**, *6*, 888–896. [CrossRef]

18. Yeung, P.S.M.; Sanders, M.E.; Kitts, C.L.; Cano, R.; Tong, P.S. Species specific identification on commercial probiotic strains. *J. Dairy Sci.* **2002**, *8*, 1039–1051. [CrossRef]

19. Hadadji, M.; Bensoltane, A. Growth and lactic acid production by Bifidobacterium longum and Lactobacillus acidophilus in goat's milk. *Afr. J. Biotechnol.* **2006**, *5*, 505–509.

20. Yuliana, N.; Rangga, A. Rakhmiati Manufacture of fermented coco milk-drink containing lactic acid bacteria cultures. *Afr. J. Food Sci.* **2010**, *4*, 558–562.

21. Bartkiene, E.; Zokaityte, E.; Lele, V.; Sakiene, V.; Zavistanaviciute, P.; Klupsaite, D.; Bendoraitiene, J.; Navikaite-Snipaitiene, V.; Ruzauskas, M. Technology and characterisation of whole hemp seed beverages prepared from ultrasonicated and fermented whole seed paste. *Int. J. Food Sci. Technol.* **2019**, 1–14. [CrossRef]

22. Wang, Q.; Jiang, J.; Xiong, Y.L. High pressure homogenization combined with pH shift treatment: A process to produce physically and oxidatively stable hemp milk. *Food Res. Int.* **2018**, *106*, 487–494. [CrossRef] [PubMed]

23. Scholz-Ahrens, K.E.; Ahrens, F.; Barth, C.A. Nutritional and health attributes of milk and milk imitations. *Eur. J. Nutr.* **2019**, 1–16. [CrossRef] [PubMed]

24. Kaniewski, R.; Pniewska, I.; Kubacki, A.; Strzelczyk, M.; Chudy, M.; Oleszak, G. Konopie siewne (*Cannabis sativa* L.)-cenna roślina przydatna i lecznicza. *Postępy Fitoter.* **2017**, *2*, 139–144. [CrossRef]

25. Mojka, K. Probiotics, prebiotics and synbiotics–characteristics and functions. *Probl. Hig. Epidemiol.* **2014**, *95*, 541–549.

26. Zaręba, D.; Ziarno, M.; Obiedziński, M. Viability of yoghurt bacteria and probiotic strains in models of fermented and non-fermented milk. *Med. Weter.* **2008**, *64*, 1007–1011.

27. Jeske, S.; Zannini, E.; Arendt, E.K. Past, present and future: The strength of plant-based dairy substitutes based on gluten-free raw materials. *Food Res. Int.* **2018**, *110*, 42–51. [CrossRef] [PubMed]

28. Akin, Z.; Ozcan, T. Functional properties of fermented milk produced with plant proteins. *LWT* **2017**, *86*, 25–30. [CrossRef]

29. Mårtensson, O.; Öste, R.; Holst, O. Lactic Acid Bacteria in an Oat-based Non-dairy Milk Substitute: Fermentation Characteristics and Exopolysaccharide Formation. *LWT-Food Sci. Technol.* **2000**, *33*, 525–530. [CrossRef]

30. Cruz, N.; Capellas, M.; Jaramillo, D.; Trujillo, A.-J.; Guamis, B.; Ferragut, V. Soymilk treated by ultra high-pressure homogenization: Acid coagulation properties and characteristics of a soy-yogurt product. *Food Hydrocoll.* **2009**, *23*, 490–496. [CrossRef]

31. Chaiwanon, P.; Puwastien, P.; Nitithamyong, A.; Sirichakwal, P.P. Calcium Fortification in Soybean Milk and In Vitro Bioavailability. *J. Food Compos. Anal.* **2000**, *13*, 319–327. [CrossRef]

32. Patel, H.; Pandiella, S.; Wang, R.; Webb, C. Influence of malt, wheat, and barley extracts on the bile tolerance of selected strains of lactobacilli. *Food Microbiol.* **2004**, *21*, 83–89. [CrossRef]

33. Vera-Pingitore, E.; Jimenez, M.E.; DallAgnol, A.; Belfiore, C.; Fontana, C.; Fontana, P.; Von Wright, A.; Vignolo, G.; Plumed-Ferrer, C. Screening and characterization of potential probiotic and starter bacteria for plant fermentations. *LWT* **2016**, *71*, 288–294. [CrossRef]

34. Navarro, R.R.; Faronilo, K.M.L.; Eom, S.H.; Jeon, K.H. Isolation, characterization, and identification of probiotic lactic acid bacterium from sabeng, a Philippine fermented drink. *J. ISSAAS* **2018**, *24*, 127–136.

35. Blasco, M. *Milk Power. Mleko roślinne 80 przepisów*; Wydawnictwo Burda: Warszawa, Poland, 2014.

36. Szakuła, M. *Konopie w kuchni*; Smiling Spoon: Warszawa, Poland, 2015.

37. Zielińska, D.; Kołożyn-Krajewska, D.; Goryl, A. Survival models of potentially probiotic *Lactobacillus casei* KN291 bacteria in a fermented soy beverage. *Zywnosc-Nauka Technol. Jakosc* **2008**, *5*, 126–134.

38. Zielińska, D.; Uzarowicz, U. Development of ripening and storage conditions of probiotic soy beverage. *Zywnosc-Nauka Technol. Jakosc* **2007**, *5*, 186–193.

39. Zielińska, D. Selecting Suitable Bacterial Strains of Lactobacillus and Identifying Soya Drink Fermentation Conditions. *Zywnosc-Nauka Technol. Jakosc* **2005**, *2*, 289–297.

40. AOAC Official Method. *18 Fat Content of Raw and Pasteurized Whole Milk Gerber Method by Weight (Method I) First Action*; AOAC International: Rockville, MD, USA, 2000.

41. AOAC. *Official Methods of Analysis of AOAC International*, 17th ed.; Horwitz, W., Ed.; AOAC International: Gaithersburg, MD, USA, 2000.

42. AOAC. *Official Method of Analysis of the Association of Official Analytical Chemists*; AOAC International: Arlington, VA, USA, 1997.

43. Miller, G.L. Use of Dinitrosalicylic Acid Reagent for Determination of Reducing Sugar. *Anal. Chem.* **1959**, *31*, 426–428. [CrossRef]

44. Czerwińska, E.; Piotrowski, W. Potential Sources of Milk Contamination Influencing its Quality for Consumption. *Rocz. Ochr. Srodowiska* **2011**, *13*, 635–652.

45. Gustaw, W.; Kozioł, J.; Waśko, A.; Skrzypczak, K.; Michalak-Majewska, M.; Nastaj, M. Physicochemical Properties and Survival of Lactobacillus casei in Fermented Milk Beverages Produced With Addition of Selected Milk Protein Preparations. *Zywnosc-Nauka Technol. Jakosc* **2015**, *6*, 129–139. [CrossRef]

46. De Man, J.C.; Rogosa, M.; Sharpe, M.E. A MEDIUM FOR THE CULTIVATION OF LACTOBACILLI. *J. Appl. Bacteriol.* **1960**, *23*, 130–135. [CrossRef]

47. Champagnem, C.P.; Rastall, R.A. Some technological challenges in the addition of probiotic bacteria to foods. In *Prebiotics and Probiotics Science and Technology*, 2nd ed.; Charalampopoulos, D., Rastall, R.A., Eds.; Springer: London, UK, 2009; pp. 763–806.

48. Yamaguishi, C.T.; Spier, M.R.; Lindner, J.D.D.; Soccol, V.T.; Soccol, C.R. Current Market Trends and Future Directions. In *Probiotics*; Springer: Berlin, Germany, 2011; Volume 21, pp. 299–319. [CrossRef]

49. Szydłowska, A.; Kołożyn-Krajewska, D. Applying potencially probiotic bacterial strains to pumpkin pulp fermentation. *Zywnosc-Nauka Technol. Jakosc* **2010**, *73*, 109–119. [CrossRef]

50. Sobczyk, M.; Ziarno, M. Characteristics of Microbial Quality and Nutrient Content (Protein, Fat, Mineral Compounds) Some Petals Muesli. *ZPPNR* **2013**, *575*, 119–129.

51. Jessa, J.; Hozyasz, K.K. Health value of coconut products. *Pediatr. Pol.* **2015**, *90*, 415–423. [CrossRef]

52. Aizpurua-Olaizola, O.; Omar, J.; Navarro, P.; Olivares, M.; Etxebarria, N.; Usobiaga, A. Identification and quantification of cannabinoids in *Cannabis sativa* L. plants by high performance liquid chromatography-mass spectrometry. *Anal. Bioanal. Chem.* **2014**, *406*, 7549–7560. [CrossRef] [PubMed]

53. Belewu, M.A.; Belewu, K.Y. Comparative Physico-Chemical Evulation of Tiger-nut, Soybean and Coconut Milk Sources. *Int. J. Agric. Biol.* **2007**, *9*, 785–787.

54. Yeo, S.-K.; Liong, M.-T.; Yeo, S.; Liong, M. Effect of prebiotics on viability and growth characteristics of probiotics in soymilk. *J. Sci. Food Agric.* **2010**, *90*, 267–275. [CrossRef] [PubMed]

55. Jurkowski, M.; Błaszczyk, M. Physiology and Biochemistry of Lactic Acid Bacteria. *Kosmos* **2012**, *61*, 493–504.

56. Nogala-Kałucka, M. *Analiza żywności. Wybrane metody oznaczeń jakościowych i ilościowych składników żywności. Wyd*; Uniwersytetu Przyrodniczego w Poznaniu: Poznań, Poland, 2016.

57. Dłużewska, E.; Nizler, M.; Maszewska, M. Technological Aspects of Obtaining Soy Beverages. *Acta Sci. Technol. Aliment* **2004**, *3*, 93–102.

58. Rozporządzenie Komisji (WE) 2073/2005 i 2019/229. Available online: https://eur-lex.europa.eu/legal-content/PL/TXT/PDF/?uri=CELEX:32019R0229&from=ES (accessed on 26 August 2019).

59. Lee, S.-Y.; Dougherty, R.H.; Kang, D.-H. Inhibitory Effects of High Pressure and Heat on Alicyclobacillus acidoterrestris Spores in Apple Juice. *Appl. Environ. Microbiol.* **2002**, *68*, 4158–4161. [CrossRef]

60. Ouwehand, A.C.; Salminen, S.J. The Health Effects of Cultured Milk Products with Viable and Non-viable Bacteria. *Int. Dairy J.* **1998**, *8*, 749–758. [CrossRef]

61. Shah, N.P. Functional foods from probiotics and prebiotics. *Food Technol.* **2001**, *55*, 46–53.

62. Gupta, S.; Abu-Ghannam, N. Probiotic Fermentation of Plant Based Products: Possibilities and Opportunities. *Crit. Rev. Food Sci. Nutr.* **2012**, *52*, 183–199. [CrossRef] [PubMed]

63. Olszewska, M.; Łaniewska-Trokenheim, Ł. Study of the nonculturable state of lactic acid bacteria cells under adverse growth conditions. *Med. Weter.* **2011**, *67*, 105–109.

64. Warmińska-Radyko, I.; Olszewska, M.; Mikś-Krajnik, M. Effect of temperature and sodium chloride on the growth and metabolism of Lactococcus strains in long-term incubation of milk. *Milchwissenschaft* **2010**, *65*, 32–35.

65. Joux, F.; LeBaron, P. Use of fluorescent probes to assess physiological functions of bacteria at single-cell level. *Microbes Infect.* **2000**, *2*, 1523–1535. [CrossRef]

66. Lahtinen, S.J.; Gueimonde, M.; Ouwehand, A.C.; Reinikainen, J.P.; Salminen, S.J. Probiotic Bacteria May Become Dormant during Storage. *Appl. Environ. Microbiol.* **2005**, *71*, 1662–1663. [CrossRef] [PubMed]

67. Zaręba, D. Fatty Acid Profile of Soya Milk Fermented by Various Bacteria Strains of Lactic Acid Fermentation. *Zywnosc-Nauka Technol. Jakosc* **2009**, *67*, 59–71.

68. Czaczyk, K.; Marciniak, A.; Białas, W.; Mueller, A.; Myszka, K. The Effect of Environmental Factors Influencing Lipopeptide Biosurfactants Biosynthesis by *Bacillus* Spp. *Zywnosc-Nauka Technol. Jakosc* **2007**, *50*, 140–149.

69. Huy, D.N.A.; Haom, P.A.; Hungm, P.V. Screening and identification of *Bacillus* sp. Isolated from traditional Vietnamese soybean-fermented products for high fibrinolytic enzyme production. *Int. Food Res. J.* **2016**, *23*, 326–331.

70. Food and Environmental Hygiene Department FEHD. Available online: https://www.cfs.gov.hk/english/whatsnew/whatsnew_act/files/MBGL_RTE%20food_e.pdf (accessed on 2 September 2019).

71. Wee, Y.J.; Yun, J.S.; Kim, D.; Ryu, H.W. Batch and repeated batch production of L(+)-lactic acid by Enterococcus faecalis RKY1 using wood hydrolyzate and corn steep liquor. *J. Ind. Microbiol. Biotechnol.* **2006**, *33*, 431–435. [CrossRef]

72. Libudzisz, Z.; Kowal, K.; Żakowska, Z. *Mikrobiologia Techniczna*; Wydawnictwo Naukowe PWN: Warszawa, Polnd, 2007.

73. Gajewska, J.; Błaszczyk, M.K. Probiotyczne bakterie fermentacji mlekowej (LAB). *Postępy Mikrobiol.* **2012**, *51*, 55–65.

74. Makarova, K.; Slesarev, A.; Wolf, Y.; Sorokin, A.; Mirkin, B.; Koonin, E.; Pavlov, A.; Pavlova, N.; Karamychev, V.; Polouchine, N.; et al. Comparative genomics of the lactic acid bacteria. *Proc. Natl. Acad. Sci.* **2006**, *103*, 15611–15616. [CrossRef] [PubMed]

75. Gupta, S.; Cox, S.; Abu-Ghannam, N. Process optimization for the development of a functional beverage based on lactic acid fermentation of oats. *Biochem. Eng. J.* **2010**, *52*, 199–204. [CrossRef]

76. Paseephol, T.; Sherkat, F. Probiotic stability of yoghurts containing Jerusalem artichoke inulins during refrigerated storage. *J. Funct. Foods* **2009**, *1*, 311–318. [CrossRef]

77. Colakoglu, H.; Gursoy, O. Effect of lactic adjunct cultures on conjugated linoleic acid (CLA) concentration of yoghurt drink. *J. Food Agric. Environ.* **2011**, *9*, 60–64.

78. Guo, Z.; Wang, J.; Yan, L.; Chen, W.; Liu, X.-M.; Zhang, H.-P. In vitro comparison of probiotic properties of *Lactobacillus casei* Zhang, a potential new probiotic, with selected probiotic strains. *LWT* **2009**, *42*, 1640–1646. [CrossRef]

79. Fenderya, S.; Akbulut, N.; Akalın, A.S. Viability and activity of bifidobacteria in yoghurt containing fructooligosaccharide during refrigerated storage. *Int. J. Food Sci. Technol.* **2004**, *39*, 613–621. [CrossRef]

80. Bonczar, G. The effects of certain factors on the properties of yoghurt made from ewe's milk. *Food Chem.* **2002**, *79*, 85–91. [CrossRef]

81. Bernat, N.; Cháfer, M.; Chiralt, A.; González-Martínez, C. Vegetable milks and their fermented derivative products. *Int. J. Food Stud.* **2014**, *3*, 93–124. [CrossRef]

82. Bernat, N.; Cháfer, M.; González-Martínez, C.; Rodríguez-García, J.; Chiralt, A. Optimisation of oat milk formulation to obtain fermented derivatives by using probiotic *Lactobacillus reuteri* microorganisms. *Food Sci. USA Technol. Int.* **2015**, *21*, 145–157. [CrossRef]

83. Blandino, A.; Al-Aseeri, M.; Pandiella, S.; Cantero, D.; Webb, C. Cereal-based fermented foods and beverages. *Food Res. Int.* **2003**, *36*, 527–543. [CrossRef]

84. Gomes, A.M.; Malcata, F.; Gomes, A.M.; Malcata, F. Bifidobacterium spp. and Lactobacillus acidophilus: Biological, biochemical, technological and therapeutical properties relevant for use as probiotics. *Trends Food Sci. Technol.* **1999**, *10*, 139–157. [CrossRef]

85. Rivera-Espinoza, Y.; Gallardo-Navarro, Y. Non-dairy probiotic products. *Food Microbiol.* **2010**, *27*, 1–11. [CrossRef] [PubMed]

86. Skrzyplonek, K.; Jasińska, M. Quality of Fermented Probiotic Beverages Made From Frozen Acid Whey and Milk During Refrigerated Storage. *Zywnosc-Nauka Technol. Jakosc* **2016**, *104*, 32–44. [CrossRef]

87. At-Otaibi, M.M. Evaluation of some probiotic fermented milk products from Al-Ahsa markets, Saudi Arabia. *J. Food Technol.* **2009**, *4*, 1–8. [CrossRef]

88. Bakirci, I.; Kavaz, A. An investigation of some properties of banana yogurts made with commercial ABT-2 starter culture during storage. *Int. J. Dairy Technol.* **2008**, *61*, 270–276. [CrossRef]

89. Pescuma, M.; Hébert, E.M.; Mozzi, F.; De Valdez, G.F. Functional fermented whey-based beverage using lactic acid bacteria. *Int. J. Food Microbiol.* **2010**, *141*, 73–81. [CrossRef] [PubMed]

90. Olson, D.; Aryana, K. An excessively high Lactobacillus acidophilus inoculation level in yogurt lowers product quality during storage. *LWT* **2008**, *41*, 911–918. [CrossRef]

91. Mousavi, Z.E.; Mousavi, S.M.; Razavi, S.H.; Emam-Djomeh, Z.; Kiani, H. Fermentation of pomegranate juice by probiotic lactic acid bacteria. *World J. Microbiol. Biotechnol.* **2011**, *27*, 123–128. [CrossRef]

92. Pereira, A.L.F.; Maciel, T.C.; Rodrigues, S. Probiotic beverage from cashew apple juice fermented with Lactobacillus casei. *Food Res. Int.* **2011**, *44*, 1276–1283. [CrossRef]

93. Salas, M.L.; Mounier, J.; Valence, F.; Coton, M.; Thierry, A.; Coton, E. Antifungal Microbial Agents for Food Biopreservation—A Review. *Microorganisms* **2017**, *5*, 37. [CrossRef]

94. Wang, Y.-C.; Yu, R.-C.; Chou, C.-C. Growth and survival of bifidobacteria and lactic acid bacteria during the fermentation and storage of cultured soymilk drinks. *Food Microbiol.* **2002**, *19*, 501–508. [CrossRef]

95. Frank, J.F.; Marth, E.M. Fermentations. In *Fundamentals of Dairy Chemistry*; Wong, N.P., Jenness, R., Keeney, M., Marth, E.H., Eds.; Springer: Boston, MA, USA, 1988; pp. 655–738. [CrossRef]

96. Medina, L.M.; Jordano, R. Survival of constitutive microflora in commercially fermented milk containing *Bifidobacteria* during refrigerated storage. *J. Food Prot.* **1994**, *56*, 731–733. [CrossRef] [PubMed]

97. Kosikowska, M.; Jakubczyk, E. Metody oznaczania bakterii probiotycznych w produktach mlecznych. *Przeg. Mlecz.* **2007**, *57*, 12–17.

98. Pereira, A.L.F.; Almeida, F.D.L.; de Jesus, A.L.T.; da Costa, J.M.C.; Rodrigues, S. Storage Stability and Acceptance of Probiotic Beverage from Cashew Apple Juice. *Food Bioprocess Technol.* **2013**, *6*, 3155–3165. [CrossRef]

 *sustainability*

Article

Effect of Press Construction on Yield and Quality of Apple Juice

Kamil Wilczyński [1], Zbigniew Kobus [2,*] and Dariusz Dziki [3]

[1]   Department of Food Engineering and Machines, University of Life Sciences postcode, 20-612 Lublin, Poland
[2]   Department of Technology Fundamentals, University of Life Sciences, 20-612 Lublin, Poland
[3]   Department of Thermal Technology and Food Process Engineering, University of Life Sciences, 20-612 Lublin, Poland
*   Correspondence: zbigniew.kobus@up.lublin.pl; Tel.: +48-81-531-97-46

Received: 31 May 2019; Accepted: 28 June 2019; Published: 2 July 2019

Abstract: The paper presents the possibility of applying different press constructions for juice extraction in small farms. The research was carried out with three different varieties of apples, namely, Rubin, Mutsu, and Jonaprince. Two types of presses were tested: a basket press and a screw press. Generally, application of the screw press makes it possible to obtain a higher yield of extraction compared to the basket press. In our study, the differences in the pressing yield among press machines also depended on the apple variety used. The juices obtained on the screw press were found to be of a higher quality characterized by a higher content of soluble solids, higher viscosity, higher total content of polyphenols, higher antioxidant activity, and lower acidity. Thus, the selection of an appropriate press is the key to producing high-quality apple juice with health-promoting properties for manufacturers of apple juice at the local marketplace.

Keywords: sustainable production; screw press; basket press; polyphenols; antioxidant activity; texture properties

---

1. Introduction

Apples (*Malus domestica*) are the most commonly used fruit for juice extraction in the European Union (EU). They are a rich source of nutrients and polyphenols and possess antioxidant properties that have beneficial effects on human health [1,2]. The yield of apples is estimated to be 12.59 million tons per year in Europe and 3.6 million tons in Poland [3].

Apples are mainly processed into concentrates, which contributes to the reduction of volume and facilitates storage. Poland is the largest producer and exporter of concentrated juices extracted from fruits grown in the temperate zone of the EU [4]. The European share of the juice yield was estimated to account for 66% in 2016.

Currently, there is a growing trend towards healthy eating and an increased interest in ecological produced food. This in turn contributes to the increasing interest of scientists in functional foods and new methods of production to preserve their quality and high level of bioactive compounds [5]. Some examples of these kinds of foods include juices that are not obtained from concentrate (NFC) and freshly squeezed non-pasteurized juices (FS). These juices are obtained from the fruit tissue by pressing and centrifugation of the pulp. Cloudy juices are classified as products obtained with a low degree of processing. They contain higher amounts of bioactive compounds, such as polyphenols or flavonoids, than clarified ones due to the omission of enzymatic and clarifying treatments. They are also richer in dietary fibre, which is necessary for the proper functioning of the digestive system, and several mineral compounds [6]. In their studies, Paepe et al. [7] and Markowski et al. [8] showed that cloudy juices contain substantially higher amounts of beneficial compounds such as polyphenols and also exhibit considerably higher antioxidant activity than clarified juices. Additionally, apple pomace obtained after low-degree processing can be used as animal feed after drying or pickling. The natural nutrients contained in the pomace can serve as a valuable source of nutrition for animals in organic farms [9].

Fruit juices are extracted on an industrial scale using different devices, depending on the type of operation (periodic or continuous) and raw material [10]. Many different press construction solutions are used in the industry for obtaining apple juice. Among them, belt press, water press [11], decanter, rack-and-frame press, hydraulic press [12], basket press [13], and screw press [14] are the most distinguished.

A basket press ensures high pressing yields of up to 60%, but it is less often used on an industrial scale. In one study, Nadulski [15] confirmed the usefulness of the basket press in juice extraction from Jonagold apples. In another study, the same author and his group [10] showed that not all apple cultivars are suitable for juice pressing under farm conditions. This suggests that the use of a suitable fruit variety for juice pressing may reduce the cost of pressing and increase the quality parameters of the juice produced.

A screw press is mainly used for oil extraction from oilseeds such as rape [16,17], flax [18], sunflower [19], cumin [20], tobacco seeds [21], and pistachio nuts [22]. Currently, screw presses are becoming increasingly popular, especially on small farms, among local entrepreneurs and for domestic applications. Based on the construction solution, screw presses are classified as single- and double-screw type. These presses are characterized by several extracting processes, where the pulp is subjected to energetic grinding and mixing. The main advantage of a screw press is that the juice obtained has a much higher amount of soluble solids and bioactive compounds [23,24].

The programmes implemented by the EU such as 'Agricultural and rural development 2014–2020' and 'Promotion of farm products' help producers develop new products based on the policy of sustainable agriculture [25,26]. These programmes allow appropriate usage of resources (raw material, energy) and increase the effectiveness of production without affecting the environment [27]. The protection and promotion of regional and traditional products is one of the most important factors supporting the sustainable development of rural areas (as it increases the income of agricultural producers, prevents depopulation and enhances the attractiveness of rural areas). In addition, this may be considered as potentially influencing the development of agricultural products. In the context of sustainable agriculture, the use of a new press construction may allow the farmers to obtain juices with higher quality and a large amount of bioactive compounds. Moreover, fresh pressing of apple juice may help open new markets.

Considering the above, it appears viable that local farms can use small presses for juice production. In this view, the aim of the present study is to compare the efficiency of screw and basket presses in the extraction of apple juice. This includes the determination of parameters affecting the juice quality such as the content of soluble solids, pH, viscosity, total phenolic content (TPC) and antioxidant activity.

2. Materials and Methods

The research material included three varieties of apples, namely Rubin, Mutsu, and Jonaprince, all obtained from the 2017 harvest. The fruits were delivered by Groups of Fruit Producers, which is related to the company Rylex Sp. z o.o with office registered in the village of Błędów (near Grójec, Poland; GPS coordinates: 51°47′N, 20°42′E), to the local Auchan store in Lublin, Poland.

The experimental flowchart is presented in Figure 1.

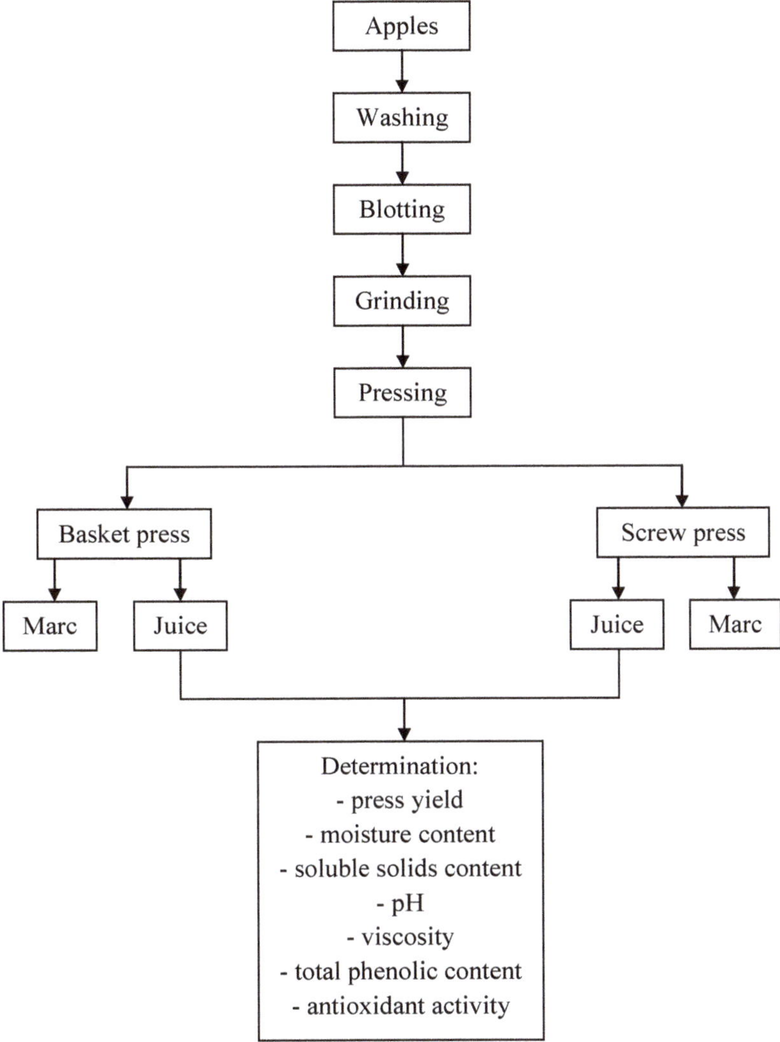

Figure 1. Experimental flowchart.

2.1. Washing and Blotting

The apples were washed in tap water and blotted dry using laboratory tissue paper.

2.2. Grinding

The apples were crushed using a shredding machine (MKJ250; Spomasz Nakło, Poland) equipped with a standard grating disc with 8 mm holes. The rotational speed of the disc was set at 170 rpm. The obtained mash was divided into portions weighing 300 g and placed in plastic containers.

2.3. Determination of Moisture Content

The moisture content of the apple mash was measured before pressing by drying 3 g of mash at a temperature of 105 °C for 3 hours [28]. The measurements were carried out in five replicates.

2.4. Analysis of Texture Properties

The texture of the apples was examined using a texture analyser (model TA.XT Plus Texture Analyzer) equipped with a measuring head that has a working range of up to 0.5 kN. A double-compression test and a cutting test were performed. The speed of the measuring head was set at 0.83 mm·s^{-1} for the double-compression test (TPA) and 1 mm·s^{-1} for the cutting test. For TPA, the apple samples with skin were cut into cylinders of 15 mm diameter and 10 mm height and were then placed with their diameter dimension parallel to the device base. Hardness was defined as the maximum force recorded during the first compression cycle (Fh). For the cutting test, a Warner–Bratzler blade was used. The samples with skin were cut into cylinders of 20 mm diameter and 20 mm height. The maximum value of the cutting force (Fc) was determined. The texture analysis was repeated ten times.

2.5. Pressing

Two kinds of presses were used to extract apple juices: basket press and screw press.

2.5.1. Types of Presses

The basket press (Figure 2) consisted of a perforated cylinder with holes of 3 mm diameter, piston, construction frame, and hydraulic system (UHJG 20/C/2; Hydrotech, Lublin, Poland), which allowed for maintaining a pressure of 4.5 MPa. The press was equipped with a tensometric sensor system for measuring pressure (EMS50; WObit, Poznan, Poland) combined with a digital recorder (MG-TAE1; WObit, Poznan, Poland).

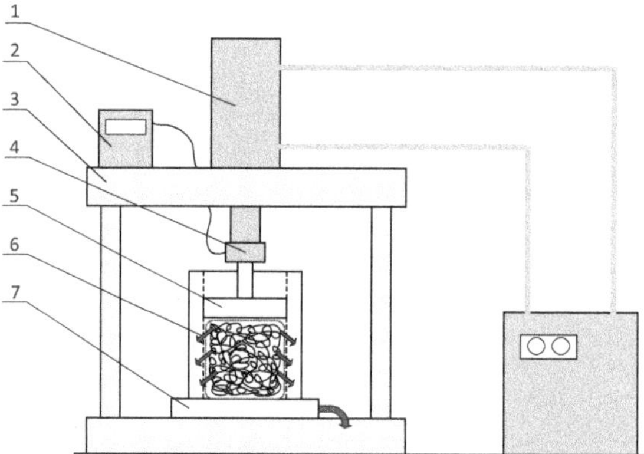

Figure 2. Construction and principle underlying the operation of the hydraulic press: 1 – hydraulic ram, 2 – measurement system, 3 – frame, 4 – tensometer, 5 – piston, 6 – cylinder and 7 – base.

The screw press used was a twin-screw type (Green Star Elite 5000; Tribest), which had a rated power of 260 W and was equipped with a sieve with holes of 0.4/0.5 mm. The press has two rotating gears. The dual stainless steel gears contain magnets and utilize bioceramic technologies that pull more nutrients into the juice. The press gears at a low 110 rpm and generates minimal heat while juicing.

2.5.2. Extraction Procedure

The sample materials weighed 300 g. They were pressed in ten replicates in each type of press.

In the case of the basket press, the material was put in a special bag made of pressing cloth, placed in the cylinder of the press, and subjected to the force applied by the piston. Once a pressure of 4.0 ± 0.1 MPa was reached, the extraction process was stopped.

In the case of the screw press, the material was directly put into the press chamber.

The extracted juice was collected in plastic containers, filtered using a Whatman No. 1 filter and stored in a refrigerator at a temperature of 4 °C.

2.5.3. Pressing Yield

The efficiency of pressing was calculated using the formula:

$$W_j(\%) = \frac{M_j}{M_i} \cdot 100 \tag{1}$$

where

W_j is the efficiency of pressing (%),
M_j is the mass of juice after pressing (kg), and
M_i is the mass of input material (kg).
The calculation was performed for all ten replicated samples.

2.6. Determination of Soluble Solids Content (Brix) and pH

After each extraction, the weight of the obtained juices was measured. The pH [29] and content of soluble solids [30] were also calculated. To determine the pH of the juices, a CP-411 pH-meter (Elmetron, Zabrze, Poland) was used. The content of soluble substances in fruit juices was measured using a PAL-3 refractometer (Atago, Tokyo, Japan). The parameters were determined in three replicates for each sample of extracted juice.

2.7. Determination of Viscosity

Viscosity was measured using a Brookfield viscometer (model LVDV-II + PRO; Brookfield Engineering Laboratories) with Rheolac 3.1 software. A 16 mL sample of juice was taken in ULA – 10EY-baker for all experiments. The spindle speed was set at 10–80 rpm, which corresponded to a shear rate of 12.24–97.84$\cdot$s^{-1}. The temperature was kept constant at 20 °C using a water bath (Brookfield TC-502P). The measurement was performed in three replicates for each sample.

2.8. Determination of TPC

The total phenolic content (TPC) of apple juices was determined according to the FC method [31] with slight modification. Gallic acid was used as the standard and was diluted with methanol (1 mg/1 mL) to give the appropriate concentrations required for plotting a standard curve. First, the sample extract (0.2 mL) was mixed with 2 mL of methanol in a 25 mL volumetric flask. Then, Folin–Ciocalteu reagent (2 mL, diluted 1:10) was added and allowed to react for 3 minutes. Next, 2 mL of Na_2CO_3 solution were added, and the mixture was made up to 25 mL with distilled water. After leaving the mixture for 30 minutes at room temperature in the dark, the absorbance at 760 nm was measured using a spectrophotometer (UV-1800; Shimadzu, Japan). The results were expressed as mg gallic acid equivalent per 100 mL of fresh juice (mg GAE 100 mL^{-1}). The measurement was performed in three replicates for each sample.

2.9. Determination of Antioxidant Activity

The antioxidant activity of apple juices was evaluated using DPPH (2,2-diphenyl-1-picrylhydrazyl) assay. For this analysis, 0.2 mL of apple juice was mixed with an aliquot of 5.8 mL of freshly prepared $6 \cdot 10^{-5}$ M DPPH radical in methanol. After allowing to stand for 30 minutes at room temperature, the spectrophotometric absorbance of the juice at 516 nm was measured using methanol as a blank. The measurement was performed in three replicates for each sample. Antioxidant activity was expressed as percentage inhibition of the DPPH radical calculated using the following equation [32]:

$$AA(\%) = \frac{Absorbanceofcontrol - Absorbanceofsample}{Absorbanceofcontrol} \cdot 100 \qquad (2)$$

2.10. Statistical Analysis

Statistical analysis of the data was performed with Statistica software (Statistica 12; StatSoft Inc., Tulsa, OK, USA) using analysis of variance for factorial designs. The significance of differences was tested using Tukey's least significant difference (LSD) test ($p \leq 0.05$) and using T-test ($p \leq 0.05$) for data in Tables 1 and 2.

Table 1. Total soluble solids (Brix) of the apple juice depending on the type of press machine.

Variety	Solid Soluble Content (in Brix)	
	Basket Press	**Screw Press**
Rubin	11.9 ± 0.00a	13.1 ± 0.00b
Mutsu	12.9 ± 0.00a	12.9 ± 0.00a
Jonaprince	11.2 ± 0.00a	11.9 ± 0.00b

a, b and c – average values in the row marked with the same letter are not statistically significantly different (T-test, $p \leq 0.05$).

Table 2. Acidity (pH) of the apple juice depending on the type of press machine.

Variety	Acidity (pH)	
	Basket Press	**Screw Press**
Rubin	3.61 ± 0.012a	3.45 ± 0.125b
Mutsu	3.72 ± 0.008a	3.62 ± 0.08b
Jonaprince	3.81 ± 0.008a	3.73 ± 0.012a

a, b and c – average values in the row marked with the same letter are not statistically significantly different (T-test, $p \leq 0.05$).

3. Results and Discussion

3.1. Moisture Content of Fresh Apples

The moisture content of the apples ranged from 84.0% to 85.27%, and no significant differences were observed between the tested varieties. Moisture content is an important parameter to be determined to carry out further investigations because it influences the yield of juice processing.

3.2. Effect of the Press Construction on the Yield of Pressing

The effect of the type of press on pressing yield is shown in Figure 3.

The yield of pressing ranged from 61.9% to 71.6%. Generally, higher pressing yields were obtained with screw press. In addition, statistical analysis showed significant differences in the efficiency of juice extraction from the pulp of the different varieties of apples. A statistically significant effect of the press type on the pressing yield was observed in the case of the Mutsu and Jonaprince varieties, whereas no significant differences were noted in the case of Rubin.

In the case of the screw press, pieces of apple are fed into the cylinder and thrown to the perforated wall by the centrifugal action of the gears. These gears crush, cut, and squeeze the pieces at the same time. This maximizes the yield as well as the quality of juice. The yield of the recovered juice depends on the diameter of the perforations, the speed of rotation of the gears and the gap between the knob and the pulp discharge casing.

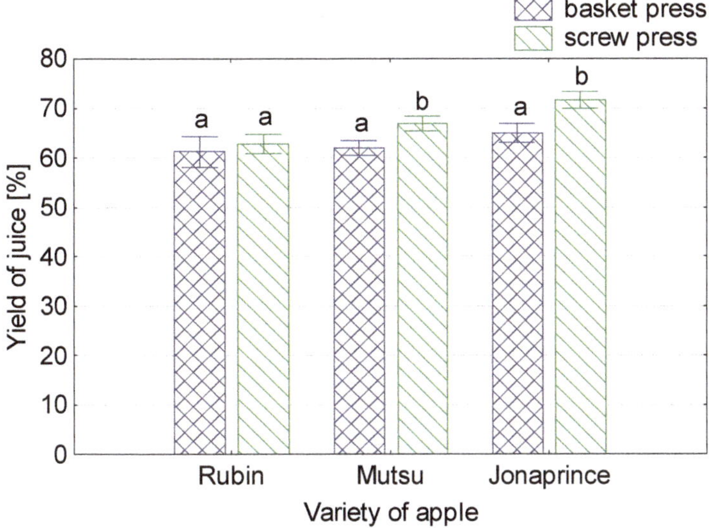

Figure 3. Effect of the type of press machine on the pressing yield.

In the case of the basket press, the crushed material is wrapped in the pressing cloth and gradually compressed by the hydraulic piston. The pressing cloth serves both as a package for the mash and as a filter for the juice.

In the case of the screw press, the correct course of the process is influenced by the texture of apples. Apples with a greater hardness are better for juice extraction on screw press, whereas those with a lower hardness are shredded to very small-sized particles, creating a layer of mousse inside the chamber, which clogs the sieve openings. This leads to a drop in the pressing yield or the passage of the mousse to the liquid phase.

To support the hypothesis about the relationship between hardness and yield of pressing, the texture properties of the apples were analysed.

Figures 4 and 5 show the hardness and cutting force of the tested apple varieties, respectively.

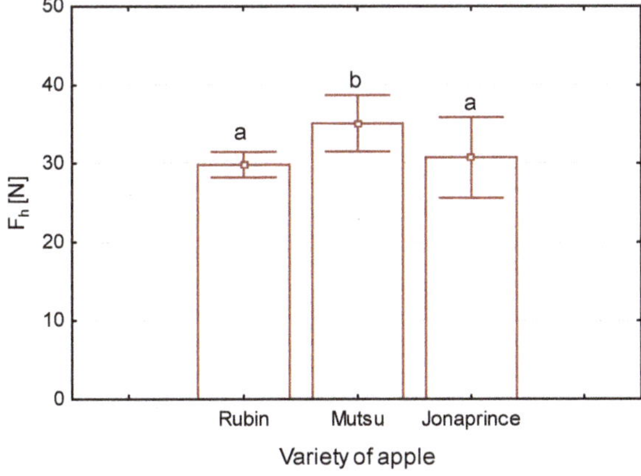

Figure 4. Hardness of the tested apple varieties.

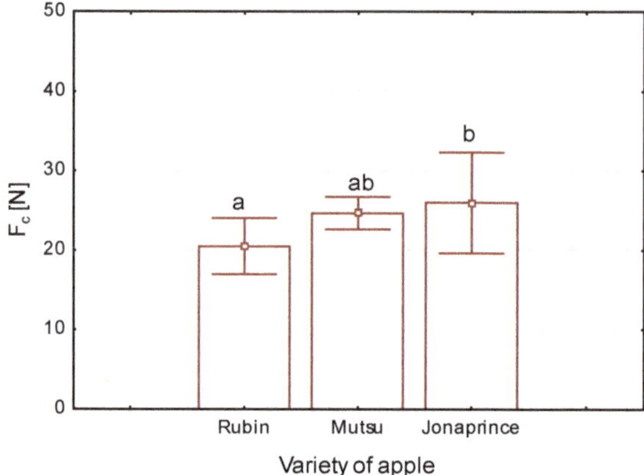

Figure 5. Cutting force of the tested apple varieties.

The texture analysis showed that, among the studied varieties, Rubin had the lowest hardness and cutting force. Probably due to these features, no statistically significant differences in the pressing yield were observed between the basket press and the screw press for this cultivar.

High efficiency of pressing was also reported by Takenaka et al. [33] when screw press was used for the extraction of citrus juice. In their experiment, the authors demonstrated that, of the three tested presses (belt, centrifugal and screw press), screw press provided the highest yield of pressing. The extraction yield obtained with the screw press was approximately 54% higher as compared to the belt press and approximately 18% higher as compared to the centrifugal extractor.

3.3. Effect of the Press Construction on the Soluble Solids Content

The content of soluble solids mainly reflects the amount of sugars and organic acids and is a very important parameter of juice quality. A study conducted by Eisele and Drake [34] showed that the content of soluble solids in juices obtained from various apple cultivars varied from 10.26 to 21.62 Brix.

In the present study, the content of soluble solids in the juices ranged from 11.2 to 13.1 °Brix (Table 1). This is similar to the content of soluble solids in the pressure-extracted apple juice produced in the Polish fruit and vegetable industry (i.e., 11.0–12.4 Brix) [35].

Generally, the juices obtained on the screw press had statistically significantly higher content of soluble solids than the ones obtained on the basket press. This was due to the wider opening of cell membranes and release of greater amounts of the deep-seated nutrients.

It was noted that the press type had no statistically significant effect on the content of soluble solids in the case of the Mutsu variety, probably due to the structure of the apple tissue. In fact, the flesh of the Mutsu apples is characterized by a coarse structure, while the Rubin and Jonaprince apples have a fine-grained structure. It was likely that the basket press enabled the breakdown of coarse-grained apples to the same extent as the screw press. Therefore, no effect of the type of press on the extract content was found in the case of the Mutsu variety. On the other hand, due to the wider opening of the cell membranes, the screw press enables a better breakdown of fine-grained structure. Therefore, there were differences in the content of soluble solids between presses in the case of the Rubin and Jonaprince varieties.

3.4. Effect of the Press Construction on Acidity

The apple juices were characterized by different levels of acidity (Table 2).

The pH of the juices obtained in this study ranged from 3.45 to 3.81. According to the literature, acidity (pH) of the apple juice varies from 3.37 to 4.24 [34]. Kobus et al. [14] found that the pH of the apple juice obtained on the screw press was in the range of 3.66 ± 0.09.

The press construction had a significant effect on the pH of the tested apple juice. It was observed that the juice extracted on the screw press had a slightly higher acidity (lower pH) compared to the juice from the basket press. The acidity was higher by 2.2%, 2.8%, and 4.6% for Jonaprince, Mutsu, and Rubin, respectively. A probable reason for the higher acidity of the juices from the screw press was increased migration of microelements, organic acids, and secondary plant metabolites such as polyphenols caused by the greater disintegration of membranes and cell walls.

Acidity is one of the most important traits of freshly squeezed apple juice. It influences the flavour, clarity, colour, aroma, and overall sensory satisfaction [36]. In addition, high acidity acts as a natural barrier against contamination by most microorganisms [37]. Consequently, it is reasonable to employ the screw press to produce juice with relatively high acidity.

3.5. Effect of the Press Construction on Viscosity

The viscosity of the juices obtained in this study ranged from 3.15 to 4.35 mPa·s (Figure 6). The viscosity of cloudy apple juice showed a wide range from 1.7 to 9.6 mPa·s, depending on the apple variety, method of processing, and concentration of soluble solids. Genovese and Lozano [38] found that the viscosity of apple juice from Granny Smith cv. ranged between 1.71 mPa·s at 10 °Brix and 3.65 mPa·s at 20 °Brix. Will et al. [39] reported that the values of viscosity ranged between 1.74 (Topaz cv.) and 2.15 mPa·s (Boskoop cv.), and the values determined by Teleszko et al. [40] ranged between 2.40 and 9.60 mPa·s.

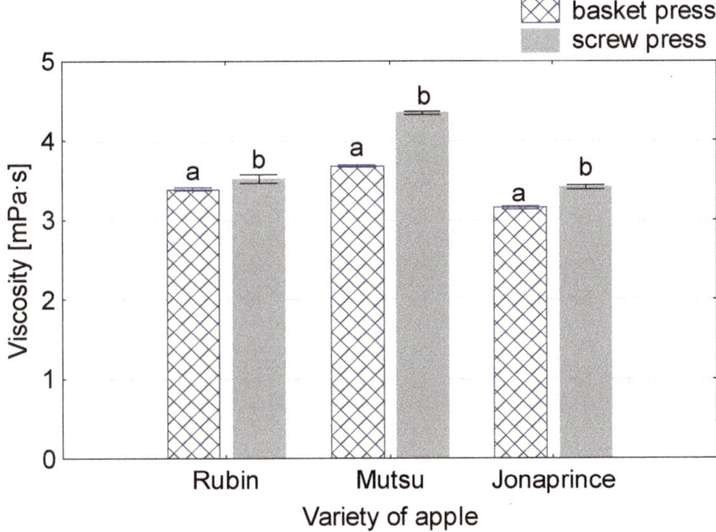

Figure 6. Viscosity of the apple juice depending on the type of press machine.

In the present study, the press construction was found to have a significant effect on the viscosity of the apple juice. In the case of Rubin, the viscosity of apple juice obtained on the screw press was 3.8% higher compared to the juice from the basket press. In the case of the Jonagold, the difference was greater and amounted to 8.3%, while the highest difference amounting to 18.2% was observed in the case of Mutsu.

The rheological behaviour of a cloudy juice is governed by both liquid viscosity and size characteristics of the solids [41].

In the case of clear juice, the viscosity depends on the content of soluble solids, with sugars playing the main role [42]. The juice from the Jonaprince variety obtained on the basket press was characterized by the lowest content of soluble solids and lowest viscosity. By contrast, the juice obtained on the screw press had a higher content of soluble solids, which probably resulted in higher juice viscosity. Thus, in the case of Jonaprince, the higher viscosity of the juice obtained from the screw press could have resulted from the greater disintegration and release of soluble phytochemical ingredients from apples.

There were no statistically significant differences found in the content of soluble solids in the juices obtained from the Mutsu cultivar between the tested presses. The higher viscosity observed in the case of this variety may be caused by the higher amount of total suspended solids in the juice obtained on the screw press, and due to the use of the pressing cloth, which serves both as a package for the mash and as a filter for the juice, for extraction on the basket press.

In the case of Rubin, the higher viscosity of the juice obtained on the screw press was mainly due to the higher content of soluble solids.

3.6. Effect of the Press Construction on TPC

Polyphenols play an important role in fruit juices because they influence the colour and flavour [11]. The content of polyphenols in juices varies depending on both the fruit variety and production technique used.

In the present study, the content of polyphenols in the juices ranged from 29.89 to 60.96 mg GAE 100 mL^{-1} (Figure 7).

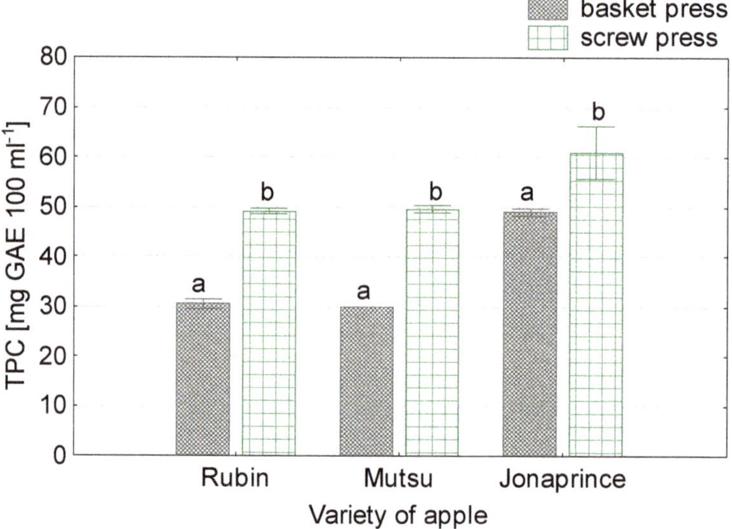

Figure 7. Total phenolic content of the apple juice depending on the type of press machine.

The TPC in juices produced in Europe varies within a broad range from 10 to 300 mg GAE 100 mL^{-1} [43]. Kobus et al. [14] showed that the content of polyphenols in juices obtained on the screw press ranged from 44.2 to 58.3 mg GAE 100 mL^{-1}.

In this study, the statistical analysis revealed a significant effect of the press type on the content of polyphenols in the obtained juices. The juices from the screw press were statistically significantly richer in polyphenols than the juices from the basket press. The higher content of polyphenols in the juice from the screw press was related to the higher percentage of solid particles and probably better grinding of apple tissue, especially skin.

It is generally known that apple skin is characterized by a higher content of polyphenols compared to the flesh. There is up to a three- to fourfold difference between these two tissue types [44]. Thus, with the complex action of the gears, the screw press ensures better disintegration of apple skin and greater release of the cell contents to the extracted juice.

It is also known that the content of phenolic compounds decreases gradually during grinding and pressing due to the oxidation of pulp and freshly extracted juices [11,45]. So, our results can also be attributed to the rate of mash oxidation. Since the apples were disintegrated directly in the screw press, the entire process of juice extraction was faster and the rate of mash oxidation was lower than in the case of the basket press.

Noteworthy, for varieties with similar contents of polyphenols in the juice extracted on the basket press (Rubin – 30.47 mg GAE 100 mL^{-1} and Mutsu – 29.89 mg GAE 100 mL^{-1}), a similar percentage of increase in polyphenols was also observed in the juice obtained on the screw press (62% for Rubin and 65% for Mutsu).

Jaeger et al. [12] found that the release of polyphenols from coarse mash was lower than that from fine mash. Since the screw press crushes apples more finely than the basket press, more polyphenols are extracted to the juice. A similar observation was reported by Heinmaa et al. [11], who noticed a higher amount of individual polyphenols in the juice extracted on a belt press (which also crushes the apples during pressing) as compared to the rack-and-frame press.

The efficiency of polyphenol extraction on individual presses may also be influenced by the textural characteristics of the apples tested. It is worth mentioning that the greatest differences were observed in the Mutsu and Rubin varieties, which are characterized by lower cutting forces.

3.7. Effect of the Press Construction on Antioxidant Activity

The antioxidant capacity of a substance is defined as its ability to scavenge reactive oxygen species and electrophiles [46]. Antioxidant activity is a very important parameter of juice quality.

The effect of press construction on the antioxidant capacity of juices is presented in Figure 8.

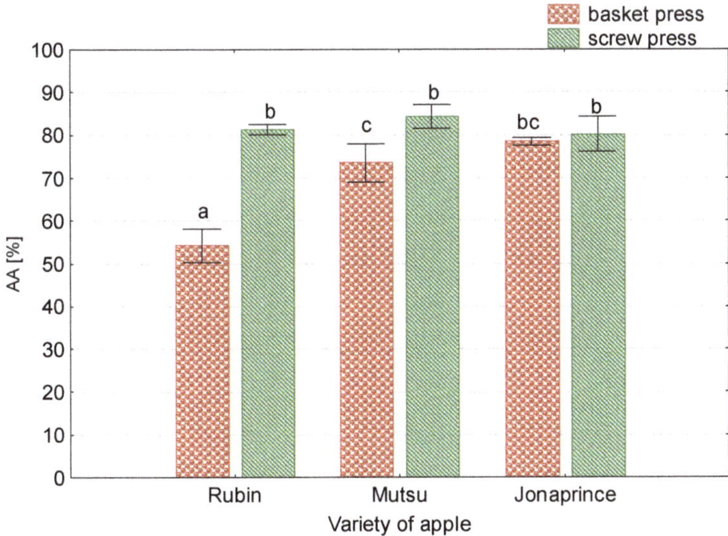

Figure 8. Antioxidant activity of the apple juice depending on the type of press machine.

The study showed that press construction had a significant effect on the antioxidant activity of the juice depending on the apple variety used. In the case of Rubin and Mutsu, the differences in antioxidant activity were statistically significant, whereas no statistically significant differences were found for Jonagold. Among the cultivars, the juice obtained from Mutsu showed the highest antioxidant activity.

The ability to scavenge free radicals was very closely related to the content of phenolic compounds in the apple juice. For the Jonagold, only small differences in the content of polyphenols in the juices were observed between the presses, which probably resulted in the absence of statistically significant differences in the antioxidant activity. Numerous studies have demonstrated a strong relationship between the polyphenols concentration and the antioxidant capacity of foods [44,47–50].

On the other hand, Wolfe et al. [51] did not find any relation between the antioxidant capacity and the TPC in apples. The lack of a correlation between these two properties may be related to the extraction process, apple varieties, percentage share of individual polyphenols, and the content of other compounds such as vitamin C, which may influence the antioxidant capacity.

Another determinant of the antioxidant activity is the degree of disintegration of the raw material. After cutting, defence metabolism is activated and synthesis and oxidation of phenolics occur simultaneously, modifying the initial phenolic composition of the fruit [46].

An effect of the processing technology on antioxidant capacity was also observed in other studies. For instance, Heinmaa et al. [11] reported that antioxidant capacity was highest in juices extracted on water press, followed by those extracted on belt press and rack-and-frame press. Their results also showed that the degrees of correlation between individual polyphenols and antioxidant activity were different.

Both antioxidant activity and TPC indicate the quality of a product with respect to its biological properties, and hence can be used in the process of quality control in the production of apple juices [52]. As indicated by our results, small-scale manufacturers should be advised to make use of a screw press to produce healthy and high-quality apple juice.

4. Conclusions

The present research demonstrated the varied efficiency of extraction of apple juice depending on the design of the press and apple variety. Generally, application of the screw press ensured higher yields of pressing compared to the basket press. The differences in the yield of extracted juice observed between the presses were related to the wider opening of cell membranes and release of greater amounts of nutrients facilitated by the screw press.

This study also showed the effect of press construction on the quality of apple juice. The juices extracted on screw press had higher TPC and antioxidant activity. This was probably due to the more intensive grinding and mixing of the raw material and the greater release of valuable bioactive components into the extracted juice. Additionally, the apple juices extracted on the screw press were characterized by a higher content of soluble solids, higher viscosity, and lower acidity. The obtained results indicate the necessity of further research on the use of screw presses for the production of juices from various varieties of apples in the farm conditions.

Author Contributions: Z.K. conceptualization; K.W. methodology; Z.K. formal analysis; K.W. investigation; K.W. data curation; Z.K. and K.W. writing—original draft preparation; D.D. writing—review and editing; Z.K. and D.D. supervision.

Funding: This research received no external funding.

References

1. Li, Z.; Teng, J.; Lyu, Y.; Hu, X.; Zhao, Y.; Wang, M. Enhanced Antioxidant Activity for Apple Juice Fermented with Lactobacillus plantarum ATCC14917. *Molecules* **2018**, *24*, 51. [CrossRef] [PubMed]
2. Kschonsek, J.; Wolfram, T.; Stöckl, A.; Böhm, V. Polyphenolic Compounds Analysis of Old and New Apple Cultivars and Contribution of Polyphenolic Profile to the In Vitro Antioxidant Capacity. *Antioxidants (Basel)* **2018**, *7*, 20. [CrossRef] [PubMed]
3. FAOSTAT. Food and agriculture organization: FAO-STAT 2016. Available online: http://www.fao.org/faostat/en/#data/QC (accessed on 23 January 2019).

4. Bugala, A. Zmiany w polskim handlu zagranicznym sokami zagęszczonymi w 2016 r. *Przemysł Fermentacyjny i Owocowo-Warzywny* **2017**, *61*, 19–21. [CrossRef]

5. Francini, A.; Sebastiani, L. Phenolic Compounds in Apple (Malus x domestica Borkh.): Compounds Characterization and Stability during Postharvest and after Processing. *Antioxidants (Basel)* **2013**, *2*, 181–193. [CrossRef] [PubMed]

6. Plocharski, W.; Markowski, J.; Groele, B.; Stos, K.; Koziol-Kozakowska, A. Soki, nektary, napoje - aspekty rynkowe i zdrowotne. *Przemysł Fermentacyjny i Owocowo-Warzywny* **2017**, *61*, 6–10. [CrossRef]

7. De Paepe, D.; Coudijzer, K.; Noten, B.; Valkenborg, D.; Servaes, K.; De Loose, M.; Diels, L.; Voorspoels, S.; Van Droogenbroeck, B. A comparative study between spiral-filter press and belt press implemented in a cloudy apple juice production process. *Food Chem.* **2015**, *173*, 986–996. [CrossRef] [PubMed]

8. Markowski, J.; Baron, A.; Le Quéré, J.-M.; Płocharski, W. Composition of clear and cloudy juices from French and Polish apples in relation to processing technology. *LWT - Food Sci. Technol.* **2015**, *62*, 813–820. [CrossRef]

9. Kruczek, M.; Drygaś, B.; Habryka, C. Pomace in fruit industry and their contemporary potential application. *World Sci. News* **2016**, *48*, 259–265.

10. Nadulski, R.; Kobus, Z.; Wilczyński, K.; Guz, T.; Ahmed, Z.A. Characterisation of selected apple cutivars in the aspect of juice production in the condition of farm. In *Farm Machinery and Processes Management in Sustainable Agriculture. Proceedings IX International Scientific Symposium Farm Machinery and Processes Management in Sustainable Agriculture, Lublin, Poland, 22–24 November 2017*; pp. 255–259. Available online: https://depot.ceon.pl/handle/123456789/14803 (accessed on 02 July 2019).

11. Heinmaa, L.; Moor, U.; Põldma, P.; Raudsepp, P.; Kidmose, U.; Lo Scalzo, R. Content of health-beneficial compounds and sensory properties of organic apple juice as affected by processing technology. *LWT - Food Sci. Technol.* **2017**, *85*, 372–379. [CrossRef]

12. Jaeger, H.; Schulz, M.; Lu, P.; Knorr, D. Adjustment of milling, mash electroporation and pressing for the development of a PEF assisted juice production in industrial scale. *Innov. Food Sci. Emerg. Technol.* **2012**, *14*, 46–60. [CrossRef]

13. Nadulski, R.; Kobus, Z.; Wilczyński, K.; Zawiślak, K.; Grochowicz, J.; Guz, T. Application of Freezing and Thawing in Apple (Malus domestica) Juice Extraction. *J. Food Sci.* **2016**, *81*, E2718–E2725. [CrossRef] [PubMed]

14. Kobus, Z.; Nadulski, R.; Anifantis, A.S.; Santoro, F. Effect of press construction on yield of pressing and selected quality characteristics of apple juice. Engineering for rural development. Jelgava, 23.-25.05.2018. Available online: https://www.researchgate.net/publication/325386311_Effect_of_press_construction_on_yield_of_pressing_and_selected_quality_characteristics_of_apple_juice (accessed on 24 Janauary 2019).

15. Nadulski, R. Ocena przydatności laboratoryjnej prasy koszowej do badań procesu tłoczenia soku z surowców roślinnych. *Inżynieria Rolnicza* **2006**, *6*, 73–80.

16. Łaska, B.; Myczko, A.; Golimowski, W. Badanie wydajności prasy ślimakowej i sprawności tłoczenia oleju w warunkach zimowych i letnich. *Problemy Inżynierii Rolniczej* **2012**, *4*, 163–170.

17. Wroniak, M.; Ptaszek, A.; Ratusz, K. Ocena wpływu warunków tłoczenia w prasie ślimakowej na jakość i skład chemiczny olejów rzepakowych. *Żywność: Nauka - technologia - jakość* **2013**, *20*, 92–104.

18. Bogaert, L.; Mathieu, H.; Mhemdi, H.; Vorobiev, E. Characterization of oilseeds mechanical expression in an instrumented pilot screw press. *Ind. Crop. Prod.* **2018**, *121*, 106–113. [CrossRef]

19. Isobe, S.; Zuber, F.; Uemura, K.; Noguchi, A. A new twin-screw press design for oil extraction of dehulled sunflower seeds. *J. Am. Oil Chem. Soc.* **1992**, *69*, 884. [CrossRef]

20. Bakhshabadi, H.; Mirzaei, H.; Ghodsvali, A.; Jafari, S.M.; Ziaiifar, A.M. The influence of pulsed electric fields and microwave pretreatments on some selected physicochemical properties of oil extracted from black cumin seed. *Food Sci. Nutr.* **2017**, *6*, 111–118. [CrossRef]

21. Sannino, M.; del Piano, L.; Abet, M.; Baiano, S.; Crimaldi, M.; Modestia, F.; Raimo, F.; Ricciardiello, G.; Faugno, S. Effect of mechanical extraction parameters on the yield and quality of tobacco (Nicotiana tabacum L.) seed oil. *J. Food Sci. Technol.* **2017**, *54*, 4009–4015. [CrossRef]

22. Rabadán, A.; Álvarez-Ortí, M.; Gómez, R.; Alvarruiz, A.; Pardo, J.E. Optimization of pistachio oil extraction regarding processing parameters of screw and hydraulic presses. *LWT - Food Sci. Technol.* **2017**, *83*, 79–85. [CrossRef]

23. Wilczyński, K.; Kobus, Z.; Guz, T. Analysis of efficiency and particles size distribution in apple juice obtained with different methods. *J. Res. Appl. Agric. Eng.* **2018**, *63*, 140–143.

24. Wilczyński, K. Charakterystyka oraz efektywność urządzeń wykorzystywanych w przemyśle do pozyskiwania soku z owoców i warzyw. In *Wybrane problemy z zakresu przemysłu spożywczego - teoria i praktyka*; Stoma, M., Dudziak, A., Kobus, Z., Eds.; Libropolis: Lublin, Poland, 2016; pp. 81–90.

25. Agriculture and rural development 2014-2020. Available online: https://ec.europa.eu/agriculture/rural-development-2014-2020_en (accessed on 28 May 2019).

26. Promotion of EU farm products. 2019. Available online: https://ec.europa.eu/info/food-farming-fisheries/key-policies/common-agricultural-policy/market-measures/promotion-eu-farm-products_en (accessed on 28 May 2019).

27. Juściński, S. The mobile service of agricultural machines as the element of the support for the sustainable agriculture. In *Farm Machinery and Processes Management in Sustainable Agriculture. Proceedings IX International Scientific Symposium Farm Machinery and Processes Management in Sustainable Agriculture, Lublin, Poland, 22–24 November 2017*; pp. 136–141. Available online: https://depot.ceon.pl/handle/123456789/14824 (accessed on 02 July 2019).

28. Horwitz, W. *Official Methods of Analysis of AOAC International*, 17th ed.; AOAC International: Gaithersburg, MD, USA, 2000.

29. PN-EN 1132:1999 - Polish version. 2019. Available online: http://sklep.pkn.pl/pn-en-1132-1999p.html (accessed on 24 January 2019).

30. PN-EN 12143:2000 - Polish version. 2019. Available online: http://sklep.pkn.pl/pn-en-12143-2000p.html (accessed on 24 January 2019).

31. Singleton, V.L.; Orthofer, R.; Lamuela-Raventós, R.M. [14] Analysis of total phenols and other oxidation substrates and antioxidants by means of folin-ciocalteu reagent. *Methods Enzymol.* **1999**, *299*, 152–178.

32. Sharma, S.; Kori, S.; Parmar, A. Surfactant mediated extraction of total phenolic contents (TPC) and antioxidants from fruits juices. *Food Chem.* **2015**, *185*, 284–288. [CrossRef] [PubMed]

33. Takenaka, M.; Nanayama, K.; Isobe, S.; Ozaki, K.; Miyagi, K.; Sumi, H.; Toume, Y.; Morine, S.; Ohta, H. Effect of extraction method on yield and quality of Citrus depressa juice. *Food Sci. Technol. Res.* **2007**, *13*, 281–285. [CrossRef]

34. Eisele, T.A.; Drake, S.R. The partial compositional characteristics of apple juice from 175 apple varieties. *J. Food Compos. Anal.* **2005**, *18*, 213–221. [CrossRef]

35. Kowalczyk, R. Wydajność tłoczenia i wskaźnik zużycia jabłek w procesie wytwarzania zagęszczonego soku jabłkowego. *Problemy Inżynierii Rolniczej* **2004**, *12*, 21–30.

36. Huang, Z.; Hu, H.; Shen, F.; Wu, B.; Wang, X.; Zhang, B.; Wang, W.; Liu, L.; Liu, J.; Chen, C.; et al. Relatively high acidity is an important breeding objective for fresh juice-specific apple cultivars. *Sci. Hortic.* **2018**, *233*, 29–37. [CrossRef]

37. Choi, L.H.; Nielsen, S.S. The Effects of Thermal and Nonthermal Processing Methods on Apple Cider Quality and Consumer Acceptability. *J. Food Qual.* **2005**, *28*, 13–29. [CrossRef]

38. Genovese, D.B.; Lozano, J.E. Contribution of colloidal forces to the viscosity and stability of cloudy apple juice. *Food Hydrocoll.* **2006**, *20*, 767–773. [CrossRef]

39. Will, F.; Roth, M.; Olk, M.; Ludwig, M.; Dietrich, H. Processing and analytical characterisation of pulp-enriched cloudy apple juices. *LWT - Food Sci. Technol.* **2008**, *41*, 2057–2063. [CrossRef]

40. Teleszko, M.; Nowicka, P.; Wojdyło, A. Chemical, enzymatic and physical characteristic of cloudy apple juices. *Agric. Food Sci.* **2016**, *25*, 34–43. [CrossRef]

41. Genovese, D.B.; Lozano, J.E. Effect of Cloud Particle Characteristics on the Viscosity of Cloudy Apple Juice. *J. Food Sci.* **2000**, *65*, 641–645. [CrossRef]

42. Juszczak, L.; Fortuna, T. Effect of temperature and soluble solids content on the viscosity of cherry juice concentrate. *Int. Agrophysics* **2004**, *18*, 17–21.

43. Gökmen, V.; Artık, N.; Acar, J.; Kahraman, N.; Poyrazoğlu, E. Effects of various clarification treatments on patulin, phenolic compound and organic acid compositions of apple juice. *Eur. Food Res. Technol.* **2001**, *213*, 194–199. [CrossRef]

44. Vieira, F.G.K.; Borges, G.S.; Copetti, C.; Gonzaga, L.V.; Nunes, E.C.; Fett, R. Activity and contents of polyphenolic antioxidants in the whole fruit, flesh and peel of three apple cultivars. *Arch. Latinoam. Nutr.* **2009**, *59*, 101–106. [PubMed]

45. Van Der Sluis, A.A.; Dekker, M.; Skrede, G.; Jongen, W.M.F. Activity and concentration of polyphenolic antioxidants in apple juice. 1. Effect of existing production methods. *J. Agric. Food Chem.* **2002**, *50*, 7211–7219. [CrossRef] [PubMed]

46. Queiroz, C.; Lopes, M.L.M.; Fialho, E.; Valente-Mesquita, V.L. Changes in bioactive compounds and antioxidant capacity of fresh-cut cashew apple. *Food Res. Int.* **2011**, *44*, 1459–1462. [CrossRef]

47. He, L.; Xu, H.; Liu, X.; He, W.; Yuan, F.; Hou, Z.; Gao, Y. Identification of phenolic compounds from pomegranate (Punica granatum L.) seed residues and investigation into their antioxidant capacities by HPLC–ABTS+ assay. *Food Res. Int.* **2011**, *44*, 1161–1167. [CrossRef]

48. Chinnici, F.; Bendini, A.; Gaiani, A.; Riponi, C. Radical scavenging activities of peels and pulps from cv. Golden Delicious apples as related to their phenolic composition. *J. Agric. Food Chem.* **2004**, *52*, 4684–4689. [CrossRef] [PubMed]

49. do Rufino, M.; Alves, R.E.; de Brito, E.S.; Pérez-Jiménez, J.; Saura-Calixto, F.; Mancini-Filho, J. Bioactive compounds and antioxidant capacities of 18 non-traditional tropical fruits from Brazil. *Food Chem.* **2010**, *121*, 996–1002. [CrossRef]

50. Drogoudi, P.D.; Michailidis, Z.; Pantelidis, G. Peel and flesh antioxidant content and harvest quality characteristics of seven apple cultivars. *Sci. Hortic.* **2008**, *115*, 149–153. [CrossRef]

51. Wolfe, K.; Wu, X.; Liu, R.H. Antioxidant activity of apple peels. *J. Agric. Food Chem.* **2003**, *51*, 609–614. [CrossRef] [PubMed]

52. Pavun, L.; Đurđević, P.; Jelikić-Stankov, M.; Đikanović, D.; Uskoković-Marković, S. Determination of flavonoids and total polyphenol contents in commercial apple juices. *Czech. J. Food Sci.* **2018**, *36*, 233–238.

 *sustainability*

Article

Biochars Originating from Different Biomass and Pyrolysis Process Reveal to Have Different Microbial Characterization: Implications for Practice

Wioletta Żukiewicz-Sobczak [1], Agnieszka Latawiec [2,3,4], Paweł Sobczak [5,*], Bernardo Strassburg [2,3,4], Dorota Plewik [1] and Małgorzata Tokarska-Rodak [1]

[1] Pope John Paul II State School of Higher Education in Biala Podlaska, Sidorska 95/97, 21-500 Biala Podlaska, Poland; wiola.zukiewiczsobczak@gmail.com (W.Z.-S.); dorotaplewik@gmail.com (D.P.); rodak.malgorzata@gmail.com (M.T.-R.)

[2] International Institute for Sustainability, Estrada Dona Castorina 124, Horto, Rio de Janeiro 22460-320, Brazil; alatawiec@gmail.com (A.L.); bernardobns@gmail.com (B.S.)

[3] Department of Geography and the Environment, Rio Conservation and Sustainability Science Centre, Pontifical Catholic University of Rio de Janeiro, Rio de Janeiro 22451-900, Brazil

[4] Department of Production Engineering, Logistics and Applied Computer Science, Faculty of Production and Power Engineering, University of Agriculture in Kraków, Balicka 116B, 30-149 Kraków, Poland

[5] Department of Food Engineering and Machines, University of Life Sciences in Lublin, Akademicka 13, 20-950 Lublin, Poland

* Correspondence: pawel.sobczak@up.lublin.pl; Tel.: +48-81-531-97-13

Received: 21 January 2020; Accepted: 14 February 2020; Published: 18 February 2020

Abstract: Sustainable technologies are increasingly promoted in various production areas. Protection of natural resources, as well as rational waste management, may lead to better optimization of technologies. Biochar, a product of pyrolysis of organic residues has found wide applications in waste management, agriculture, energy and construction industry. In the present study biochar samples produced in Poland and in Brazil were analysed for microbial content using three substrates: Plate Count Agar, Malt Agar, and Potato Agar. Both qualitative and quantitative measurements were done. Microscopic analysis of the biochar structure was also performed. We found that microbial cultures in both biochars represented a wide range of biodiversity of microorganisms genera and species. We demonstrate that the biochar samples differ depending on the botanical origin as well as on the production technology. Structure of the tested samples also varied depending on the botanical origin. Sample 1-PL (pine) was characterised by a compact and regular structure, while sample 2-PL (oak) showed porous and irregular structure. Sample from Brazil (1-BR) showed a more delicate structure than Polish biochars. Obtained properties may suggest a range of implications for practice.

Keywords: biochar; microbiological analysis; structure; implications for practice

---

1. Introduction

Biochar, carbonous substrate originating from the process of pyrolysis [1] has been demonstrated to have various potential benefits such as increasing agricultural yields [2–4], soil and sediment remediation [5] or improving animals health in livestock farming [6]. There has been also a recent interest in using biochar in construction [7] with potential benefits for diminishing humidity inside buildings, thermal isolation and fungus development control [8–12]. Biochar or microporous carbon is also used for humidity control, a property that has been extensively researched by Nakano et al. (1996) and Abe et al. (1995). Humidity control by microporous carbon can reduce asthma and dermatitis occurrences by controlling the growth of mould and ticks [10–13]. However, there is little research that looks into a range of biochars, identifying and quantifying microbial colonies. Although the number and the type of microbial colonies in biochar have various implications, including crop interactions and safety, the microbial composition remains an under-investigated aspect of biochar research. Of particular interest, are the bacteria from family Bacillus. They are common throughout a range of environments and tolerate acid and alkaline conditions. On the account of the possibility to create endospores they can survive in extreme environments and are commonly isolated from unfresh and spoiling food products. Majority of the Bacillus species are safe for people with the exception of B. anthracis and B. cereus. Due to their widespread persistence and resistance, these bacteria are used in commercial production of enzymes, antibiotics, insecticides, vitamins and metabolites such as hyaluronic acid (used for example in cosmetics) [14,15]. Enzymes such as α-amylases, able to hydroliseα-1,4-glycoside bonds in starch, originating from Bacillus sp. are commonly used in food, chemical and textile industry as well as for the production of pharmaceutics [16–18].

Another important characterization of biochar is their fungus content, such as Alternaria, Aspergillus, Candida, Cladosporium, Penicillium and Rhizopus. Although favouring humid environments, Aspergillus and Penicillium, are able to survive in dry conditions. A range of fungus metabolites are antibiotics, such as the metabolites of Penicillium puberulum. An important feature of fungus is their mycotoxin production (such as alphatoxins), on the account of their cancerogenic and teratogenic properties, even in small concentrations [19,20]. They are degraded in an alkaline environment and under UV radiation [21–25]. The implications of microbial and fungus content in biochar can be far-reaching. Even though some of these organisms may remain inactive due to its limitations in accessibility, the presence of microbes and fungus may have both positive and negative implications, and their use will vastly depend on destination. Biochars rich in desired bacteria may be used as a substrate in in-vitro production whereas those with limited undesired organisms may be used for construction or in the food industry. However, the microbial characterization is not yet commonplace and the properties of different biochars from different biomass and different pyrolysis processes have been rarely reported while the microbial community composition associated with biochar is poorly understood [26,27].

Here we present the qualitative and quantitative comparison of two dissimilar biochars, focusing on the microbial content, and discuss the implications for the use of these biochars in practice. The biochars were produced in different pyrolysis processes: biochar from Brazil was produced in home-made stove adopted usually by the farmers in their on-site production for their agricultural use or for charcoal sale. Biochar from Poland was industrially produced from a large-scale energy plant. Our results have important implications for the use of biochar in practice and are the first that discuss in comprehensive manner biochars from dramatically different production processes and different biomass with different potential applications.

The aim of the study was to compare Brazilian and Polish biochar samples in terms of the microbial content and the structure of the surface. In addition, an attempt was made to assess the properties of test samples in terms of their potential use in practice.

2. Materials and Methods

The research material consisted of two types of biochar coming from Brazil and Poland:

1. biochar from Brazil obtained from Gliricidia plant (1-BR),
2. biochar from Poland obtained from Pine woodchips (1-PL),
3. biochar from Poland obtained from Oak woodchips (2-PL).

Biochar samples provided for the research were placed in sterile conditions. Similarly, throughout the study, extensive precautions were put in place to limit microbial contamination.

1. Brazilian biochar characterization

The Brazilian biochar was produced in home-made drum stove from Gliricidia sepium at the temperature of approximately 350 °C. Gliricidia is a commonly used plant in organic farming for nitrogen-fixing. It is also used as 'living fence' at agricultural farms and pasturelands. Despite its benefits, it grows fast and may provide excessive shadow, limiting plant growth. In certain circumstances, there is, therefore, a need for cutting off the upperground branches that do not have an alternative use. Moreover, at the site where the biochar was produced (Brazilian Agricultural Research Station, Embrapa Agrobiologia, Seropedica km 47), Gliricidia become an invasive species that disseminated rapidly entering native forest fragments. The source of biomass for biochar production in Brazil, therefore, did not compete with alternative biomass uses, as at times criticised in literature [28]. Branches of Gliricidia were dried for two weeks before pyrolization. Cut and dry biomass was put into the stove, fired up and closed with drum cup and isolated with a layer of sand. The pyrolization time was 24 h. After that time biochar was cooling for another 24 h (with open stove) and sieved at 4 mm (Figure 1).

biochar	pH H2O	C	H	N	H/C
1. Brazil	8.5	60.4	2.47	1.18	0.04 1

Figure 1. Steps of production: Rio de Janeiro state where biochar was produced, Gliricidia tree, dry Gliricidia before being cut at put into the stove, drum stove, biochar (before being sieved), table with biochar properties [4].

2. Polish biochar characterization

Biochar was produced according to Fluid SA company technology. The process consisted in thermal refining of plant biomass and other post-production biomass residues through their autothermal roasting at the temperature 260 °C in reduction atmosphere and without the use of additional energy, catalysts, and chemical additives.

Carbonising products consisted mainly of biochar (from 65% to 70% of energy compared to the energy of the feed) and process gases (from 20% to 35% of energy compared to the energy of the feed). During the process, a significant increase of the carbon element (C) in relation to the biomass of the feed was recorded (1.5 to 2.0 times) as well as an increase of energy density, on average by 4.0 times, reduction of the amount of hydrogen (H), on average by 2.5 times, and reduction of the amount of oxygen (O_2), on average by 3.0 times. Polish samples of biochar were produced from pine woodchips (1-PL) and oak woodchips (2-PL).

The research was carried out at the Innovation Research Center (CBNI) and Regional Center of Agriculture, Environmental and Innovative Technology (EAT) Pope John Paul II State School of Higher Education in Biała Podlaska, Poland.

2.1. Microbiological Analysis of Biochar Samples Consisted of Qualitative and Quantitative Determination of Bacteria and Fungi Species

For the research three standardized substrates were used in the assessment process:

(a) The total number of microorganisms cultivated on PCA substrate (PLATE COUNT LAB-AGAR™) in temperature of 30 °C and time period 72 h.

(b) The total number of fungi colonies cultivated on PDA substrate (POTATO DEXTROSE LAB-AGAR™) in temperature of 30 °C and time period 72 h, after that time the temperature was decreased to 21 °C for the next 72 h.

(c) The total number of fungi colonies cultivated on MA substrate (MALT EXTRACT LAB-AGAR™) in a temperature of 24 °C and time period 144 h.

The substrates used had been purchased from BIOMAXIMA S.A. The dilution method PN-EN ISO 7218 [29] was used with the dilutions of 10^{-1}, 10^{-2}, 10^{-3}, 10^{-4}, and 10^{-5}.

The qualitative and quantitative assessment was done on the basis of microbial flora species composition using macroscopic and microscopic methods, as well as taxonomic keys and atlases. It was expressed in CFU/g units [30–35].

2.2. Microscopic Examination of the Structure

Examination of the biochar sample structure was done using a research microscope Nikon Eclipse E-200 with fluorescence attachment and SCA image analysis.

3. Results

In the sample, 1-BR cultured on PCA substrate dominated strains of Bacillus sp. (the average number of colonies from UNC to 204 colonies for dilutions ranging from 1 to 10^{-5}. The plates were overgrown on the entire surface with the colonies of Bacillus sp. On two plates confluent growth was observed, which made it impossible to perform a quantitative count (Table 1).

Table 1. Microbiological analysis of the culture from the biochar sample from Brazil (1-BR) cultivated on PCA (PLATE COUNT LAB-AGAR™) substrate (CFU/g).

		I	II	III	Average Number of Colonies
	1	Bsp. (UNC)	Bsp. (UNC)	Bsp. (UNC)	UNC
	10^{-1}	Bsp (UNC)	Bsp (UNC)	Bsp (UNC)	UNC
Dilution	10^{-2}	Bsp. (214)	Bsp. (194)	Bsp (UNC)	204.0
	10^{-3}	Bsp. (36)	Bsp. (62)	Bsp (UNC)	49.0
	10^{-4}	Bsp. (3)	Bsp. (6)	Bsp. (23)	10.7
	10^{-5}	Bsp. (2)	0	Bsp. (4)	2.0

I, II, III—repetitions; (UNC)—in brackets the uncountable colonies; (NC)—in brackets the number of colonies (NC); Bsp.—Bacillus sp.

On malt substrate (Table 2) strains of Bacillus sp. were identified for the dilution of 1 and 10^{-1}. Furthermore, in the second repetition for the dilution of 10^{-1}, the following strains were isolated: Aspergillus versicolor (UNC), Aspergillus flavus (UNC), Aspergillus fumigatus (UNC), Aspergillus niger (3), Aspergillus candidus (2), Aspergillus nidulus (7), Penicillium sp. (UNC), Rhizopus sp. (2), while during the third repetition: Aspergillus versicolor (UNC), Aspergillus fumigatus (8), Aspergillus niger (8), and Penicillium sp. (UNC). For the dilution of 10^{-2}, in the first repetition, besides Bacillus sp. the following were isolated: Aspergillus versicolor (UNC), Aspergillus nidulus (1), Aspergillus flavus (1), Aspergillus niger (5), Rhizopus sp. (UNC), in the second repetition: Aspergillus versicolor (18), Aspergillus fumigatus (1), Penicillium sp. (8), while in the third repetition: Aspergillus versicolor (3), and an overgrowth of Bacillus sp. on the entire surface plate was recorded. In the case of dilution of 10^{-3}, on average 7.3 colonies of Aspergillus sp., Aspergillus versicolor, Aspergillus flavus, and Penicillium sp. were isolated, as well as Bacillus sp., in the case of which confluent growth was observed.

Table 2. Microbiological analysis of the culture from the biochar sample from Brazil (1-BR) cultivated on MALT substrate (CFU/g).

		I	II	III	Average Number of Colonies
	1	Bsp (*)	Bsp (*)	Bsp (*)	-
Dilution	10^{-1}	Bsp (*)	Aver (UNC) Afl (UNC) Afum (UNC) Anig (3) Acan (2) Anid (7) Psp. (UNC) R (2) Σ(UNC)	Aver (UNC) Afum (8) Anig (8) Psp (UNC) Σ(UNC)	UNC
	10^{-2}	Aver (UNC) Afl (1) Anig (5) Anid (1) R (UNC) Bsp (*) Σ(UNC)	Aver (18) Afum (1) Psp (8) Σ(27)	Aver (36) Asp. (1) Psp (6) R (2) Σ(45)	36.0
	10^{-3}	Aver (6) Asp. (1) Bsp (*) Σ(7)	Aver (5) Afl (1) Asp. (1) Psp (1) Σ(8)	Aver (5) Psp (1) Σ(7)	7.3
	10^{-4}	Bsp (*)	Aver (2) Σ(2)	Aver (1) Σ(1)	1.0
	10^{-5}	0	0	0	0

I, II, III—repetitions; (UNC)—in brackets the uncountable colonies; (NC)—in brackets the number of colonies (NC); Σ (NC)—total numbers of colonies; Aver—Aspergillus versicolor; Afl—Aspergillus flavus; Afum—Aspergillus fumigatus; Anig—Aspergillus niger; Acan—Aspergillus candidus; Anid—Aspergillus nidulus; Asp.—Aspergillus sp; Psp.—Penicillium sp; R—Rhizopus sp; Bsp (*)—plate overgrown by Bacillus sp.

For the dilution of 10^{-4}, in the first repetition, Bacillus sp. was isolated, characterised by a confluent growth, while in the second and third repetition Aspergillus versicolor species were recorded with the average number of colonies equal to 1.0. After averaging the readings the average number of fungi colonies ranged from UNC to 45. On POTATO substrate (Table 3) for the dilution of 1, no colonies of microorganisms were recorded.

In the case of dilutions from 10^{-1} to 10^{-5}, in all repetitions, strains of Bacillus sp. overgrown on the entire surface of the plate were observed, which made a quantitative count impossible to perform. Additionally, the following genera or species were isolated: Aspergillus sp., Aspergillus niger, Aspergillus flavus, Aspergillus versicolor, Aspergillus candidus, Aspergillus fumigatus, as well as Penicillium sp. (the average number of colonies ranging from 0.3 for the dilution of 10^{-4} to 3.7 for the dilution of 10^{-3}). However, for the dilution of 10^{-5}, no colonies of microorganisms were isolated. In the case of sample no. 1-PL cultured on PCA substrate (Table 4) dominated strains of Bacillus sp.

Table 3. Microbiological analysis of the culture from the biochar sample from Brazil (1-BR) cultivated on POTATO substrate [CFU/g].

		I	II	III	Average Number of Colonies
	1	0	0	0	0
	10^{-1}	Bsp (*)	Anig (2) Afl (2) Asp. (2) Bsp (*) Σ(6)	Aver (UNC) Acan(2) Anig (1) Afl (UNC) Σ(UNC)	3.0
	10^{-2}	Bsp (*)	Aver (2) Acan(1) Afl (1) Σ(4)	Aver (3) Bsp (*) Σ(3)	2.3
Dilution	10^{-3}	Bsp (*)	Bsp (*)	Aver (6) Acan(1) Anig (1) Afum (1) Afl (1) Psp (1) Bsp (*) Σ(11)	3.7
	10^{-4}	Bsp (*)	Bsp (*)	Psp (1) Bsp (*) Σ(1)	0.3
	10^{-5}	0	0	0	0

I, II, III—repetitions; (UNC)—in brackets the uncountable colonies; (NC)—in brackets the number of colonies (NC); Σ (NC)—total numbers of colonies; Aver—*Aspergillus versicolor*; Afl—*Aspergillus flavus*; Afum—*Aspergillusfumigatus*; Anig—*Aspergillus niger*; Acan—*Aspergillus candidus*; Anid—*Aspergillus nidulus*; Asp.—*Aspergillus sp*; Psp.—*Penicillium sp*; R—*Rhizopus sp*; Bsp (*)—plate overgrown by *Bacillus sp*.

Table 4. Microbiological analysis of the culture from the biochar sample from Poland (1-PL, 2-PL) cultivated on PCA substrate (CFU/g).

		I		II		III		Average Number of Colonies	
	Biochar	**Pine 1-PL**	**Oak 2-PL**	**Pine 1-PL**	**Oak 2-PL**	**Pine 1-PL**	**Oak 2-PL**	**Pine 1-PL**	**Oak 2-PL**
	1	Bsp. (UNC)	0	Bsp. (UNC)	Afl (1)	Bsp. (UNC)	0	UNC	0
Dilution	10^{-1}	Bsp (421)	0	Bsp (443) Micro (1) Σ(444)	0	Bsp (771)	0	545.3	0
	10^{-2}	Bsp. (61)	0	Bsp. (54)	0	Bsp (124)	0	79.7	0
	10^{-3}	Bsp. (6)	0	Bsp. (3)	0	Bsp (10)	0	6.3	0
	10^{-4}	0	0	0	0	Bsp. (1)	0	0.3	0
	10^{-5}	0	0	0	0	0	0	0	0

I, II, III—repetitions; (UNC)—in brackets the uncountable colonies; (NC)—in brackets the number of colonies (NC); Σ (NC)—total numbers of colonies; Bsp.—*Bacillus sp*; Micro—*Micrococcus sp*; Afl—*Aspergillus flavus*.

The average number of colonies ranged between 545.5 for the dilution of 10^{-1} and for the dilution of 1 also uncountable growth of *Micrococcus sp*. For the dilution of 10^{-5}, no colonies of microorganisms were isolated. On MALT substrate (Table 5), for the dilution of 1, strains of *Aspergillus flavus* were isolated, and the average number of colonies count was 1.7.

Table 5. Microbiological analysis of the culture from the biochar sample from Poland (1-PL, 2-PL) cultivated on MALT substrate (CFU/g).

		I		II		III		Average Number of Colonies	
	Biochar	Pine 1-PL	Oak 2-PL	Pine 1-PL	Oak 2-PL	Pine 1-PL	Oak 2-PL	Pine 1-PL	Oak 2-PL
Dilution	1	Afl (2)	0	0	0	Afl (1) # (2) Σ(3)	0	1.7	0
	10^{-1}	0	0	0	0	0	Psp. (1)	0	0.3
	10^{-2}	0	0	0	0	0	0	0	0
	10^{-3}	0	0	0	0	0	0	0	0
	10^{-4}	0	0	0	0	0	0	0	0
	10^{-5}	0	0	0	0	0	0	0	0

I, II, III—repetitions; (UNC)—in brackets the uncountable colonies; (NC)—in brackets the number of colonies (NC); Σ (NC)—total numbers of colonies; Afl—Aspergillus flavus; Psp.—Penicillium sp; #—unidentified fungi colony.

Within the range of dilutions between 10^{-1} and 10^{-5}, no colonies of microorganisms were isolated. On POTATO substrate (Table 6) strains of Bacillus sp. overgrown on the entire plate were observed (average number of colonies from UNC for the dilution of 1 to 1 colony for the dilution of 10^{-4}). For the dilution of 10^{-5}, no colonies of microorganisms were isolated. In the case of sample no. 2-PL on PCA substrate for the dilution of 1, only in the second of the three repetitions, 1 colony of Aspergillus flavus was isolated. On the plates, with the dilutions ranging from 10^{-1} to 10^{-5}, no colonies of microorganisms were isolated. On the MALT substrate for the dilution of 10^{-1} 1 colony of Penicillium sp. was isolated. Meanwhile, for the dilutions ranging from 10^{-2} to 10^{-5} no colonies of microorganisms were isolated. On POTATO substrate, for the dilution of 10^{-1}, 1 colony Aspergillus versicolor was isolated, while for the dilution of 10^{-2} 1 colony of unidentified fungi was recorded. On the plate, with the dilution of 10^{-3}, no microorganisms were recorded. In contrast, on the plate with the dilution of 10^{-4} one yeast colony was isolated. On the plate, with the dilution of 10^{-5}, no microorganisms were recorded. After the microbiological analysis of biochar samples, it was noted that sample 2-PL was the least microbiologically polluted. However, it should be stated that in Polish samples biologically different microorganisms were present. In the sample, 1-PL dominated strains of Bacillus sp. While in sample 2-PL dominated, though in small amount, strains of fungi species: Aspergillus flavus, Aspergillus versicolor, and Penicillium sp. In the case of biochar sample from Brazil, the same microorganisms were present as the ones isolated and identified in Polish samples.

Table 6. Microbiological analysis of the culture from the biochar sample from Poland (1-PL, 2-PL) cultivated on POTATO substrate (CFU/g).

		I		II		III		Average Number of Colonies	
	Biochar	Pine 1-PL	Oak 2-PL	Pine 1-PL	Oak 2-PL	Pine 1-PL	Oak 2-PL	Pine 1-PL	Oak 2-PL
Dilution	1	Bsp (*)	0	Bsp (*)	0	Bsp (*)	0	0	0
	10^{-1}	Bsp (*>100)	0	Bsp (*>100)	Aver (1)	Afl (1) Bsp (*>100)	0	0.3	0.3
	10^{-2}	Bsp (*32)	# (1)	Bsp (*43)	0	Bsp (*54)	0	0.3	0
	10^{-3}	Bsp (*1)	0	Bsp (*5)	0	Bsp (*9)	0	0	0
	10^{-4}	0	0	Bsp (*1)	0	0	## (1)	0.3	0
	10^{-5}	0	0	0	0	0	0	0	0

I, II, III—repetitions; (UNC)—in brackets the uncountable colonies; (NC)—in brackets the number of colonies (NC); Σ (NC)—total numbers of colonies; Aver-Aspergillus versicolor; Afl—Aspergillus flavus; Bsp (*)—plate overgrown by Bacillus sp; #—unidentified fungi colony; ##—unidentified yeast colony.

The microscopic analysis, carried out using research microscope Nikon Eclipse E-200 with fluorescence attachment and SCA image analysis, shows the structure of biochar 1-BR (Figure 2), biochar 1-PL (Figure 3) and biochar 2-PL (Figure 4). Biochar 1-BR presents irregular structure, but this is not as dense as in the case of biochar 2-PL (oak). The structure of biochar sample 1-BR (Brazil) is undulating and porous. However, this sample seems to have the most irregular structure when compared with samples 1-PL (pine) and 2-PL (oak).

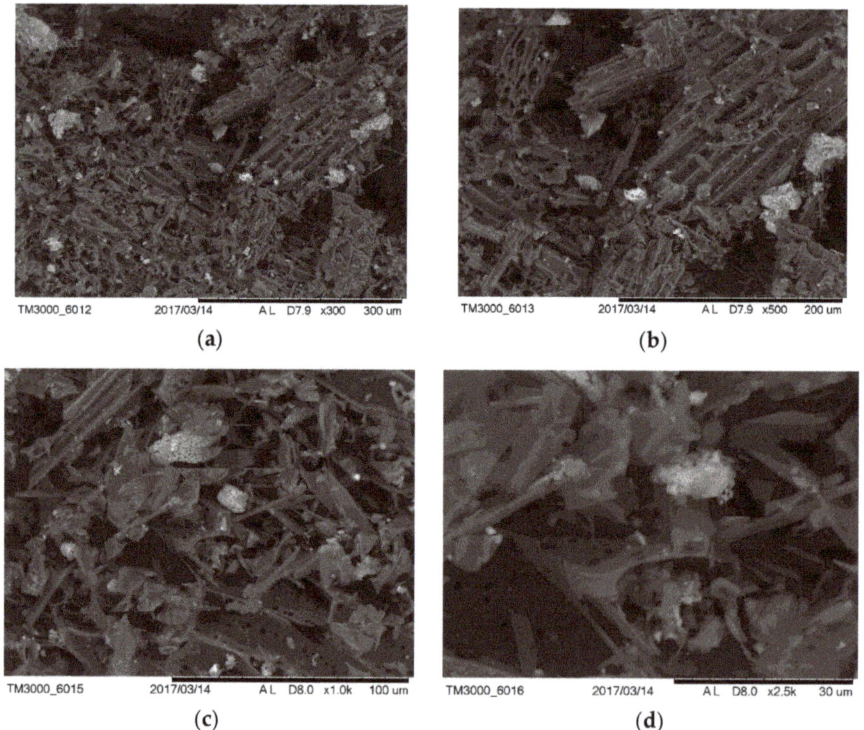

(a)　(b)

(c)　(d)

Figure 2. Microscopic images of biochar from Brazil (1-BR). (**a**) magnification 300×, (**b**) magnification 500×, (**c**) magnification 1000×, (**d**) magnification 2500×.

Figure 3. Macroscopic images of biochar from Poland (1-PL). (**a**) magnification 300×, (**b**) magnification 500×, (**c**) magnification 1000×, (**d**) magnification 2500×.

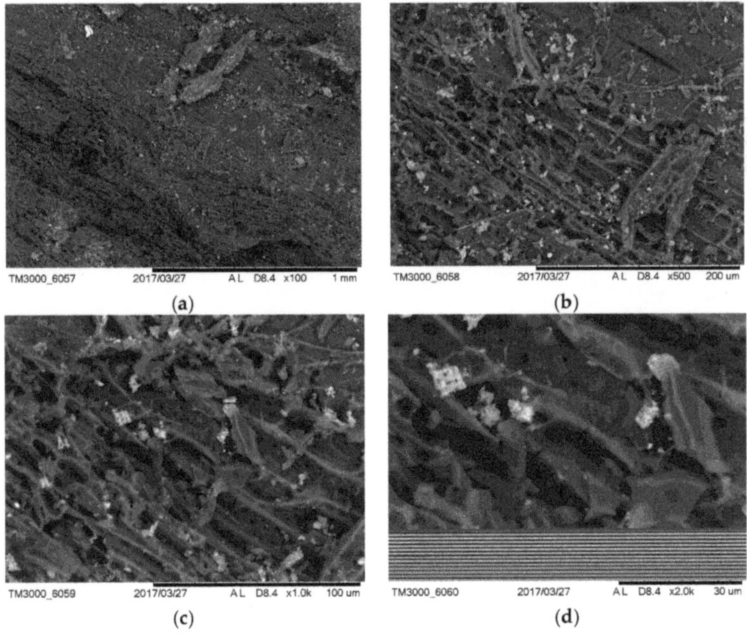

Figure 4. Microscopic images of biochar from Poland (2-PL). (**a**) magnification 100×, (**b**) magnification 500×, (**c**) magnification 1000×, (**d**) magnification 2000×.

Oak (sample 2-PL), being hardwood, in the microscopic images shows a hard and porous structure, while pine (sample 1-PL) has regular, slightly porous, and delicate structure.

4. Discussion

The microbiological analysis of the three biochar samples demonstrated that samples 1-PL and 2-PL were less microbiologically contaminated than sample 1-BR. In the case of sample 1-BR, similar microorganisms were isolated as from samples 1-PL and 2-PL. In sample 1-BR, among others, the following were present: Bacillus sp. Aspergillus sp. Aspergillus niger, Aspergillus flavus, Aspergillus versicolor, Aspergillus candidus, Aspergillus fumigatus, as well as Penicillium sp. and Rhizopus sp. Such a broad range of microorganisms was isolated neither from sample 1-PL nor 2-PL. At the same time, there is a certain similarity between the biochar from Poland and Brazil, namely the strains isolated from Brazilian biochar were present in two types of biochar from Poland, originating from two botanically different sources. When analysing the two samples from Poland opposite observations should be made, namely that biologically different microorganisms were present in these two samples. In sample 1-PL (pine) strains Bacillus sp. dominated. In sample 2-PL (oak) dominated, though in small amounts, strains of fungi: Aspergillus flavus, Aspergillus versicolor, as well as Penicillium sp. In the sample from Brazil strains Bacillus sp., Aspergillus sp., and Rhizopus sp. were the most numerous ones. After the microbiological and microscopic analysis, it can be concluded that the two samples have very interesting characteristics that can be used depending on the purpose. It was observed that sample 2-PL was less microbiologically contaminated. However, it should be stated that in the tested samples biologically different microorganisms were identified. In the sample 1-PL strains Bacillus sp. dominated. In the sample 2-PL dominated, thought in small amount, strains of fungi: Aspergillus flavus, Aspergillus versicolor, as well as Penicillium sp. Additionally, differences were observed in the microscopic images, where sample 1-PL (pine) was characterised by a compact and regular structure, while sample 2-PL (oak) had irregular and porous structure. At the same time, the significantly differing structure was observed in the case of biochar 1-BR (Brazil). It was more delicate than the other test samples. The porous structure and specific surface of biochar are its most important physical properties and they are responsible for the course of various processes in the soil. Biochar has high internal porosity, which affects the water absorption, sorption capacity, and retention of nutrients in the soil. Therefore, it improves physicochemical properties of the soil, facilitates the use of nutrients by plants, and prevents nutrients leaching. Biochar may improve soil structure (water-air properties) [36,37]. Presence of certain genera and species in the test samples may be associated with the generic traits of the raw materials from which they were derived. In addition, the manner, in which test samples were produced, might also be important. The biochar from Poland came from controlled, patented, and repeatable production. On the other hand, the biochar from Brazil was produced in a simple way reflecting the possibility to be repeated by the farmers, wherein not every production parameter can be controlled. This type of dependency is directly proportional to the microbiology of test samples. Summing up, all the samples have very interesting properties that may be used depending on the intended purpose. Implications for practice are now widely used in many research centres around the world. Herrman et al. (2019) concluded that the addition of biochar from rubber tree caused a raise of the soil pH value as well as its nutrient content. In the research cited higher sensitivity to the activity of applied biochar was recorded in the case of fungi occurring in the soil as compared to bacteria present there [38]. Similar studies were performed in Bulgaria [39] using biochar in the cultivation of wheat and maize in crop rotation. The stimulating effect of biochar on soil microflora was observed, in particular on the number of bacteria. The research was considered to be a promising method for preserving soil fertility. Azis et al. (2019) concluded that in the case of soil on which wheat is cultivated addition of biochar, in experimentally determined amount, improves the physicochemical properties of the soil and soil microflora, which in turn has a positive effect on the productivity of the soil and increased yields [40]. In other studies, Hardy et al. investigated the long-term application of biochar in order to improve the quality of soil microorganisms. For this purpose, specialized models imitating soil conditions were designed. Biochar was shown to modify the quantity and quality of soil microorganisms, however, in order to properly design a successful modification of the soil using biochar its properties should be considered in conjunction with the soil conditions [41]. Unfortunately, in the available literature of the subject, there is no similar research with which one could compare the results obtained in this work. Based on the definition of sustainable development, strategies of environmental protection should consist of continuous, integrated and anticipatory actions undertaken in production processes. Such actions lead to an increase in production efficiency and reduce risks to humans and the natural environment. Sustainable production aims at the manufacture of goods which uses processes limiting environmental pollution. In this context, biochar production based on waste-free use of organic biomass has the traits of sustainable development.

5. Conclusions

1. In the biochar sample from Brazil (1-BR), both bacteria: Bacillus sp., and fungi: Penicillium sp., Aspergillus versicolor, Aspergillus flavus, Aspergillus fumigatus, Aspergillus niger, Aspergillus candidus, Aspergillus nidulus, and Rhizopus sp. were observed and identified.

2. In biochar sample from pine woodchips (1-PL) colonies of Bacillus sp. and Micrococcus sp. bacteria were observed and identified.

3. In biochar sample from oak woodchips (2-PL) fungi: Aspergillus flavus, Aspergillus versicolor, and Penicillium Sp. were observed and identified.

4. On the basis of microscopic images, diverse structures of the samples were recorded. Sample 1-PL (pine) is characterised by a compact and regular structure, while sample 2-PL (oak) shows porous and irregular structure. Sample from Brazil (1-BR) is characterised by a structure more delicate than these of the remaining test samples.

5. Samples coming from controlled production (1-PL, 2-PL) were less microbiologically contaminated than biochar sample from Brazil (1-BR).

6. The results obtained can be used as relevant utilitarian data for implications for practice.

Author Contributions: Conceptualization, W.Ż.-S. and P.S. and A.L.; methodology, W.Ż.-S. and D.P.; prepared the materials, D.P.; W.Ż.-S.; analysed the data, W.Ż.-S.; M.T.-R.; wrote the paper, W.Ż.-S.; P.S., A.L.; B.S. visualization, P.S. All authors have read and agree to the published version of the manuscript.

Funding: This research received no external funding.

Acknowledgments: The authors would like to cordially thank the President of Fluid SA, eng. Jan Gładki and International Institute for Sustainability in Rio de Janeiro, Brazil for help in this work.

Conflicts of Interest: The authors declare no conflict of interest.

References

1. Lehmann, J.; Joseph, S. *Biochar for Environmental Management: Science and Technology*; Routledge: Abingdon, UK, 2012.
2. Filiberto, D.M.; Gaunt, J.L. Practicality of biochar additions to enhance soil and crop productivity. *Agriculture* **2013**, *3*, 715–725. [CrossRef]
3. Latawiec, A.E.; Strassburg, B.B.N.; Junqueira, A.B.; Araujo, E.; de Moraes, L.F.D. Biochar amendment improves degraded pasturelands in Brazil: Environmental and cost-benefit analysis. *Sci. Rep.* **2018**, *9*, 1–12. [CrossRef]
4. Castro, A.; da Silva Batista, N.; Latawiec, A.; Rodrigues, A.; Strassburg, B.; Silva, D.; Araujo, D.; de Moraes, L.F.D.; Guerra, J.G.; Galvao, G.; et al. The effects of *Gliricidia*-derived biochar on sequential maize and bean farming. *Sustainability* **2018**, *10*, 578. [CrossRef]
5. Hale, S.E.; Cornelissen, G.; Werner, D. Sorption and remediation of organic compounds in s oils and sediments by (activated) biochar. In *Biochar for Environmental Management Science, Technology and Implementation*; Ohannes Lehmann, J., Stephen, J., Eds.; Earthscan: London, UK, 2015.
6. Schmidt, H.; Hagemann, N.; Draper, K.; Kammann, C. The use of biochar in animal feeding. *PeerJ* **2019**, *7*, e7373. [CrossRef]

7. Schmidt, H.P. Novel uses of biochar. In Proceedings of the USBI North American Biochar Symposium, Center for Agriculture, University of Massachusetts, Amherst, MA, USA, 13–16 October 2013.

8. Abe, I.; Hitomi, M.; Ikuta, N.; Kawafune, I.; Noda, K.; Kera, Y. Humidity-control capacity of microporous carbon. *J. Urban Living Health Assoc.* **1995**, *39*, 333–336.

9. Nakano, T.; Haisbi, T.; Mizuno, T.; Takeda, T.; Tokumoto, M. Improvements of the under floor humidity in woody building and water content of wood material. *Mokuzai Kogyo* **1996**, *51*, 198–202.

10. Morita, H. The effects of humid controlling charcoal on the environ-mental antigenic allergy. In *Proceeding of the 35th Annual Meeting of Japanese Society for Dermatoallergology, Yokohama, Japan, 16–17 July2005*; Japanese Society for Dermatoallergology: Wakayama, Japan, 2005; p. 115.

11. Taketani, T. Evaluation of the effect of humid controlling charcoal on the infantile bronchial asthma. *Allergy* **2006**, *55*, 467.

12. Riva, L.; Nielsen, H.K.; Skreiberg, Ø.; Wang, L.; Bartocci, P.; Barbanera, M.; Bidini, G.; Fantozzi, F. Analysis of optimal temperature, pressure and binder quantity for the production of biocarbon pellet to be used as a substitute for coke. *Appl. Energy* **2019**, *256*, 113933. [CrossRef]

13. Souradeep, G.; Harn, W.K. Factors determining the potential of biochar as a carbon capturing and sequestering construction material: Critical review. *J. Mater. Civ. Eng.* **2017**, *29*, 04017086. [CrossRef]

14. Schallmey, M.; Singh, A.; Ward, O.P. Developments in the use of *Bacillus species* for industrial production. *Can. J. Microbiol.* **2004**, *50*, 1–17. [CrossRef]

15. Harwood, C.P.; Cranenburgh, R. Bacillus protein secretion: An unfolding story. *Trends Microbiol.* **2008**, *16*, 73–79. [CrossRef]

16. Aiyer, P.; Modi, H. Isolation and screening of alkaline amylase producing bacteria *Bacillus Licheniformis*. *Asian J. Microbiol. Biotechnol. Environ. Sci.* **2005**, *7*, 895–897.

17. De Souza, P.M.; de Oliveira e Magalhães, P. Application of microbial A -amylase in industry—A review. *Braz. J. Microbiol.* **2010**, *41*, 850–861. [CrossRef]

18. Sailas, B.; Smitha, R.B.; Jisha, V.N.; Pradeep, S.; Sajith, S.; Sreedevi, S.; Priji, P.; Unni, K.N.; Sarath Josh, M.K. A monograph on amylase from *Bacillus* spp. *Adv. Biosci. Biotechnol.* **2013**, *4*, 227–241.

19. Hussein, H.S.; Brasel, J.M. Toxicity, metabolism and impact of mycotoxins on humans and animals. *Toxicology* **2001**, *167*, 101–134. [CrossRef]

20. Kołaczyńska-Janicka, M. Mikotoksyny—Realne zagrożenie. *Kukurydza* **2006**, *1*, 59–62.

21. Bittencourt, A.B.F.; Oliveira, C.A.F.; Dilkin, P.; Correa, B. Mycotoxin occurrence in corn meal and flour traded in São Paulo, Brazil. *Food Control* **2005**, *16*, 117–120. [CrossRef]

22. Cegielska-Radziejewska, R.; Lesnierowski, G.; Kijowski, J. Antibacterial activity of hen egg white lysozyme modified by thermochemical technique. *Eur. Food Res. Technol.* **2009**, *228*, 841–845. [CrossRef]

23. Cegielska-Radziejewska, R.; Leśnierowski, G.; Kijowski, J.; Szablewski, T.; Zabielski, J. Effects of treatment with lysozyme and its polymers on the microflora and sensory properties of chilled chicken breast muscles. *Bull. Vet. Inst. Pulawy* **2009**, *53*, 455–461.

24. Ghali, R.; Hmaissia-Khlifa, K.; Ghorbel, H.; Maaroufi, K.; Hedili, A. Incidence of aflatoxins, ochratoxin A and zearalenone in tunisian foods. *Food Control* **2008**, *19*, 921–924. [CrossRef]

25. Łozowicka, B. Zanieczyszczenia chemiczne w żywności pochodzenia roślinnego. *Prog. Plant Prot.* **2009**, *49*, 2071–2080.

26. Dai, Z.; Barberán, A.; Li, Y.; Brookes, P.C.; Xu, J. Bacterial community composition associated with pyrogenic organic matter (biochar) varies with pyrolysis temperature and colonization environment. *mSphere* **2017**, *2*, e00085-17. [CrossRef]

27. Russo, G.; Verdiani, G.; Anifantis, A.S. Re-use of agricultural biomass for nurseries using proximity composting. *Contemp. Eng. Sci.* **2016**, *9*, 1151–1182. [CrossRef]

28. Rodriguez, A.; Latawiec, A.E. Rethinking organic residues: The potential of biomass in Brazil. *Mod. Concep. Dev. Agron.* **2018**, *1*, 1–5. [CrossRef]

29. *PN-EN ISO 7218Microbiology of Food and Animal Feeding Stuffs—General Rules for Microbiological Examination*; Polish Committee for Standardization: Warsaw, Poland, 2013.

30. Ramirez, C. *Manual and Atlas of the Penicillia*; Elsevier Biomedical Press: Amsterdam, The Netherlands, 1982.

31. Baran, E. *Zarys Mikologii Lekarskiej*; Volumed: Wrocław, Poland, 1998.

32. Larone, D.H. *Medically Important Fungi. A Guide to Identification*, 5th ed.; ASM Press: Washington, DC, USA, 2011.

33. Kwaśna, H.; Chełkowski, J.; Zajkowski, P. *Grzyby*; Instytut Botaniki PAN: Kraków, Poland, 1991; Volume XXII.

34. Krzyściak, P.; Skóra, M.; Macura, A.B. *Atlas Grzybów Chorobotwórczych Człowieka*; MedPharm: Wrocław, Poland, 2011.

35. Samson, R.A.; Hoekstra, E.S.; Frisvad, J.C.; Filtenborg, O. *Introduction to Food- and Airborne Fungi*, 6th ed.; Centraalbureau voor Schimmelcultures: Utrecht, The Netherlands, 2002.

36. Bis, Z. Czy węgiel będzie dzieckiem Europy? *Nowa Energ.* **2013**, 2–3. Available online: http://kie.is.pcz.pl/images/wegiel_dzieckiem_europy.pdf (accessed on 15 September 2019).

37. Bis, Z. Biowęgiel-powrót do przeszłości, szansa dla przyszłości. *Czysta Energ.* **2012**, *6*, 28–31.

38. Herrmann, L.; Lesueur, D.; Robin, A.; Robain, H.; Wiriyakitnateekul, W.; Bräu, L. Impact of biochar application dose on soil microbial communities associated with rubber trees in North East Thailand. *Sci. Total Environ.* **2019**, *689*, 970–979. [CrossRef]

39. Petkova, G.; Nedyalkova, K.; Mikova, A.; Atanassova, I. Microbiological characteristics of biochar amended alluvial meadow soil. *Bulg. J. Agric. Sci.* **2018**, *24* (Suppl. S2), 81–84.

40. Aziz, S.; Laiba, Y.; Asif, J.; Uzma, F.; Zahid, Q.; Isfahan, T.; Syed, K.H.; Ali, M.I. Fabrication of biochar from organic wastes and its effect on wheat growth and soil microflora. *Pol. J. Environ. Stud.* **2020**, *29*, 1069–1076. [CrossRef]

41. Hardy, B.; Sleutel, S.; Dufey, J.E.; Cornelis, J.-T. The long-term effect of biochar on soil microbial abundance, activity and community structure is overwritten by land management. *Front. Environ. Sci.* **2019**, *7*, 110. [CrossRef]